BioMethods Vol. 10

Series Editors

Dr. T. Meier
MyoContract
Biozentrum, Room 388
Universität Basel
Klingelbergstr. 70
CH-4056 Basel
Switzerland

Prof. Dr. H.P. Saluz
Hans-Knöll-Institut
für Naturstofforschung e.V.
Beutenbergstr. 11
D-07745 Jena
Germany

Microsystem Technology: A Powerful Tool for Biomolecular Studies

Edited by
J. M. Köhler
T. Mejevaia
H. P. Saluz

Springer Basel AG

Editors

Dr. J.M. Köhler
Institut für Physikal.
Hochtechnologie Jena
Hemholtzweg 4
D-07743 Jena
Germany

T. Mejevaia
Rosenstr. 7
D-07749 Jena
Germany

Prof. Dr. H.P. Saluz
Hans-Knöll-Institut
für Naturstofforschung e.V.
Beutenbergstr. 11
D-07745 Jena
Germany

Library of Congress Cataloging-in-Publication Data

Microsystem technology : a powerful tool for biomolecular studies /
edited by J.M. Köhler, T. Mejevaia, H.P. Saluz.
p. cm. – (Biomethods)
Includes bibliographical references and index.
ISBN 978-3-0348-9784-6 ISBN 978-3-0348-8817-2 (eBook)
DOI 10.1007/978-3-0348-8817-2
1. Nanotechnology. 2. Biotechnology. 3. Combinatorial chemistry.
4. Molecular biology–Automation. I. Köhler, M. (Michael),
1956–. II. Mejevaia, T. III. Saluz, H. P., 1952–. IV. Series.
TP248.25.N35M53 1999
572– dc21
98–47629
CIP

Die Deutsche Bibliothek – CIP-Einheitsaufnahme

Microsystem technology : a powerful tool for biomolecular studies /
ed by J. M. Köhler ... - Springer Basel AG, 1999
(BioMethods ; Vol. 10)
ISBN 978-3-0348-9784-6

Originally published by Birkhäuser Verlag in 1999
Softcover reprint of the hardcover 1st edition 1999
Printed on acid-free paper produced from chlorine-free pulp. TCF ∞

Cover design: Markus Etterich, Basel

ISBN 978-3-0348-9784-6

9 8 7 6 5 4 3 2 1

Contents

List of contributors IX

Preface XIII

Colour Plates XV

1 Molecular building principles in nature
J. Michael Koehler and Hans Peter Saluz 1

2 Robotic equipment and microsystem technology in biological research
Holger Eickhoff, Igor Ivanov, Markus Kietzmann, Elmar Maier, Markus Kalkum, David Bancroft and Hans Lehrach 17

3 Components and systems for microliquid handling
Steffen Howitz 31

4 Nanotiterplates for screening and synthesis
Günther Mayer, Klaus Wohlfart, Andreas Schober and J. Michael Köhler 75

5 Chip technology for micro-separation
Andrew J. de Mello and Andreas Manz 129

6 Rapid multisample PCR in miniaturized ultrathin-walled microwell plates
Alexander N. Tretyakov, Rimma A. Pantina, Oleg K. Kaboev and Hans Peter Saluz 179

7 Generation of large libraries
Carola Burgtorf and Hans Lehrach 189

8 Generation and screening of solution-phase synthetic peptide combinatorial libraries
Andrew Wallace, Kenneth S. Koblan, Riccardo Cortese, Jackson B. Gibbs and Antonello Pessi 225

9 Using oligonucleotide probe arrays to access genetic diversity
R.J. Lipshutz, D. Morris, M. Chee, E. Hubbell, M.J. Kozal, N. Shah, N. Shen, R. Yang and S.P.A. Fodor 241

10 Generation of libraries by print technologies
Eugen Ermantraut, Stefan Wölfl and Hans Peter Saluz 255

11 Synthesis and screening of bead-based libraries
Dirk Vetter 265

12 Sensors for biomolecular studies
Jan Rickert, Thomas Wessa and Wolfgang Göpel 279

13 Fluorescence detection of single molecules applicable to small volume assays
Jörg Enderlein, W. Patrick Ambrose, Peter M. Goodwin and Richard A. Keller 311

14 Fluorescence correlation spectrometry (FCS): Measuring biological interactions in microstructures
Gabriele Gradl, Rolf Guenther and Silvia Sterrer 331

15 Scanning force microscopy: A microstructured device for imaging, probing, and manipulation of biomolecules at the nanometer scale
Wolfgang Fritzsche 353

16 DNA resequencing, mutation detection and gene expression analysis of oligonucleotide microchips
Andres Metspalu and John M. Shumaker 371

17 Miniaturized arrays for DNA analysis
Maryanne J. O'Donnell and Hubert Köster 399

18 Manipulation of particles, cells and liquid droplets by high frequency electric fields
Thomas Schnelle, Torsten Müller and Günter Fuhr 417

19 Optical trapping and manipulation
Karl Otto Greulich 453

20 Computer modeling of protein, nucleic acid, and drug structures
Jürgen Sühnel 475

21 Optimizing structure and function relationship of nucleic acid molecules
Michael Famulok, Andres Jäschke and Stefan Wölfl 507

22 Optimisation of molecular function
Christian V. Forst 529

23 Appendix
Udo Luhmann 555

Index 575

List of contributors

Ambrose, W.P.
CST-1, MS M888,
Los Alamos National Laboratory,
Los Alamos, NM 87545, USA

Bancroft, D.
GPC Aktiengesellschaft
Genome Pharmaceuticals Corporation
Lochhamer Str. 29
D-82152 Martinsried/Munich, Germany

Burgtorf, C.
Max-Planck-Institut
für molekulare Genetik
Ihnestr. 73
D-14195 Berlin-Dahlem, Germany

Chee, M.
Affymetrix
3380 Central Expressway
Santa Clara, CA 95051, USA

Cortese, R.
Istituto di Ricerche di Biologia Moleculare
(IRBM) P. Angeletti,
Via Pontina KM 30.600,
J-00040 Pomezia (Roma), Italy

Enderlein, J.
Institut für Analytische Chemie
Chemo- und Biosensorik
Universität Regensburg
Postfach 101042
D-93040 Regensburg, Germany

Ermantraut, E.
CLONDIAG CHIP TECHNOLOGIES GmbH
CCT, Löbstedter Str. 105
D-07743 Jena, Germany

Eickhoff, H.
Max-Planck-Institut für molekulare Genetik
Ihnestr. 73
D-14195 Berlin-Dahlem, Germany

Famulok, M.
Laboratorium für molekulare Biologie –
Genzentrum München
Feodor-Lynen-Str. 25
D-81375 München, Germany

Fodor, S.P.A.
Affymetrix
3380 Central Expressway
Santa Clara, CA 95951, USA

Forst, C.V.
Beckman Institute
University of Illinois
405 North Mathews Avenue
Urbana, IL 61801, USA

Fritzsche, W.
Microsystems Department
Institute of Physical High Technology
Helmholtzweg 4
D-07743 Jena, Germany

Fuhr, G.
Humboldt-Universität zu Berlin,
Institut für Biologie,
Membranphysiologie,
Invalidenstr. 43
D-10115 Berlin, Germany

Gibbs, J.B.
Department of Cancer Research,
Merck Research Laboratories
West Point, PA 19486, USA

Goodwin, P.M.
CST-1, MS M888,
Los Alamos National Laboratory,
Los Alamos, NM 87545, USA

Göpel, W.
Institute of Physical and Theoretical Chemistry
and Center of Interface Analysis and Sensors
University of Tübingen
D-72076 Tübingen, Germany

Gradl, G.
EVOTEC BioSystems GmbH,
Humboldt-Universität zu Berlin
Institut für Biologie
Membranphysiologie
Invalidenstr. 42
D-10115 Berlin, Germany

Greulich, K.O.
Institut für Molekulare Biotechnologie e.V.
Postfach 100813
D-07708 Jena, Germany

Guenther, R.
EVOTEC BioSystems GmbH,
Schnackenburgallee 114
D-22525 Hamburg, Germany

Howitz, S.
GeSiM Gesellschaft für Silizium-
Mikrosysteme mbH
Rossendorfer Technologiezentrum
Bautzner Landstr. 45
D-01454 Großerkmannsdorf/Rossendorf,
Germany

Hubbell, E.
Affymetrix
3380 Central Expressway
Santa Clara, CA 95051, USA

Ivanov, I.
GPC Aktiengesellschaft
Genome Pharmaceuticals Corporation
Lochhamer Str. 29
D-82152 Martinsried/Munich, Germany

Jäschke, A.
Institut für Biochemie, Freie Universität Berlin
Thielallee 63
D-14195 Berlin, Germany

Kaboev, O.
St. Petersburg Nuclear Physics Institute,
Molecular and Radiation Biophysics Division,
188350, Gatchina, Russia

Kalkum, M.
Max-Planck-Institut für molekulare Genetik
Ihnestr. 73
D-14195 Berlin-Dahlem, Germany

Keller, R.A.
CST-1, MS M888,
Los Alamos National Laboratory,
Los Alamos, NM 87545, USA

Kietzmann, M.
GPC Aktiengesellschaft
Genome Pharmaceuticals Corporation
Lochhamer Str. 29
D-82152 Martinsried/Munich, Germany

Koblan, K.S.
Department of Cancer Research,
Merck Research Laboratories,
West Point, Pennsylvania 19486, USA

Köhler, J.M.
Institute of Physical High Technology
Helmholtzweg 4
D-07743 Jena, Germany

Köster, H.
Sequenom Inc.
11555 Sorrento Valley Road
San Diego, CA 92121, USA

Kozal, M.J.
Affymetrix
3380 Central Expressway
Santa Clara, CA 95051, USA

Lehrach, H.
Max-Planck-Institut für molekulare Genetik
Ihnestr. 73
D-14195 Berlin-Dahlem, Germany

Lipshutz, R.J.
Affymetrix
3380 Central Expressway
Santa Clara, CA 95051, USA

Luhmann, U.
Hans Knöll Institut für Naturstoff-Forschung,
Beutenbergstr. 11
D-07746 Jena, Germany

Manz, A.
Zeneca/SmithKline Beecham Centre
for Analytical Sciences,
Imperial College of Science, Technology
and Medicine,
London, SW7 2AY, UK

Maier, E.
GPC Aktiengesellschaft
Genome Pharmaceuticals Corporation
Lochhamer Str. 29
D-82152 Martinsried/Munich, Germany

Mayer, G.
Institut für Physikalische Hoch-
technologie (IPHT)
Abt. Mikrosysteme
Postfach 100239
D-07702 Jena, Germany

de Mello, A.J.
Zeneca/SmithKline Beecham Centre
for Analytical Sciences,
Imperial College of Science, Technology
and Medicine,
London, SW7 2AY, UK

Metspalu, A.
Institute of Molecular and Cell Biology
and Children's Hospital
University of Tartu, Estonian Biocentre,
23 Riia St.
Tartu 51010, Estonia

Mejevaia, T.
Rosenstr. 7
D-07749 Jena, Germany

Morris, D.
Affymetrix
3380 Central Expressway
Santa Clara, CA 95051, USA

Müller, T.
Humboldt-Universität zu Berlin,
Institut für Biologie, Membranphysiologie,
Invalidenstr. 43,
D-10115 Berlin, Germany

O'Donnell, M.J.
Sequenom Inc.
11555 Sorrento Valley Road
San Diego, CA 92121, USA

Pantina, R.
St. Petersburg Nuclear Physics Institute,
Molecular and Radiation Biophysics Division,
188350, Gatchina, Russia

Pessi, A.
Instituto di Ricerche de Biologia Molecolare
(IRBM) P. Angeletti,
Via Pontina KM 30.600
I-00040 Pomezia (Roma), Italy

Rickert, J.
Institute of Physical and Theoretical Chemistry
and Center of Interface Analysis and Sensors
University of Tübingen
D-72076 Tübingen, Germany

Saluz, H.P.
Cell and Molecular Biology,
Hans Knöll Institut für Naturstoff-Forschung
Beutenbergstr. 11
D-07746 Jena, Germany

Schnelle, Th.
Humboldt-Universität zu Berlin
Institut für Biologie, Membranphysiologie,
Invalidenstr. 43,
D-10115, Berlin, Germany

Schober, A.
Institut für Physikalische Hoch-
technologie (IPHT)
Abt. Mikrosysteme
Postfach 100239
D-07702 Jena, Germany

Shah, N.
Affymetrix
3380 Central Expressway
Santa Clara, CA 95051, USA

Shen, N.
Affymetrix
3380 Central Expressway
Santa Clara, CA 95051, USA

Shumaker, J.M.
Baylor College of Medicine
Department of Molecular and Human Genetics
One Baylor Plaza
Houston, TX 77030, USA

Sterrer, S.
EVOTEC BioSystems GmbH,
Schnackenburgallee 114
D-22525 Hamburg, Germany

Sühnel, J.
Biocomputing,
Institut für Molekulare Biotechnologie e.V.
Postfach 100813
D-07708 Jena, Germany

Tretyakov, A.
St. Petersburg Nuclear Physics Institute,
Molecular and Radiation Biophysics Division,
188350, Gatchina, Russia

Wallace, A.
Centre for Peptide and Protein Engineering
School of Biology and Biochemistry,
The Queen's University of Belfast,
97 Lisburn Road
Belfast BT9 7BL, Northern Ireland, UK

Wessa, T.
Institute of Physical and Theoretical Chemistry
and Center of Interface Analysis and Sensors
University of Tübingen
D-72076 Tübingen, Germany

Wölfl, S.
Hans-Knöll-Institut für Naturstoff-Forschung
Beutenbergstr. 11
D-07745 Jena, Germany

Wohlfahrt, K.
Institut für Physikalische Hoch-
technologie (IPHT)
Abt. Mikrosysteme
Postfach 100239
D-07702 Jena, Germany

Vetter, D.
Graffinity Pharmaceutical Design GmbH
Im Neuenheimer Feld 515
D-69120 Heidelberg

Yang, R.
Affymetrix
3380 Central Expressway
Santa Clara, CA 95051, USA

Preface

Biomolecular studies are the trial of Man to understand how Nature manages information at the molecular level. The understanding of molecular information handling in nature is essential for the molecular optimization in chemistry, molecular biology, molecular pharmacology and therefore – as an example – for the development of specifically acting drugs.

The famous recent method of technical information management is digital electronics. Over the past few years, evidence has arisen that computerized and molecular information managements have many similar and overlapping aspects. For example, both technology and nature use digitized information and both use small structures for the efficient handling of information. Furthermore, they optimize their processes in order to gain a maximum of information with a minimum of invested energy.

During the last two decades, novel experimental techniques in biomolecular sciences have paved the way for artificial biomolecular optimization. In the same time interval, the progress of micro system technology has been extended from the field of digital electronics and sensing to micro liquid handling, and the field of chip-supported substance handling began. It appears that the "marriage" of physical micro technology and molecular processing will be consummated soon. The contact of both fields has been realized in for example DNA chips. Such connections will also become relevant in additional fields in the near future. Biomolecular investigations are the first to profit from these fast growing scientific and technical connections between micro systems and molecular sciences.

Microsystem technology occupies increasingly more a key position for the application and further development of molecular sciences. It opens novel markets with an enormous developmental potential for almost all key indications, such as early diagnosis, diagnostics and therapy monitoring of infectious and cancer diseases, prenatal diagnostics of inherited diseases, predisposition diagnostics or therapy monitoring of novel therapeutic principles, especially in gene therapy.

The aim of this multi-author book is to provide a comprehensive collection of micro system technologies used in modern molecular biology, biotechnology, combinatorial chemistry, molecular pharmacy and other related fields. It is directed toward readers from many disciplines and anyone interested in this promising topic. This book supports research work in this field and will help natural scientists and engineers to use the most current knowledge for development of other miniaturized technologies and devices, and to study naturally occurring molecular processes as well as use them for technical operations,

J. M. Köhler
T. Mejevaia
January 1999 *H. P. Saluz*

Colour Plates

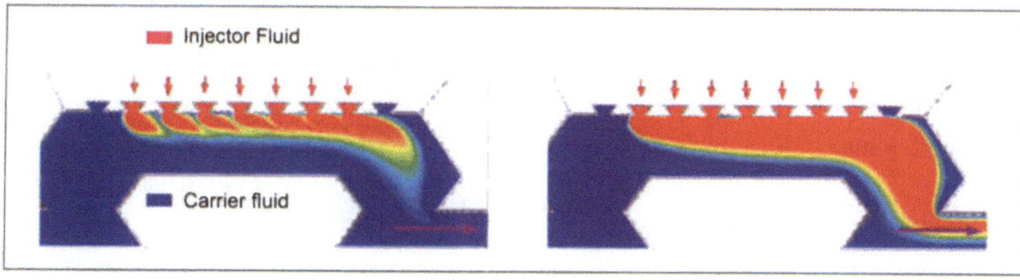

See page 65 1

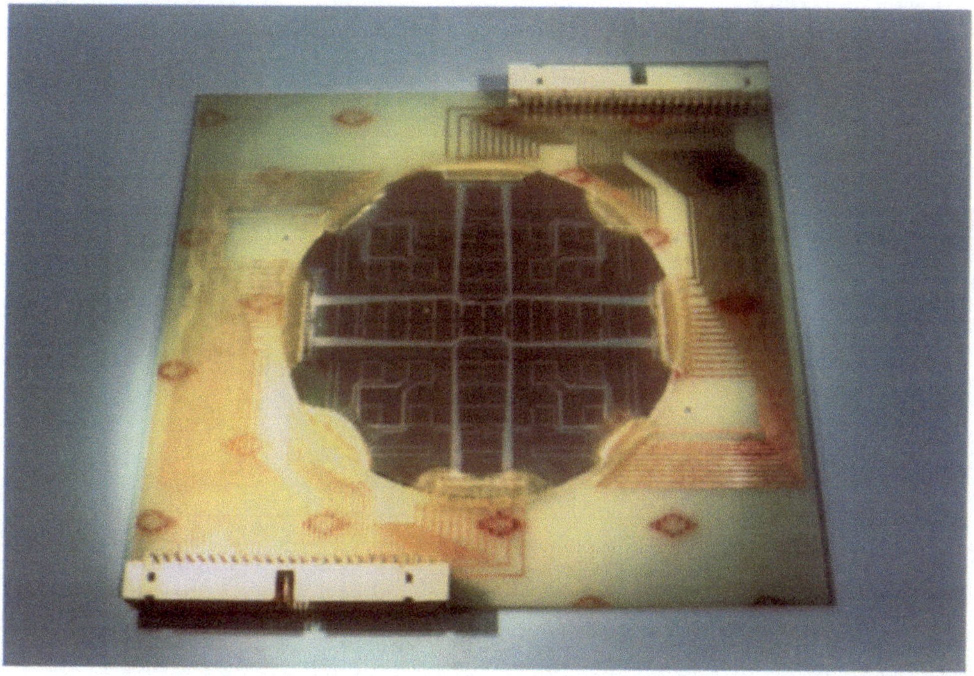

See page 110 2

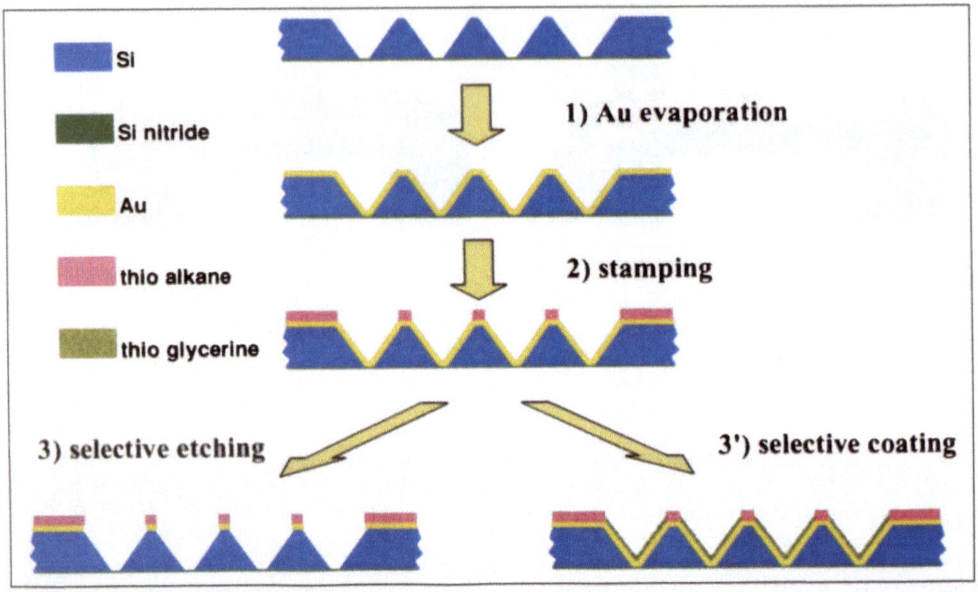

3

See page 117

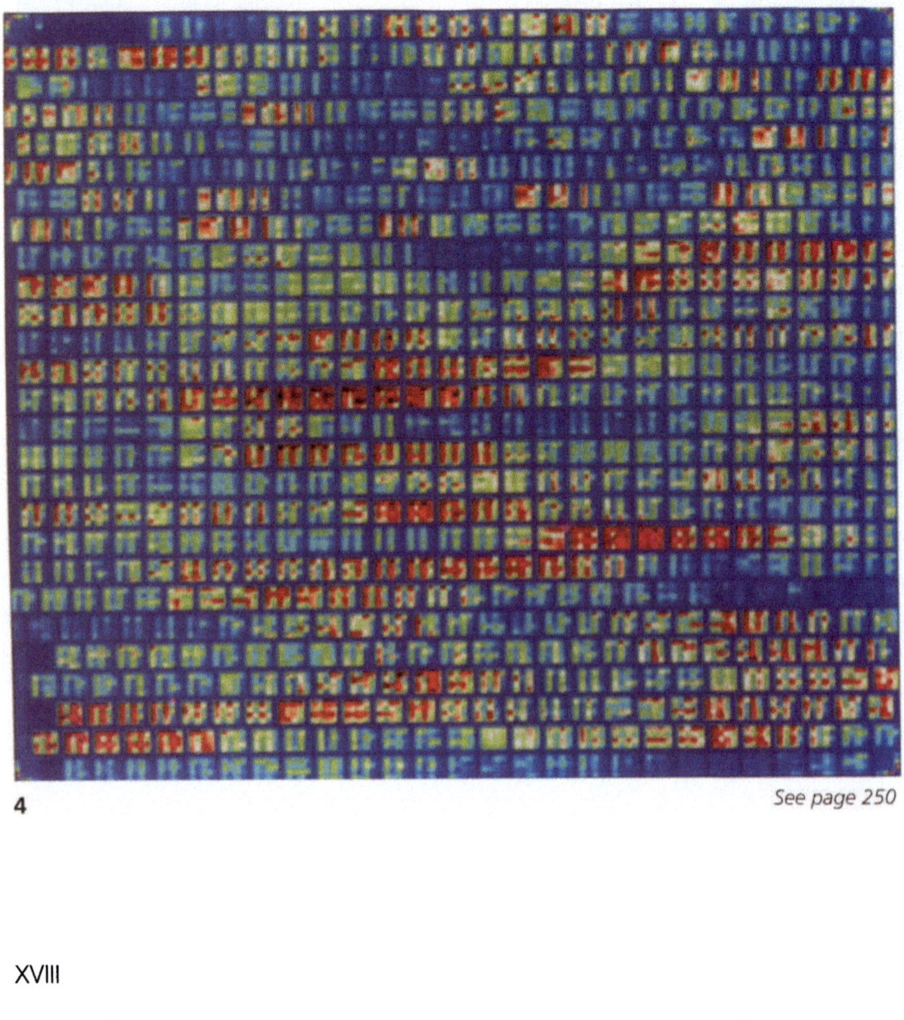

4

See page 250

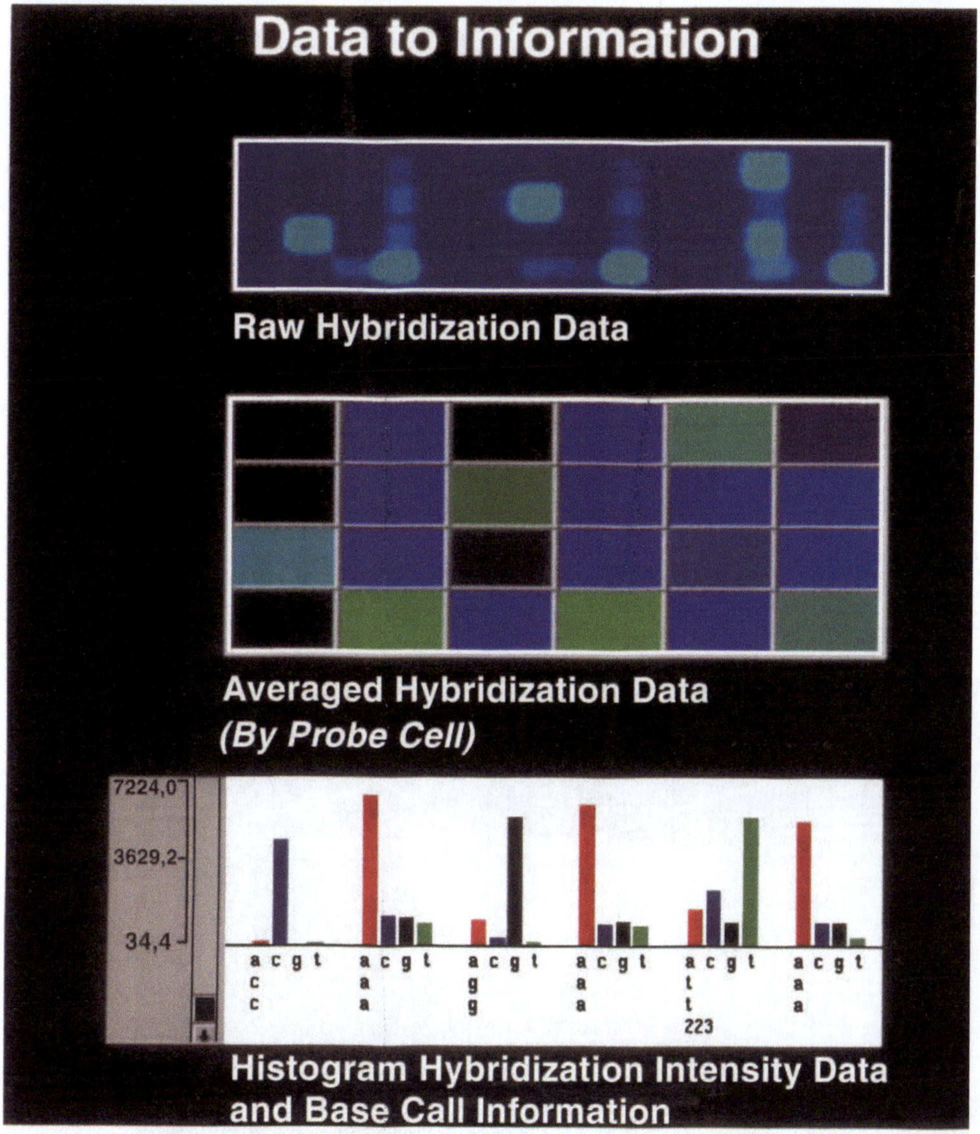

See page 251

5

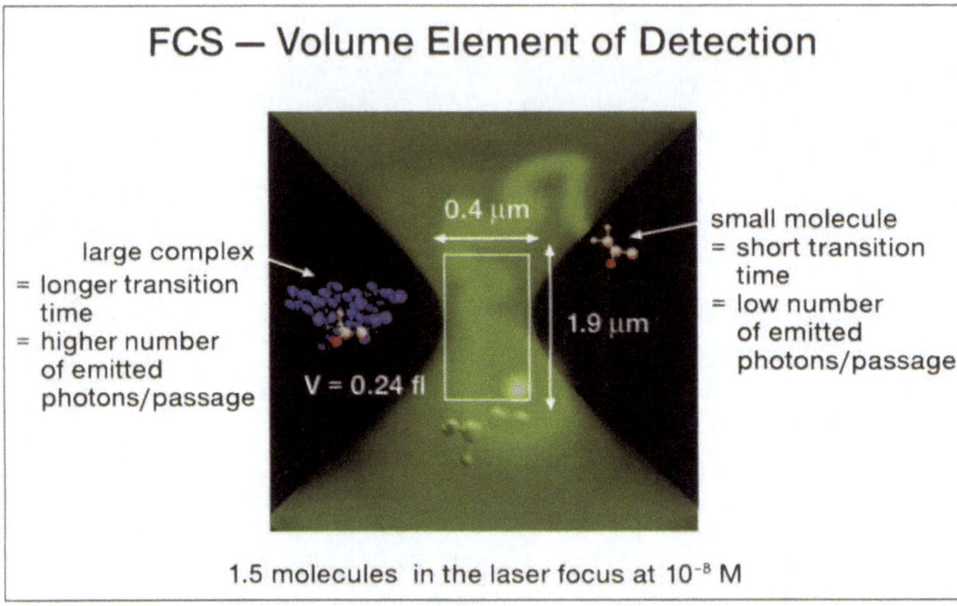

6

See page 335

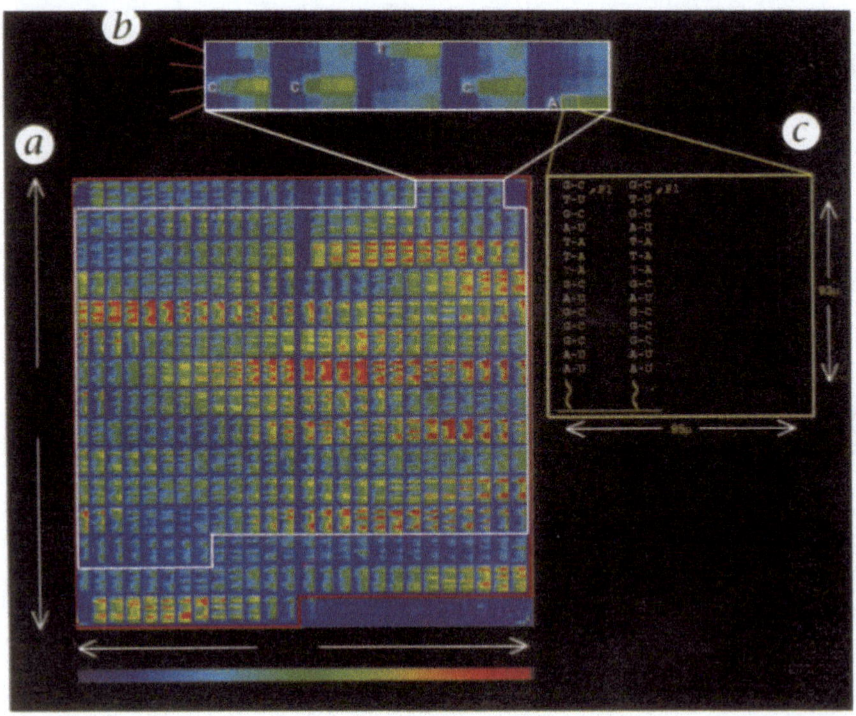

7

See page 381

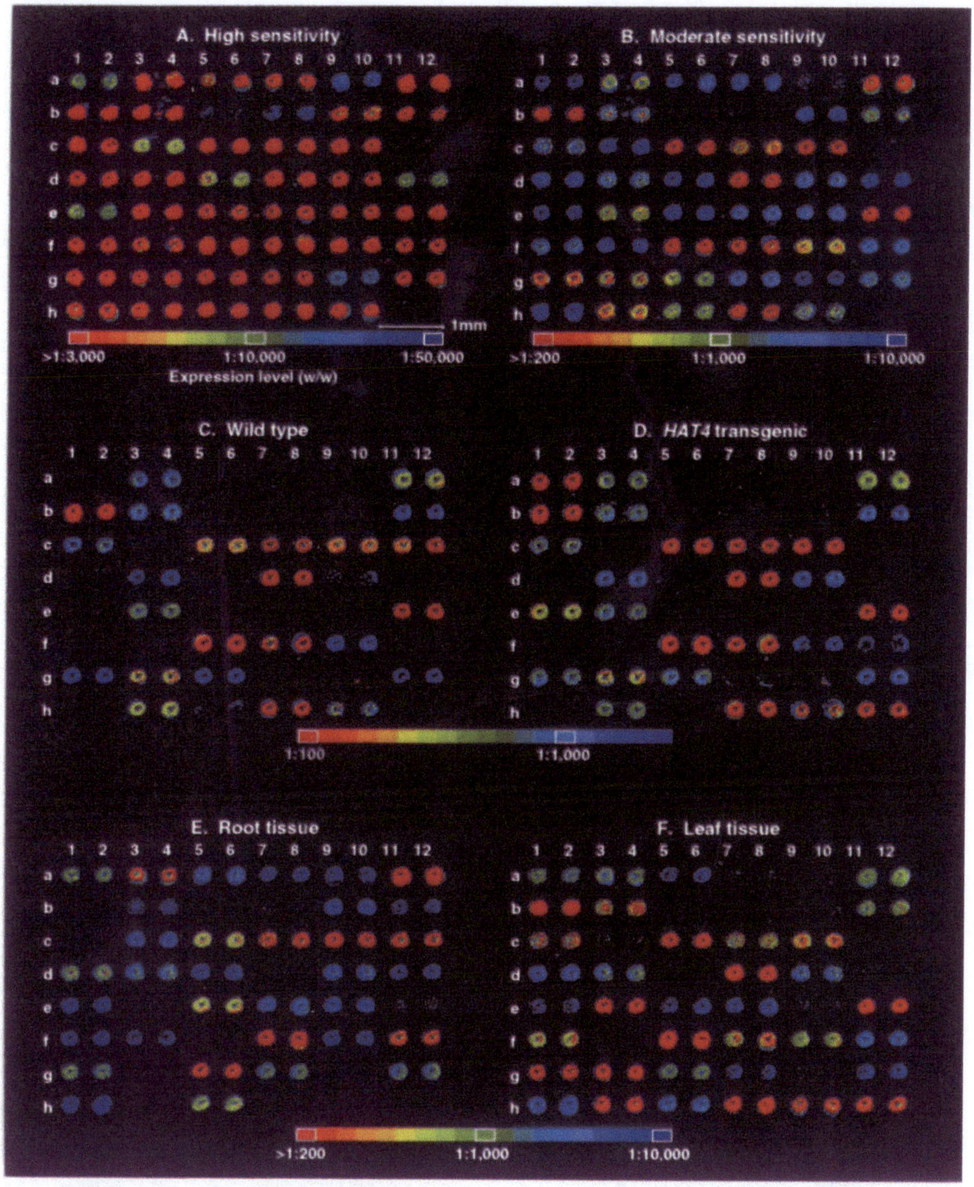

See page 385

8

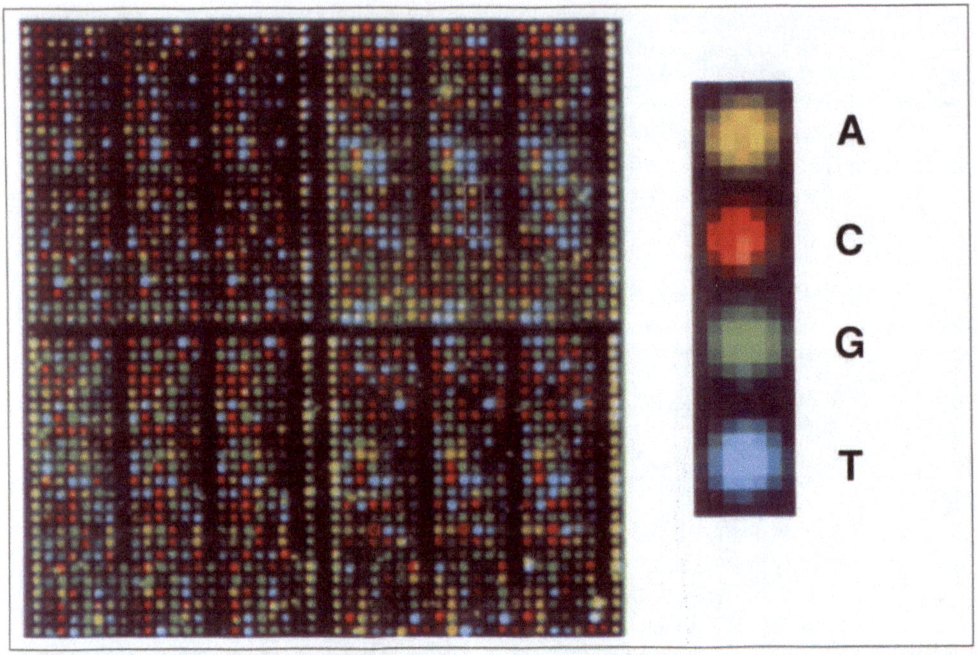

9

See page 391

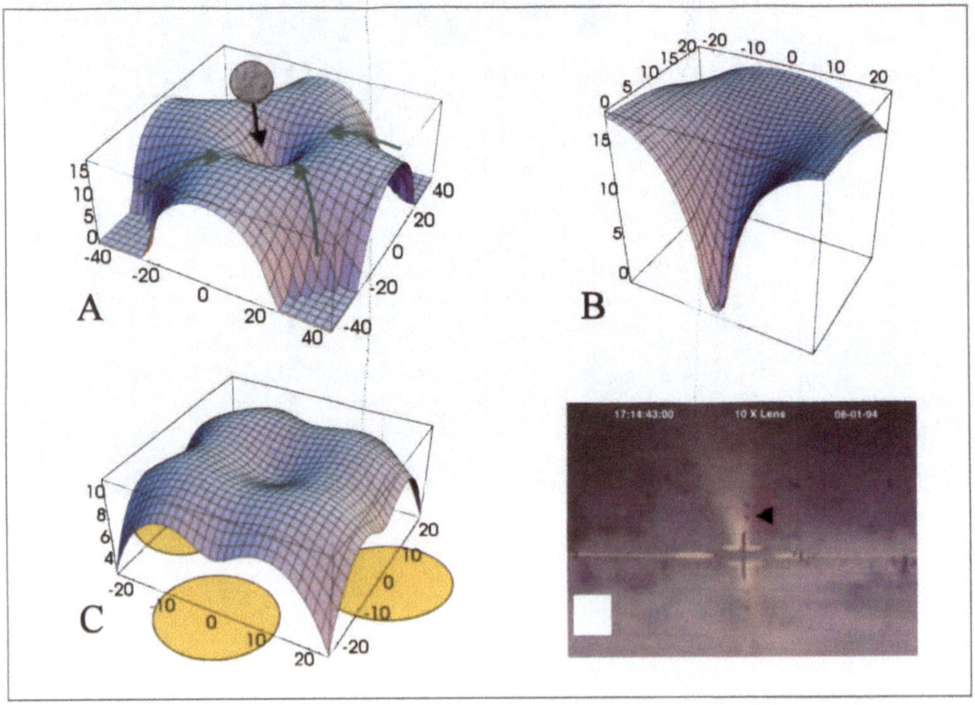

10

See page 430

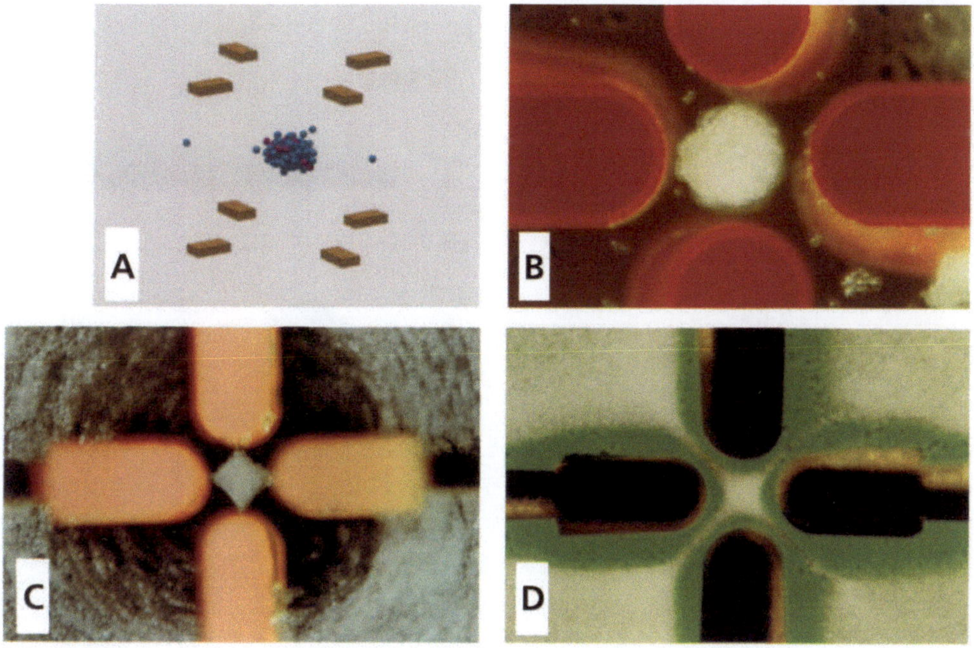

See page 434

11

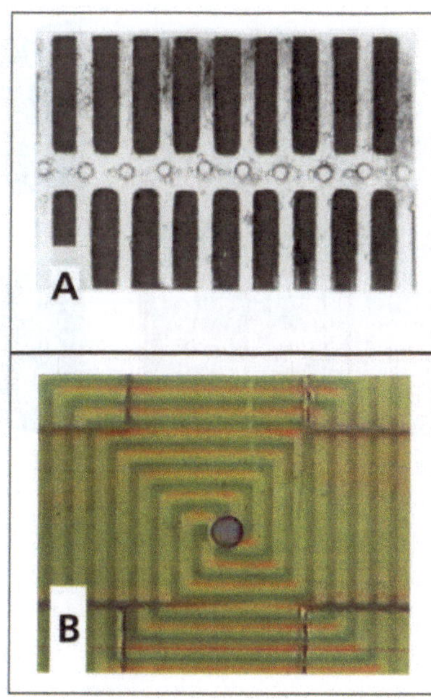

See page 438

12

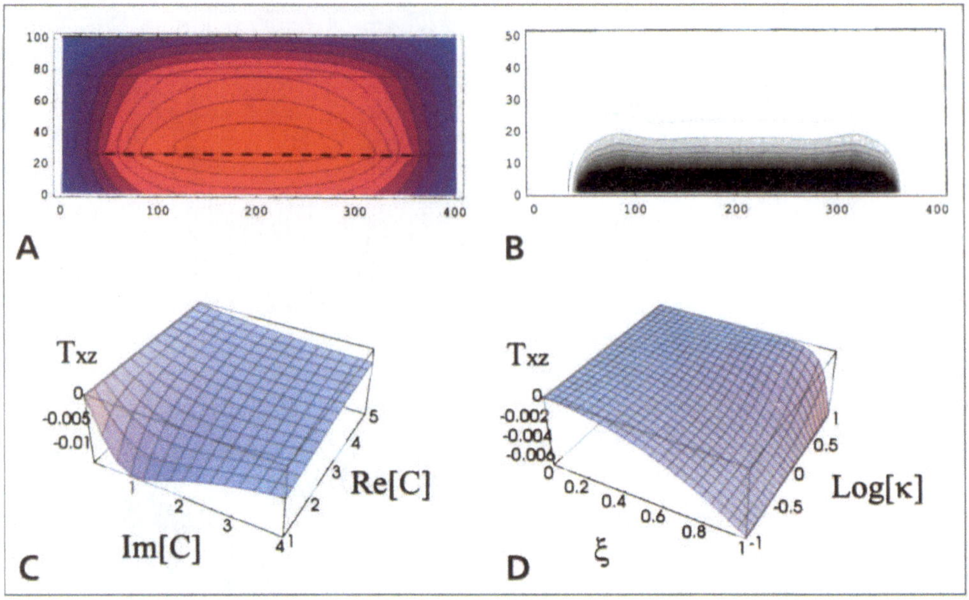

13

See page 441

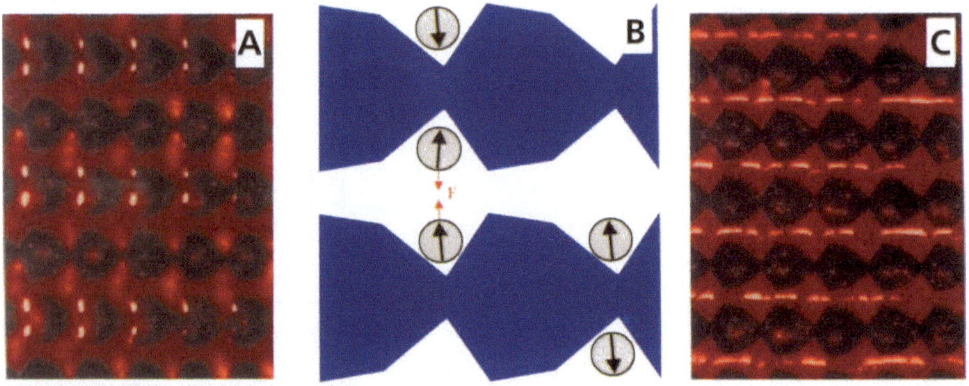

14

See page 449

1 Molecular building principles in nature

J. Michael Koehler and Hans Peter Saluz

1.1 Introduction

Almost all matter of living beings consists of molecules. Discrete molecules are characterized by strong bond forces between their atoms and weak bond forces between the atoms or groups of atoms of different molecules. Only passively functioning elements such as bones of vertebrates or shells of mussels are not exclusively based on discrete molecules, i.e. they mainly consist of inorganic solid-state compounds forming a three-dimensional network of strong chemical bonds.

The determining matter of the living world is composed of high-molecular weight molecules, such as peptides and nucleic acids precisely defined in size and construction. The large molecules of living cells are surrounded by characteristic ensembles of smaller molecules. The complex molecular composition of living cells corresponds to a supermolecular arrangement of a great set of completely different molecules varying in time and space. Thereby, the extended plurality of molecules forms a topological and functional network. Molecules within the cells act as building elements in this network. In addition, the key molecules act as tools for chemical operations, i.e. for biomolecular processing. A subtle molecular and supermolecular construction of living cells is absolutely required to enable the key molecules to act as well controlled tools in the molecular management of basic life processes on the nanometer-scale.

The scene changes using the objects of traditional technical fields where solid materials, such as metals, alloys, solid crystalline or glassy compounds dominate. Synthetic molecular materials were introduced to a great extent by means of the technical development of synthetic organic polymers. However, these polymers are not used in a subtly constructed molecular surrounding, but as large ensembles of only one or a very few types of chemical substances.

The innersphere structure of technically produced macromolecules is rather simple in comparison with enzymes. They are formed by only one or a few building units. The units in the chain of macro molecules are randomly distributed. The length of macro molecules, i.e. the number of units in a chain, is not fixed but deviates more or less from a medium chain length. For technical applications the molecules are exclusively used as materials and not as nanometer-sized tools. The only exception is chemical catalysis.

Biochemistry teaches that enzymes work not only in living beings as molecular tools but also in chemical surroundings, the composition of which can be much less complex than the innersphere of living cells. Molecular biology dealt originally with the complex functional network of biomolecules in living systems. Today it demonstrates, that even highly complex biomolecular processes can be realized outside of the living cells, i.e. in a reduced biochemical surrounding. Recent molecular biology claims that the complex molecular management of basic processes of life is not an unexplainable mystery. It progresses to solve two crucial questions: First, what are the basic principles of molecular organization that enable nature to manage molecules in such a surprising manner? Second, what should we do to use these principles in future technical applications based on molecular biology? The answers to these questions should promote the development of molecular biology from an analytic science to a synthetic one and should help to elucidate the question of what micro system technology can contribute to this process.

1.2 Molecular diversity

Living organisms use only a very small part of the potential diversity of chemical compounds. Most of the approximately 90 chemical elements found in natural terrestrial matter do not appear in living cells or can be found in tiny traces only. The six light elements C, H, N, O, P and S form more than 99% of living matter. What is more, the overwhelming part of the mass of biomolecules which is naturally formed by the six chemical elements mentioned above, represents only a very small part of all classes of chemical compounds. Each of the considered atoms has an approximate size of $(0.2\ \text{nm})^3$. The number of possible arrangements of six elements of this size increases tremendously with the size of a given volume. In a one-nm^3 cube, there are $5 \cdot 5 \cdot 5 = 125$ positions. A number of 6^{125} possible arrangements can be calculated, if six different elements could be arbitrarily placed at each position. This huge number is hardly reduced if different arrangements are assumed to be equivalent due to a possible symmetry of some of the considered cubes. This number is increased to 6^{125000} if a cube of 10 nm-edge length is taken for the arrangement, and to a value of about $(10^{10})^{10}$, a number of 10 billions Os, if a cube of 1 µm is regarded, corresponding to the size of a small cell. Although the biochemical synthesis is limited to a relatively small set of substance classes, the diversity of biochemical molecules and their biomolecular functions remain unimaginably high. This fact can be illustrated by regarding the number of possible biomacromolecules, which can be build up by a small set of chemical building units. The building units in a macromolecular chain can be presented as letters in a string. The number of molecular building units in the whole molecule (n) corresponds to the length of the string. The number of building unit types (k) corresponds to the number of letters. The number of different molecules (z) increases with the length (z) and the number of types (k):

$$\mathbf{z = k^n} \qquad \textbf{(1)}$$

Even for small (k)-values, the number (z) increases dramatically with the increasing length (n) of the molecule. For an average enzyme of a length of 300 amino acid residues, approximately 10^{300} different molecules could be formed if only the 20 naturally occurring amino acids are used. This number is much greater than the estimated number of protons present in the universe (10^{80}) and also much greater than the product of protons and the estimated elementary time intervals since the "Big Bang" (10^{120}). This superastronomical number of possibilities is characteristic for combinatorial relations. The strongly increased number of possible types by variation of building units in the macromolecules is typical for the so-called combinatorial explosion. How did nature learn to optimize molecules in such an unimaginable space of possibilities? Trial and error alone cannot explain the evolution of living cells where thousands of functionally optimized biomolecules occur. Therefore, mechanisms must exist to manage the optimization problems of data sets of smaller numbers of possibilities. Both variation and selection in populations are the ecological aspects of natural optimization during evolution. The development of a reasonable set of rules for the management of individuals in populations, specialization and diversity in complex systems should be the rational aspect in this process. So far the well-determined management of single molecules cannot be realized and used in technical optimization of bio-analogous molecular information. Therefore, we have to use the advantage of natural intelligence in molecular managing by combining the handling of populations or similar entities of molecules by suitable sets of rules of handling procedures. The knowledge of these rules should support the study and reproduction of elementary structures and functional operations of biomolecular processes. If this biological know-how will be used in future, techniques will be available to develop highly complex chemical systems and operational rules. By this means technical systems of the future could not only realize the same efficiency as the complex biomolecular systems of living cells, but probably be definitely superior to all that nature developed during evolution.

1.3 Biomolecular construction

Hierarchy of molecular interactions

Life is built up in space and time. Even on a molecular scale the control of structure and dynamics in space is one of the most important preconditions of life. This control of space is realized in nature by the system of connections between atoms and molecules in a well-defined spatial arrangement. Nature controls the space by a management of topologies of interactions between bulding units. On a molecular scale, a hierarchy of molecular interactions is used for the organization of significant topologies in order to control the self-organization of complex chemical systems in space and time. The management of molecular diversity in nature is based on the handling of a hierarchy of chemical interactions between particles. Covalent binding forces between high-energy atoms, such as the C-C-bonds in aliphates guarantee the long-time stability of many smaller molecules and molecular building units. This type of bond is also very stable in aqueous solutions, and almost independent of pH and ionic strength. Strongly oxidizing agents or special chemical tools, such as redox enzymes, are necessary to cleave such bonds.

The second class of bonds in the hierarchy of strength is represented by strong covalent bonds of a given polarity. The C-O-, C-N-, C-S and P-O-bonds are important examples of this type. These bonds are comparatively stable. However, they are more easily attacked by water and its dissociation products H^+ and OH^-. The reason is their asymmetric distribution of electrons between more and less electronegative atoms. In nature this type of bond is used in stable, but more variable molecular configurations. This class of bond is found in the central chemical function of hard molecular interactions. They are responsible for the functionalization and realization of a strong coupling of elementary molecular building units.

Almost all important classes of biomolecules are constructed of such elementary building elements which consist of a significant combination of these two classes of bonds. All more stable macromolecules are composed of building units which are coupled by the more flexible polarized covalent bonds. The formation and positioning decision of these bonds are mainly realized by condensation and hydrolysis reactions. In an aqueous environment it was shown that hydrogen is particularly suited to saturize the end valences in functional groups. The O-H-, N-H-, S-H- and C-OH-bonds are small units that enable the molecular building units to guarantee strong interactions by condensation reactions. Atoms, such as C and N, usually undergo bonds in more than two directions. Nevertheless, all strong covalent bonds used for the connection of the elementary building units in biomolecules, lead to a linear arrangement, to a first, one-dimensional organization. The reactive interaction leads to bonds that enable the elementary molecular units to rotate, i.e. only one dimension of the molecular topology is determined by the primarily strong covalent bonds. The coordinative bonds represent the next higher hierarchical group of binding strength with a very broad spectrum of binding forces. They are characterized by the donation of one or a few electron pairs of a so-called ligand to a metallic cation. Usually, one central cation is surrounded by several ligands forming a coordinative (or "complex") compound.

More stable, naturally occurring coordinative bonds are realized by ligands with two or more coordinative binding electron pairs, thus forming "chelate complexes." Single valent ligands such as halide or hydroxide ions have more or less an assisting function. The complex compounds have a characteristic shape due to the multiple valence of coordinative bonds to the central ion. Nature preferably uses complexes with a set of stronger coordinative bonds realized by both chelate ligands in one plane and weaker coordinative interactions in the axial direction. By this means, coordinative bonds become a classification principle to manage the second dimension in molecular arrangements. In many molecules, the different chelate ligands of a complex compound are connected by a chain of rotary covalent bonds. This freedom of rotation can be limited by the formation of a complex with the central metal atom, thus leading to a topological determination in more than one dimension. Both of the first mentioned kinds of covalent bonds are so strong that molecules of linear chains of tens of thousands of atoms remain stable, even if each unit is connected by one bond only. In order to manage the molecular topologies and their dynamics in higher dimensions, weaker bonds are used. These bonds stabilize the whole arrangement by both their individual binding force and the cooperative action of a plurality of such bonds. In addition to the coordinate bonds that involve a large ensemble of weaker bonds, the even weaker H-bridge bonds are involved in the molecular organization. However, the character of a more spacious molecule is practically not influenced by a single H-bridge bond. The number, the local density and the distribution of multiple H-bridges determine the three-dimensional topology of many biomacromolecules and their supermolecular arrangements.

Hierarchy of structures

The hierarchy of interatomar and molecular interactions is a precondition to control the huge fundamental diversity of molecules. However, this is not sufficient to explain the control of the superastronomical number of the random positioning of particles in space. There is an additional principle required enabling nature to reduce the molecular diversity to a controllable level. This principle was found in a very simple set of rules of structure formation. This set uses the possibilities of well-stepped interaction forces:

(1) Biomolecular systems are organized by hierarchies of chemical units.
(2) Strong chemical bonds form the basic elements of the hierarchy.
(3) The stability of the characteristic chemical bonds of a structural level decreases with the increasing level of the hierarchical molecular systems. The number of these typical bonds increases with the increasing organization level.

(4) Macromolecules are built by standard building elements.
(5) The basic standard building elements consist of both, a variable and a constant region.
(6) The constant region of molecular building elements is responsible for standard coupling reactions between the elementary building units. Usually, this constant region possesses two complementary reactive centers which enable the building units to form macromolecular linear chains.
(7) The variable region of building elements is responsible for specific properties of the building unit.
(8) A limited set of variable regions is used in the building elements of a class of biomolecular macromolecules.
(9) Macromolecules are first constructed as linear chains. By this means, the order of coupled building units represents primarily a one-dimensional object (string). The stability of a chain is guaranteed by comparatively strong binding forces between the building units.
(10) Second, the linear chain is folded to a three-dimensional object. This process is governed by the order of functions in the variable regions of the different building units within the chain. It is supported by the formation of medium and lower chemical bonds. (Probably, in this folding process, there is at least one additional level of structural hierarchical organization. This level has to determine block regions of macromolecules. We assume that the binding forces within the blocks are higher than between the blocks. See below!)
(11) Third, the folded macrobiomolecules are assembled in a supermolecular arrangement by low binding forces.

1.4 Management of topologies – supermolecular arrangements

The hierarchy of chemical structures in biomolecular systems corresponds well with the hierarchy of topologies. First, the topological connection between atoms within the building elements is fixed. Second, the linear topological interconnection in the macromolecular chain is defined. There are several aspects of a third step of an intramolecular organization in form of submolecluar units of a limited number of elementary building units in the construction of macromolecules. Loops, sheets and helical structures of domains or stiff chain elements and more flexible parts in polypeptides and nucleic acids are typical examples for such a submolecular structuration of macromolecules. Fourth, the three-dimensional topology of the molecular parts in folded molecules is fixed, and finally, the supermolecular topology is defined by the arrangement of biomacromolecules in a spatial organization to realize specific chemical functions.

1.5 Management of the random- and time-nanomachinery by chemical kinetics

Following the ideas of chemical kinetics, it becomes clear that the most stable bonds guarantee the longest time scales of stabilities. This fact is responsible for the long time stability of the elementary building units in biomolecular systems. The bonds between the building units are a little bit less stable, resulting in a slightly reduced lifetime of the macromolecular chains. Most of the single weak bonds stabilizing the three-dimensional shape of a folded macromolecule show much shorter lifetimes, i.e. the lifetime of a certain three-dimensional state can be very short. This small time constant of the three-dimensional entities and configurations enables the molecules to undergo changes in shape and motions of molecular parts. Such motions and changes in shape are crucial properties of biomolecular functions in the living world. The hierarchy of lifetimes in the molecular organization is the kinetic basis for the functional organization of life. Moreover, in molecular terms, this hierarchy of time constants forms the bridge between the kinetically controlled chemical world of smaller molecules and building units and the nanomachinery of the supermolecular biological world with their well determined mechanisms of individual molecular or supermolecular changes, respectively. Whereas small molecules and building units follow mainly the statistical laws of conventional kinetics, their reactions and transport become support for the more individually constructed macromolecules and supermolecular arrangements. Controlling the time scales by the structural and binding force hierarchy, the small molecules become "slaves" in the growth and movement of macrobiomolecules and by this means support their function as biomolecular tools in the even much more complex system of living cells.

1.6 Management of information

All properties of a living being can be understood as information. From an evolutionary point of view, information is the set of facts which enables a being to survive and produce offspring. Both definitions are subjective, but they illustrate how complicated the description of the information of living systems is. A more formal way to the definition of information came from informatics. It describes the loss of uncertainty H by the definition of characters in a set of data with the individual probability p_i (Shannon and Weaver, 1971):

$$H = \text{Sum}\ (p_i \cdot \ln(1/p_i)) \qquad \textbf{(2)}$$

or in case of equal probabilities of all possible states W with

$$p_i = 1/W \qquad \textbf{(3)}$$

$$H = \ln\ (W) \qquad \textbf{(4)}$$

This corresponds with the well known Boltzmann-Planck-Hartley formula for entropy in quantum systems:

$$S = k_b \cdot \ln\ (W) \qquad \textbf{(5)}$$

We are currently unable to analyze the information H (Eq. 4) for a living system as a whole, but we can look at parts of cells in which information can be described formally, that is, in the sequence of macromolecules. Therefore in the following, we assume for simplicity that the molecular information of sequences in genomes is regarded as a representative for the organism as a whole.

Regarding the above-mentioned numbers of possible arrangements of building units in macromolecules, it is absolutely impossible that meaningful biomolecular systems were optimized for whole macromolecules or even

Table 1 Possible states of free configurable amino acid sequences for different genome sizes

Biological organism level/organism	Number of amino acids represented by genome (n) approx.	Calculated number of possible states (k^n) approx.	H (with $W = k^n$) approx.
Protein	400	10^{520}	1200
Simple virus	5000	10^{6505}	15000
Simple prokaryote*	150000	10^{195000}	448500
Eukaryotic cell (yeast)**	2500000	$10^{3250000}$	7500000
Invertebrate (drosophila)***	55000000	$10^{71500000}$	165000000
Vertebrate (man)***	1000000000	$10^{1300000000}$	3000000000

* *Number taken from Moor and Sitte, 1971, Fig. 24.*
** *Number estimated following A. Hinnen, personal communication, 1997.*
*** *Data taken from Darnell et al., 1994, Table 4–3.*

whole cells by a purely trial and error process of replication, variation and selection. This fact is the reason for the following speculative look at another way for the management of information in the optimization of biological macromolecules.

There are several possibilities that the optimization of biological information, structures and functions proceeded in a process, in which each level in the hierarchically organized biological system was a target of selection. In this manner not all possible combinations of elements for the whole system had to be varied, but only the information on the next lower level of a regarded subsystem. Some simple reflections about boundaries of evolutionary optimization should help to understand the principle of biomolecular optimization. We regard the optimization between the single amino acids and the level of a simple procaryontic cell with a genome corresponding to about 150000 amino acid residues. We suppose that only a limited number of individual species (types) could be optimized at each level during evolution, and say that a number N of about 1025 tested individuals in a population changing by variation and selection is realistic to assume an evolutive optimization during a geological time period:

$$N = 10^9 \text{ years} \cdot 10^4 \text{ generations/year} \cdot 10^{12} \text{ individuals/generation} \\ = 10^{25} \text{ individuals} \quad \mathbf{(6)}$$

Further, it is assumed for this number that the individual mutation rate must not grow over the reciprocial value of string length in order to avoid information loss by error accumulation. In addition, reasonable volumes, concentrations and generation times are not in contradiction to the assumed number. Next, we need to know the multiplicity at each organization level. Therefore, we assume for simplification that all levels have nearly the same multiplicity, which means that the same number of subunit types k, and the same length m of the subunits formed by the next lower level are valid (principle of self similarity of the hierarchy). So, it can be formulated for each level i:

$$k_i = \text{const.} = k \tag{7}$$

$$m_i = \text{const.} = m \tag{8}$$

At each level and in each subsystem, the number of possible states should be equal to the maximum value of individuals tested in the evolution:

$$N_i = k_i^{mi} = 10^{25} \tag{9}$$

It is assumed that the variation and the selection proceed in all levels approximately independently, meaning that N is assumed to be equal to all N_i. The number of levels in the hierarchy z can be estimated from the length of the whole genome n (amino acid related data set) and the node strength m (number of subunits in each unit):

$$n = m^z \quad \text{or} \quad z = \ln(n)/\ln(m) \tag{10}$$

Under this assumption, the node strength m should be the same for the lowest level and can be calculated additionally from the number of basic building element types k, which is 20 (native amino acids):

$$N = N_1 = k_1^{m1} \quad \text{or} \quad m = \ln(N)/\ln(k) \tag{11}$$

For the example (simple prokaryote):

$$m = \ln(10^{25})/\ln(20) = 19.2 \tag{12}$$

and the number of levels z (following Eq. 10):

$$z_{prokaryot} = \ln(n)/\ln(m) = \ln(150000)/\ln(19.2) = 4.033 \tag{13}$$

Both estimated data m and z can be interpreted in a very simple model of hypothetical genome subdivision (Tab. 2):

Table 2 Hypothetical formal structure of the genome of a simple prokaryote

Level No.	Length (amino acide residues)	Interpretation
4	150000/19.2	Organism (cell)
3	7813 /19.2	Supermolecular arrangement (multiprotein complex)
2	409 /19.2	Protein
1	21	Oligonucleotide building block
0	1	Amino acid

Levels 0, 2, 3 and 4 can be very simply interpreted as functional units. They are evolved as specific types in natural evolution and observable in all cells. But the interpretation of level 1 offers a very interesting parallel to cell and early evolution, too, if this level was compared with the nucleic acid molecules of corresponding length, i.e., length in the range of about 63 bases (compare Orgel, 1986). In the t-RNA we find a molecule of about such a length (76 bases), which is essential for the translation of genetic code in protein sequences and which was identified as key molecule in the molecular evolution of early life (Eigen, 1980). Perhaps it is not accidental that the number of basic elements (20 amino acids) is very close to the estimated length of subunits ($m = 19$). It can be speculated that the rational selection and variation principle has been worked already at the level of the amino acids, and “20” is the general magic number of natural evolution. But following the explained model, this magic number can be understood. It is explained by the gain of information of a certain size and structure in a set of individuals living in time (generations) and space (populations). It seems to be possible to extend the above speculative considerations, which can be interpreted for the simple prokaryote, to higher organisms, too. This schematic way reflects the reality only in a rough approximation, but may help to understand how rational nature might be in handling information. The information which must be managed in populations is compared by measuring the information of a evolutionary product in the case of free combinations (H0), and for the model of hierarchically organized variation and selection H1 (Tab. 3):

Table 3 Hypothetical structure of genomes and aspects of information managment in evolution (with k1 = 20 and m = 19)

Levels z	Level interpretation	$n (= 19^z)$	$\lg(k^n)$	$H0 = \ln(k^n)$*	$H1 = \ln(k \cdot m^{m \cdot z})$
0	Amino acid	1	1	2.3	3
1	Oligo building block	19	25	57.5	59
2	Protein	361	470	1093	115
3	Supermolecular complexes simple viruses	6859	$9 \cdot 10^3$	$2 \cdot 10^4$	171
4	Complex viruses, simple procaryonts	130 321	$17 \cdot 10^4$	$4 \cdot 10^5$	227
5	Eucaryotic cells	2 476 099	$3.2 \cdot 10^6$	$7.5 \cdot 10^6$	283
6	Simple organisms	47 045 881	$61 \cdot 10^6$	$142 \cdot 10^6$	339
7	High organisms like vertebrata	893 871 740	$1.1 \cdot 10^9$	$2.7 \cdot 10^9$	395

* *corresponding to Shannon's information (Shannon and Weaver, 1971).*

The values for n in Table 3 deviate only slightly from the data of Tables 1 and 2, suggesting the general importance of the discussed model. In reality, not all levels should have exactly the same values mi, but from the economical point of view of hierarchical evolutionary processes, the m_i-values for the various levels should not deviate too much from each other. The comparatively low values of H1 in comparison with H0 suggest that the evolution of complex organisms is controlled less by random walks in a very large sequence space but by extraction of fewer paths in a hierarchically clustered sequence space, which is given by comparatively small sets of ordering parameters. These ordering key parameters might be represented by the organization of main levels of biological complexity.

1.7 Molecular optimization technology

In the laboratory we do not have geological time scales for the optimization of molecular information. Therefore, the rational design and the rational planning of experimentation is a decisive precondition for successful optimization in the field of the great systemic multiplicity discussed above. The handling of great numbers of molecules (10^{20}–10^{30}) alone does not solve the problem. The modeling of examples in a superastronomical number of possible cases alone cannot satisfy the challenge for well-understood and well-controlled molecular optimization. But the integration of models on the molecular dynamics and the subcellular nanomachinery of living systems on the one hand, and the technical work with artificial populations of chemical species (substance libraries) on the other hand could enable us to understand better the natural organization of dynamic complexity and open the way for intelligent optimization of complex bioanalogous molecular functions and artificial nanomachineries.

We can learn from nature that not only systematic screening and rational design help to find the way to optimized molecules. It seems that the clustering of sequence spaces and the adaptation of library sizes and structures to optimize problems can be decisive in accelerating the development of new substances for materials or drugs. The technical problem consists in the finding of the right way to connect rational information and screening strategies. This calls for computer controllable systems with integrated substance handling and information handling. It is assumed that the technical development of new substances will in future become a problem of the cross linking between substance handling, reaction control and digital information control. Chemical recipes will no longer remain as a fixed data set, but they will be improved during automated cycles of mixing, reaction initiation, reaction and product measurement under computer control using sets of basic rules.

Whereas conventional laboratory techniques are barely able to solve the problems of handling substance populations in systems requiring rapid optimization, micro system technology was invented and developed during the last few years as a powerful tool for biomolecular investigations. It pairs functional advantages like large surface-to-volume ratios, fast diffusion and fast heat transfers with logistic advantages like complex channel systems and highly parallel substance carriers in small device volumes.

The marriage between micro systems and molecular systems offers a very promising additional development beside the use of physical micro techniques for molecular studies and substance development. Traditionally, micro system technology uses non-biological matter as material, preferably inorganic solids. The use of biological and much more technically adapted bio-analogous materials and molecular tools in technology opens an exciting prospect for a new technical culture. The study and the development of the use of biogenic and bioanalogous materials in micro system technology and the application of micro systems and micro components in biomolecular handling is more than a recent challenge to a technical improvement. They open a completely new world for technology and science.

References

Darnell J, Lodish H, Baltimore D (1994) Molekulare Zellbiologie. Walter de Gruyter, Berlin, New York

Eigen M (1980) Das Urgen. Johann Ambrosius Barth, Halle

Mohr H, Sitte P (1971) Molekulare Grundlagen der Entwicklung. Akademie-Verlag, Berlin

Orgel LE (1986) Molecular Replication and the Origins of Life. *In*: J. De Boer, E. Dal and O. Ulfbeck (eds): The Lesson of Quantum Theory. Elsevier Science Publishers B.V, 283–293

Shannon C, Weaver W (1971) The Mathematical Theory of Communication. Urbana, Chicago, London

2 Robotic equipment and microsystem technology in biological research

Holger Eickhoff, Igor Ivanov, Markus Kietzmann, Elmar Maier, Markus Kalkum, David Bancroft and Hans Lehrach

2.1 Introduction

Each cell of a living organism contains the whole genetic information in form of DNA molecules. The size of the DNA from a single cell, or genome, of human beings is 3×10^9 nucleotide base pairs. Although the DNA information is usually identical in each cell there are several hundred different cell types. This is due to the fact that genetic information is read out from genes and transcribed from DNA into a cell-specific population of mRNA molecules, which itself can be further translated into different types of proteins. Every step of these cellular processes includes complex interactions of DNA, RNA and protein (Alberts, Bray et al., 1994). To understand these interaction mechanisms scientists started to decode the genetic information (Dulbecco, 1986). This task became finally a major goal of the Human Genome Project (Cantor, 1990).

The benefits of this project are visible already before one-tenth of the genome sequence has been revealed. Genomic databases have enabled scientist to access, retrieve and process biological information (Zehetner and Lehrach, 1994). At the same time, the Human Genome Project has changed the attitude and direction of biological research (Tilghman, 1996). Currently, the interest of researchers is focused on finding genes, analysing their expression patterns and their *in vivo* functions as well as further features of the corresponding proteins. The order and the expression profile of biological information is another level of complexity even more important for the understanding of organisms. Genes whose expression is highly specific to a tissue, organ, cell type or disease may be attractive as targets for the development of highly specific therapeutics and diagnostic (Maier, Meier-Ewert et al., 1997).

Since there are approximately 100000 genes predicted for the human genome, new methods and reliable techniques for processing many samples in parallel and at high throughput are needed.

Here we describe how automated robotic systems can facilitate biological research. Robots have been developed mainly for the parallel analysis and the characterization of large DNA array (Meier-Ewert, Maier et al., 1993, 1994). These automated techniques allow the examination of tens of thousands of clones in parallel by hybridisation-based approaches. We also show how to implement the principles of these robotic systems for other biological tasks including protein analysis and the characterisation of gene expression.

2.2 Hybridisation based approaches to genome analysis

Most of the methods for DNA characterisation are based on the fundamental fact that DNA is able to form a full or a partially complementary double helix hybrid from two separate single stranded nucleotide chains (Watson and Crick, 1953). Hybridisation is the interaction of two DNA strands. To detect hybridisation events one strand (target) is usually immobilised on a solid support, e.g. nylon membranes, whereas its counterpart (probe) is fished out by the target from a hybridisation solution. The probe is labelled and the hybridisation is detected by measuring the signal on the solid surface in the region of immobilised target (Wetmur, 1991; Meinkoth and Wahl, 1984).

Hybridisation approaches are important tools for large scale DNA characterisation and require, among other things, upstream clone picking and spotting, the probe hybridisation itself, and downstream image and computer analysis (Lehrach, Bancroft et al., 1997). First of all, a pool of DNA molecules to be analysed is prepared for insertion into bacteria such as *E. coli.* Randomly spread colonies of bacteria are grown on agar plates. Each colony carries a unique DNA fragment or clone. Since bacteria can carry only a relatively short DNA fragment, a large number of clones, or a clone library, is needed for a full coverage of a genome or even a tissue-specific cDNA library. A typical size of a cDNA library is a hundred thousand clones. After picking, selected clones can be grown and kept in microtitre plates. This allows long-term storage, analysis and subsequent retrieval of individual clones. Clones from microtitre plates can be used for DNA amplification by PCR or they can be arrayed on nylon membranes for subsequent hybridisation with specific probes (Meier-Ewert, Maier et al., 1993).

Automated clone picking

The pattern of colonies randomly grown on agar plates is checked by an image analysis system to address the position of the colonies for picking. The randomly grouped, proportioned and shaped bacterial colonies are automatically selected on the basis of given criteria: colour, shape, size. The image analysis software is able to recognise clones as small as 0.5 mm in diameter and select for blue/white genetic systems in *E. coli.* After defining the colonies' position, software translates coordinates into robot movement for picking. One pin of the picking head (Fig. 1) touches the colony. Then the 96-pins of the picking head transfer and inoculate colonies in a microtitre plate for growth and storage.

We have integrated the picking feature in a flat-bed robot being capable of picking and spotting. In the past 8 years we designed and tested several

Figure 1 *Picking Head. The picking head consists of three major parts: a CCD camera for taking a picture of an agar colony tray (left side), an x, y moving table that guides the pressure line to the chosen pins and the picking tool with 96 pins*

generations of clone picking systems. The system is capable to pick and inoculate approximately 3000 clones per hour into 384-well microtitre plates, (Maier, 1995).

Robotics systems for automated clone arraying

After picking and growing thousands of colonies in microtitre plates, the colonies are arrayed with a 384-pin head (Fig. 2) onto nylon membranes. The spotting head is moved on a servo-controlled, three-axis, linear drive system with an accuracy of 25 µm. A complete spotting run includes the handling of up to 72 microtitre plates, bar-code reading, lid lifting, 384 parallel clone transfers, pin sterilisation and pin drying. The volume of liquid transferred with a pin depends on the tip diameter, which varies from 150 µm to 450 µm which corresponds from 5 nl to 50 nl liquid volume. The smaller the pin diameter, the higher the spotting density that can be achieved. For routine ope-

Figure 2 *Spotting Tool. The 384 individually spring-loaded pins are mounted with 4.5 mm spacing in the area of a microtitreplate. The stainless steel pin-tip size varies between 150 µm and 450 µm*

rations 27648 clones are spotted in a duplicate pattern as a 5×5 box format around 2304 guide dots per $22 \text{ cm} \times 22 \text{ cm}$ nylon membrane. The duplicate spotting simplifies detection and identification of positive clones after each hybridisation. The nylon membranes have been reused at least 20 times without significant loss of hybridisation information. With the system described here it is possible to immobilise and analyse up to 147456 clones on a single $22 \text{ cm} \times 22 \text{ cm}$ nylon surface.

Large scale thermocycling

In addition to growing arrayed colonies on membranes, suitable DNA amounts can be generated by DNA amplification from colonies, e.g. by the polymerase chain reaction (PCR). PCR techniques (Saiki, Scharf et al., 1985; Saiki, Bugawan et al., 1986) have been developed for a long time and they play a central role in large scale genome analysis programs. Commercially available cycling devices can handle up to four times 384 probes in parallel. We have built a laboratory thermocycling prototype for high throughput DNA amplification based on large water baths (Maier, 1995). A basket filled with 135 different 384-well microtitre plates (51840 reactions) is moved with a pneumatically driven x/z sliding stage between three 220-liter water baths at three eligible temperatures. The microtitre plates are heat sealed with a plastic foil in a commercially available heat sealer to prevent cross contamination. After the amplification step the DNA product is sufficiently pure for spotting (Meier-Ewert, Maier et al., 1993).

Non radioactive hybridisation and detection

The high throughput experiments based on hybridisation require ideally non-radioactive detection methods. At present only a few articles have been published that describe the use of directly labelled fluorescent probes for hybridisation with DNA on solid supports. This is due to the low signal to noise ratio that can be obtained with directly labelled probes. We use a hybridisation protocol which utilises an enzymatic signal amplification (Maier, Crollius et al., 1994). The DNA probe has a tag, e.g. digoxigenin, which is recognised by an antibody conjugated with alkaline phosphatase (anti-digoxigenin-AP Fab fragment, Boehringer-Mannheim). The hybridisation is visualised with a non-fluorescent substrate for alkaline phosphatase called Attophos. This substrate becomes highly fluorescent after a phosphate group in the Attophos molecule is removed. The detection is quite sensitive since each active centre of alkaline phosphatase can process about 10^4 to 10^5 substrate molecules per minute (Cherry, Young et al., 1994). This labelling method is reliable for a wide range of hybridisation probes, ranging from short oligonucleotides to long PCR products. For documentation and analysis of hybridisation the positive signals are detected by excitation by UV light (365 nm) and photographed with a high resolution CCD camera (Photometrics PXL, KAF 1400 chip) through a fluorescence emission filter (589 nm bandpass filter, bandwidth 80 nm, Herolab, Germany). The pictures are digitised into a Macintosh PowerPC 8100. The obtained spatial resolution is drastically increased in other detection systems that use Time Delay Integration Linescan cameras or laser scanning principles.

Automated image analysis

An important feature in high throughput laboratories is the automated, large scale characterisation of positive clones in the investigated libraries. The main requirements to such an automated analysis system are firstly the automatic grid finding on an array and secondly the determination of positive clones. Unfortunately, different hybridisation arrays show different qualities. The quality of the picture can be affected by several factors, e.g. uneven distribution of spots on a picture due to non flatness of nylon membranes or a high hybridisation background. Human judgement is so far the best method for the decision of whether a clone is positive or not. Nevertheless, the algorithms for an automatic spot finding are quite well developed (Geman and Geman, 1984; Lehrach et al., 1997). At the current stage nearly 80% of all positive clones can be scored automatically.

2.3 New technologies in high throughput screening: Miniaturisation is a driving force

Integrated circuits made personal computers possible that have revolutionised the world. In addition, the semiconductor industry has been able to double the complexity of a chip every 5 years with reducing cost.

We now witness the development of modern chip-biology, which adopts methods and technologies from the semiconductor industry. High density arrays with integrated solid-phase oligonucleotide synthesis for rapid multiplex analysis of nucleic acid samples have been introduced (Chee et al., 1996; Hacia et al., 1996; Schena et al., 1996). Microlithographic etching of silicon wafers enables the creation of precisely controlled structures, for example "obstacle courses", which might be a good substitution of common gels for separation of long polymer molecules (Volkmuth et al., 1995). Over the last few years, miniaturisation became a driving force in molecular biology and genome analysis.

High throughput screening methods of clone libraries would benefit from further automation and miniaturisation as well. For example, increasing the density of arrays and therefore the number of DNA targets would increase the analysis speed and the information flow. As an alternative to conventional arraying with pins, a microdrop spotting on demand of technology was developed. With the aim of reducing the size of hybridisation arrays by one or two orders of magnitude, the genetic samples are pipetted with a piezoelectric multi-channel microdispensing robot. The piezoelectric dispensing system was originally developed for use in ink-jet printers. The major part is a piezoelectric element or piezoelectronic actor in tube shape, which expands and contracts in the process of the applied AC-voltage. This actor is connected to a tapered glass capillary with an outlet nozzle size of 25 µm to 50 µm. The liquid that has to be dispensed can be filled into the glass capillary in two ways. The first possibility is to aspirate the liquid through the dispensing nozzle by applying a gentle vacuum. The second one is to fill the capillary from a reservoir in the back. Once a glass capillary is filled with several microliters, liquid droplets can be shot out by applying alternating voltage. The piezoactor expands and then contracts so that a droplet is fired out of the glass capillary (Fig. 3). The microdrop system is able to dispense 25 µm to 100 µm droplets

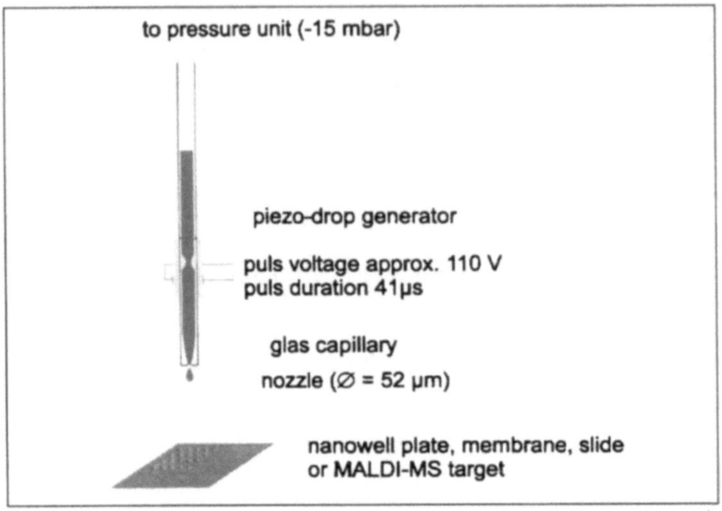

Figure 3 *Scheme of a Piezo-Jet dispenser. A minimum of 11 µl of liquid has to be aspirated. Application of the right voltage together with the right pulse shape to the piezoelectric element results in a drop with 60 µm diameter, corresponding to 100 pl. The drop velocity is approx. 0.58 m/s*

(10 pl to 100 pl). The frequency of a commercially available single nozzle system is approximately 2000 drops per second. The droplet shape is influenced by the diameter of the capillary and the cleanliness of the nozzle edge. Small crystals or a soiled tip surface result in poorly shaped droplets or no droplets at all.

An eight-nozzle head has been recently implemented. It moves in the x, y, z positions with 5 µm resolution using a servo-controlled, linear drive system. A 9-mm spacing between nozzles enables aspiration of solutions directly from the wells of a 384-well or 96-well microtitre plate. After aspirating the samples, dispensing nozzles move to a camera to check whether a suitable droplet is formed or not. The camera captures an image of a droplet in a stroboscopic light. Integrated image analysis system scans the image to verify the quality of the droplets. If a droplet is poorly formed, the image analysis system directs the head to clean the edge of the nozzle. The piezo-dispensing parameters, e.g. the voltage and impulse length for each of the nozzles, are independently controlled (Fig. 4).

With the system described here it is also possible to perform precise filling procedures for nanotiter plates or silicon wafers. Therefore, a second video

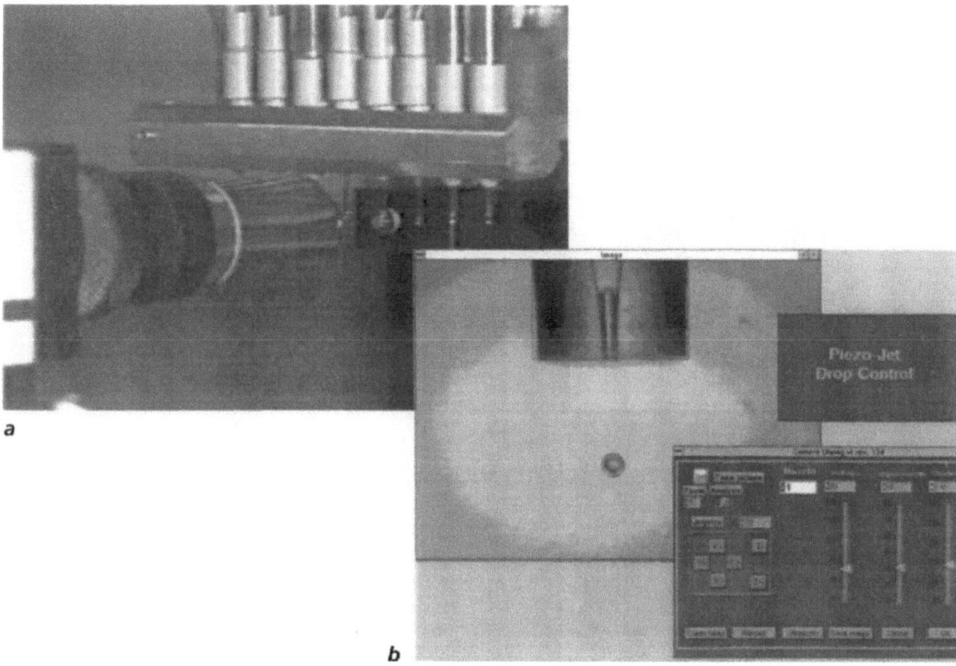

Figure 4 *Automatic drop control. For quality control of drops a stroboscopic image is acquired. (a) Shows the stroboscopic light and the microscope objective connected to the CCD camera. The eight dispensing capillaries can be inspected and adjusted. The drop parameters can be adjusted online (b) using user-friendly buttons for voltage, pulselength and stroboscope delay*

camera is attached to the dispensing head. It enables via software to identify certain cavities on a silicon surface and to dispense a chosen number of droplets independently into each of the chosen cavities (Fig. 5).

The spot size of a microdrop system on a nylon membrane varies between 50 µm and 120 µm and the array density is approximately 2000 spots/cm^2. The functionality of the system allows to dispense on the fly and it takes less than 3 min to array 100 × 100 spots in a square with dot size of 100 µm diameter and 230 µm distance between centres (Fig. 6). At this density it is possible to immobilise a small cDNA library consisting of 14000 clones on a microscope-slide surface.

Another application of the microdispensing technology is the preparation of probe plates for MALDI (Matrix Assisted Laser Desorption Ionisation)-Mass-spectrometry, which has the potential for high throughput applications in DNA analysis. A commercially available MS instrument has a potential to

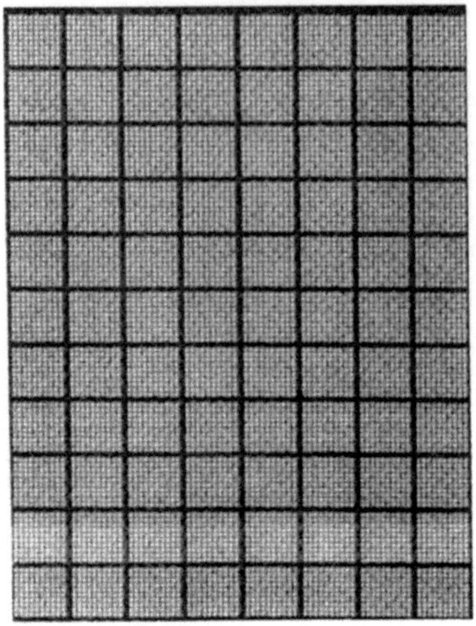

Figure 5 *Piezo Drop Arraying. The picture shows a density of 2150 clones per square centimeter arrayed by the microdispensing device. The spots are 100 μm in diameter and the spacing between the spots is 230 μm. Every spot contains the same DNA fragment. The picture has been taken with a laser scanning device developed at the Max Planck Institute for Molecular Genetics in Berlin*

analyse one probe in a few milliseconds. We have developed a prototype that uses microdispensed arrays of several thousand DNA fragments or proteins on a MALDI-MS target with a size of approx. 2×2 cm.

Unfortunately, there are some drawbacks limiting the use of MALDI-MS in genome analysis. For proteins and genome projects the probes have to be purified. Salts and detergents, e.g. SDS out of SDS-PAGE gels or staining reagents, drastically increase the background during the measuring. The background noise overlays the overall spectrum, making it very often impossible to interpret the obtained peaks. Other limitations include the mass range of the investigated species. At the current stage in MALDI-MS DNA sequencing a maximum of 80 bp can be resolved at a one basepair level (Murray, 1996; Kirpekar, Nordhoff et al., 1995).

With the ongoing miniaturisation process in genome analysis, new tools have to be developed for all necessary handling steps in, e.g. a miniaturised

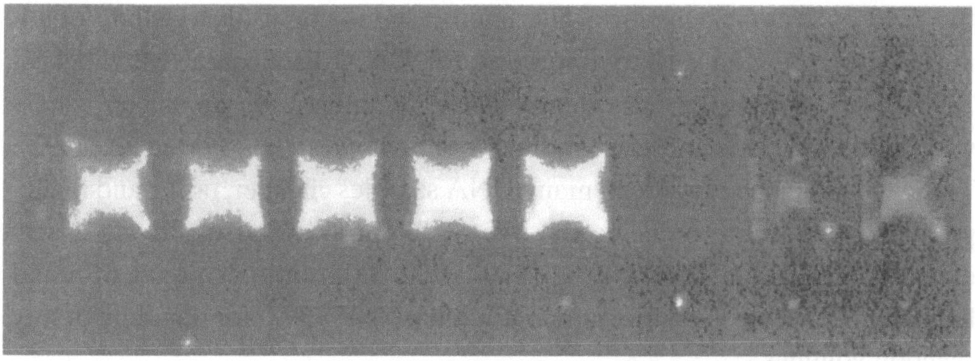

Figure 6 Filling of silicon wafer cavities. Each cavity was anisotropically etched to a size of 0.5 mm × 0.5 mm. The picture shows a row. Five cavities on the lefthand side were filled with 5 nl (50 drops each containing 100 pl) of a fluorescent dye solution (Cy5, Amersham), one cavity was not filled and two cavities on the righthand side were filled with a 10-fold lower concentrated solution of the dye. The total amount of Cy5 in the cavities varies between 1 femtomol (bright signals) and 100 attomoles (weak signals). Picture has been taken with the same detection device as in Figure 5

hybridisation approach. In addition, the detection systems have to be improved. Smaller spot sizes result in less amounts of targets and sensitive detection systems are required with an increased spatial resolution, e.g. using light-optical principles. Optical methods like laser scanning devices for large areas or different microscopy methods including confocal laser scanning microscopes for areas smaller than a few mm^2 are the methods of choice in the near future. For the analysis of single molecules in small cavities other optical methods like scanning nearfield optical microscopy (SNOM) (Moers et al., 1996; Iwabuchi et al., 1997) or fluorescence correlation spectroscopy (FCS) (Rigler, 1995; Oehlenschlager et al., 1996) as well as non optical methods like atomic force microscopy (AFM) (Hansma et al., 1996; Lyubchenko and Shlyakhtenko, 1997) have successfully been tested for individual experiments.

2.4 Conclusions

In the future there will be more and more "Lab-On-A-Chip" devices (Service, 1995), powerful integrations of microfluidics, micromechanics and detection systems (Kovacs et al., 1996). Some application of these devices could be in the field of faster diagnostics. The microsystems might also include oligonucleotide arrays on different surfaces for DNA diagnostics (Mirzabekov, 1994). There are promising examples of PCR integration on silicon wafers with online detection and/or analysis (Woolley et al., 1996; Taylor et al., 1997). These technologies can screen many DNA samples cheaper, faster and highly parallel. In addition, reducing the size of the processes can provide a new experimental design (Burke et al., 1997), e.g. different techniques of liquid handling like electroosmotic pumps (Freaney et al., 1997) for handling of tiny reaction volumes.

For these scenarios to become reality soon, molecular biologists must work more closely with engineers, physicists and chemists to remove the gap between the macroscopic "laboratory world" and the microsystems "chip world".

References

Alberts B, Bray D, Lewis J, Raff M, Roberts K, Watson JD (1994) Molecule Biology of the Cell. Garland Publishing, New York, London

Burke DT, Burns MA, Mastrangelo C (1997) Microfabrication technologies for integrated nucleic acid analysis [Review]. *Genome Research* 7(3): 189–197

Cantor CR (1990) Orchestrating the human genome project. *Science* 248(4951): 49–51

Chee M, Yang R, Hubbell E, Berno A, Huang XC, Stern D, Winkler J, Lockhart DJ, Morris MS, Fodor SP (1996) Accessing genetic information with high-density DNA arrays. *Science* 274(5287): 610–614

Cherry JL, Young H, Di Sera LJ, Ferguson FM, Kimball AW, Dunn DM, Gesteland RF, Weiss RB (1994) Enzyme-linked fluorescent detection for automated multiplex DNA sequencing. *Genomics* 20(1): 68–74

Dulbecco R (1986) A turning point in cancer research: sequencing the human genome. *Science* 231(4742): 1055–1056

Freaney R, McShane A et al (1997) Novel instrumentation for real-time monitoring using miniaturized flow systems with integrated biosensors. *Annals Clin Biochem* 34(Part 3): 291–302

Geman S, Geman D (1984) Stochastic Relaxation, Gibbs distributions and the Bayesian restoration of images. *IEEE Transactions in Pattern Analysis and Machine Intelligence*

Hacia JG, Brody LC, Chee MS, Fodor SP, Collins FS (1996) Detection of heterozygous mutations in BRCA1 using high density oligonucleotide arrays and two-colour fluorescence analysis [see comments]. *Nature Genetics* 14(4): 441–447

Hansma HG, Revenko I, Kim K, Laney DE (1996) Atomic force microscopy of long and short double-stranded, single-stranded and triple-stranded nucleic acids. *Nucleic Acids Res* 24(4): 713–720

Iwabuchi S, Muramatsu H, Chiba N, Kinjo Y, Murakami Y, Sakaguchi T, Yokoyama K,

Tamiya E (1997) Simultaneous detection of near-field topographic and fluorescence images of human chromosomes via scanning near-field optical/atomic-force microscopy (SNOAM). *Nucleic Acids Res* 25(8): 1662–1663

Kirpekar F, Nordhoff E, Kristiansen K, Roepstorff P, Hahner S, Hillenkamp F (1995) 7-Deaza purine bases offer a higher ion stability in the analysis of DNA by matrix-assisted laser desorption/ionization mass spectrometry. *Rapid Commun Mass Spectrometry* 9(6): 525–531

Kovacs GTA, Petersen K, Albin M (1996) Silicon micromachining – sensors to systems. *Analyt Chem* 68(13): A407–A412

Lehrach H, Bancroft D, Maier E (1997) Robotics, computing, and biology: an interdisciplinary approach to the analysis of complex genomes. *Interdisciplinary Science Reviews* 22: 37–44

Lyubchenko YL, Shlyakhtenko LS (1997) Visualization of supercoiled DNA with atomic force microscopy *in situ*. *Proc Natl Acad Sci USA* 94(2): 496–501

Maier E (1995) Robotic technology in library screening. *Lab Robotics Automation* 7(3): 123–132

Maier E, Crollius HR, Lehrach H (1994) Hybridisation techniques on gridded high density DNA and *in situ* colony filters based on fluorescence detection. *Nucleic Acids Res* 22(16): 3423–3424

Maier E, Meier-Ewert S, Ahmadi AR, Curtis J, Lehrach H (1994) Application of robotic technology to automated sequence fingerprint analysis by oligonucleotide hybridisation. *J Biotech* 35(2-3): 191–203

Maier E, Meier-Ewert S, Bancroft D, Lehrach H (1997) Automated array technologies for gene expression profiling. *Drug Disc Today* 2(8): 315–324

Meier-Ewert S, Maier E, Ahmadi A, Curtis J, Lehrach H (1993) An automated approach to generating expressed sequence catalogues. *Nature* 361(6410): 375–376

Meinkoth J, Wahl G (1984) Hybridization of nucleic acids immobilized on solid supports. *Analyt Biochem* 138(2): 267–284

Mirzabekov AD (1994) DNA sequencing by hybridization – a megasequencing method and a diagnostic tool. *Trend Biotechnol* 12: 27–32

Moers MH, Kalle WH, Ruiter AG, Wiegant JC, Raap AK, Greve J, de Grooth BG, van Hulst NF (1996) Fluorescence *in situ* hybridization on human metaphase chromosomes detected by near-field scanning optical microscopy. *J Microscopy* 182(Pt 1): 40–45

Murray KK (1996) DNA sequencing by mass spectrometry. *J Mass Spectrometry* 31(11): 1203–1215

Oehlenschlager F, Schwille P, Eigen M (1996) Detection of HIV-1 RNA by nucleic acid sequence-based amplification combined with fluorescence correlation spectroscopy. *Proc Natl Acad Sci USA* 93(23): 12811–12816

Rigler R (1995) Fluorescence correlations, single molecule detection and large number screening. Applications in biotechnology. *J Biotech* 41(2–3): 177–186

Saiki RK, Bugawan TL, Horn GT, Mullis KB, Erlich HA (1986) Analysis of enzymatically amplified beta-globin and HLA-DQ alpha DNA with allele-specific oligonucleotide probes. *Nature* 324(6093): 163–166

Saiki RK, Scharf S, Faloona F, Mullis KB, Horn GT, Erlich HA, Arnheim N (1985) Enzymatic amplification of beta-globin genomic sequences and restriction site analysis for diagnosis of sickle cell anemia. *Science* 230(4732): 1350–1354

Schena M, Shalon D, Heller R, Chai A, Brown PO, Davis RW (1996) Parallel Human Genome analysis – microarray-based expression monitoring of 1000 genes. *Proc Natl Acad Sci USA* 93(20): 10614–10619

Service RF (1995) The incredible shrinking laboratory. *Science* 268 (5207): 26–27

Taylor TB, Winn-Deen ES, Picozza E, Woudenberg TM, Albin M (1997) Optimization of the performance of the polymerase chain reaction in silicon-based microstructures. *Nucleic Acids Res* 25(15): 3164–3168

Tilghman SM (1996) Lessons learned, promises kept – a biologist's eye view of the Genome Project. *Genome Research* 6(9): 773–780

Volkmuth WD, Duke T, Austin RH, Cox EC (1995) Trapping of branched DNA in microfabricated structures. *Proc Natl Acad Sci USA* 92(15): 6887–6891

Watson JD, Crick FHC (1953) Molecular structure of nucleic acids. A structure for deoxyribose nucleic acid. *Nature* 171: 737–738

Wetmur JG (1991) DNA probes: applications of the principles of nucleic acid hybridization. *Crit Reviews Biochem Molecular Biol* 26 (3-4): 227–259

Woolley AT, Hadley D, Landre P, Demello AJ, Mathies RA, Northrup MA (1996) Functional integration of PCR amplification and capillary electrophoresis in a microfabricated DNA analysis device. *Analyt Chem* 68(23): 4081–4086

Zehetner G, Lehrach H (1994) The reference library system – sharing biological material and experimental data. *Nature* 367(6462): 489–491

3 Components and systems for microliquid handling

Steffen Howitz

3.1 Introduction

The present work gives an overview of some selected microfluidic components. It presents some components which are already in use and others which are still in a stage of development. These components allow the solution of microfluidic problems of an entirely new dimension. Examples are dispensing and pipetting systems for the submicrolitre range as well as complex microsystems. Such microsystems allow self-supporting microliquid handling in manifolds containing capillaries which have a length and height of some 100 micrometres. The author intends primarily that readers who are not familiar with microsystem technology may form an idea of this modern technological development. It is intended that the possibilities and limits of microfluidics shall be imparted to the reader with the aim that he will be able to develop and formulate his own visions of new applications. To achieve this within the limited scope of this contribution, the subject will be treated by examples of the possibilities of a system philosophy selected by GeSiM [1].

The following chapter gives priority to microfluidic components and systems which are produced using silicon and glass to allow an active or passive handling of fluids in the range of some picolitres to some microlitres. This seems reasonable and justified, as today microfluidic components and systems made of silicon are mostly advanced. Irrespective of this fact, there are a number of other developments which use molded plastics [2], glass [3] and ceramic materials [4].

Microfluidic components are discrete miniaturized components which can carry out one or more specific functions. Each microfluidic component has at least one fluidic input interface and at least one fluidic output interface. Depending on the function, a microfluidic component can also have electric and mechanical interfaces. Microfluidic components of practical relevance

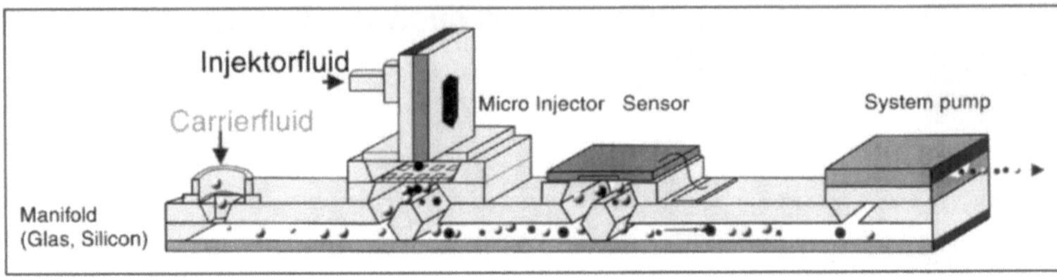

Figure 1 *Cross view of Micro Liquid Handling System [1, 5]*

include chemical microsensors, micropumps, microvalves, micromixers, microrestrictors and fluidic micro-to-macro connecting hardware. Interlinking of microfluidic components in a microchannel system produces complex microliquid handling systems. Figure 1 shows an example of an arrangement as a schematic sectional view, the manifold being made of silicon and glass.

The geometry of the manifold can be varied. Integration of microfluidic components uses suitable layout and bonding methods. The handling of real fluids in microfluidic components or systems causes a number of new problems. For example, surface and interface properties determine the behaviour of fluids to a much larger extent than in macrosystems.

3.2 Materials for the manufacture of microfluidic components

Basic materials for microfluidic components and microliquid handling systems described below in detail are silicon and glass. It is mainly these materials which the fluids come into contact with.

As regards its electronic properties, the type of silicon used for microfluidic applications is not different from that used in microelectronics. The low-price ⟨100⟩ silicon is typically used for wet-chemical microfabrication. The specific resistance of this silicon is in the range of 0.5 to 50000 Ωcm. The electronic properties of silicon form an interesting point of view for the design of microfluidic components with multiple functions, particularly if microfluidic and microsensoric functions are to be combined in the same component. Examples are flow sensors combined with pressure sensors [6], or applications requiring microchannels with heater and temperature sensors in an integrated form [7]. Silicon is easy to pattern in the course of 3D micromachining. It has a high mechanical strength and can be directly combined with glass or silicon. In addition, it has a high chemical resistance to organic solvents, strong acids and weak bases. These properties are the reason why silicon has gained acceptance as one of the most important materials of microfluidics.

Glass is the second important material for the production of microfluidic components and systems. Mostly it is used as an optically polished borosilicate glass of the Pyrex 7740 type (Na_2O-B_2O_2-SiO_2) or Borofloat type [8]. The coefficient of thermal expansion of this glass is adjusted to that of silicon. Therefore, the glass can be bonded at temperatures of 400° to 450°C and at voltages of 400 to 2000 V by the atomic bonding method, anodic bonding, directly to the silicon without using additional bonding materials.

In addition to the materials silicon and glass, bonding materials are significant to the joining of different components. They are required for the assembly, building-up and bonding of microcomponents and they determine the overall behaviour of microsystems to a very large degree. Bonding materials make a major contribution to the contact between the liquid and the system. They have an influence on the mechanical strength and therefore they have to be paid attention to. Typical building-up and bonding techniques are glueing, eutectic alloying and soldering using metallic solders or solders containing glass. Today glueing has gained general acceptance for many bonding processes. This is due to the easy processing, the short time required for the pretreatment of the bonding surfaces and the wide range of available bonding agents.

Alternative materials used for the microsystem-to-macrosystem junction in the manufacture of fluidic connection components are PVC, polymethyl methacrylate (PMMA), polyether ether ketone (PEEK) and special steel.

In microfluidics, bulk material is not the only material used. To achieve special surface properties, silicon and glass may be coated in the course of CVD or PVD processes. Insulator films such as SiC, SiO_2, Si_3N_4 and Ta_2O_5 [9] or metallizations such as TiPt [10]. Another method of influencing the surface properties is the deposition of organically modified polymers [11] with thicknesses of 3 to 5 µm. This material may gain particular importance if used in the liquid form for the "internal lining" of the completely assembled component or microsystem.

3.3 Technologies of manufacturing microfluidic components and systems

The technologies of manufacturing microfluidic components from silicon and those of the production of microelectronic integrated circuits are similar in many aspects. Functional layers are deposited on the surface of the silicon by CVD or PVD processes, exactly like in the production of electronic circuits. This deposition process is followed by photolithography and layer patterning to generate pattern-delineated masks on the silicon surface. These processes are, in turn, followed by three-dimensional micromachining of the silicon as a basic method of manufacturing microfluidic components. Ion implantation, diffusion and metallization are processes which may also be used for the manufacture of microfluidic components. As in the IC production, the silicon wafers go through these procedures several times while they are processed. Each process step is followed by wet- or dry-chemical cleaning steps proceeding in plasma which require a relatively long time. These procedures are carried out in clean rooms. The clean rooms have appropriate venting systems and are air-conditioned, depending on the requirements. Clean rooms used for photolithography are additionally illuminated with yellow light. Many of the process media used are toxic and therefore they have to be disposed of after use in such a way that they do not pollute our environment, requiring a considerable amount of time and money.

The above demonstrates that the technology of IC production resembles that of microfluidic components in many aspects. Now the special features of the production of microfluidic components will be treated. The differences between IC and microfluidic component production result from the different requirements which integrated circuits and microfluidic components have to meet.

The effects used in IC's are essentially based on the fact that electric charges in the semiconductor crystal are transferred, stored or are changed by the influence of an external energy flow. Therefore, mainly the electronic properties of the semiconductor crystal are important to IC's. There is almost no mass transfer in the microelectronic component.

In addition to the transfer of electric charges, the semiconductor crystal in the microfluidic component has to allow mass transfer. This proceeds in three-dimensional structures which can be manufactured to the micrometre to provide extremely defined prerequisites for the moving fluid. In contrast to IC production, it is not the surface of the semiconductor substrate but the three-dimensionally structured bulk material which determines the component. Typical structures are channels, reactors, membranes, breakdowns, micro-bridges, mesa structures and microlattices. Microfluidic components become particularly complex if they require a co-existence of electronic and micro-fluidic functions. Here the pure techniques of IC technology and microfluidics fail. Exactly at this point, new ways have to be searched for, representing the challenges to the technologist in the field of microfluidics.

What are the major technological features of the manufacturing process of microfluidic components? First, it is the process of three-dimensional fine line patterning itself. Second, it is the necessity of carrying out sequential steps in a three-dimensionally patterned silicon substrate. It is exactly at this point where conventional planar technology is left.

The generation of micropatterns of a tight packing density on a silicon wafer requires an anisotropic technique of three-dimensional styling. Basically, microfluidic structures are generated subtractively, that is, by a material-removing etching technique. An etching technique is referred to as anisotropic if there is exactly one direction of propagation of the etch front having a propagation velocity which exceeds that in the directions of propagation of the other etch fronts by several orders.

All anisotropic wet etching techniques, e.g., in a basic batch of a 30% KOH at a working temperature of 80°C, utilize the fact that the most tightly packed principal plane ⟨111⟩ of the silicon crystal is dissolved particularly slowly. Therefore, the principal planes ⟨111⟩ function as particularly resistant and structure-determining etch-stopping planes in the wet etching of single-crystal silicon. All the other principal planes are dissolved faster by up to two orders.

The inlet and outlet regions of a micropump after wet etching are shown in Figure 2.

The principal planes ⟨111⟩ of the silicon block form the lateral boundaries of the channels and grooves which are still open to the top. Uniform steps on the left boundary of the chamber can be seen, particularly on the right side of Figure 2b. These steps reproduce on that pattern features of the layout where there is a deviation from the orthogonality of the principal planes ⟨111⟩ in the lattice. The photo shows a structure of a ⟨100⟩ silicon substrate which inter-

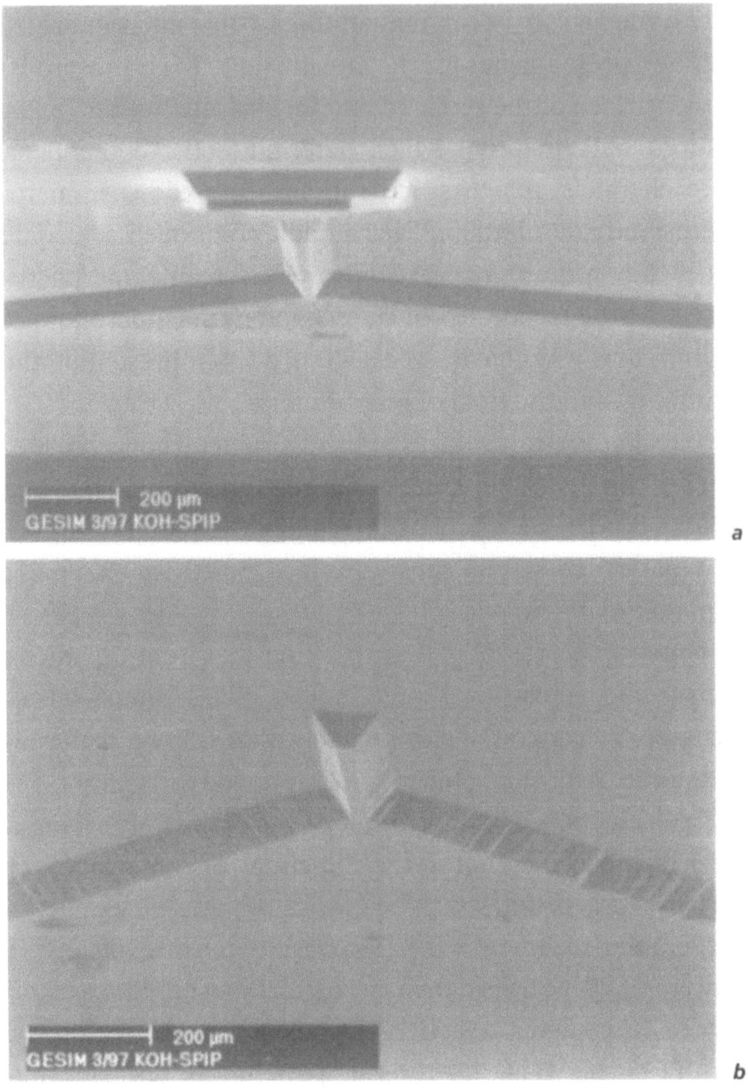

Figure 2 *SEM-picture of a micromachined pump structure (wet etched). a) Inlet pad; b) orifice [1]*

sects the main phase at 45°. The steps are exit points of the principal planes ⟨111⟩ and result in a process-specific roughness. The lateral boundary surfaces of the channels and reactor are formed by principal planes ⟨111⟩. They push through the wafer surface of the ⟨100⟩ silicon substrate at an angle of 54.7°, that is, the microchannels also have those slopes.

Such effects caused by the crystal may have an adverse effect on the performance of fluidic components. Therefore, it is important to develop anisotropic etching techniques which do not have that dependence. Such techniques include all dry etching techniques. They differ in the working gas, working temperature, etch rate and masking system used. Today the technological possibilities and the process costs involved in the etching equipment vary greatly. Just the requirements of microfluidics promoted or at least stimulated these developments.

The microfluidic components presented in this article were patterned by advanced silicon etching (abbreviated to ASE) [12, 13, 15]. This dry etching technique utilizes SF_6 without the presence of oxygen as a working gas. The directional dependence in ASE plasma etching is achieved in a process which constantly proceeds in two stages. At the first stage, all surfaces are coated with a polymer film. At the second stage, this polymer film is partially reopened. If in this process of polymer opening silicon surfaces are exposed, atomic fluorine can etch the silicon. The slopes and boundaries produced in the course of the etching process are also coated with the passivating polymer. However, this polymer cannot be opened, preventing the etching action of fluorine in these areas. All silicon areas which are not perpendicular to the

Figure 3 *SEM-photo of an etched pump structure (dry etching: ASE) [12]*

Table 1 Advantages and disadvantages of etching technologies for the manufacture of deep micropatterns in silicon

	Advantages	Disadvantages
Anisotropic wet etching of silicon	– Allows simultaneous processing in lots. – Etch rate is independent of the pattern and is very homogeneous. – Allows simultaneous processing of both wafer sides. Otherwise masking is necessary. – Electrochemical etch stop allows the manufacture of ultrathin membranes. – Equipment costs are relatively low. – Operating cost are relatively low. – Disposal of residual chemicals is without problems.	– Dependence on the crystal orientation of the silicon block. Only those of the layouts which correspond to the location of the crystal planes can be implemented. Round and deep pattern features cannot be generated. – Only SiO_2, Si_3N_4, Ta_2O_5, Pt, Au and NiCr can be used for the etch mask. – The aspect ratio is determined by the crystal. – Steep profiles can be generated only in wafers with a steep behaviour of the principal plane ⟨111⟩. – Is not compatible with the technology of microelectronics.
Anisotropic dry etching of silicon	– The technique is compatible with the IC technology of microelectronics. – Electronic and fluidic functions can be combined. – A width-to-height aspect ratio of 1:50 can be reached on a base area of few square micrometres. – The anisotropic effect is independent of crystal orientation. Any two-dimensional layout can be implemented. There is no need to take account of the location of the principal planes. – In addition to the usual oxide and nitrite layers, photoresists annealed at 90°C can also be used as a material for etch masking. – Depth patterning of Si can also be carried out on surfaces with patterns of metallization.	– Wafers are always processed on one side only. This may be desirable in selected cases, but the technological processes require much time. – Just a single-slice reactor involves high equipment costs. The costs of a multislice reactor are extremely high. – High operating and maintenance costs – The etch rate depends on the layout and has to be optimized individually. – The disposal of the residual chemicals is expensive and risky.

plane normal of the wafer to be processed are passivated. It is important that the effect of anisotropic removal of material is not dependent on the crystal orientation of the single-crystal silicon. This can be seen from Figure 3 which shows an example of round pattern features. Avantages and disadvantages of dry and wet etching technologies for the manufacture of deep micropattern in silicon are given in Table 1.

3.4 Microfluidic components and systems

Complex fluidic microsystems gain increasing importance whenever diagnostic and analytical problems have to be miniaturized. In future, a main application will be molecular biology. We here refer to the DNA Chip Program of the USA [16]. Molecular diagnosis, clinical diagnosis, active substance research, high-through-put screening and DNA sequencing techniques are examples of applications in molecular biology [17].

The layout of fluidic microsystems orientates itself by the advanced designs of hybrid technology. This orientation is useful, as the building-up and bonding techniques of hybrid technology are particularly suitable just for the crosslinking of components strongly differing in material and geometry. When having a look at the large number of microfluidic components which have been developed so far, we see the significance of the fluidic microsystem technology as an technology which is able to interlink these components within the microsystem efficiently, robustly and reliably.

Many developments towards the fluidic microsystem have been classed under the topic "μ-total analysis systems" (abbreviated to μ-TAS) since 1994 [18, 19].

Manifolds as a basis of fluidic microsystems

A manifold is the system basis of a fluidic microsystem. In a manifold, fluidic components can be crosslinked to form a microsystem. It is the task of the manifold to keep individual components mechanically defined, to interconnect them fluidically and to couple them electrically if required. If the function of the manifold in microfluidics is to be explained, the comparison

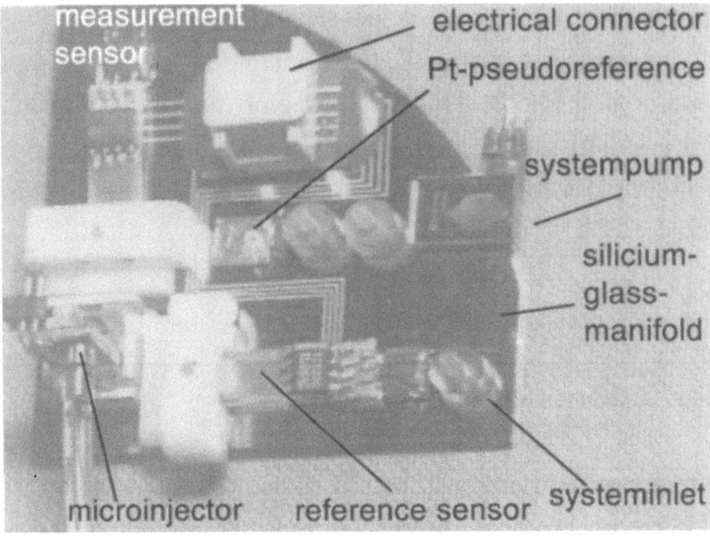

Figure 4 *Detail of a microsystem with a manifold and assembled components [1, 20]*

with a printed circuit board used in electronics may be helpful. Just in the same way as electronic assemblies and systems are created in printed circuit technology, microsystems are created by the combination of sensors, actuators and electronic components on the manifold. For illustration, a manifold of a microsystem used for determining the nitrate concentration in aqueous solutions is shown as a detail in Figure 4 [20].

Figure 4 shows a glass-silicon manifold, parts of the conductor system, chemical microsensors and an injection module for the injection of the calibration liquids. These fluidic components were mounted on the manifold as modules in a hybrid manner. The manifold may comprise one or several fluidic planes as well as electrical functional layers, depending on its material. When designing new microsystems, we proceed on the assumption that the geometric requirements of the components to be integrated are defined and we adjust the manifold to the microsystem. Therefore, the manifold can be varied and defined components are integrated.

Today the service "customized manifold of a silicon-glass construction for use in microfluidics" is offered on the market [1]. The material to be selected is determined by the application and the technological know-how. Major developments of this technology are given in Table 2.

Table 2 Summary of manifold technologies

Material	Advantages	Disadvantages	Ref.
Silicon-glass	– Extremely short iteration cycles – The modular concept is easy to implement – Highly resistant to reagents – Micropatterns can be generated up to the submicron range – Anodic bonding produces an extremely reliable and robust channel system – Easy to link by conventional techniques	– The material is relatively expensive – It requires a lot of time to manufacture multilevel manifolds	[1, 21–24]
Glass surface with titanium/platinum micropatterns	– Simple design to be used for fundamental research – Extremely short iteration cycles if laser ablation is used for fine line patterning	– This technique does not produce a flow system with microcapillaries. They would have to be attached if required.	[10]
Quartz-glass	– Ideal combination for capillary electrophoresis	– Expensive materials – Bonding of materials is critical	[25]
Glass-glass		– Bonding technique is problematic	[26]
Silicon-silicon	– Extremely short iteration cycles – The modular concept is easy to implement – Ideal resistance to reagents – Micropatterns can be generated up to the submicron range	– Risky and time-consuming bonding technique – Expensive materials	[27]
PMMA	– Cheap material – Micropatterns can be generated up to the submicron range – The modular concept is easy to realize	– Low resistance to organic solvents – The technology is very expensive – Very long iteration cycles	[28, 29]

Table 2 (continued)

Material	Advantages	Disadvantages	Ref.
PDMS	– Cheap material – Micropatterns can be generated up to the submicron range	– Low resistance to organic solvents – Manifolds are manufactured by covering from the top and bottom. The technique requires a great deal of time and is unsuitable for mass production.	[30]
Photoresist	– Simple bonding technology – Extremely cheap material – Short iteration cycles can be implemented	– Swelling if aqueous fluids are used – Low resistance to organic solvents	[31]
LTCC-Ceramic	– Crossing fluid channels can be produced – Consistent bonding technology by lamination – Fluidic and electronics components can be integrate in one manifold using screen printing and laser drilling – Cheap and very robust material	– 100 µm minimal structure width – Roughness of LTCC ceramic 3 µm – Shrinking of LTCC in the burn out process of about 15 %	[4, 68]

With regard to the concept and selection of a manifold, some questions have to be answered beforehand. What liquids are to pass the manifold for what time? What components are to be integrated in what manner? Do the desired detecting methods require special materials? What building-up and bonding techniques can be used for the integration of the components? What loading occurs at the fluidic manifold-component interfaces? Are the bonding materials sufficiently resistant to the fluid? What requirements do the electrical interfaces have to meet? How strong is the mechanical loading of the overall system? Can this loading be taken up by the manifold? What do the peripherals look like? Should disposable components be integrated into the manifold and may this be realized by combining macro- and micro-techniques?

Micropump, microdispenser and micropipette components for the fluid drive

There are relatively few suppliers of miniaturized fluid drives on the market in comparison with the large number of fluid drives which have so far been studied in scientific laboratories. Those who are interested in the use of microfluidic components inquire, above all, about the performance parameters and about the availability of the components. As a rule, the potential users do not care to what extent techniques of microsystem technology or just conventional techniques are used for the production of the components. In order that microfluidic components may gain user's acceptance, they have to meet a number of severe criteria. These criteria include cost effectiveness, precision and accuracy in handling of liquids of volumes in the microlitre, nanolitre and even picolitre range, the possibility of volume validation, long-term stability in case of permanent loading, practical handling in priming and cleaning, the robustness of the system, the availability of adapted PC-aided electronic control units, the resistance of the component to organic solvents, acids and bases, the suitability of the component for the transport of suspensions and biologically active substances and the possibility to couple the microcomponent to the state-of-the-art apparatus of liquid handling.

Micropumps

For the purpose of directional transport of liquids, miniaturized fluid drives also have to combine functional components for volumetric displacement and flow restriction. These functional components have to be included in the smallest area. Typical dimensions are some 100 µm × 100 µm. These functional components have to be manufactured with a tolerance of a few micrometers. They have to be assembled to each other accurately. Their construction as a whole should be as simple and robust as possible. As this task is among the most demanding tasks of microfluidics, it is no wonder that today a great number of quite different solutions exist. A selection of microfluidic drives which have so far been realized is shown in Table 3. It is intended to give the reader an idea of the variability of usable principles.

Among the fluid drives to be dealt with below, priority will be given to the presentation of different types of micropumps capable of producing an open

Table 3 Overview of micropumps for the transport of liquids and suspensions

Material	Fluid structure	Type of drive	Features	Ref.
Silicon glass	Anisotropically etched silicon substrate, two passive valves	Piezoelectric diaphragm drive	– delivery rate: 8 µl/min at $V_w = 100$ V, $f_w = 1$ Hz – max. delivery pressure: 10^4 Pa – for closed-loop systems, no open jet – area requirement: several cm^2, large dead volume – simple silicon-glass composite	[32]
Silicon glass	Anisotropically etched silicon substrate, two passive valves	Thermoelectric diaphragm drive	– delivery rate: 34 µl/min at $V_w = 6$ V – max. delivery pressure: 5×10^4 Pa – for closed-loop systems, no open jet – area requirement: several cm^2, large dead volume – complex glass-Si-Si-Si-glass composite	[33]
Silicon	Several anisotropically etched silicon substrates, two passive valves	Electrostatic diaphragm drive	– bidirectional delivery characteristic – delivery rate: forward: 250 µl/min at a pressure of 3×10^4 Pa, backward: 350 µl/min at a pressure of 5×10^3 Pa, $V_w = 200$ V – for closed-loop systems, no open jet – area requirement: several (100 × 100) μm^2 – complex Si-Si-Si-Si composite	[34]
Silicon	Two anisotropically etched micro-lattices	Ions generated and accelerated in an electrostatic field	– bidirectional delivery characteristic – extremely high delivery rate: 30000 µl/min, pressure: 2.5×10^3 Pa – required area in the size of the microchannel and therefore almost no dead volume, easy to fill – only electrically non-conductive fluids can be delivered	[35]

Silicon glass	Anisotropically etched silicon with diaphragms and isotropically etched glass, two active valves	External pneumatic drive for valves and pump	– bidirectionality due to the control of the valves – cannot run without the external drive – delivery rate: several µl/min, leakage rate: 0.1% – area requirement: several (100 × 100) µm² – small dead volume – used in closed-loop systems	[36]
Glass	Glass capillary with discharge nozzle, without valve	Piezoelectric diaphragm drive, hollow glass cylinder forms the diaphragm	– unidirectional open-jet pump cannot be used for closed-loop systems – max. delivery rate: 120 µl/min, drops of 50 pl to 1.5 nl can be emitted	[37, 41]
Glass platinum	Thin-film metal structure on glass, without valve	Travelling-wave dielectrophoresis, the dielectrophoretic and translatory force are superimposed on each other	– defined particle drive, in particular cell drive – bidirectionality due to polarity reversal of the field – efficient transport through smallest gaps – transport, fixation and positioning of cells and particles suspended in a solution are effected by the electric field – flow velocity 5 to 500 µm/s, V_w = 4 to 15 V	[40, 41]
Silicon glass	Anisotropically etched silicon with diaphragm, without valve	Piezoelectric diaphragm drive, silicon forms the diaphragm	– unidirectional open-jet pump cannot be used for closed-loop systems – max. delivery rate: 660 µl/min at f_{wmax} = 5.5 kHz, V_w = 30 to 120 V, drops of 300 pl to 2.5 nl can be emitted – area requirement: min. 1.8 mm², dead volume: 300 to 1000 nl	[1, 38, 39]

jet which works on the drop-on-demand principle. The capability of a fluid drive to produce an open jet means that the liquid passes through an opening into the free atmosphere after it has been accelerated in the active part of the drive. As a rule, the opening has the form of a nozzle. The characteristic of the microfluid drives capable of producing an open jet which works on the drop-on-demand principle is that defined quanta of fluids in the volume range of several picolitres to nanolitres are released. Today this principle is used particularly in ink-jet printers. It is also realized in the drives mentioned in references [1, 37–39] given in Table 3. Fundamental differences between the drive of an ink-jet printer and a microfluid drive are the modified materials used for the components and the modified interfaces, the possibility to arrange customized components in a discretely and complexly integrated manner and the fact that these components cannot only be used for printing but also for dispensing and pipetting of liquids. These developments were necessary in order to be able to use microfluid drives for applications in the fields of life science, molecular genetics, combinatorial chemistry and clinical diagnosis [43].

It is demonstrated below by the example of a microejection pump developed by GeSiM how modified components can meet differentiated criteria of use.

Microejection pump, type MEP

Let us first consider the basis element of the future modifications, the microejection pump MEP. An example of the MEP (external dimension 10 mm × 5 mm × 1.5 mm) is shown in Figure 5. The MEP has a fluid inlet (cross section 700 µm × 700 µm), a pump chamber (dead volume 300 to 800 nl), a silicon diaphragm (excursion by means of a piezoelectric actuator), a discharge nozzle (cross section 100 µm × 100 µm) and a microchannel (200 µm × 150 µm) interconnecting these features within the silicon substrate. The numerical values given in brackets are typical examples.

The MEP can meet three criteria of use. First, the MEP can be used for the fluid transport, as a suction pump is mounted at the end of the channel. Second, it can be used for the delivery of defined quanta of liquids or suspensions onto any substrate. Third, it can be used for printing precise printouts. The MEP can be operated manually or by means of x-y-z motion modules.

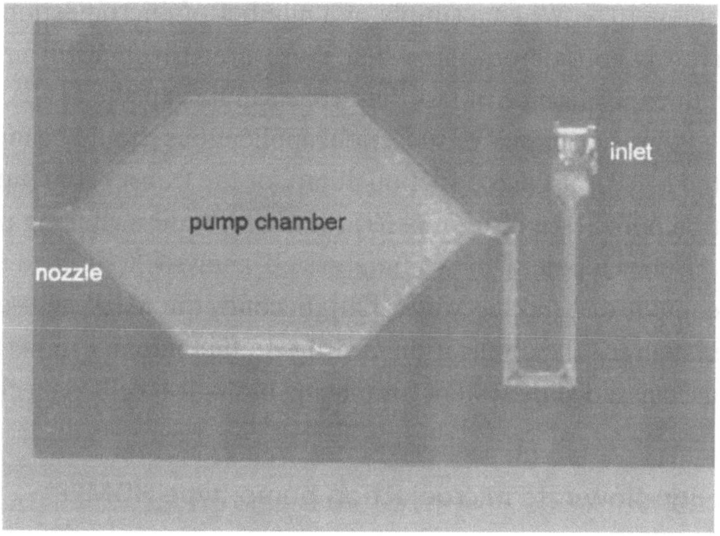

Figure 5 *Microejection pump (MEP)*

The MEP cannot be used in closed-loop systems. The pressure difference between the pump inlet and the discharge nozzle may amount to a water column of only a few centimetres.

Typical operating parameters of the MEP are a voltage of 35 to 120 V, a maximum frequency of 1.5 kHz, a drop diameter of 70 to 130 μm and a maximum delivery rate of 120 μl/min. As regards the parameter ranges, it should be pointed out that the individual parameters are not independent of each other and are not continuously variable. The MEP is a component in which fluid-dynamic processes proceed and the delivery behaviour of which varies due to frequency-dependent resonances. Therefore, the manufacturer determines a separate triplet of values for each type of pump which comprises pulse duration, frequency and voltage. A user always receives a micropump with the relating triplet of values. Parameter determination is done using undegassed and filtered water. If liquids deviating from aqueous fluids are to be delivered, additional tests will have to be carried out.

Today the MEP concept is already accepted by many users and it is confirmed by interesting applications. The following details demonstrate how precisely these micropumps work. When 100 drops of 1 nl each are dosed, the coefficient of variation is below 2%. When 10 drops are dosed, it is still below 10%. When a supply of a number of microejection pumps of the same type is considered and when these are driven with an identical triplet of values, a variance of less than 10% can be expected [1].

Three examples of successful applications should be mentioned here. First, the MEP is used as a suction pump for the transport of liquids and as an injection pump (see also on p. 59) for microfluidic switching processes, options in the development of a miniaturized analysis system to determine the NO_3 content of drinking water [20]. Second, the MEP as used as a microdosing system for the application of drugs to the human eye [44]. Third, the MEP is used as a dosing system for placing biological cells on a chilled substrate [45].

High-flow-rate microejection pump, type HDMEP

The development of a micropump of the HDMEP type was aimed at providing to users a fluid drive with a significantly increased delivery rate [39]. The modification of the base component MEP made in comparison with the HDMEP relates to the micropattern in the region of the fluid inlet. An additional feature, a diffuser, was integrated into the inlet channel directly in front of the entry into the pump chamber. This development utilizes the known fluid-dynamic effect that the flow resistance of a diffuser used for flowing fluids is directional. The diffuser was arranged in the inlet channel of the HDMEP micropump in such a way that its flow resistance is in the loading mode lower and in the unloading mode higher than that of the standard inlet channel of an MEP. A fluid drive of the HDMEP design which has external dimensions of 14 mm × 2.8 mm × 1.5 mm is shown in Figure 6.

Except for the diffuser arranged in the inlet channel of the pump chamber, HDMEP-type micropumps have no specific features distinguishing them

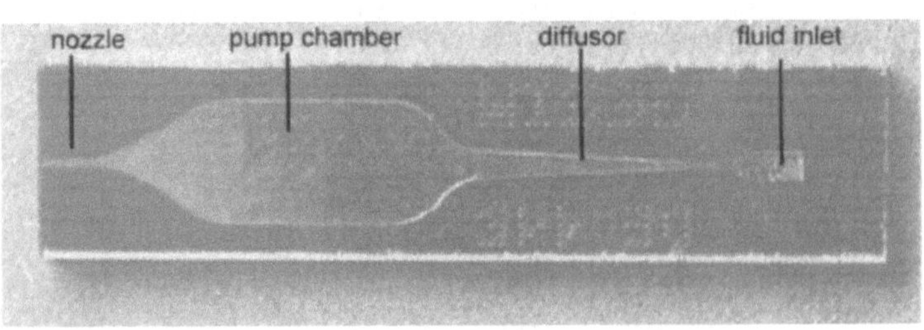

Figure 6 *Microejection pump HDMEP [1, 39]*

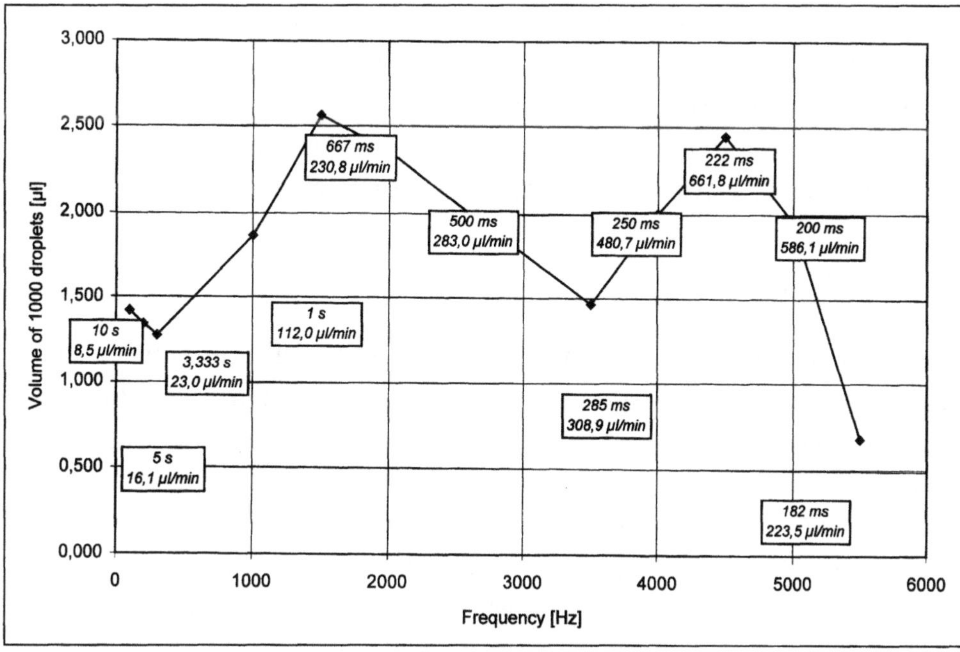

Figure 7 *Relation between frequency and flow rate of an HDMEP [1]*

from MEP-type micropumps. Their criteria of use also correspond to those of the MEP pumps.

The essential difference lies in the operating parameters. The maximum working frequency of the MEP is 2 to 3.5 kHz, depending on the form of the slopes of the voltage pulses, and the highest delivery rate measured at this frequency was 180 µl/min. The HDMEP can still accurately deliver even at 5.5 kHz and it reaches a delivery rate of 660 µl/min at this frequency. The micropumps of the HDMEP type make particularly clear that in microfluidic drives the relations between working frequency and delivery rate or working voltage and delivery rate are by no means generally linear. On the contrary, the delivery rate of an HDMEP varies significantly in the working frequency band. To give you an impression, the relation between working frequency and delivery rate is shown in Figure 7. This figure makes particularly clear how useful it is when the manufacturer does not give the ranges of the operating parameters but connects the function of the individual component with fixed operating parameters.

Where can the HDMEP-type micropump be used? Their use is interesting if particularly large volumes have to be realized within short dosing times. An example of GeSiM's everyday development work concerned a dosing system for two drugs. It was required to concurrently apply two active substances with a strong difference in the final volume. Both volumes had to be discharged within a dosing time of less than 1 s. The task was solved by combining MEP- and HDMEP-type micropumps on the same chip.

Multichannel dispensers

The MEP- and HDMEP-type micropumps described above provide as base components ideal conditions of the manufacture of dispensers for the nanolitre range. The variability of the number of channels, channel design, array of discharge nozzles and external dimensions make the dispenser flexible. There are two ways to realize multichannel dispensers. First, discrete micropumps of the MEP or HDMEP type can be mounted on a special adapter for the multichannel dispenser as is shown by an example in Figure 8.

The adapter shown in Figure 8 carries a glass-silicon manifold which, in turn, carries discrete micropumps. They are held on the adapter and are connected fluidically to the system peripherals and electrically to the control unit. The Multidos-type control unit (1) allows the programming of up to

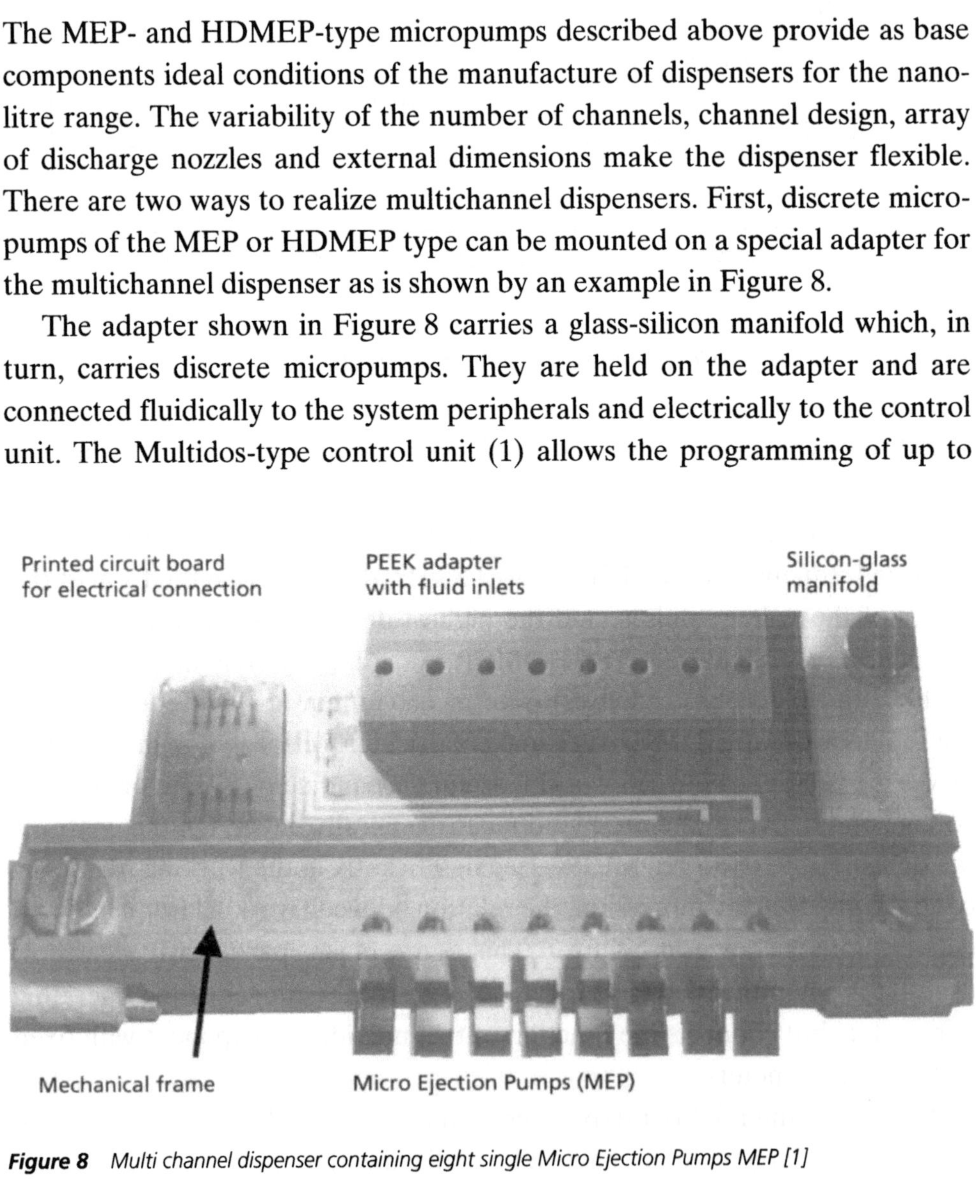

Figure 8 *Multi channel dispenser containing eight single Micro Ejection Pumps MEP [1]*

250 micropumps according to the individual requirements. The design of Figure 8 has the disadvantage that the distance between the discharge nozzles can never be smaller than the width of the discrete chip used. However, an advantage is the flexible construction of this arrangement.

A second way is to integrate several active elements into the same chip. Prototypes of double-channel dispensers are shown in Figure 9.

This solution allows a substantially higher concentration of many discharge nozzles. If a silicon-glass-silicon composite instead of a silicon-glass one is realized, the number of channels can even be doubled. For example, at GeSiM two- and four-channel dispensers are built and dispensers with far more channels are being developed. The discharge nozzles are arranged at a distance of 330 µm.

The individual pump channels in multichannel dispensers, irrespective of their design, have the same features as the above discrete micropumps. There is a significant extension of the uses. It lies in the possibility to disperse drop arrays sequentially or in parallel. The distance between the drop emission plane and the substrate may amount to several millimetres and determines

Figure 9 *Double-channel dispenser type MEP [1]*

the accuracy of the resultant printouts. Multichannel dispensers are useful when large amounts, small volumes with a high local resolution have to be discharged into defined arrays. The coating of any planar substrate surfaces but also the coating of micro- and nanotitre plates are examples of realized uses with the GeSiM system "Nanoplotter" [1].

Another interesting application of such microsystem will result if the multichannel dispenser is combined with an ultrasonic levitator and an x-y-z manipulator unit [46]. The starting point is the known fact that liquids, suspensions and solid particles can be contactlessly held in the sound pressure maximum of a standing ultrasonic wave [47]. If a multichannel dispenser is adjusted to the range of the sound pressure maximum, microdrops from different channels can be held in this maximum. The principle of the technical solution is shown in Figure 10. The combination of levitator, manipulator and dispenser allows the realization of various functions. Such functions regard the contactless holding, the mixing without carry-over, the thermal treatment of the drop at the site of holding, the concentration and dilution of the drop at the site of holding, the realization of optical methods of analysis in the drop at the site of holding, the taking of the drop by means of a pipette and the discharge of the drop into any sample holders.

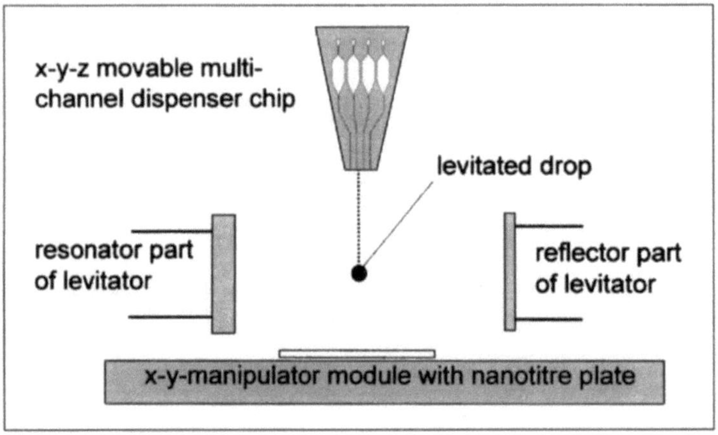

Figure 10 *Example of the arrangement of a multichannel dispenser, levitator and nanotitre plate with different manipulator modules*

Pipettes

The procedures of pipetting and dispensing have a great deal in common and only one essential difference. Therefore, it is obvious that micropumps of the MEP and HDMEP type are used for both applications.

Let us first roughly compare the procedures of dispensing with those of pipetting. In case of dispensing, the fluid to be dispersed is fed exclusively through the fluid inlet into the pump chamber and is then discharged through the nozzle region. In case of pipetting, the fluid to be dispersed is sucked through the nozzle region and is than discharged through the nozzle region again. A pipette can also be used as a dispenser if required. But conversely, a dispenser cannot be used as a pipette. A pipette is an element that can meet more complex criteria of use than a dispenser.

The procedure of discharging liquids or suspensions in the pipette or dispenser is identical with that described above for fluid drives of the MEP and HDMEP types. For the use of microfluidic components, it has also proved useful to couple the pipette or dispenser to the selector valve and injection pump. The coupling of micro- and macrofluidic components makes it possible to have the filling, flushing, dispensing and pipetting proceed automatically. Therefore, the process of liquid handling can be made efficient, reproducible and reliable. The coupling of micropipettes to injection pumps and selector valves results in an extension of the criteria of use which can be fulfilled. On the one hand, this extension regards pipetting using an air gap, that is, a defined air cushion is arranged between the system liquid and the liquid to be pipetted in order to prevent a diffusion of both fluids into each other. On the other, it regards the extension of the volume range that can be handled by means of microfluidic components when pipetting. This range is upwards limited by the maximum delivery rate of the micropipette. This restriction can be eliminated by coupling to the injection pump.

If the fluid is to be taken in through the nozzle region of an MEP- or HDMEP-type pipette, the base components shown in Figures 5 and 6 have to be modified. The carry-over of separated fluids which are taken in and are then discharged at another place is problematic. When different fluids are pipetted successively, the quality of this procedure is expressed by the so-called carry-over effect. The geometry and the properties of the effectively wetted surface of the micropipette determine the carry-over effect in the pipette. The tip of the pipette is shaped into a point to reduce the effectively wetted surface when sampling is done by means of the pipette directly through the nozzle region. In addition, the width of the pipette is dimensioned only as large as actually required for the special purpose. The surface of the micropipette, and in particular the part which is actually wetted when the pipette is immersed, has a hydrophobic design. SPIP- and PPIP-type pipette chips manufactured by GeSiM are shown in an unassembled condition in Figure 11. In each of these cases, the fluidic layout of the MEP-type micropump is the basis. The SPIP type is a wet etched type of pipette. The PPIP type is partly etched in the SF_6-plasma of an ASE reactor.

Pipettes can be given a number of specific features using their shapes and fluidic layouts. These features have a significant effect on the performance and the criteria of use of each pipetting system.

By varying their external dimensions, pipettes can be designed for pipetting onto 96-well or 384-well microtitre plates. The pipette may also have

Needle pipette (type NPIP)

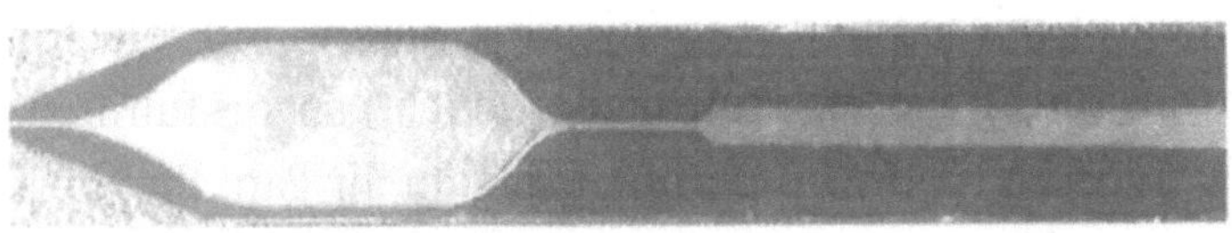

Standard pipette (type PPIP), inlet at the right edge

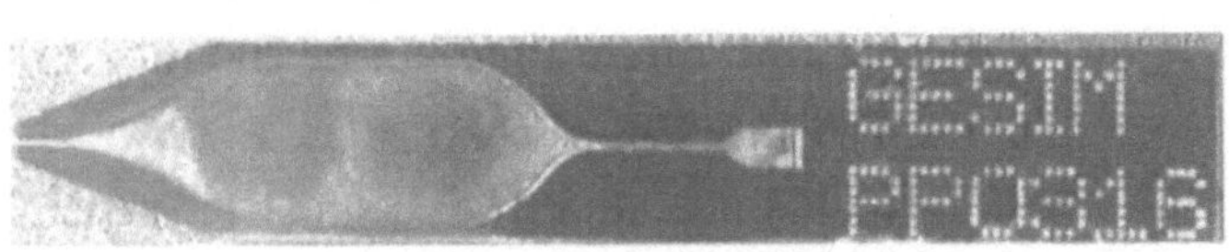

Standard pipette (type PPIP), inlet at the backside

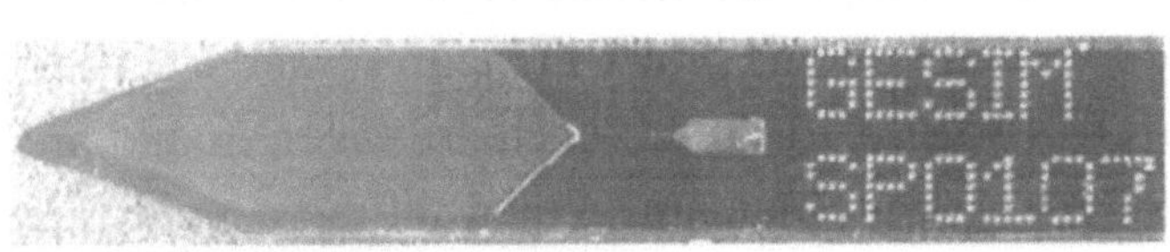

Standard pipette (type PPIP), inlet at the backside

Figure 11 *Micro pipette before assembly [1]*

extremely extended chip dimensions (e.g., L × W × H: 50 mm × 2.8 mm × 1.5 mm). When designed in such a way, they are suitable for pipetting into deep vessel systems. After the pipette has been used for taking up samples from standardized vessel systems, it can be used for dispensing onto any substrates. Figure 12 shows a GeSiM pipette emitting 500 drop packages of 1 nl each into the wells of a glass-silicon nanotitre plate. Each well of the nanotitre plate can accomodate a volume of 600 nl.

If specific features are needed, variations of the fluidically active patterns are also useful in pipettes. For example, pipettes with one or several fluid inlets are manufactured at GeSiM. Significant differences result in dependence on where the inlet is connected, that is, whether it is connected on the face or on the back of the pipette chip.

As mentioned above for the dispenser, pipettes can also be designed with several channels. Multichannel pipettes are always composed from discrete pipettes. The arrangements can be grouped into those where the pipettes are rigidly connected to an adapter (see Fig. 13) and those where the pipettes are

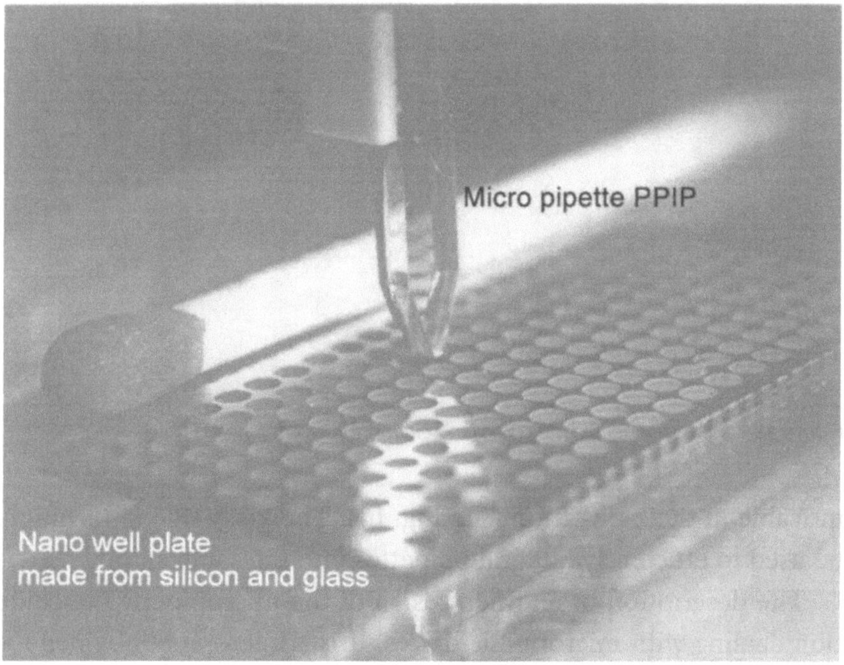

Figure 12 *Micro pipette PPIP dispensing sample into a nano well plate [1]*

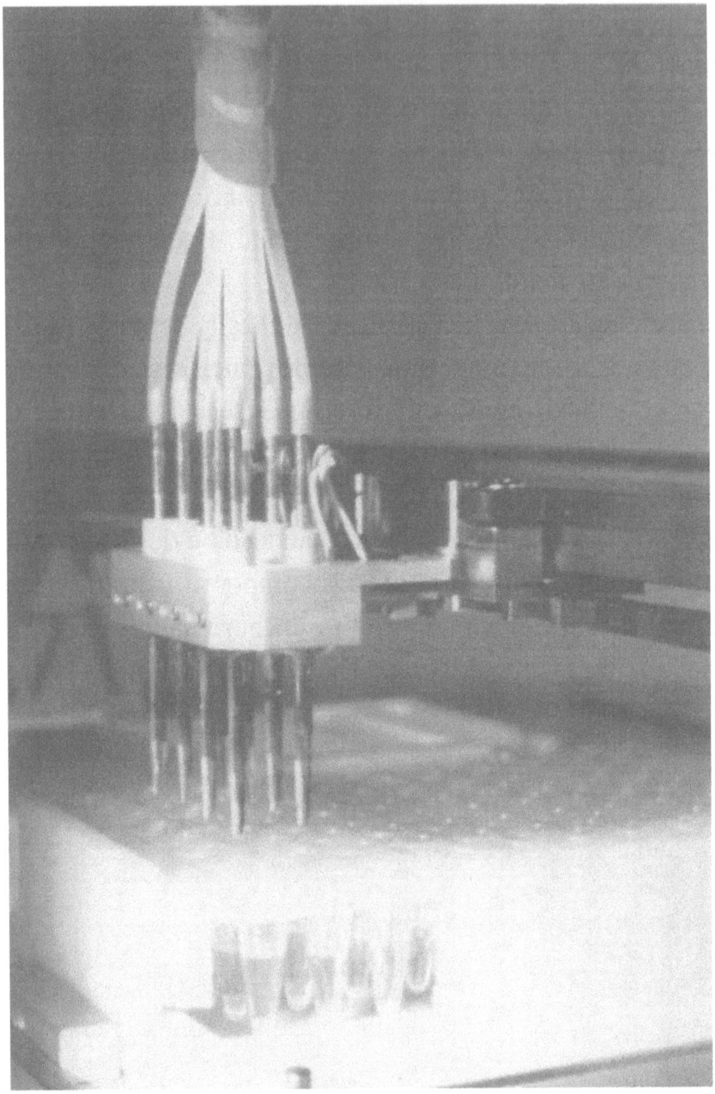

Figure 13 *Multi channel pipette containing eight single micro pipettes [1]*

movable in the z-direction (see Fig. 14). The glass-silicon technique can also be used to efficiently integrate pipettes into the same chip.

The description of another design of the micropipette concludes the section dealing with microfluidic drives. It is a micropipette which has an additional feature, besides those described above, namely a facility for the controlled heating of the pipette. The development of this feature was stimulated

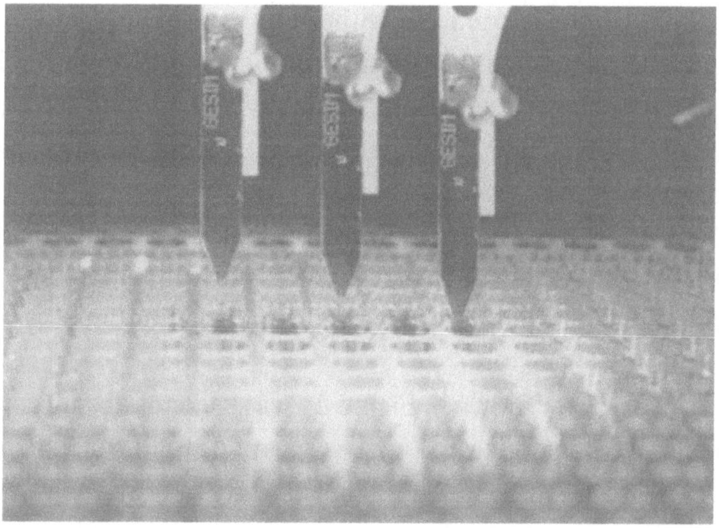

Figure 14 *Micro pipettes (Type PPIP) adjusted to a microwell plate (384). Every pipette can be moved up and down separately*

by applications in which higher-viscous fluids were to be pipetted in a defined manner. Basically, the pipetting of liquids up to a maximum viscosity of 5 mPas causes no problems, whereas this microfluidic component is not suitable for higher viscosities. If fluids of a higher viscosity are to be handled by the micropipette, defined heating can be used in many cases as a means to reduce the viscosity. However, thermal activation is not applicable to liquids reaching at such temperatures the range where their viscosity cannot be handled any longer, that is, where the liquids begin to destructure irreversibly. A pipette that can be heated in a defined manner which is based on an MEP-type micropump is presented in [48]. It has, besides the fluidic structure described above, a heater integrated into the silicon bulk, a platinum temperature sensor and an electronic control. In this micropipette, the liquid also comes into contact with silicon and glass only. Thermal activation is by impressing joule heat throughout the entire silicon bulk material of the micropipette. The temperature sensor is directly attached to the back of the silicon body of the micropipette by means of thin-film technology. The electronic control of the pump and heater parameters is done in the "Multidos" controller [1].

It is an interesting application when the heater function is combined with dispensers and pipettes having one or several fluid inlets. The following procedure can be carried out without any problems. Two fluids, A and B, flow into the pump chamber. In the pump chamber, both fluids are heated to the reaction temperature. Both fluids react according to the equation A + B + heat = C. The fluid C is then discharged from the micropipette in the desired drop packages.

A PPIP-type micropipette with heater, platinum temperature sensor and PEEK case is shown in Figure 15. Electrical contacting is done via a flexible printed circuit board of polyimide and the fluid inlet is realized via a PEEK capillary.

The pipette shown in Figure 15 was used for dosing experiments with ethylene glycol and glycerol. The viscosity of both fluids had to be reduced to a value of 5 mPas by heating. The viscosity of ethylene glycol at room temperature is approx. 16.1 mPas and it could be delivered from 75°C up. The viscosity of glycerol at room temperature is 934 mPas and it could be delivered at 122°C. The electrical parameters of the heatable pipette are as follows: The resistance of the heater was 220 ohm. The resistance of the temperature sensor was 280 ohm at room temperature and 320 ohm at 100°C. The electric power consumption of the heating of the pipette chip was 360 mW at an operating voltage of 10 V in the heating resistor. The time required for heating a filled micropipette from room temperature to 100°C at this power consumption was 110 s. This time has to be regarded as the warm-up period required to reach readiness for operation. When the micropipette has reached its operating temperature, the fluid flowing into the pump chamber can be delivered continuously.

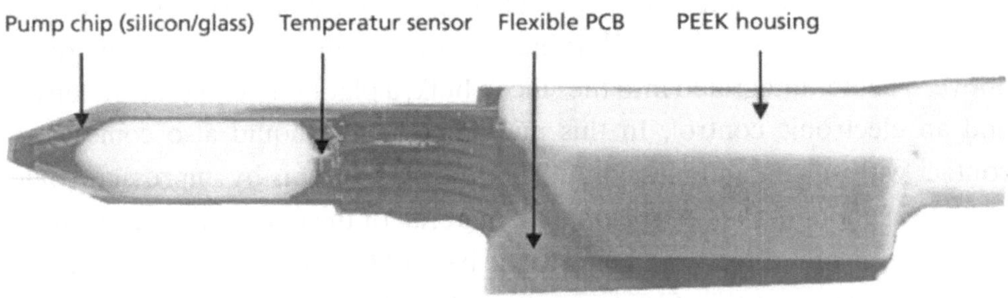

Figure 15 *Micro pipette (Type SPIP) with heating structure and temperature sensor [1, 48]*

Microvalves, microinjectors, micromixers and components for the completion of microfluidic systems

The construction of microsystems of the complexity of a µ-TAS requires a number of additional components, besides those described above in sections on p. 39 and p. 43. When considering the problems of the µ-TAS developments from the angle of microfluid handling, the components which have not yet been dealt with in this chapter can be classed under the term "fluidic switching elements". Most of the requirements of today's comfortable fluidic microsystems can be met by the combination of manifolds, fluid drives and fluidic switching devices. Microfluidic components required for the realization of switching processes will be mentioned and partly described in this section.

To explain what the author understands by a microfluidic switching process, let us have a look at the diagram of an elementary fluidic structure shown in Figure 16.

A system channel with an inlet and a suction pump arranged at the end of the channel is the basis of the elementary fluidic structure. An inlet channel runs into the system channel via a nodal point. This layout of two interconnected microchannels is often used in complex fluidic microsystems. Let us assume that a microfluidic switching device, e.g., a valve, is arranged directly upstream from the nodal point in both the inlet channel and the system channel. Let us further assume that in the initial state valve IC is closed and the valve SC is open. When the system pump at the end of the system channel sucks, only fluid 1 flows through the system channel to the outlet. If only valve IC is opened and valve SC is closed, fluid 2 will flow from the moment of switching from the inlet channel to the outlet of the system channel via the

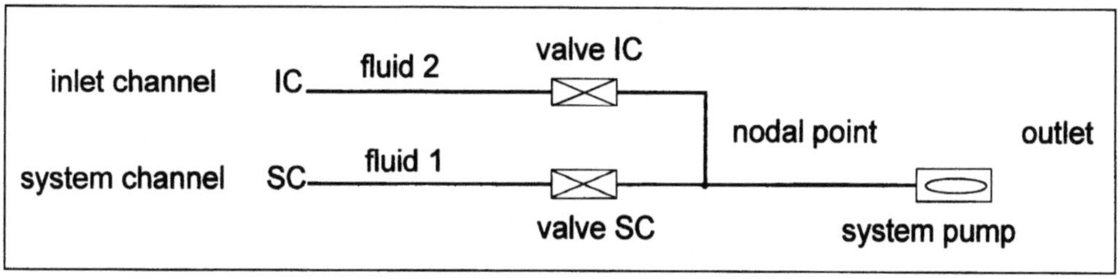

Figure 16 *Elementary fluidic structure*

nodal point. The change from fluid 1 to fluid 2 at the nodal point is intended as an example of a fluidic switching process which can be realized by the actuation of a microfluidic switching device.

Switching processes of liquids in microchannels can be realized by the use of valves, microinjectors and switching devices utilizing auxiliary hydraulic energy. Switching devices utilizing auxiliary hydraulic energy, e.g., described in [49, 67], will not be discussed in this contribution.

Microvalves

Let us first consider the microvalves. Operating principle, materials and technical parameters of the microvalves which have been developed so far may be very different. Table 4 is a comparison of several microvalves.

Those who plan the practical use of microvalves are faced with the fact that Redwood Microsystems are the only commercial supplier of such components with internal diaphragm drive. Microvalves with external drive have an almost ideal closing behaviour and are sufficiently small. The disadvantage of

Table 4 Some of the realized microvalves used for the switching of fluids

Material	Fluid structure	Drive	Features	Ref.
Glass silicon epoxy r.	Silicon diaphragm bonded anodically to a pattern-delineated Pyrex glass substrate	External, pneumatic	– to construct multiway valves, valve diaphragm requires little space – pneumatic drive used for opening and improved closing – flow rate 1 µl/min, leakage rate 1% – the concept allows the combination of valve, mixer and pump in the same manifold by the same technology	[36]
Silicon epoxy r.	Silicon composite patterned in three planes, directly bonded	Internal, thermo-electric	– valve and flow sensor are manufactured in the same component by the same technology – can be used for pressure surges up to 100 kPa – flow rate 5 ml/min, leakage rate does not depend on ambient temperature – time constant: valve 15 ms, time constant of flow sensor 0.2 ms	[50]

valve concepts using external drives is the necessary combination of the valve with macroscopic system elements. They include hose connections, compressors, vacuum pumps and switching valves. Most of the components mentioned in Table 4 have not yet been launched. However, this situation will change soon. In the opinion of the author, valve concepts using external drives will probably be the first to find their way to the user. The greatest advantage of the use of microvalves can be realized wherever fluidic switching processes between microchannels with pressure differentials are required. In such cases,

Table 4 (continued)

Material	Fluid structure	Drive	Features	Ref.
Silicon silicone rubber epoxy r.	one silicon substrate with two separate channels and one diaphragm of silicone rubber	External, pneumatic	– excellent switching behaviour, area requirement: 225 µm × 125 µm – complete sealing at a differential pressure between inlet and outlet of 30 kPa – flow rate 300 µl/min, no leakage flow	[51]
Silicon epoxy r.	Silicon composite patterned in two planes, directly bonded	No drive, passive flap valve	– one-way valve based on flap design – valve action is dependent on the existence of a flow – flow rate 15 ml/min, leakage flow not specified	[52, 53]
Silicon glass	Anodically bonded glass-silicon-glass composite	Internal, thermo-electric	– time constant 1000 ... 500 ms – normally closed and normally open is commercial available – flow rate 300 µl/min – leakage rate depends on ambient temperature and differential pressure – valve dimensions: 6 mm × 7.4 mm × 4.4 mm – can be integrated as a module	[54], [67]
PMMA polyimide	PMMA body produced by micromoulding	No drive, passive diaphragm valve	– one-way double-diaphragm valve, diaphragms of titanium 2.7 µm thick and polyimide several µm thick – valve diameter: 0.46 to 1.4 mm – flow rate 36 ml/s, reseat pressure < 18 kPa, leakage flow 20 µl/s – requires a relatively small area	[55, 67]

there is no alternative to microvalves so far. Microinjectors break down at pressure differentials between system and environment of above 6×10^2 Pa.

Microinjectors

Microinjectors are an alternative of realizing fluidic switching processes by means of microvalves. Microinjection of liquids is already in use in microcapillary electrophoresis [56, 57] and in flow-injection analysis [58]. The common feature of these microinjectors is that no valves are used in the fluidic switching process.

The injector principle in the microcapillary electrophoresis of microsystems using glass manifolds is based on the electrophoretic drive of liquid molecules which are influenced by impressed electric fields. Due to their excellent dielectric properties, glass manifolds are indispensable for these planar electrokinetic systems with field strengths of about 800 V/cm. Microinjectors which organize the fluidic switching process according to this principle will not be treated in this section.

Let us now proceed to the second group of realized injectors comprising MEP-type micropumps and specially patterned silicon-glass manifolds. The general operating principle is again explained by an elementary fluidic structure (see Fig. 17).

Here again, we have a fluidic switching process, that is, the injection of fluid 2 from the inlet channel into the system channel where fluid 1 flows is to be realized. Again, an MEP-type micropump sucking if required is arranged at the outlet of the system channel. However, there are no microvalves in this elementary structure and the nodal point is shown differently. It is not con-

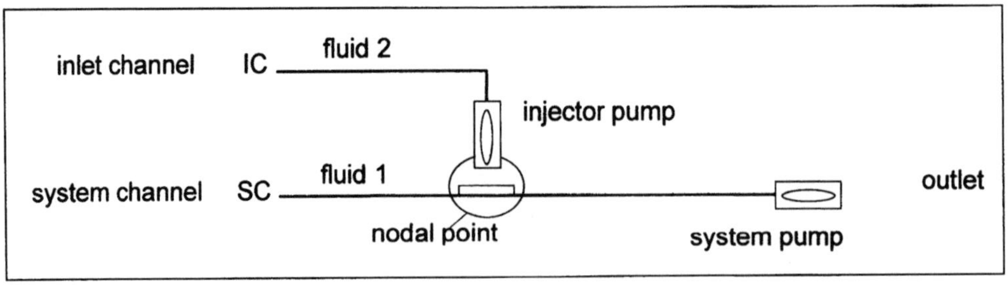

Figure 17 *Elementary structure of a microinjector based on MEP-type micropump*

nected to the system channel. The nodal point is now formed by the arrangement of an MEP-type micropump and a fluid diode.

The fluid diode is a component which has not yet been described in this chapter [59]. It is a silicon chip which has screen pores on one side and a short channel on the other. The screen pores are either wet etched or are etched in plasma (see Fig. 18). The fluid diode is mounted on a manifold with the channel side down in such a way that the system channel runs exactly beneath

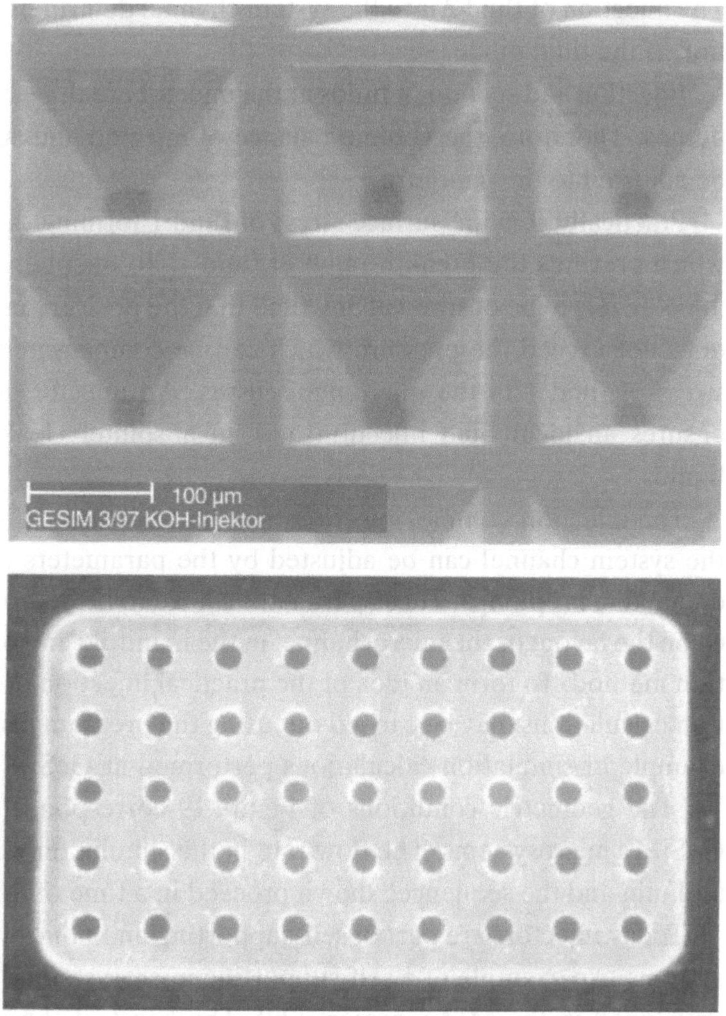

a

b

Figure 18 SEM photo of the screen pores of the fluid diode (detail) [1]. a) Screen pores are wet etched; b) screen pores are dry etched in the ASE process

the fluid diode. When fluid 1 in the system channel passes the screen pores, the screen pores are blocked due to the formation of a large number of micromeniscusses. This blockade from inside determines the blocking direction of the fluid diode. The screen pores of the fluid diode are visible on the surface. This design ensures that the system channel in the region of the nodal point is always accessible through the screen pores.

When drops of fluid 2 are fired through an air gap onto the micromeniscusses, they wet the screen pores and can be sucked into the system channel. The injection of fluid 2 into the system channel determines the forward direction of the fluid diode.

Injection and suction of fluids at the injector are done by MEP-type micropumps. Therefore, the volume balance of injected and sucked-off fluids can be adjusted to the nanolitre.

Practically, it is the surface stress of fluid 1 forming the micromeniscusses which prevents the breakthrough of fluid 1. To maintain the function of the injector, it is to be ensured at any time that the pressure in the system channel does not exceed the pressure which can be compensated for by the surface stress of fluid 1 in the micromeniscusses. An injector adjusted in this way ensures an ideal fluid injection and an absolutely leakproof switching of liquids.

The injection volume, injection time and flow rate of fluid 2 injected into the system channel can be adjusted by the parameters of the two drop-on-demand micropumps. These parameters have to be determined in dependence on the layout of the microchannel in the manifold and on the desired detection method. To form an idea of the practical injection process, FEM simulation calculations may be carried out using the program package Flowtran. An example of simulation calculations performed at GeSiM is shown in Figure 19. The geometric conditions of Figure 19 correspond to those realized in GeSiM's microsystems. The flow rate in the simulation was determined to be 5 µl/min and the sequences shown proceed in a time of 3.5 s.

Today injectors are part of self-supporting and non-self-supporting microsystems. An example of a self-supporting microsystem is the nitrate monitor developed in the MIAS project [20]. The heart of the nitrate monitor is a microsystem comprising two coated film electrodes, a pseudoreference electrode, an injector and a silicon-glass manifold. A non-self-supporting application might be an injector reduced to a manifold and a screen pore array. In this

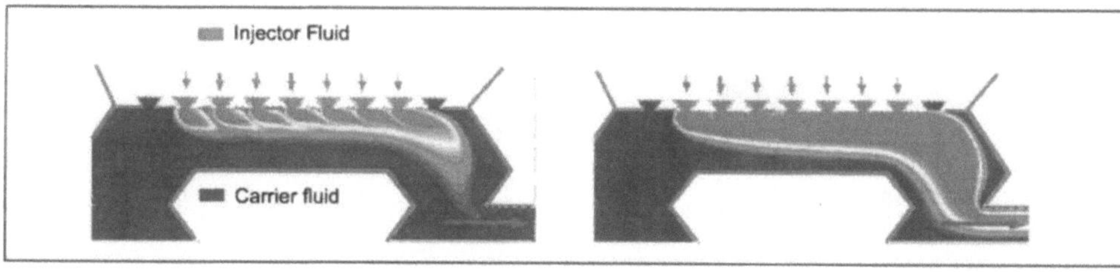

Figure 19 *FEM – simulation of the injection of sample in the Fluidic Diode (see colour plate 1)*

case, injection can be taken over by a micropipette coupled to a high-speed x-y-z pipetting system. The diagram of an arrangement of the injector is shown on the upper side and the results of a series of calibrations for six NO_3 concentrations measured at the injector are shown on the lower side of Figure 20.

The flow rate of the solution in the system channel as measured using the nitrate monitor was 4 µl/min. Each of the four NO_3 concentrations was injected into the system channel at a volume of 3.5 µl over an injection period of 3 s. Experiments carried out with this system showed that the optimum flow rate of the fluid in the system channel was 5 µl/min. To characterize the injection process, a coated film electrode with a polymer membrane sensitized to NO_3 ions was used. When the flow rate was increased five-fold, the peak height of the sensor signals fell to 50% of that of the values measured at 4 µl/min. At an injection volume of 3.5 µl, stable measuring signals were already recorded. When, however, the injection volume was reduced to below 2.5 µl, the peak height decreased significantly.

The numerical values show that the volume balance of the fluids injected or carried off at the nodal point has a considerable effect on the performance of the injector. The definition of the volume balance is always a time-consuming procedure. In an MEP-type micropump, the volume cannot be determined only by counting the emitted drops. The determination of variations in the drop volume using this method is impossible. This problem can be solved by the integration of a flow sensor. A suitable flow sensor must be miniaturized and must have a high resolution in the range of lowest flow rates. Only those of the described injectors which have a facility for flow measurement can be used for really demanding analytical measuring tasks over a long measuring time, using fluids with different values of viscosity.

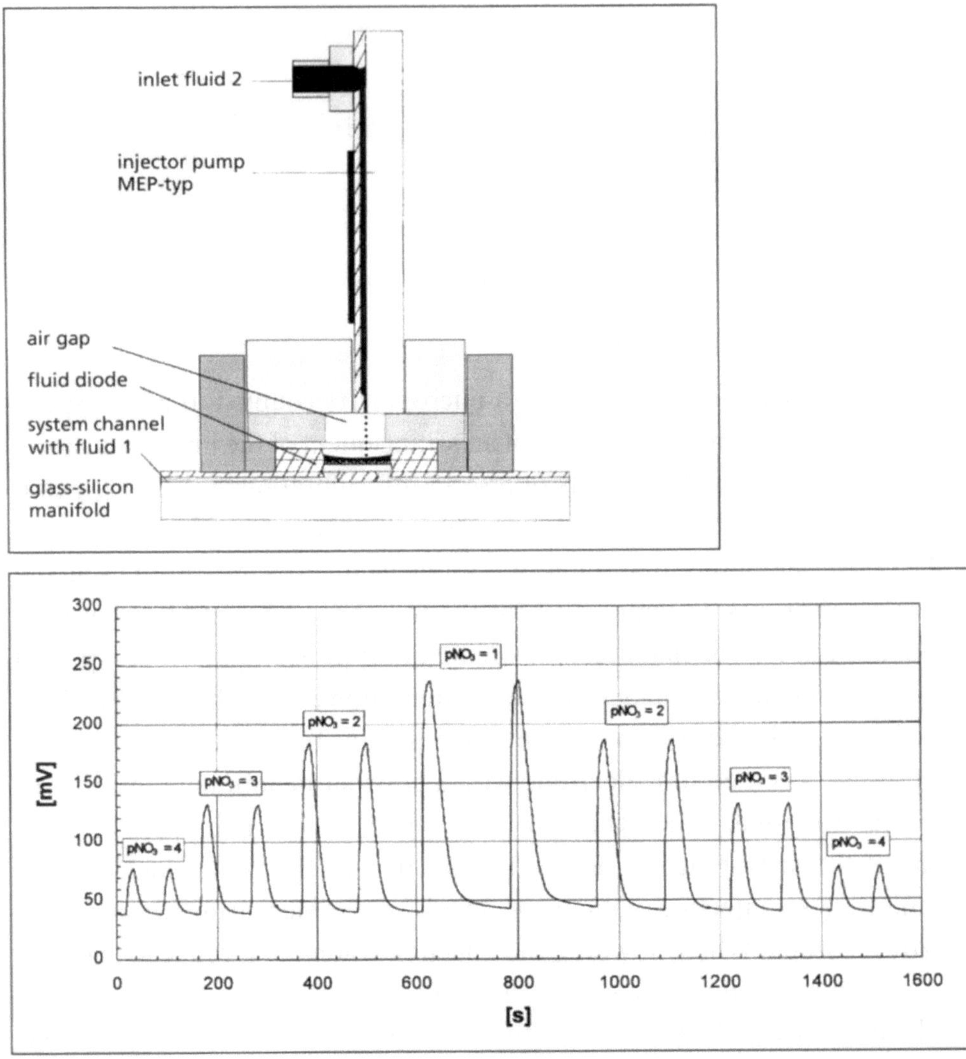

Figure 20 *Construction diagram of the injector and results of a series of nitrate calibrations [1]*

The demands made on the sensitivity of a relevant flow sensor are extremely heavy. This will become clear when we imagine that the transport of a few nanolitres has to be defined. When, for example, a volume of 10 nl is discharged at the micropump, the liquid level in the microchannel (width × height: 500 μm × 200 μm) varies by a mere 100 μm.

An overview of miniaturized flow sensors is given by Nguyen and Dötzel in [60]. In particular the flow sensors developed in Wien [61] and Twente [62] seem to have been just made for applications in the field of flow measurement in MEP-type micropumps. As these flow sensors were not on the market, an electrocaloric flow sensor was constructed by GeSiM. Production technology as well as building-up and bonding techniques were adapted for the control of MEP-type micropumps [63]. The flow sensor allows the measurement of continuous flows in the range of 1 to 100 µl/min with a measurement error of less than 0.5%. The dimensions of the sensor chip are 8 mm × 5 mm. The sensor comprises a diaphragm, resistors for heaters and temperature sensors, a microchannel and several channels for the fluid flow. The diaphragm is an Si_3N_4 film 1 µm thick which is deposited through PVD in a stress-free manner. It is directly wetted by the fluid on both sides. This sensor is able to qualitatively detect the emission of a single drop. Quantitative determination begins at drop packages of 4 nl. Even more exact values can be obtained by faster electronic sampling rates in signal processing. A detail of the layout of the flow sensor as well as the flow characteristic for flow rates of 1 to 10 µl/min are shown in Figure 21.

The flow sensor is provided to users to meet two criteria of use. First, it is the flow sensor as used as a discrete component. Second, it is the flow sensor as part of the flow control system for micropump applications, e.g., MEP-type micropumps.

Micromixers

The mixing of fluids is a process step which is important to many applications of chemical analysers. This also applies to microsystems. With regard to µ-TAS, it seems reasonable to manufacture micromixers as modules which can be integrated as a component into a microsystem if required. Operating principle, materials and technical parameters of the micromixers which have been developed so far may be very different. Table 5 is a comparison of several micromixers.

In microsystems with typical Reynolds numbers below 50, the flow conditions are strictly laminar. Therefore, dynamic mixers, that is, such mixers able to utilize defined turbulences, are not suitable for an approach to the problem [66]. It is no wonder that almost all of the mixers which have been realized for µ-TAS so far are static mixers. Static mixers are exclusively based on the diffusion between the liquids to be mixed. Diffusion requires time and this time

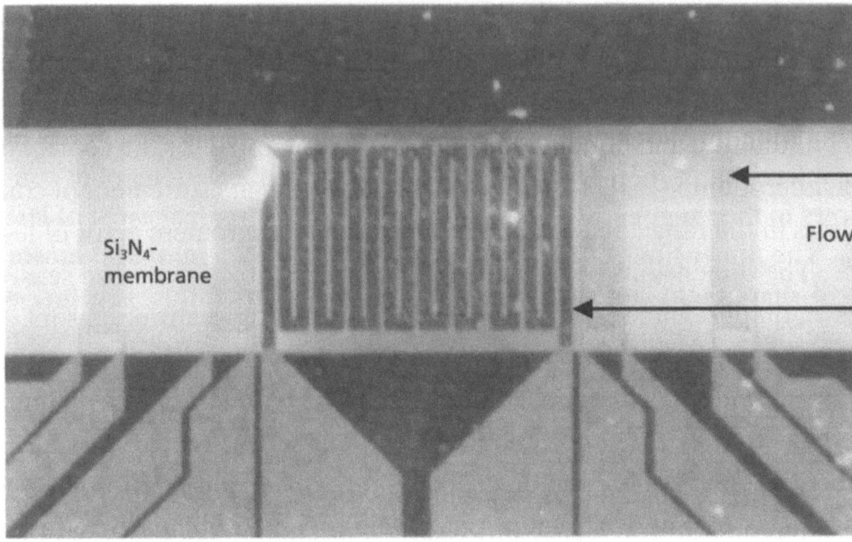

a

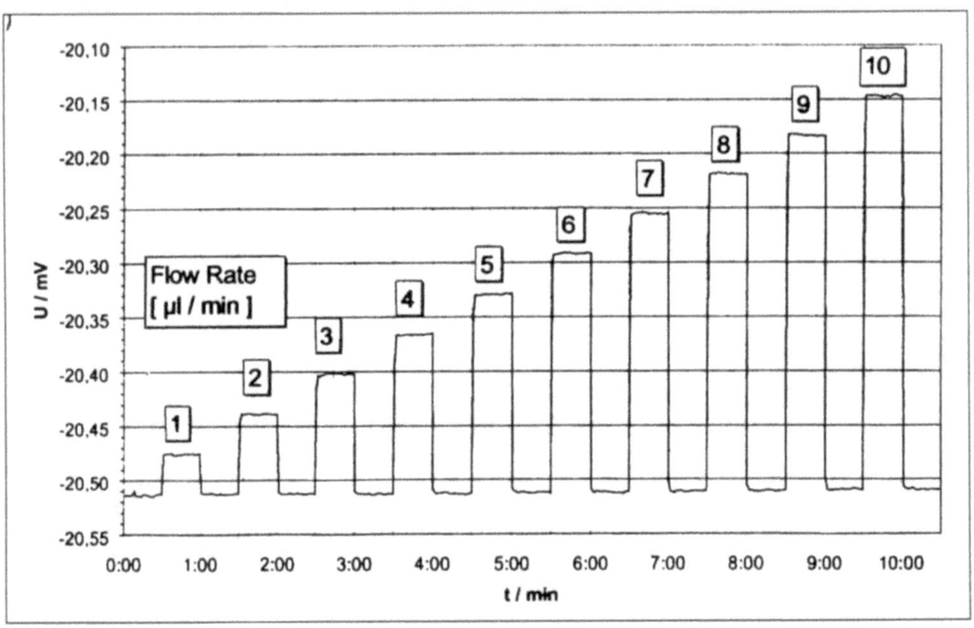

b

Figure 21 *Flow sensor [1, 63]. a) Layout detail; b) flow characteristic areal plott of measurement*

Table 5 Overview of several selected micromixers for μ-TAS concepts

Material	Mixer type	Features	Ref.
PMMA	Möbius-type static mixer	– Manufacture by laser ablation in polymer – Mixing is done by the repeated passive passing through a large number of similar structures where the fluid strand is split, rebonded together and so on	[64]
Solid resist silicon	Möbius-type static mixer	– Four-step mixer can mix accurately even at the low flow rate of 285 μl/min – mixing channel dimensions: W 600 μm, H 100 μm, L 170 mm – capacity of the mixer: 6.5 μl – the mixer has one basic element per mixing stage where the fluid strand is turned by 90° several times	[31]
Glass silicon	Multiparallel lamellas-type static mixer	– Reduction of the diffusion ways by splitting a large strand into many parallel lamellas (so-called "multiple flowing liquid layers") – manufacture by SF_6O_2 plasma etching and KOH wet etching – area requirement of the mixer structure: approx. 500 μm × 500 μm – dead volume less than 500 nl	[65]
Glass silicon	Dynamic MEP-type mixer	– MEP-type micropump with T-shaped outlet channel – diffusion occurs after the fluids in the flying microdrop have been mixed batchwise, therefore the diffusion distances are extremely short – dead volume less than a few nanolitres – the mixing ratio can be controlled by the layout of the component and by electrical operating parameters	[1]

can be provided in the microsystem either by low flow rates or by long diffusion channels. Further, [31] points out that the shape selected for the microchannel is also important to the formation of a large interface between the liquids to be mixed.

The micromixer described by Larsen [65] is characterized by a particularly high efficiency, little space required and a small dead volume. Due to its construction by glass-silicon technique, it is resistant to organic solvents, strong acids and weak bases.

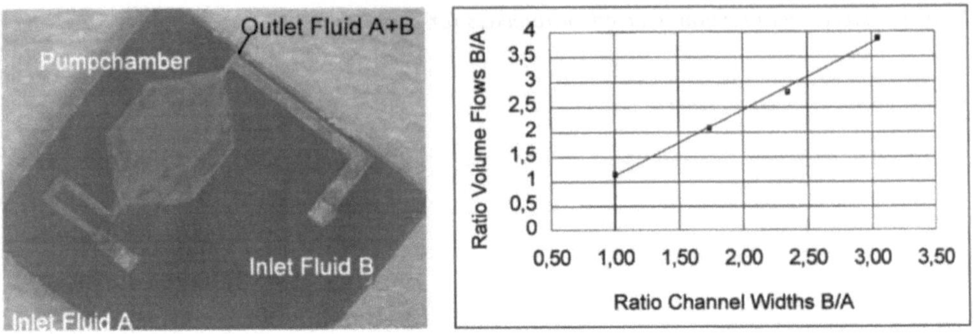

Figure 22 *Active mixer based on a MEP-type micropump [1]*

The active micromixer based on an MEP-type micropump which was developed by GeSiM is an interesting solution for a dynamic mixer. An example of a mixer as well as its characteristics are shown in Figure 22.

The micromixer is comprised of two fluid inlets, a pump chamber and a T-shaped outlet channel with a lateral channel coupling. The layout of the fluid channels, that is, the hydraulic cross-sections of the channels where fluid A and fluid B flow through, determines the mixing ratio. A second possibility to vary the mixing ratio for a given layout is to vary the working frequency. Active mixing is always done in the loading mode of the micromixer. Fluid A and fluid B are resucked into the pump chamber. This is an intermittent process leading to turbulences in the mixing zone limited to the outlet nozzle and the outlet region of the pump chamber. Despite the typical channel geometry of 100 µm × 50 µm (width × height), turbulences occur in this fluid-dynamic process. We can proceed on the assumption that the drop discharged from the pump is a mixture of both fluids. Further, we can assume that the process of mixture formation due to diffusion also proceeds in this drop of about 1 nl.

Let us consider the criteria of use of the presented micromixers. The micromixers reported in [31, 64 and 65] are suitable for use in closed-loop microsystems. Their function is realized by the action of a fluidic drive in the system. The micromixer of Figure 22 works as a system which is capable of producing an open jet and which actively drives the fluids to be mixed. It is a drive and a mixer in the same microfluidic component. However, it is not suitable for use in a closed-loop system. If this active micromixer is combined with the above microinjector, defined mixtures of different fluids can also be injected into closed-loop microsystems.

References

1 Produktkatalog der Firma GeSiM mbH, Großerkmannsdorf (1997)

2 Menz W, Mohr J (1997) Mikrosystemtechnik für Ingenieure, Kapitel 7, 2. erw. Aufl. – Weinheim, VHC, pp 231–318

3 Harrison DJ et al. Chemical and biochemical reaction and separations within microchips; book of abstract: 1st Int Conf on Microreaction Technology; Feb 23–25, 1997

4 Bauer R, Howitz S. Projekt Keramische Mikrosystemkomponenten für die Mikrofluidik, PT-Nr. 3511/547, SMWA, Dresden, Feb 24, 1997

5 Fischer-Frühholz S. Das Labor auf dem Chip Teil 1 und 2; BioTec 1/97 und BioTec 2/97, Media Tec Verlag, pp 40–44

6 Nguyen N-T, Dötzel W (1996) Mikromechanische Strömungssensoren im Überblick, F & M 104: 644–648

7 Wolf G et al. Supermikrokalorimeter für thermochemische Anwendungen, Abstract der 12. Ulm-Freiberger Kalorimetrietage 19.–21.03.1997 Bergadkademie Freiberg, S 5, 70

8 Produktkatalog Borofloat™ Glas; Jenaer Glaswerk GmbH, Jena, 1994

9 Drost A (1996) Seminar Mikrodosiersysteme in IFT-München, Vortrag Technologien der Mikrosystemtechnik, Dec 3 1996, München

10 Fuhr G. Pyrex-Surface für Dielektrophoretische Pumpen und Feldkäfige; Examples of three dimensional micro-structures for handling and investigation of adherently growing cells and sub-micron particles, Proc of the 2nd Int Symposium µ-TAS '96, Basel, Nov 21–22, 1996, pp 39–54

11 Haas K-H. Annual Report FhG Institut für Silicatforschung, Abteilung ORMOCERe, pp 46–63

12 Richter K, Orfert M, Howitz S, Thierbach S, Deep plasma silicon etch for microfluidic applications, 6. Int. Conference on Plasma Surface Engineering 1998, Garmisch-Partenkirchen

13 Bhardway JK, Ashraf H. Micromachining and microfabrication process technology, SPIE-proceedings series, Vol 2639, Oct 23, 24, 1995, Austin, Texas, pp 224–233

14 Bhardway JK, Ashraf H, McQuarrie A. Symposiumon microstructures and microfabricated systems, Annual Meeting of the Electrochemical Society, Montreal, Quebec, Canada May 4–9, 1997

15 mst-news ZASE Advances MEMS Technology 20/97, p 23

16 Regalado A (1996) The DNA-chip in diagnostics, Start-UP

17 The Genesis Report, Miniaturized Diagnostics Grow in Importance, USA-NJ, pp 2–6

18 van den Berg A, Bergveld P. Proc of the µ-TAS '94 Workshop, MESA Research Institute, University of Twente, The Netherlands, Nov 21–22, 1994, Twente

19 Widmer HM. Proc of the 2nd Int Symposium on Miniaturized Total Analysis Systems, µ-TAS 96, Nov 21–22, 1996, Basel

20 Projektbeschreibung; Miniaturisiertes Analysesystem zur Bestimmung der NO_3-Konzentration im Trinkwasser, SMWA 1996, PT-Nr: 1676/355, Dresden, 1995

21 Manz A et al. µ-TAS: Miniaturized total chemical analysis systems, Proc of the µ-TAS '94 Workshop, MESA Research Institute, University of Twente, The Netherlands, Nov 21–22, 1994, Twente, pp 5–57

22 Sbiaa Z et al. 3D integrated micropump and microvalves for a µ-TAS, Proc of the 2nd Int Symposium on Miniaturized Total Analysis Systems, µ-TAS 96, Nov 21–22, 1996, Basel µ-TAS 96, p 239

23 Woias P et al. A micromachined open tubular reactor for heterogeneous immunoassays, Even There, µ-TAS 96, p 256
24 Meckes A et al. Concept and design considerations for a miniaturized gasanalyser Even There µ-TAS 96, p 126–128
25 Nakanisi H et al. Micromachined quartz and pyrex chips for capillary dlectrophoresis, Even There µ-TAS 96, p 236
26 Gretillat MA et al. A new fabrication method of borosilicate glass capillary tubes with lateral inlets and outlets, ebenda µ-TAS 96, p 214
27 Roeraade J et al. Nanochemistry and nanoseparations of biomelecules, Even There µ-TAS 96, pp 34–38
28 Rapp R. Performance of a modular µ-TAS with electrochemical detection; Even There µ-TAS 96, p 237
29 Bley P. 1. Statuskolloquium des Projektes Mikrosystemtechnik, KfK-5238; Sept 23–24, 1993
30 Effenhauser CS et al. Detection of single DANN melecules and DANN fragment analysis in molded silicon elastomer microchips, Proc of the 2nd Int Symposium on Miniaturized Total Analysis Systems, µ-TAS 96, Nov 21-22, 1996, Basel, p 124
31 Svasek P et al. Dry film resist based fluid handling components for µ-TAS fast mixing by parallel multilayer lamination, Even There µ-TAS 96, pp 78–80
32 van Lintel HTG et al. (1988) A piezoelectric micropump based on micromachining of Silicon, *Sensors and Actuators* 15; pp 153–167
33 van de Pol FCM et al. (1990) A thermopneumatic micropump based on microengineering techniques, *Sensors and Actuators* A21–A23, pp 198–202
34 Zengerle R et al. (1995) A bidirectional silicon micropump, *Sensors and Actuators* A 50, pp 81–86
35 Richter A et al. (1991) A microfabricated electrohydrodynamic EHD pump, *Sensors and Actuators* A 29, pp 159–168
36 Branebjerg J, Rye Nielsen C. Design and properties of system components for liquid-handling, MCM technical report, Danfoss A/S DK-6430 Nordborg, Denmark, in preparation
37 Döring M (1991) Flüssigkeiten mikrofein dosieren, F & M 99, 11: S459–463
38 Howitz S, Wegener T, Bürger M. Electrically controllable micro-pipette, Patent PCT/DE 96/00139
39 Howitz S, Wegener T, Bürger M, Gehring T. Mikroejektionspumpe mit dynamisch variabler Eingangsrestriktion, Patent, D 661152 DE, München, Dec 11, 1996
40 Fuhr G et al. (1991) Asynchronous travelling-wave includes linear motion of living cells, Studia Biophysica, Vol 140, No 2, pp 97–102
41 Fuhr G, Shirley SH (1995) Cell handling and characterisation using micron and submicron electrode arrays, state of the art and perspectiver of semiconductor microtools, *J Micromechanics and Microengineering*, 5, pp 77–85
42 Microdrop GmbH; Produktkatalog, Hamburg, November, 1996
43 Lemmo A, Fisher JT, Geysen HM, Rose G (1997) Characterisation of an inkjet chemical microdispenser for combinatorial library synthesis, Analytical Chemistry, Vol 69, No. 4, February 15
44 Seiler T, Howitz S, Wegener T, Bürger M. Verfahren und Vorrichtung zur kontaktfreien Applikation von flüssigen diagnostischen oder therapeutischen Pharmaka am menschlichen Auge, Patent, D 66051 DE, München, Sept 24, 1996
45 Fuhr G, Hornung J, Hagedorn R, Müller T, Howitz S. Kryokonservierung und Tieftemperaturbearbeitung von biologischen Zeilen, Patent, PCT/DE 95/01490, München, Okt 25, 1994
46 Fischer S, Howitz S. Vorrichtung zur Zuführung von Fluiden in eine stehende Ultraschallwelle, D 670048 DE, München May 16, 1997

47 Dantec Invent Measurement Technology GmbH, Produktkatalog US-Levitationssystem, Erlangen

48 Howitz S, Wegener T, Bürger M (1997) Mikropipette mit integriertem Heizer und Temperatursensor zur Nanoliterdosierung höher viskoser Flüssigkeiten, 4. Dresdner Sensor-Symposium, Dresden, Dez 1997, in preparation

49 Blankenstein G et al. Flow switch for analyte injection and cell/particle sorting, Proc of the 2nd Int Symposium µ-Tas '96, Basel Nov 21–22, 1996, pp 82–84

50 Franz J et al. A silicon microvalve with integrated flow sensor, Proc Eurosensors IX, Stockholm, Sweden, pp 313–316

51 Vieider C et al. A pneumatically actuated micro valve with silicon rubber membrane for integration with fluid handling systems, Proc Eurosensorx IX, Stockholm, Sweden, pp 284–286

52 Ulrich J et al. Static and dynamic flow simulation of a KOH-etched micro valve, Proc Eurosensors IX, Stockholm, Sweden, pp 17–20

53 Koch M et al. Simulation and fabrication of micromachined cantilever valves, Proc Eurosensors X, Leuven, Belgium, pp 849–852

54 Redwood MicroSystems, Inc. 959 Hamilton Avenue, Menlo Park CA 94025, Technical data, normally open gas valve NO-1500 & NO-3000; Normally Closed Gas Valve NC-1500-TFluistor Microvalves

55 Rapp R et al. Konzeption, Entwicklung und Realisierung einer Mikromembranpumpe in LIGA-Technik; KfK Bericht 5251; Karlsruhe, October 1993

56 Jed Harrison D, Nghia Chiem. Microchip Lab for Biochemical Analysis, Proc of the 2nd Int Symposium µ-Tas '96, Basel, Nov 21–22, 1996, p 31

57 Manz A et al. (1993) Advances in chromatography, 33, pp 1–66

58 Howitz S et al. A fluidic-ISFET Micro-System, Proc of the 5th International Meeting on Chemical Sensors, Rome, July 11–14, 1994, p 257.

59 Howitz S, Pham MT. Mikro-Fluiddiode, Patent, PCT/DE/95/00200, München Feb 17, 1995

60 Nguyen N-T, Dötzel W (1996) Mikromechanische Strömungssensoren im Überblick, Carl Hanser Verlag München, F & M 104/9, pp 644–648

61 Kohl F et al (1994) A micromachined flow sensor for liquid and gaseous fluids, *Sensor and Actuator* A41–42, pp 293–299

62 Lammerink TSJ et al. (1993) Mico-liquid flow senor, *Sensor and Actuator* A37–38, pp 45–50

63 Köhler JM, Baier V, Schulz T, Dillner U, Gehring T, Howitz S, Caloric liquid flow sensor for lowest flow rates, Eurosensors XII, 13.–16. Sep. 1998, pp 765–768

64 Mensinger H et al. Microreactor with integrated static mixer and analysis system, Proc of the µ-Tas '94 Workshop, MESA Inst Univ of Twente, Nov 21–22, 1994, pp 237–243

65 Larsen UD, Blankensteins G et al. Fast mixing by parallel multilayer lamination Proc of the 2nd Int Symposium µ-Tas '96, Basel, Nov 21–22, 1996, µ-TAS 96, pp 228–230

66 Branebjerg J et al. Application of miniature analyzers: From microfluidic components to µ-TAS, Proc of the µ-Tas '94 Workshop, MESA Inst Univ of Twente, Nov 21–22, 1994, pp 141–151

67 Volmer J, Hein M. Miniaturisierte fluidische Schaltelemente in LIGA-Technik mit integrierter elektrischer Steuerung, Dissertation Universität Karlsruhe; KfK Bericht Nr. 5375

68 Gimsa J, Rebenklau L, Howitz S, A LTCC chamber for characterization of internal electric particle properties by a laseroptical method, Microsystemtechnologies 1998; Potsdam 12/1998

4 Nanotiterplates for screening and synthesis

Günther Mayer, Klaus Wohlfart, Andreas Schober and J. Michael Köhler

4.1 Introduction

Miniaturization of parallel processing of liquids

Biomolecular processing includes the crosslinked handling of substances and information. Analytical information has to be extracted from solutions and from a lot of different sources. Structural information has to be extracted from molecules, kinetic information from interaction or decomposition processes. The need of conversion between matter and information consists in the opposite direction, too. Mechanistic concepts and kinetic calculations have to be converted in new synthesis, structural concepts have to be realized in new molecules, new materials, and particularly, sequence information must be converted in biomolecular tools, like substances for new diagnostic tests, for new therapeutics, for new bioanalogous types of materials, and for synthetic enzymes for technical processes.

Today, we are beginning to understand how nature organizes the interfaces between substance flows, energy flows and information flows. But, a lot of basic mechanisms and many details are not well understood. We know that a hierarchical network of kinetic and spatial couplings is used for the management of the complex molecular processes. The micro compartmentation of cells is one of the most basic aspects of spatial organization. It enables the cell to realize substance management in parallelization. Typical size of micro compartments in nature is in the dimension of about 10^{-20} l (molecular membrane wells), 10^{-18} l (cell organelles) or $10^{-15} \ldots 10^{-12}$ l (complete bacteria cells).

Parallelization of substance handling is a basic method for optimization of laboratory work, too [1–3]. Today, the typical volumes in automated laboratory operations are in the range between about 1 ml and 5 µl. But, how large is the minimum volume that is absolutely necessary for receiving information or introducing information in molecules? In a lot of operations like binding tests with luminometric assays only very small amounts of substances are necessary, between about 10^3 to 10^6 molecules, and even single molecule detection is possible, for example by fluorescence correlation spectroscopy (FCS) [4–6]. For a discussion of FCS see also the article in this book by G. Gradl et al. [7]. Moreover, even at small concentrations, only small volumes are needed. At a concentration of nMol solutions (10^{-9} mol/l) 1 µl contains about 600000000 molecules, 1 nl liquid contains 600000 molecules, and 1 fl liquid contains 600 molecules. The limitation of the miniaturization of liquid operations is now given not by the basic limits of measurement methods, but by the techniques of micro liquid handling. One problem of micro liquid handling consists in the separation of miniaturized reaction vessels in arrays below the µl-range. This corresponds to the extension of the micro titer technique into the nanoliter range, from the micro titer plate to the nano titer plate and perhaps down to the picoliter range.

Potential of microfabrication technology for nano titer plates

Microfabrication is well suited for the definition of compartments with small volumes. Line widths below 100 nm can be produced by the use of the most advanced lithographic procedures like focused electron beam lithography and ion beam etching. Structural details in the range of al (attoliter = 10^{-18} l) can be prepared. The advanced well established lithographic techniques for the fabrication of integrated electronic circuits work with structure sizes in the order of about 0.3 µm, corresponding to a volume of about 30 al. Well established standard processes produce lines of 1 µm, corresponding to 1 fl, more simple optical proximity lithographic procedures supply minimal line widths of about 5 µm, corresponding to 0.1 pl. These numbers impressively demonstrate that lithographic methods can supply measures much smaller than needed for the next steps of micro liquid handling.

But, microfabrication is bound up with a dimensional problem. Besides patterning techniques as deep-UV or x-ray lithography [8], most technological steps are of planar nature, primarily. Patterns are generated in one plane only. The third dimension is the stepchild of the pattern transfer in micro lithography and given in most cases by second parameters like the thickness of material layers or etching times.

The largest measurement in the third dimension is presented by the thickness of microtechnical substrates. The thickness of standard 4″ silicon wafers is about 0.5 mm. This value can be used as an approximate mark. The introduction of microtechnical formed vessels for liquid handling becomes reasonable in the volume range of 1 µl. For the potential of microtechnology the sub-µl and the nl-ranges have to be investigated.

Recent developments and applications

The micro titer technique is mainly based on plastic material. It is well established in a lot of laboratories and supported by various devices. It is applied preferentially in the volume range between about 20 and 200 µl.

Microfabricated carriers for parallel processing with lower volumes were proposed about one decade ago [9]. An interesting way to supply microsized compartment arrays in polymer material uses the LIGA process, in which a master of the carrier is produced by x-ray deep lithography and subsequent electroforming in metal and forming in plastics [10–12]. Micro compartments from PMMA, made by moulding of a master formed by mechanical micro engineering, were successfully used for growing *in vitro* hepatocyte cell cultures [13].

But also silicon itself can be used for the fabrication of compartment arrays [14]. For the fabrication of smaller reaction vessels, silicon is a material which offers a lot of advantageous properties: it is comparatively insensitive against organic solvents and has a high biological and environmental compatibility. Also the silicon surface can be chemically modified if necessary to meet for special requirements in modern molecular biology. Silicon also offers a very low fluorescent background in a large range of the visible spectrum in contrast to a broad band of plastic materials, which is important if optical detection techniques have to be applied for screening, e.g., laser-induced fluorescence (LIF) or fluorescence correlation spectroscopy (FCS). Last but not least, silicon benefits from highly developed and sophisticated technology of micro electronics and micro system technology [15]. This and the versatile and reproducible properties make silicon a widely used material for a lot of classes of micro systems. The possible integration of functional elements is a big advantage of micro fabricated compartment arrays, particularly on the basis of silicon, which is well introduced in the development of micro sensors and microactuators. These advantages make silicon a good candidate for the development of highly integrated arrays of reaction vessels matching forth-coming requirements of biochemists and molecular biologists.

The usefulness and suitability of micro compartment arrays made from silicon has already been shown in a number of impressive experiments. A sort of micro compartment arrays has been used as a micromechanically defined filter plate and sample holder in a system for the fusion of single lettuce cells [16]. Nano titer plates with about 6000 chambers of 0.1 µl with integrated membranes and/or micro sieves were prepared and tested for chemical reactions and even the growth of cells and difficult biomolecular reactions like 3SR in silicon compartments have been demonstrated [17].

In the following we will present insight into the technique of nano titer plates. Mainly we will concentrate on nano titer plates made from silicon in various variants because we hold this as the most promising system at least at the moment. Besides the rare fabrication techniques we will also show how to adapt silicon made nano titer plates to special requirements by techniques for modification of surface properties and/or integration of sensor elements demonstrated at the example of temperature control. We will close this paper with some examples of applications of our nano titer plates in combinatorial chemistry and evolutive biotechnology.

4.2 Design considerations for micro compartments

Miniaturization of reaction vessels not only has the potential of savings in space and material, i.e. pocketable biochemical laboratories with rational dealing with agents, but also offers the possibility for enhanced processing methods, such as optimized heat transfer due to the large surface to volume ratio [18–20]. Also the mass transport can be greatly improved in properly designed systems.

Let us consider a chemical reaction taking place in a miniaturized reaction vessel. The total number N of compartments, the total area A of the array and

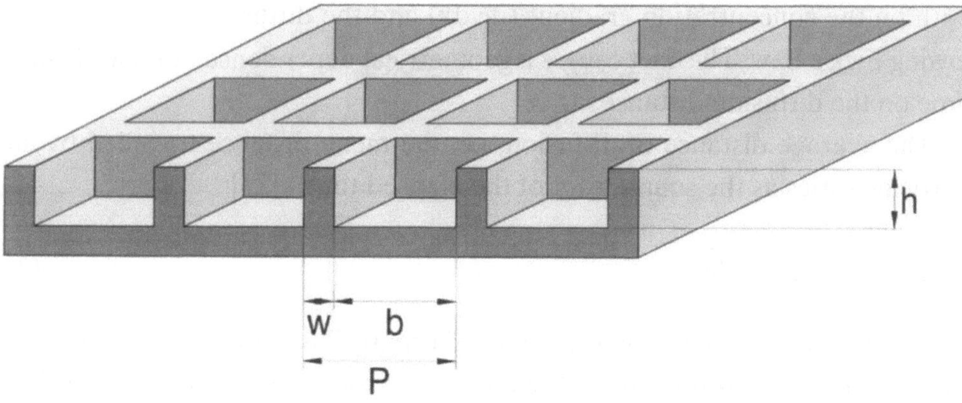

Figure 1 Geometry of micro compartment arrays. See text for explanation

the volume V of each microcompartment in a nano titer plate are the most simple parameters, which have to be considered in the design. Without a restriction to generality we can concentrate on cuboid compartments with quadratic cross-section of width b and depth h (Fig. 1).

The areal density F of compartments is simply given by the period P of the arrangement:

$$F = \frac{N}{A} = \frac{1}{P^2} \quad (1)$$

The volume of a single compartment can be expressed in terms of the width b and the aspect ratio r, i.e. the ratio of depth h to width b of the compartment:

$$V = rb^3 \quad (2)$$

The optimization of the geometry of a micro compartment depends on the processes which have to been performed in it. Particularly, the time consumption is one important parameter. Frequently, the rates of chemical reactions are controlled by the transport of reaction agents (educts) for fast chemical processes. If no convection takes place the transport of educts is controlled by diffusion. If the educts are filled serially into a reaction vessel, the reaction agents first have to diffuse from the interior of the solution in the upper part of the compartment to the lower one before any reaction can take place. The time needed for the movement of the diffusion front in the liquid depends both on the concentration gradient (dc/dx) and the diffusion distance x, following Fick's Law. This fact causes a geometrical dependence of the diffusion time on the diffusion distance.

The average distance of diffusion, i.e. the mean distance travelled by all particles varies as the square root of the elapsed time t [21]:

$$x_D = \langle x \rangle = 2\sqrt{Dt/\pi} \tag{3}$$

In case of small molecules in water, the diffusivity D is in the order of 10^{-9} m²/s (for example, the diffusion coefficient of sucrose in water at T = 298 K is 0.521×10^{-9} m²/s [21]). Resulting in a diffusion time τ_D:

$$\tau_D \cong 7{,}8 \times 10^8\, x_D^2 \tag{4}$$

with x_D in [m] and t_D in [s].

In case of two equal volume parts the average diffusion distance is half the depth of the compartment. In order to compare different shapes of the compartment it is very informative to express the diffusion time in terms of the compartment volume and the aspect ratio:

$$\tau_D \cong 2 \times 10^8\, \sqrt[3]{(r^2\, V)^2} \tag{5}$$

This relation is displayed in Figure 2 for some typical volumes as parameter.

The figure shows that, depending on the compartment volume and shape, the interdiffusion time can take from ms to hours and days. In standard micro titer plates a fast and complete reaction in general cannot be achieved without additional steps for intermixing as stirring or shaking. For nano titer plates the problem is released. Fast reactions in the order of seconds or below can be achieved with no additional means if the compartments are properly designed.

However, it should be noted once more that these considerations are valid only in the absence of other intermixing processes such as stirring, shaking or convection. In real life there is normally turbulence already introduced by the filling process, which by itself will shorten reaction times. This is true in the case of micro titer plates but also for nano titer plates (standard dispensing devices suitable for nano titer plates such as microdrops® eject droplets with a speed of some m/s [22]).

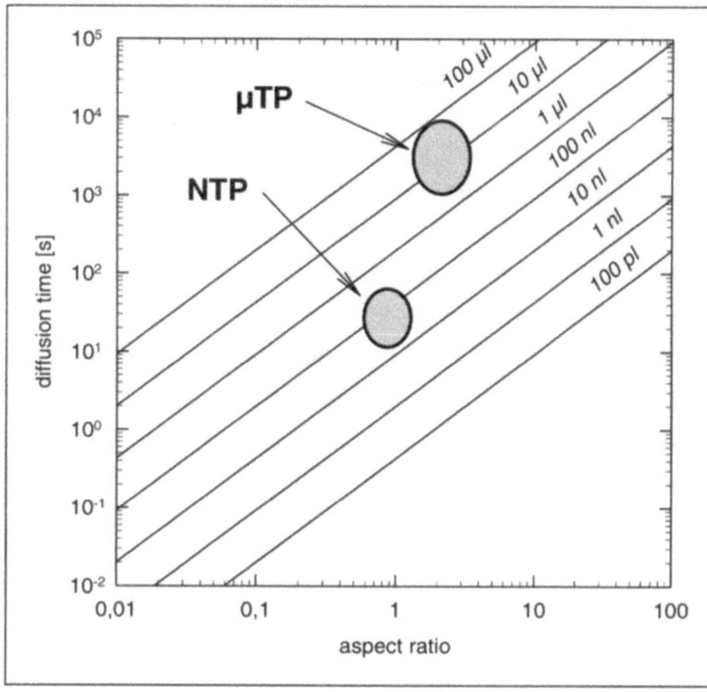

Figure 2 *Estimated interdiffusion times of two agents serially filled in micro compartments of constant volumes as a function of the compartment shape. Low aspect ratios mean shallow compartments, high aspect ratios are deep and narrow compartments. Typical geometries of micro titer plates and nano titer plates are shown*

4.3 Fabrication of nano titer plates by silicon deep etching

Silicon deep etching

Our micro compartment arrays are fabricated by means of a chemical wet etch technique. This technique is in principle known since the early 1950s. For an overlook see, for example, refs. [23, 24]. The etch rate of silicon in alkaline solutions like KOH, NaOH, LiOH but also NH_4OH, tetramethylammonium-hydroxide (TMAH) or a mixture of ethylenediamine, pyrocatechol and water (EDP) exhibit a strong dependence on the crystal orientation of the exposed crystal planes. Therefore this technique is often called anisotropic etching or orientation-dependent etching. In general the etch rate of the {111} planes is very small compared to the etch rates of the other crystal planes and the ratio of the etch rates of the main crystal planes $r_{100}:r_{110}:r_{111}$ can be as large as 300:600:1 [25] or even higher. This is why {111} planes can be treated in the design of micromechanical elements as natural etch stops. The etch rate further depends on the type of the etchant, the concentration and exponential rises with temperature. In general, temperatures of 60°–80°C are used. Under these circumstances the etch rates for {100} planes are in the order of some 100 nm/min to ≈ µm/min.

The chemical mechanism of orientation-dependent etching of silicon is not yet completely understood. Agreement exists that the dissolution of silicon consists of two main oxidation and reduction steps where hydroxide ions and water react with the silicon surface:

$$Si + 2\,OH^- \rightarrow Si(OH)_2^{++} + 4\,e^- \tag{6}$$

$$Si(OH)_2^{++} + 4\,e^- + 4\,H_2O \rightarrow Si(OH)_6^{--} + 2\,H_2 \tag{7}$$

The four electrons in the first step are injected into the conduction band of the silicon and later consumed in the reduction step (7). Concerning the anisotropy of the etch rates an important role is played by the fact that the number of dangling bonds per unit cell on a silicon surface depends on the crystallographic orientation of the surface plane [24, 26]. For a {111} surface there is only one dangling bond and three backbonds, whereas on {100} and {110} planes there are both two dangling bonds and backbonds [27]. However, this difference in the areal density of dangling bonds is not sufficient to explain differences in the etch rates of two orders of magnitude [24]. Kendall [25] proposes a model which states that {111} planes get oxidized very rapidly and could be covered by a thin oxide layer immediately after contact with the etchant, while Palik assumes that the anisotropy is caused by differences in activation energies of the backbonds [28]. However, a model that explains the anisotropy quantitatively from first principles has not been developed yet.

Crystallographic limitations

For chemical and biochemical reactions to be carried out in micro compartments, the properties of the compartment surfaces play an important role, because the ratio of surface area to volume increases rapidly for small volumes in the range of µl and below. Therefore smooth surfaces will be preferred in cases where adsorption and immobilization of substances is undesired. This will be especially true for small concentrations where in extreme cases there are only a few hundred or even only a few molecules present.

Anisotropic chemical wet etching of silicon produces very smooth and exactly defined surfaces if carried out to the end, i.e. the structures are defined completely by {111} planes intersecting solely at concave angles forming no convex edges. If convex edges or wedges are present, the shape of the resulting structure is difficult to control, because higher index planes are generated like {211}, {311}, {322} and others. The etch rate of high index planes depends in a complex way on the etch parameters like temperature, concentration and composition of the etchant.

For micro compartments defined by {111} planes the shape and also the volume of the compartment and the achievable integration density of compartments is completely defined by the shape of the etch mask and the geometry of the silicon substrate.

Silicon wafers can be fabricated in almost every crystallographic direction. The most used and therefore the cheapest are wafers sawed along a {100} plane. If such standard (100)-Si wafers are used for preparation of the micro compartment arrays, quadratic or rectangular compartment orifices are possible. The geometry of the reaction chambers is an inverted pyramid with side wall angles of 54.74° relative to the wafer surface (part a) of Fig. 3). The bottom width of the pyramids W_{bottom} is given by the top width (the opening of the etch mask) W_{top} and the wafer thickness d:

$$W_{\text{bottom}} = W_{\text{top}} - \sqrt{2}\, d \tag{8}$$

This shape limits the total number of the chambers, which can be achieved on a given area. Standard backside polished 4-inch silicon wafers have a thickness of ≈ 490 µm. (SEMI-standard for 4-inch wafers is 525 µm, ≈ 35 µm are removed during the backside polishing process). Using standard (100) wafers and designing a width of the bottom membrane of $W_{\text{bottom}} = 100$ µm, the maximum chamber density is ≈ 150/cm^2, which is equivalent to a compartment number of 12000 per 4-inch wafer.

(110) silicon is a material which is increasingly used in micro system technologies [29]. Here micro compartments with vertical sidewalls are possible. Four of the eight {111} planes intersect the surface at an angle of 90° and

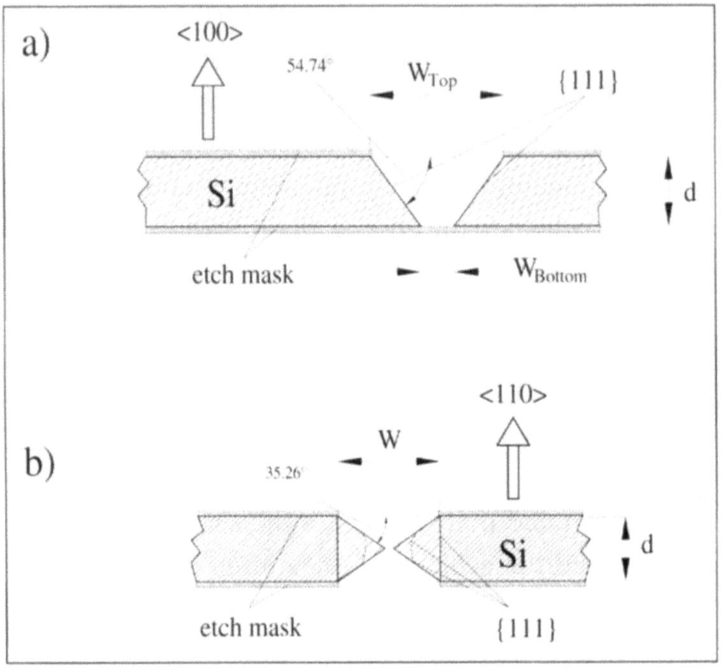

Figure 3 *Position of the {111} planes relative to the surface of silicon. a) (100) silicon; b) (110) silicon. Four planes are perpendicular to the surface, the other planes intersect the surface under an angle of 35.26°*

build angles to each other of 70.5° and 109.5°, respectively. So, compartments with rhomboid orifices or, in general, parallelogram shapes can be built. The remaining {111} planes intersect the surface under an angle of 35.26° and build a wedge which looks out into the chamber (Fig. 3b). This can be completely removed if the chamber is etched from both front and back sides of the wafer and the projection of the chamber width on the ⟨001⟩-direction exceeds the minimal width of

$$W_{min} = \sqrt{2}\, d \tag{9}$$

Perpendicular to this direction the compartments can be made as small as technologically possible, but should be at least 150–200 µm to allow for convenient liquid handling.

Integration of membranes and micro sieves

In the fabrication of micromechanical elements by deep etching the introduction of a special etch mask is necessary. While in standard surface patterning techniques with thin films the selectivity and resistance of photo resists against the etchant is often sufficient, this is not true for silicon deep etching. The etching often takes several hours in a very aggressive medium at high concentrations and elevated temperatures. For this purpose often SiO_2 is used. SiO_2 shows etch rates of several nm/min under standard deep etching conditions.

For the fabrication of our nano titer plates we use a special sandwich layer of SiO_2/Si_3N_4. This has a number of advantages. If the thicknesses of the layers are carefully adjusted, the averaged temperature expansion coefficient of the layer system meets that of the silicon substrate. With such stress compensated layer systems it is possible, to build large free standing membranes with a total thickness below 1 μm. Membranes spanning several cm^2 have been fabricated as a support for thermopile sensors of very high sensitivity [30–33]. Experience has shown that it is advantageous to have a small remaining stress in the layer system. In this way the membrane is tight and smooth and ripples and corrugations are avoided.

The SiO_2/Si_3N_4 layer system is transparent in a broad range of the optical spectrum. Due to this it is possible to use optical detection techniques for the analysis of the compartment contents. It is also possible to make the bottom membrane partly opaque or reflective if needed, for example, for reference purposes. Techniques that can be applied are photo luminescence or laser induced fluorescence (LIF), absorbance or transparency measurements, both spectrally resolved or integrated, and also special techniques like fluorescence correlation spectroscopy (FCS). In combination with position sensitive detectors such as CCD-cameras very efficient screening setups can be built at a high parallelization level.

In multi step solid phase synthesis techniques, e.g., combinatorial chemistry, the possibility for washing and filtering superfluous reagents out of the reaction vessel is essential. Otherwise the reaction products will not be unambiguous. Therefore a technique is needed to combine the membranes with sieves. With the SiO_2/Si_3N_4 sandwich layer system it is possible to pattern the bottom membranes to form micro sieves by only one additional lithographic step. The standard silicon deep etching procedure can be used without any modification.

Of course, there are also some drawbacks using a SiO_2/Si_3N_4 layer system as an etch mask. First, the technological expenditure to fabricate the SiO_2/Si_3N_4 layer system is higher compared to SiO_2. Second, the patterning of the layer system is carried out in hot phosphoric acid or by means of dry etching techniques. This makes a sacrificial layer necessary which serves as an etch mask.

The Si_3N_4/SiO_2-membranes used in our nano titer plates have to be carefully handled to avoid damages. Depending on the experimental requirements the stability of the membranes can be enhanced by several ways. If optical transparency is no criterion the membranes can be combined with metal layers which can be carried out by means of standard sputter techniques. The large ductility of metals like Ti, Au, or NiCr makes micro sieve membranes robust enough to withstand relatively hard load in sucking reagents from the compartment through the micro sieve by applying vapour to the nano titer plate.

However, this technique is not applicable if optical transparency is necessary. For a large class of applications micro sieves are not necessary. In this case the nano titer plates can be combined with glass carriers by means of anodic bonding as described on p. 93.

Technological procedure

For the fabrication of nano titer plates we use standard 4 inch (100) or (110) silicon wafers with polished backside. After cleaning and removal of the native oxide the wafers are covered on both sides with the SiO_2/Si_3N_4 sandwich layer system by means of low pressure chemical vapour deposition (LPCVD)*. Last, the sacrificial layer of 1.2 µm SiO_2 is deposited. The wafers are now ready to be structured.

For the photo lithographic process the wafers are covered with photo resist (Hoechst AZ 1514-H) on both sides and a soft bake completes the prepara-

* The deposition of the SiO_2/Si_3N_4 layer system was done by the “Institut für Halbleiterphysik (IHP)” in Frankfurt/Oder, Germany.

tion. The exposure is done using a mask aligner AL 6-2 (electronic visions, Austria) at $\lambda = 365$ nm (i-line). This mask aligner allows for the precise adjustment of the micro sieves on the back side of the wafer relative to the compartments on the front side. The resist is developed in an alkaline solution. After the hard bake the resist pattern is transferred to the SiO_2 sacrificial layer by wet etching.

The next step is the preparation of the silicon deep etch mask. The transfer of the SiO_2 pattern to the SiO_2/Si_3N_4 layer system is carried out in concentrated phosphoric acid at 180°C. Because of the low etch rate of the Si_3N_4 even in this aggressive medium (1...2 nm/min) this step takes several hours.

Finally the actual silicon deep etching is carried out. We use NaOH at 80°C. Again this step takes several hours. In case of the (110) silicon micro compartments with integrated micro sieves there is an additional isotropic etch step necessary [34]. This must be carried out prior to the anisotropic etching. The reason is that the micro pits that are built under the micro sieve etch mask must be expanded to build one single pit. Otherwise the anisotropic etching from the back side would stop.

Results

Nano titer plates etched from (100) silicon

With standard (100) wafers we have fabricated nano titer plates both in a full wafer design for a maximum number of compartments and a chip design for easy handling. Figure 4 shows a photograph of a 4-inch wafer with about 6000 compartments. The top width of a single compartment is 800 µm × 800 µm. With a spacing between the compartments of 200 µm the areal utilization is 64%.

For the handling of the nano titer plates in an automaton we introduced some marks to enable an automatic position recognition and to make the addressing of a distinct compartment easy and reliable. The marks simply consist of some omitted compartments that form a pattern, that can be recognized by an image processor.

A SEM graph of a single compartment is shown in Figure 5. The width of the bottom membrane is 150 µm and the total volume of the compartment amounts to 140 nl.

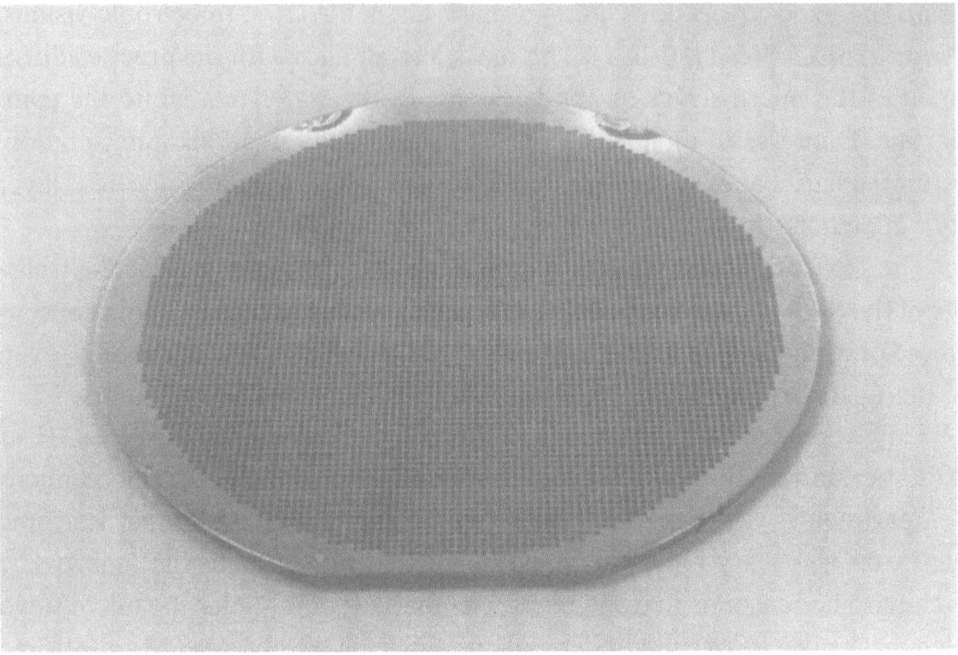

Figure 4 Silicon wafer containing 6000 compartments on a 4-inch diameter

The bottom membrane of the compartment in Figure 5 is built by a micro sieve. The micro sieve consists of 7×7 pores with a pore diameter of 10 μm and a period of 20 μm. Pore sizes down to 7 μm have been successfully prepared by means of standard wet etching techniques. Smaller pore sizes can be realized in combination with dry etching techniques and a modified deep etching procedure using a special protection technique based on silicone. Pore sizes down to 800 nm have been realized, which can be used as carriers for electron holography of DNA [35].

We have also developed a new wafer design with the compartments arranged in a segmented chip-like manner. The wafer contains 37 groups of 10×10 compartments. Non-periodic compartments on the rim of the wafer are used as marks for the automatic position recognition system. This modified layout primarily serves as a basis for the integration of elements for thermo control, which is discussed in detail in Section 4.6. It also has advantages for emptying the compartments during rinsing steps. This is done by applying vacuum to the bottom sieve membranes and sucking out the solution in the compartment. By subdividing the wafer in groups the vacuum can be

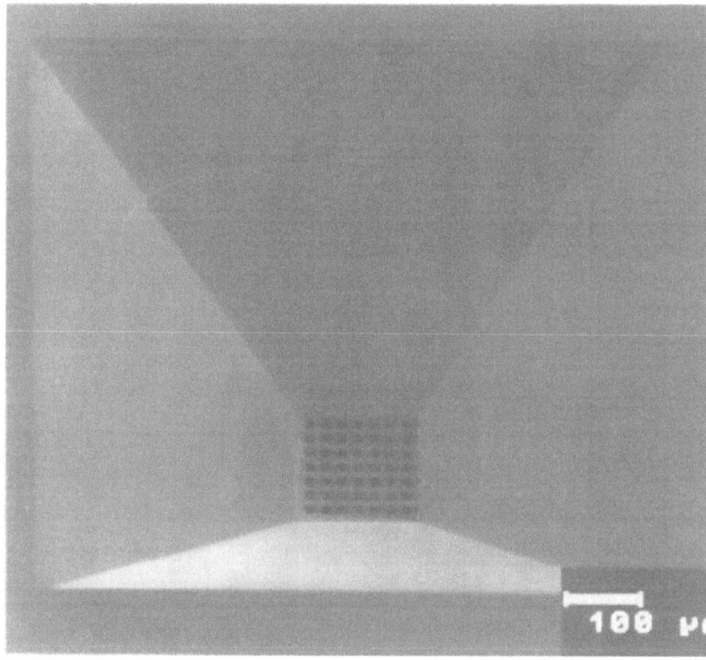

Figure 5 *SEM graph of a single compartment with micro sieves on the bottom membrane. The size of a single micro pore is ~ 10 µm*

applied in a more defined and reproducible way and the force on the nano titer plate is reduced.

To meet different experimental requirements we have also prepared nano titer plates in a chip-like manner with various numbers of compartments (Fig. 6). Table 1 gives an overview on the nano titer plates we have fabricated. The volume of a single compartment can be varied in a certain range by adjusting the width of the opening. However, the bottom membrane becomes more fragile for large compartment volumes because of the larger bottom width, especially if micro sieves are integrated.

For the fabrication of the 20 nl-compartments (NT 784/20) we have used special thinned wafers with a thickness of 275 µm. Because the top width of the compartments is coupled to the bottom width by the thickness of the wafer (Eq. 8), using thinner wafer is a straight forward way to increase the areal density of compartments with optically transparent bottom membranes. We have achieved integration levels of 400/cm^2, which would correspond to as much as 25000 compartments on a 4-inch wafer. In the chip-like layout the

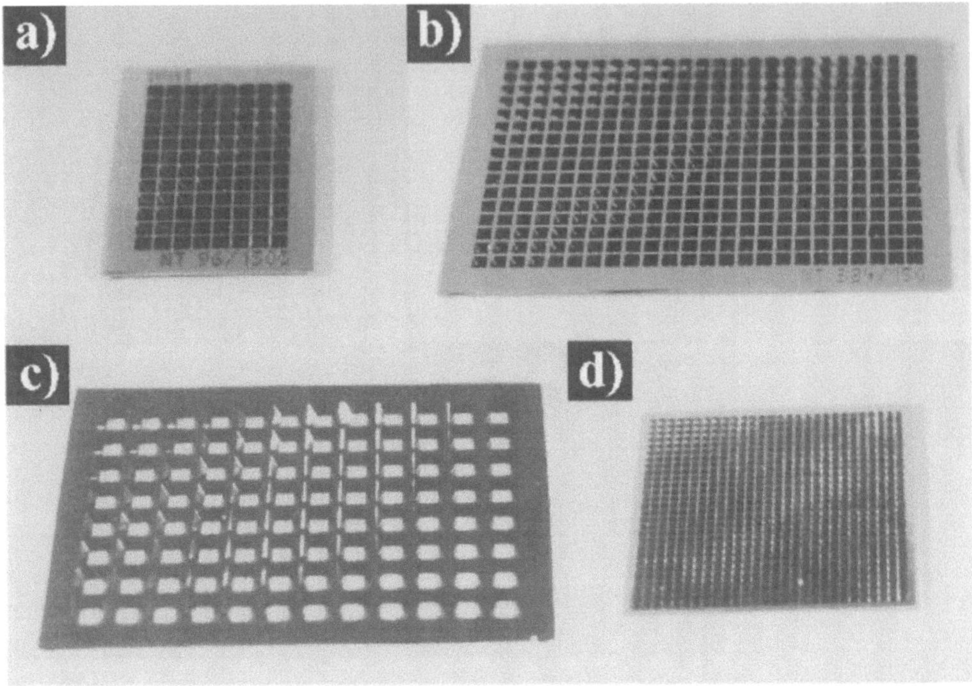

Figure 6 *Chip sized variants of nano titer plates: a) NT 96/150: chip with 96 wells of 140 nl; b) NT 384/150: chip with 384 wells of 140 nl; c) NT 96/1000: chip with 96 wells of 1 µl; d) NT 784/20: chip with 784 wells of 20 nl. For the physical dimensions see Table 1*

Table 1 Geometrical data of nano titer plates made from (100) silicon

Variant	Size [mm]	No. of compartments	Volume [nl]	W_{Top} [µm]	W_{Bottom} [µm]	Period [mm]
NT 6000/150	⌀ 100	6000	140	800	150	1
NT 3700/150	⌀ 100	3700	140	800	150	1
NT 96/150	11 × 15	8 × 12	140	800	150	1
NT 100/150	12 × 12	10 × 10	140	800	150	1
NT 384/150	19 × 27	16 × 24	140	800	150	1
NT 96/400	12 × 18	8 × 12	400	1200	550	1.5
NT 320/400	24 × 30	16 × 20	400	1200	550	1.5
NT 96/1000	19 × 27	8 × 12	950	1700	1050	2
NT 784/20	15 × 15	28 × 28	20	450	60	0.5

total number of compartments on the wafer amounts to 16000, because the need for some edges around the chip reduces the averaged integration level.

However, the volume of a single compartment is:

$$V = \frac{d}{3}\left[W_{\mathrm{Bottom}}^2 + W_{\mathrm{Bottom}}\,W_{\mathrm{Top}} + W_{\mathrm{Top}}^2\right] \tag{10}$$

By inserting Eq. (8) the volume becomes a third order function of the wafer thickness. Therefore, if the bottom width W_{Bottom} of the compartment is small compared to the wafer thickness d, the volume of the compartments gets very small and fluid handling and evaporation will become a severe problem.

Nano titer plates etched from (110) silicon

A better utilization of the wafer can be achieved using (110)-Si. Because the side walls are vertical, the volume of the compartments shrinks only linearly with the wafer thickness. We have prepared micro compartments from (110)-Si. SEM graphs are shown in Figure 7. The back side of the compartments is covered by a sieve membrane with a pore size of $12 \times 20\ \mu m^2$. The shape of the compartments is a parallelogram with side lengths of $280 \times 700\ \mu m^2$ and a wafer thickness of 485 µm, which calculates to a total volume of 90 nl. Chips consisting of an array of 448 micro compartments have been prepared, which amounts to ~ 380/cm² or a total of ~ 15000 per 4-inch wafer. Using (110)-Si wafers with a smaller thickness an integration level of ~ 50000 per 4-inch wafer is feasible.

Besides the higher achievable compartment density, the (110)-Si nano titer plates offer a higher aperture, which will result in a higher S/N-ratio in transmission spectroscopy experiments. Also the ratio of the *free* surface to the compartment volume is smaller for (110)-compartments compared to (100)-compartments. If the meniscus due to wetting properties is neglected, the ratio is 1.85 mm^{-1} and 4.27 mm^{-1}, respectively, for completely filled compartments. This will be reflected in different evaporation rates, as will be shown on page 112ff. On the other side, compartments with micro sieves made from (100)-Si due to their pyramidal shape can exhibit funnel effect in filtering processes, which can be advantageous if working with few or single beads.

Figure 7 *Array of micro compartments etched in (110)-Si. a) view from the front side. There are 448 micro compartments on a single chip. Note that the compartment opening is not rectangular; b) view from the back side through the micro sieves*

Modification of nano titer plates by anodic bonding

Anodic bonding is a technique to connect glass and silicon wafers and was first discovered by Wallis and Pomerantz in 1969 [36]. The key effect is the formation of stable covalent chemical Si-O-Si bonds between the Si-wafer and the glass [37, 38]. This is accomplished by applying a voltage to the wafer pair with the Si-wafer set on a positive (anodic) potential, which serves a twofold purpose: first a space charge region is built at the glass-Si interface, because the small and relatively mobile alkali ions present in the glass (e.g., Na^+) move to the opposite side of the glass and are neutralized at the cathode. The arising electric field in the space charge region causes a strong attraction force between the silicon and the glass, which brings them in intimate contact. Then the remaining immobile O^--ions can form bindings to dangling bonds on the silicon surface. The result is a durable hermetic and irreversible seal between the Si- and the glass wafer. The bond strength usually even exceeds the destruction threshold of the materials. To obtain a sufficient ion conductivity the procedure is carried out at elevated temperature.

For successful anodic bonding some prerequisites must be fulfilled:

- to allow a good contact the surfaces must be smooth (polished).
- to avoid mechanical stress and cracks the thermal expansion coefficients must match.

For the silicon glass combination most often borosilicate glasses as Pyrex (Corning #7740) or Borofloat® (Schott) is used as glass, but also special types with optimized thermal expansion coefficients like SD-2 (Hoya) [39, 40] or glasses with high ion mobility [41]. Typical process parameters are temperatures of 400°–500°C and voltages of some hundred V to KV, partly depending on the thickness of the glass wafer, where lower temperatures can be compensated by higher voltages to a certain degree. Bonding times of a couple of minutes are usually sufficient.

The silicon wafer can also be covered by a thin film of poly-silicon, SiO_2 or Si_3N_4 [42]. Even silicon can be anodically bonded to silicon if a thin intermediate layer of sputtered or evaporated glass is used. In this case very low voltages and temperatures can be used, which reduces the load on sensitive materials [43–45]. However, quality inspection of the bonded interfaces is difficult and must be carried out with IR systems.

Material combinations are not restricted to the silicon glass system, but also other semiconductor (Ge, GaAs) glass combinations have been demonstrated or even certain metals (Ta, Ti, Kovar) and ceramics can be anodically bonded to glass. For an overview on possible material combinations see, for example [36, 46, 47].

We have fabricated various types of nano titer plates anodically bonded to glass carriers. We used commercially available Borofloat with thicknesses of 1.1 mm, 500 µm and for special applications 200 µm. Bond parameters were 800 V and 450°C.

Nano titer plates bonded to glass carriers can be used if suction or filtering is not necessary and optical transparency and mechanical stability is a criterion. They are especially useful for screening techniques like confocal microscopy or FCS, where only small detection volumes of some μm^3 are investigated. The transparent glass carrier could also be an ideal support for the integration of micro lens arrays with enhanced performance compared to conventional optics. The micro lens arrays can in principle be fabricated by the ion exchange technique [48, 49]. Another way for integration of micro-optical elements is replication of a master by moulding, which could be done by LIGA [50]. However, because of the high temperature load during the anodic bonding process, the moulding has to be carried out after the bonding. We have used an elastomere (polydimethylsiloxane, PDMS) for the moulding, which has a refractive index of 1.41 [51]. For the preparation of the master we have used a very simple and low-cost approach by embossing a thermoplast with an array of steel balls used for ball bearings as stamping mould*. With this technique we have achieved micro lens arrays of reasonable quality with lens diameters of ~ 800 µm and a focal length of approx. 1.3 mm.

Anodic bonding can also be used to produce micro compartment arrays with small openings to reduce evaporation of solvent as described on p. 112. By using sputtered intermediate glass layers it should also be feasible to combine nano titer plates with small openings with such micro sieve membranes in a stacked arrangement.

* This technique has been suggested by E.B. Kley, FSU-IAP, Jena.

4.4 Highly integrated nano titer plates from glass and polymers

In principle, it is no problem to etch small grooves in silicon, which can then be filled with liquid as fl- or pl-compartments. There is a simple geometric rule of the number of micro compartments *n*, which can be arranged on a given area *A* in the case of the orientation dependent etching of (100) silicon. This compartment density *D* depends on the volume and is maximal in the case of square pyramidal compartments of a given volume *V*:

$$D = \frac{n}{A} = \sqrt[3]{\left[\frac{\tan(54.7°)}{6\,V}\right]^2} = \frac{1}{\sqrt[3]{18\,V^2}} \qquad (11)$$

About 175 compartments of 100 nl, about 17500 compartments of 100 pl or 380000 compartments of 1 pl can be arranged at 1 cm^2. In reality, the maximum density is a few percent below this value in order to provide the space for the walls between the micro compartments.

Higher area densities can be achieved if true anisotropic techniques with free control of the aspect ratio can be used. However, the integration of optically transparent membranes is a technological challenge. We have used two different techniques to fabricate high integrated nano titer plates. Using photo structurable glass as substrate material with inherent optical transparency quite large aspect ratios can be realized. Anisotropic dry etching techniques on the other side seem to be a practicable way to fabricate ultra high integrated nano titer plates or plates in the sub-nl or pl-range.

Nano titer plates from photosensitive glass

Photosensitive glass as a material for micro machining

Photosensitive glass is a new material that has gained interest in micro system technologies in the last years [52, 53]. The effect of photosensitivity of some kinds of glass has in principle been known since a long time, however the exact mechanism has been unclear. The key point is the photo induced transfer of an electron from a Ce-ion to a Ag-ion, which are added in small amounts to the glass melting. The cerium is incorporated in the glass matrix as a positive ion, which can be further oxidized by UV radiation:

$$Ce^{3+} + h\nu \rightarrow Ce^{4+} + e^- \tag{12}$$

$$Ag^+ + e^- \rightarrow Ag \tag{13}$$

During the subsequent tempering process the atomic silver serves as a seed to form small glass crystallites. The form giving process is built by an ultrasonic-assisted etching in HF, where the glass crystallites have a higher solubility compared to the amorphous regions. The dissolution of exposed regions is therefore 30 times faster than that of not exposed regions.

Results

The micro compartment arrays were fabricated using a photosensitive glass substrate provided by the Sensorglas Ilmenau, Germany. For the UV-light exposure the glass substrate was covered with a metallic thin film mask. The micro compartments are arranged with a period of 0.1 mm, resulting in an area density of 10000/cm^2 (Fig. 8). The structural sizes can be as small as 20 µm if necessary [54]. The shape of the compartments is a small tube with a circular opening and the total volume amounts to ~2 nl.

The compartments can be arranged comparatively close to each other, which leads to a significant increase of compartment density in comparison with wet etched silicon compartments of the same volume. However, the wet etching process is not selective enough to supply a bottom membrane of high optical quality. Also, the production of micro sieve membranes seems to be difficult.

The optical quality of the bottom of the compartments etched in glass is not very high because the roughness of the etched surface of the glass is in the order of some µm. The high roughness is caused by the micro crystallites that are built during the tempering process. The size of these micro crystallites is in the order of 1–5 µm. The high roughness is also unfavourable for processes with very low concentrations, where adsorption of molecules could be a severe problem, but could on the other hand be advantageous for catalytic processes if it is possible to deposit a catalyst on the large inner surface of the compartments.

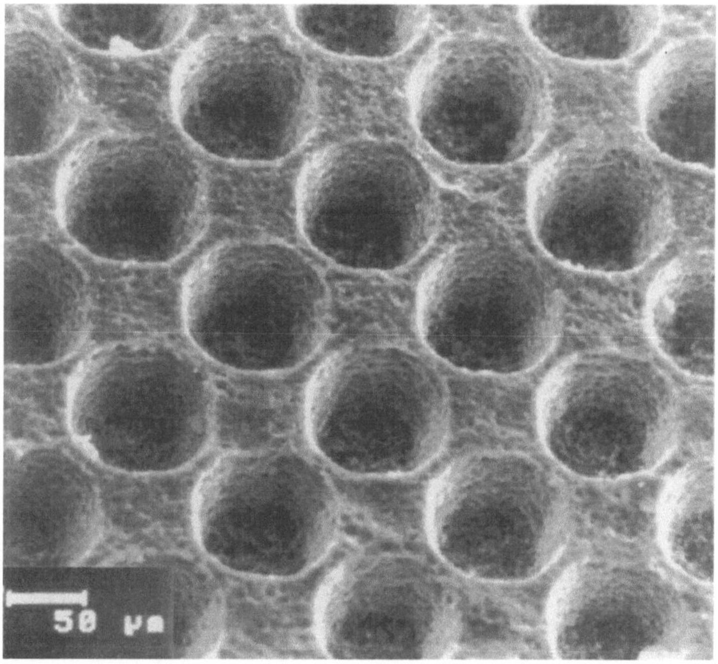

Figure 8 *Micro compartments fabricated from photosensitive glass. The volume of a compartment amounts to ~2 nl.*

Nano titer plates with transparent bottom membranes from polymers

Arrays of small compartments with high aspect ratios can be fabricated by anisotropic dry etching. This process is comparatively expensive for carriers of compartments with single volumes in the range of sub-µl and upper nl-range, because the etch time is high despite the great progress in high rate reactive ion etching. But, smaller compartments with volumes in the lower nl- and the pl-region can easily be produced by the application of dry etching processes. Functional elements, like sieve bottom membranes or optical membranes of high quality can be integrated without any problem.

Fabrication technique

We have used a standard photo resist to produce micro compartments with an optically transparent bottom membrane. The main technological steps are depicted schematically in Figure 9.

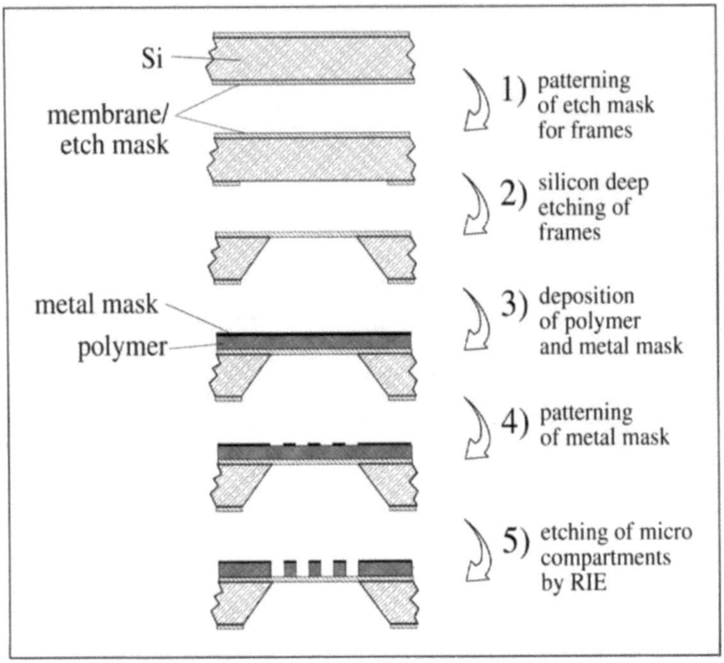

Figure 9 *Main technological steps for the preparation of polymer micro compartments with transparent bottom membranes*

For the carriers we use the standard 4-inch (100) silicon wafers that are used for the fabrication of the silicon nano titer plates. Front and back sides of the wafer are covered with the SiO_2/Si_3N_4 layer system. This layer system is used as the optically transparent bottom membrane. The membrane is supported by a silicon frame which is prepared by silicon orientation dependent etching (steps 1 and 2 in Fig. 9). Then the polymer film is deposited by spin coating. Using standard photo resist layer thicknesses of 10 µm were realized, but with special high viscosity resists structures with thicker layers and higher aspect ratios are possible [8]. At last the polymer layer is covered with a metallic mask (about 300 nm in thickness, step 3). The metal mask is patterned by means of photo lithography and dry etching techniques, but also chemical wet etching can be used (step 4). Finally, the mask pattern is transferred by reactive ion etching (RIE) into the polymer layer (step 5).

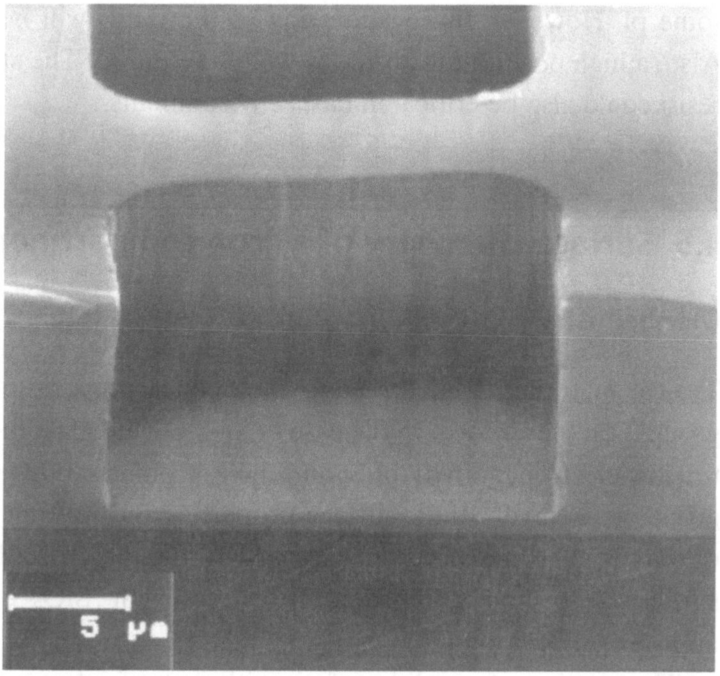

Figure 10 *Highly integrated micro compartments prepared in polymer film on a optically transparent membrane (SEM graph). The volume of a compartment is 4 pl*

Results

The micro compartment arrays fabricated in this way are shown in Figure 10. The size of a single compartment is $20 \times 20\ \mu m^2$ on a 25 µm grid, which calculates with the layer thickness of 10 µm to a compartment volume of 4 pl. The areal density of these pico titer plates is $160000/cm^2$. Chips with four fields of compartment arrays each containing 6400 compartments were fabricated. The size of a single chip is as low as $6 \times 6\ mm^2$, and it is straight forward to design larger chips with more than 10^6 compartments each with the possibility for optical read out.

The fabrication of the high integrated polymer film nano titer plates has been carried out as a kind of a demonstration of the feasibility. Limiting for the usability up to now is the handling of corresponding fluid amounts. Modern dispenser systems such as Microdrop are able to produce drops with some pl. However, the evaporation of the fluids will be a severe problem. Also, a high positioning accuracy will be required. The stability of the photo resist could also be critical in basic environments.

4.5 Surface chemistry of silicon compartments

OH-chemistry

Typical materials used for nano titer plates preparation show many OH-groups* on the surface. In the case of Si, SiO_2 and glass this property is due to the low electronegativity of Si and the low polarizability of Si(IV). The OH-groups cause hydrophilic surface properties and wetting. The contact angles of water can be increased or decreased by making hydrophobic or hydrophilic surfaces, respectively, considerably by treating Si-, SiO_2- or glass-surfaces with appropriate compounds.

The wetting of a solid surface with water is dependent on the relation between the interfacial tensions σ between solid, liquid and vapour/air:

water/air: σ^{wa}
solid/water: σ^{sw}
solid/air: σ^{sa}

Typical values for interfacial tensions between solid surfaces and water are in the range between $5–50 \times 10^{-5}$ N/cm. The ratio between these tensions determines the contact angle Θ of a water droplet in the atmosphere on a given surface. The relation is described by Young's equation:

$$\sigma^{sa} - \sigma^{sw} = \sigma^{wa} \cos\Theta \quad (14)$$

An illustration of this relation on surfaces of different wetting characteristic to water in ambient air is given in the following figure.

For $\theta = 90°$ the interfacial tensions surface/air and surface/liquid are equal. According to the definition, surfaces with measured contact angles of more than 90° are characterized as non-wetting, whereas a surface with less than 90° is called wetting.

* For fully hydrated silica surfaces a density of silanol groups of up to 5×10^{14} cm^{-2} was found [55]. This compares to the number of surface atoms of ~ 18×10^{14} cm^{-2} for a {111}-Si surface.

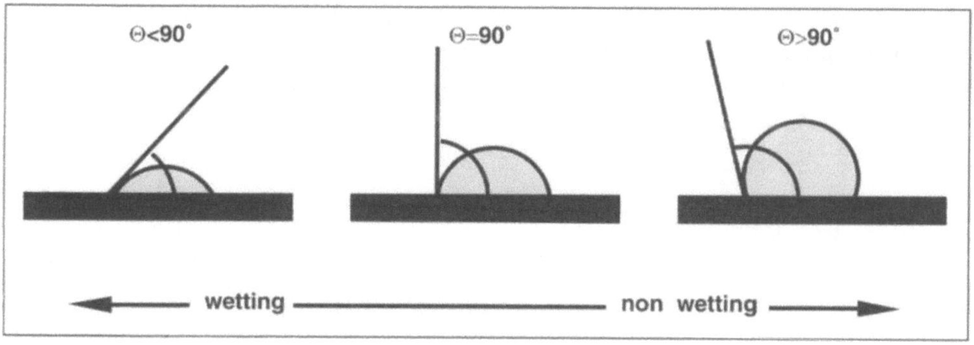

Figure 11 *Wetting characteristics of surfaces with different interaction energies*

Within the micro meter range surface forces are dominant to other forces like gravitation in solid-liquid interaction. Therefore, with appropriate modification of surfaces in the micro meter range it is possible to manage solid-liquid interaction in small dimensions.

Since interfacial tensions cannot be measured directly, the measurement of the contact angle between surface and droplet of a liquid is widely used to characterize wetting characteristics of surfaces [56]. The interfacial tensions are usually calculated from contact angle measurements of two different liquids. Due to a hysteresis effect the advancing and receding contact angle have to be distinguished.

Some typical compounds suitable for hydrophobic and hydrophilic surface treatments of various substrates with measured contact angles to water (advancing type) are given in Table 2.

In addition to the chemical treatment of surfaces, wetting characteristics of surfaces can further be shifted by minimizing the contact area between liquid and surface by increasing the surface roughness. This is already realized in nature, for example, in the cuticular micro structure of leaf surfaces [58] and in fractal surfaces [59]. Such super water-repellent surfaces have self-cleaning properties (Lotus-effect), since particles on micro structured hydrophobic surfaces are more readily wetted and washed off immediately.

Microstructured pads of polydimethylsiloxane (PDMS) exhibit an increased contact angle against water of up to 140° compared to a flat PDMS surface of 100°, as shown in Figure 12. Structured surfaces of such properties might have various applications in technology and micro system design in future.

***Table 2** Contact angles of common substrates and frequently used modification agents for hydrophobizing and hydrophilizing. Because small contact angles below 15° are difficult to determine, only upper limits are given in this case. The variation of the other angles was < ± 3°*

Substrates	Modification agent	Contact angle to water (advancing)
Hydrophobic surface modification:		
silicon, glass	octadecyltrichlorosilane (OTS)	96°
polydimethylsiloxane (PDMS)	–	100°
gold substrates	hexadecanethiol	100°
Hydrophilic surface modification:		
silicon, SiO_2 (*)	–	62°
silicon, SiO_2	hexamethyldisilazane (HMDS)	65° [56]
glass substrates (quartz)	–	42°
gold sputtered on silicon	–	82°
gold substrates	thioglycerine	25°
gold substrates	mercaptohexadecanoic acid	< 15° [57]
gold substrates	mercaptoundecanol	< 15° [57]
polydimethylsiloxane (PDMS)	O_2-plasma treatment	< 10°

* *The wettability of silicon depends strongly on the pretreatment and the history of the surfaces. For freshly prepared Si and SiO_2 contact angles as low as 2° have been reported [56].*

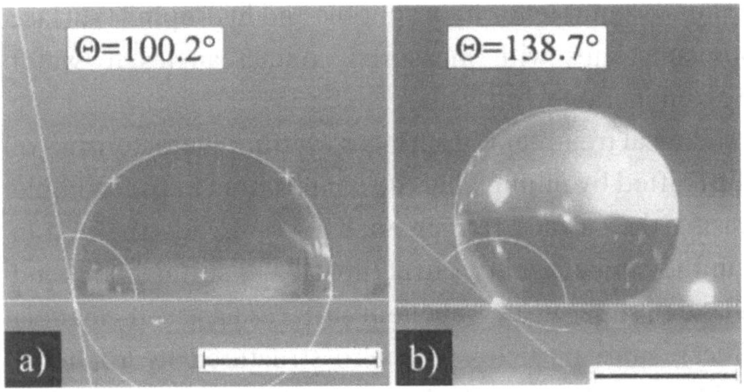

***Figure 12** Water droplet on a flat PDMS surface. a) and a microstructured PDMS surface; b) and corresponding contact angles. The "broken line" in the right image is the side view on the chequerboard-type pattern used. The structure size was 8 µm and the height 5 µm. In the left side of the image the structure is partly hidden by the zero line for the determination of the contact angle. The black bars represent 500 µm*

Surface modification and chemistry with alkyl silanes

Prior to chemical modification surfaces should be cleaned and activated. In case of glass and silicon substrates this is achieved by ultra sonic treatment and washing in solutions of hydrochloric acid/hydrogen peroxide or sulphuric acid/hydrogen peroxide. To avoid contamination until treatment, samples should be stored in an exsiccator or under distilled water.

Surface modifications are typically one-step reactions taking place at room temperature. In case of silanisation with reactive halogen alkanes like octadecyltrichlorosilane (OTS) the reaction should be preferably carried out in an inert gas atmosphere in order to avoid reaction with water vapour. A typical silanisation protocol for hydrophobizing of glass and silicon surfaces is given with the following procedure.

A cleaned and activated substrate is washed with chloroform and put in a Petri dish with a freshly prepared 1 vol% solution of OTS in chloroform and stirred for 5 min. Then the substrate is again washed first in chloroform then in methanol and finally in water. Silicon surfaces treated in this way show contact angles of 96° to water.

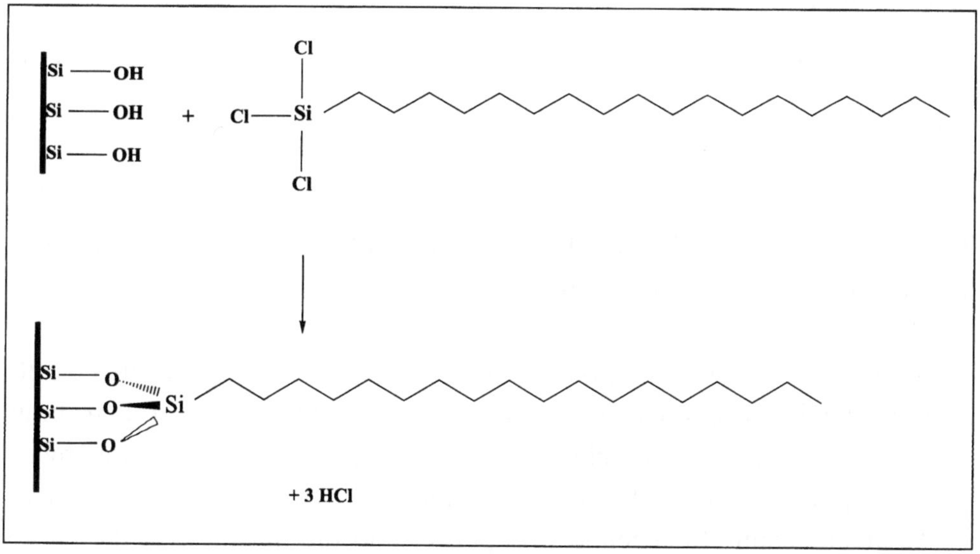

Figure 13 *Hydrophobizing reaction of silicon and glass surfaces with octadecyltrichlorosilane*

Gold thiol chemistry

The properties of surfaces modified by surface reactions depend on the density of binding sites. The substitution of H in OH-groups by alkyl groups, alkylsilyl groups or others leads to more or less dense distributed organic groups at the surface. The task of this type of modifications consists mainly in the conversion of local binding sites.

The formation of closed monomolecular films on microtechnical surfaces is another demand for the preparation of chip carriers for nanochemical handling despite the conversion of more or less dense and randomly distributed binding sites. Self assembled monolayers (SAM) are well suited for such a purpose. SAM films guarantee chemically homogeneous surface properties [60]. In case of secondary functional groups, it is possible to bind other substances and construct surfaces with very different chemical properties. SAMs are characterized by a comparatively strong ordered arrangement of molecules in a surface film which are chemically bonded to the surface.

Fast surface reaction of high selectivity and yield can be realized using the binding of thiols on gold surfaces [61]. Thin sputtered films of Au (10–100 nm) are sufficient for the immobilization of a dense molecular film of thiols. The exact mechanism of the formation reaction is not yet clear. For the formation of RS^--Au^+ the loss of the SH hydrogen is required [62], where the proton can be lost as H_2:

$$RS-H+Au_n(0) \rightarrow RS^- - Au^+ + {}^1/_2\,H_2 + Au_n(0) \qquad (15)$$

or as H_2O:

$$RS-H+Au_n(0)+\text{oxidant} \rightarrow RS^- - Au^+ + {}^1/_2\,H_2O + Au_n(0) \qquad (16)$$

The bonding between the Au (111) surface and the sulfur atom has both σ- and π-character [62] with bonding strength close to that of typical covalent bonds.

Alkanethiols on gold comprise the most robust and best characterized self-assembled mono layers. The polymethylene chains of thiols longer than about 10 carbons form a crystalline-like array on the gold surface and are tilted by an angle of approx. 27° [57]. By variation of the functional group and the length of the alkane chain, the chemical properties and the thickness of the mono layer of the exposed surface can be precisely controlled [63]. Alkanethiols

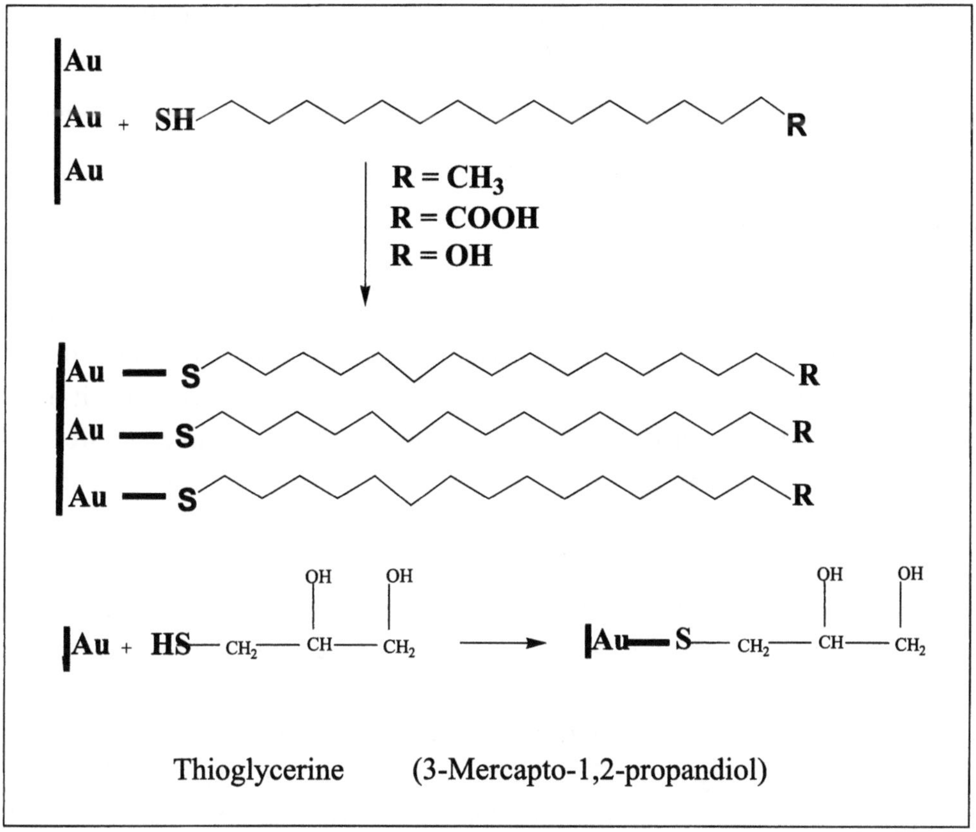

Figure 14 *Reaction of gold exposed to an ethanolic solution of a long-chain alkanethiol (upper sketch) or short-chain thioglycerine (lower). See also text for a discussion of the formation mechanism*

with long (12–18 carbon atoms) aliphatic chains give a dense mono layer and a strong bond to gold surfaces. By using an alkanethiol like hexadecanethiol $CH_3(CH_2)_{15}SH$ the resulting mono layer shows strong hydrophobic properties whereas the same compound with a terminal carboxyl-(COOH) or hydroxyl-(OH) group will show hydrophilic properties and ensure wetting tendency.

4.6 Integration of temperature control

In biochemical and molecular biological experiments there are often thermally activated reactions or reactions that require a well defined temperature. In some cases the optimum reaction temperature is not known at all. We have therefore developed a nano titer plate which allows the precise determination and adjustment of the temperature in the compartments.

Design considerations

The thermo controlled nano titer plate was designed to have regions of different temperature that enables the experimenter to carry out biochemical experiments simultaneously at different temperatures on one single chip or wafer. This is made possible by adjusting a temperature gradient generated by a series of thin film heaters. Temperature can be controlled by another series of thin film resistors, which act as thermo sensors. The heaters should be capable to provide temperatures of up to 100°C to the compartments and the temperature gradient should be as smooth as possible. As the basic geometry a wafer was chosen with the hot zone at the center and the temperature uniformly decreasing to the rim.

To estimate the necessary electrical power we have performed a FEM analysis based on a very simplified model containing a single central heater and cooling by the surrounding air. The result was that the heater should be capable to produce at least 4 W of thermal power. Because the resulting temperature gradient was very steep most of the compartments would be at ambient or only slightly increased temperature. With some additional heaters displaced from the wafer center the temperature gradient can be smoothed.

As a basis for the layout of the temperature controlled nano titer plate we have used the type of nano titer plate with 3700 compartments which was described earlier on p. 88. The thermal elements are to be integrated on the backside of the nano titer plate. The borders between the chip-like groups of compartments offer enough space for the heating elements. As material for the heaters and sensors we have chosen platinum because of its very linear temperature coefficient of the resistance and its biochemical compatibility. For the geometrical design of the heaters one has to consider the resistivity and the critical current density of the material. For bulk material with a rectangular cross-section the resistance R is related to the resistivity ρ and the resitor's geometry by the well known formula:

$$R = \rho \frac{l}{A} = \rho \frac{l}{bd} \tag{17}$$

(l: length, b: width, d: layer thickness)

However, the resistivity of thin films depends in a complex manner on the fabrication conditions (fabrication technique, pressure, temperature, target type, substrate, adjacent layers, tempering…) and the thickness of the evaporated or sputtered layer. The reason is mainly caused by crystal imperfections of evaporated or sputtered thin films. Therefore, it is common practice to introduce the so-called square resistance R_{sq}, which is the resistance of a square of arbitrary size:

$$R_{sq} = \frac{\rho}{d} \tag{18}$$

R_{sq} has in principle to be determined for every combination of fabrication parameters and layer thickness. The total resistance of a given structure of constant cross-section (e.g., a meander or a supply) is the square resistance times the ratio of the length and the width of the structure, which is equal to the number n of squares of size b the structure can be divided into:

$$R = nR_{sq} = R_{sq} \frac{l}{b} \tag{19}$$

The concept of the square resistance is very helpful in designing new patterns like meanders and supplies.

For platinum layers with a thickness of 200 nm we use a square resistance of 3.5 Ω and a critical current density of 2×10^5 A/cm^2 (we use this as a conservative value. In highly integrated microelectronic devices current densities up to 10^6 cm^{-2} are applied even to Al interconnections [64]). The design parameters for the thin film heaters are summarized in Table 3.

Table 3 Design parameters for the thin film heaters of the thermo controlled nano titer plate

width	300 µm
total length	19 mm
resistance	220 Ω
max. current	100 mA
max. power	2 W

For the fabrication of the sensors we use also platinum meanders with a resistance of ~800 Ω. The layout for the heaters and sensors is displayed in Figure 15. The small dots represent the bottom membranes of the compartments arranged in groups of 100. Four heaters are placed around the central chip of the wafer and additional eight heaters are placed in the outer region of the wafer in a symmetrical arrangement. The arrangement is designed to allow the generation of up to 100°C at the central part of the wafer. The temperature can be controlled by 49 platinum resistors used as thermo sensors, 37 of them at the center of each chip area. The remaining are placed on the four axes between the chip areas to allow for a precise determination of the temperature gradient. The layout also shows the supplies and the bond pads which are built by an aluminum layer. All the elements are independently accessible.

Technology

The fabrication of the thermo control elements starts with the ready fabricated standard nano titer plate. The thermo control elements are built by standard thin film techniques: first the layers of platinum (thickness ~200 nm) and aluminum (thickness ~800 nm) are deposited by sputtering. Then the aluminum supplies and bond pads are patterned by means of photolithography and chemical wet etching. The platinum for the heater and sensor meanders is patterned by using photolithography and ion beam etching (IBE). Then a layer of SiO_2 (~1 μm) is deposited by PECVD (plasma enhanced chemical vapour deposition) for electrical isolation and to protect the aluminum against oxidation and aggressive media. The bond pads are opened by means of photolithography and chemical wet etching.

The nano titer plate is mounted to a printed circuit board (PCB) which also serves as the carrier for the wafer. The electrical interconnections are made by ultrasonic wire bonding. At last the bond wires are encapsulated with a epoxy resin to protect them against mechanical and chemical damage.

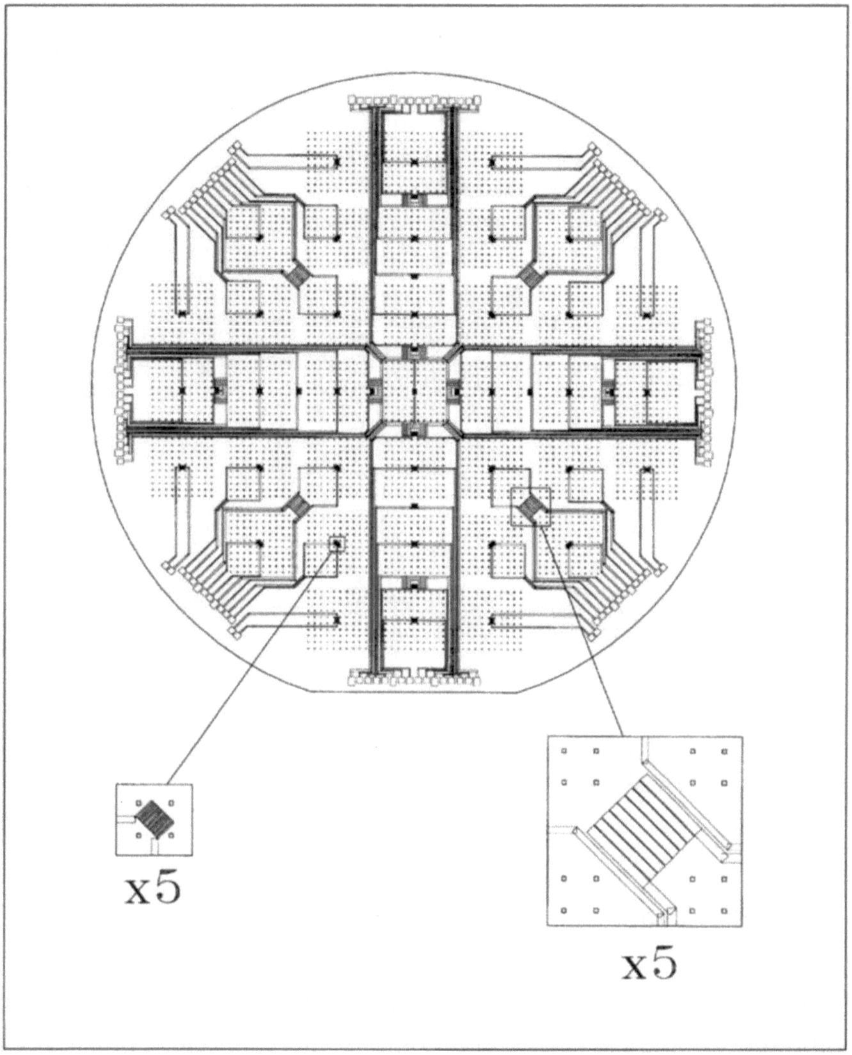

Figure 15 *Arrangement of the thin film elements for heating and temperature measurement on a micro compartment array including bond pads and supplies. Small dots represent the micro compartments, as seen from the back side of the wafer (bottom membranes). The layout contains a total of 12 heaters (large meanders) and 49 thermo sensors (small meanders; see x 5 enlargements). The wafer diameter is 100 mm*

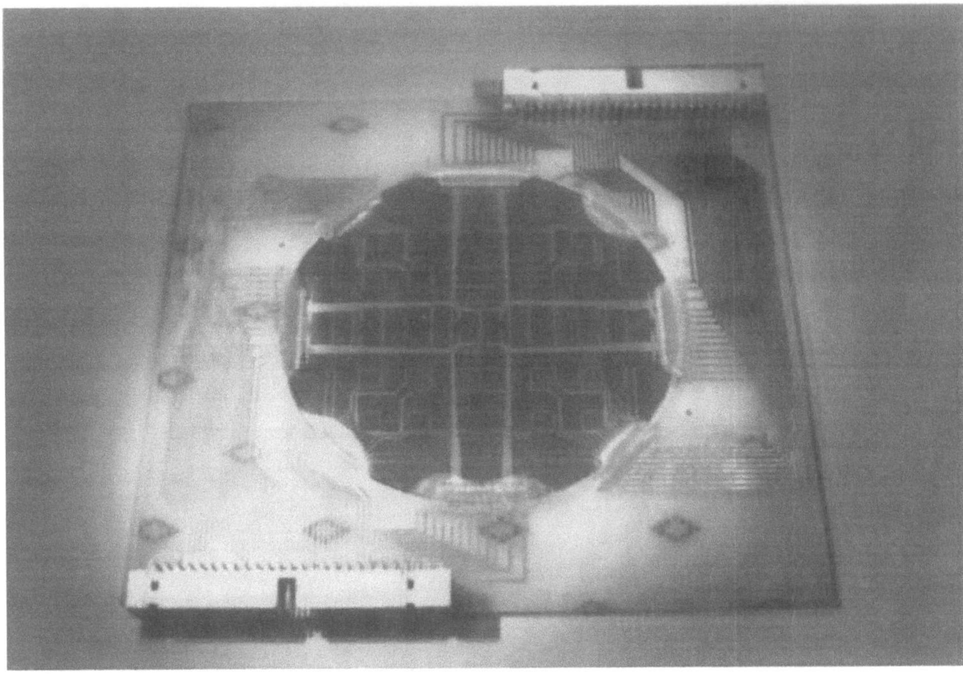

Figure 16 *Photograph of the backside of temperature controlled nano titer plate with 3700 wells, mounted on a PCB (see colour plate 2)*

Characterization

Before any measurements can be done, all the heaters and sensors have to be calibrated, i.e. the resistance and temperature coefficient of the resistance T_K have to be determined which is defined by:

$$T_K = \frac{\Delta R}{R_0} \frac{1}{\Delta T} \tag{20}$$

We find slight deviations from design parameters for the resistance of the heaters and sensors of 10–15%, which can be explained by deviations of the layer thickness from the nominal value. The temperature coefficient was found to be $1.8 \ldots 1.9 \times 10^{-3}\ K^{-1}$.

First results of this thermo-controlled micro compartment array are given in Figure 17, which shows the temperature distribution across the wafer. The

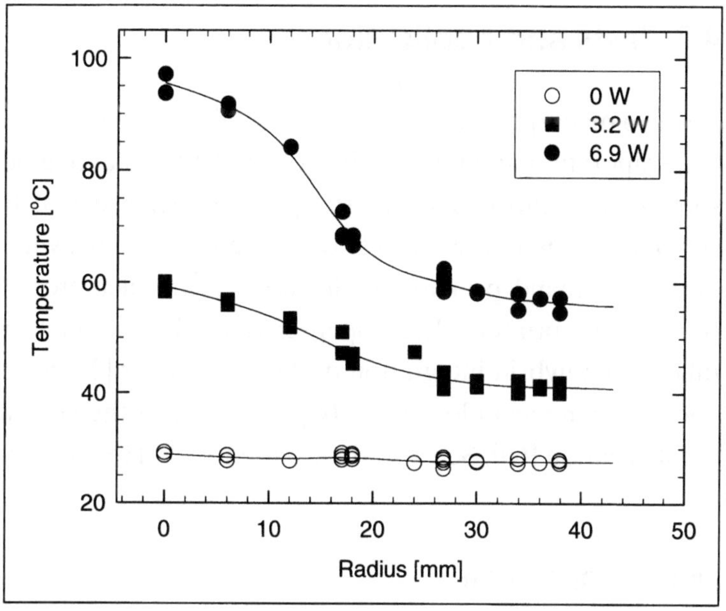

Figure 17 *Temperature distribution across the micro compartment array, if driven with the four central heaters. Electrical power supplied was 6.9 W (circles) and 3.2 W (squares). Zero position is at the center of the wafer. The variation of the data at a given radius arises from measurements in the four quadrants of the wafer. The lines only serve as a guide to the eye*

system was driven by the four central heaters with an electrical power of up to 6.9 W (circles). The variation of the data at a given radius arises from measurements in the four quadrants of the wafer. Cooling was accomplished via the surrounding air ($T = 27°C$), no active cooling was required. As can be seen temperatures exceeding 95°C can easily be generated in the central region. More important, temperature gradients of more than $\Delta T = 40°C$ across the wafer radius are possible.

The temperature difference relative to the surrounding scales almost linear with the supplied electrical power. With an electrical power of 3.2 W (squares) the temperature ranges from 60°C at the wafer center to 40°C in the outside diameter. Experiments using the additional outermost heaters have not been performed so far. We expect a smoother decay of the temperature distribution, so more of the micro compartments can be set to a higher temperature. In this way the peak value at the center could be lowered and liquid evaporation will be reduced.

4.7 Tests and applications

The two major problems of handling small liquid volumes on sub-*mm* areas are the evaporation and the avoiding of fluidic crosstalking between neighbouring spots or compartments. For the precise dispensing of liquids in the *nl* and sub-*nl* range there are now various systems available, which most often use the drop-on-demand principle with piezo-electric actuators as it is known from some ink jet printers. The dispenser can either be built in form of a thin glass tube with a hybrid integration of the piezo actor [22, 65, 66] or made by using micro system technologies [67, 68], which offers the advantage of possible high parallelization [69]. See also the Chapter by S. Howitz, this volume [70].

Liquid evaporation

Concerning the evaporation of liquids the most important influence arises from the temperature and the composition of the surrounding atmosphere (e.g., the relative humidity in the case of aqueous solutions). There exist a number of provisions one can take to reduce evaporation from micro compartments, e.g. reducing the liquid temperature (if applicable), layering with oil or wax or working in saturated atmospheres or, of course, putting a cover glass over it. Most of these are hardly applicable if the micro compartment has to be accessible during the processes. Because evaporation consists of two steps, i.e. the actual evaporation and the transport of the vapour, the shape of the micro compartment and especially the shape and size of the orifice could also reduce evaporation.

Experimental

We have examined the evaporation from micro compartments using three different categories of compartments [71]: type A are compartments made from standard (100)-Si, type B with vertical sidewalls, made from (110)-Si and type C is the inverted form of type A with the small bottom hole as opening. In all cases the bottom membranes/sieves were removed and the bottom hole sealed with Pyrex glass by means of anodic bonding as described on p. 93. Parameters are summarized in Table 4.

Table 4 Parameters of micro compartments used for evaporation tests. The height was ~ 490 µm in all cases

Type	Compartment shape	Volume [nl]	Opening shape	Opening [mm²]
A	inverted pyramidal	140	square	0.64
B	vertical walls	90	parallelogram	0.165
C	pyramidal	140	square	0.0225

Test liquid was pure water, testing environment T = 22°C and relative humidity r.h. = 50%. The compartments were filled using a Microdrop® (piezo-ceramic [65, 66] injekt principle, drop diameter = 100 µm, V = 0.5 nl) and observed through a microscope. After a certain time of evaporation the compartments were refilled to the same level, which gives us the amount of evaporated liquid.

Results and Discussion

Results of the evaporation tests are shown in Figure 18. The total volume of compartments of type A evaporates in 7 min. As a comparison: the same amount of liquid, i.e. 140 nl of pure water dropped on a flat silicon surface evaporated in ≈2.5 min. For a better comparison of the compartments with different volumes we have scaled the data of the type B compartments to a volume of 140 nl (open circles). This is justified by the fact that the evaporation in compartments of this type exhibits an almost perfect linear time dependence. The original data are also depicted (solid circles). Despite the smaller liquid volume of type B the time for a complete evaporation is longer compared to type A.

As a first result, the total evaporation time for type C is about 20 min and therefore 3.5 times longer as for type A. If the size of the opening would play a negligible role, evaporation times for type A and C would have to be exactly equal because the volume of the compartments is the same.

Both for type A and C the evaporation rates reduce at the bottom of the compartments. For type A we explain this by the free surface of the liquid, which gets smaller during the evaporation, while for type C the explanation

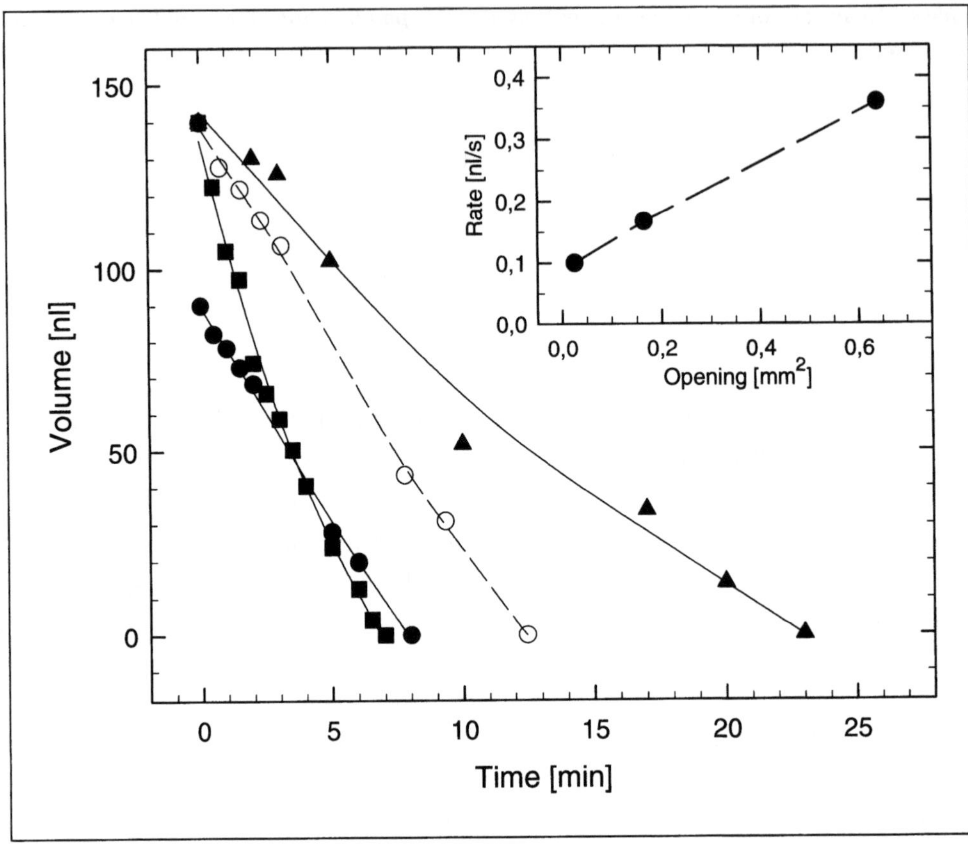

Figure 18 *Evaporation of pure water from micro compartments of three different geometries. Compartment type A (squares), type B (solid circles) and type C (triangles). Open circles are compartments of type B scaled to the volume of type A and C for better comparability. The lines are just to guide the eye. Inset: Initial evaporation rate versus the opening area of the micro compartments. The line serves as a guide for the eye*

might be given by the small clearance, which prevents an effective escape of the vapour.

In order to clarify the mechanism responsible for the evaporation, we have calculated the absolute evaporation rate r_{evap} from the initial slopes of the evaporation curves. The inset of Figure 18 shows this versus the clearance, i.e. the opening area of the micro compartments. There is a clear dependence, which is almost linear. This implies that the bottleneck for the evaporation is given by the opening area, at least at the major part of the evaporation.

The rate of the evaporation is an important parameter if one wants to compensate the evaporation by refilling solvent. In this case the system must be fast enough to refill the total amount of solvent which evaporated during the time needed for one refilling round-trip. This can be expressed in the relation:

$$r_{\text{evap}}\left(\frac{n_{\text{drop}}}{f_{\text{drop}}}+\tau_{\text{ww}}\right)n_{\text{w}} \leq V_{\text{drop}}n_{\text{drop}} \tag{21}$$

where r_{evap}: evaporation rate; n_{w}: number of wells; V_{drop}: volume of a drop; f_{drop}: frequency of drop dispenser, n_{drop}: number of dispensed droplets; τ_{ww}: average well-to-well position time of the dispensing system. Of course, the amount of evaporated solvent must be smaller than the volume of the well. The right-hand side of Eq. 21 simply gives the volume dispensed in each well per round-

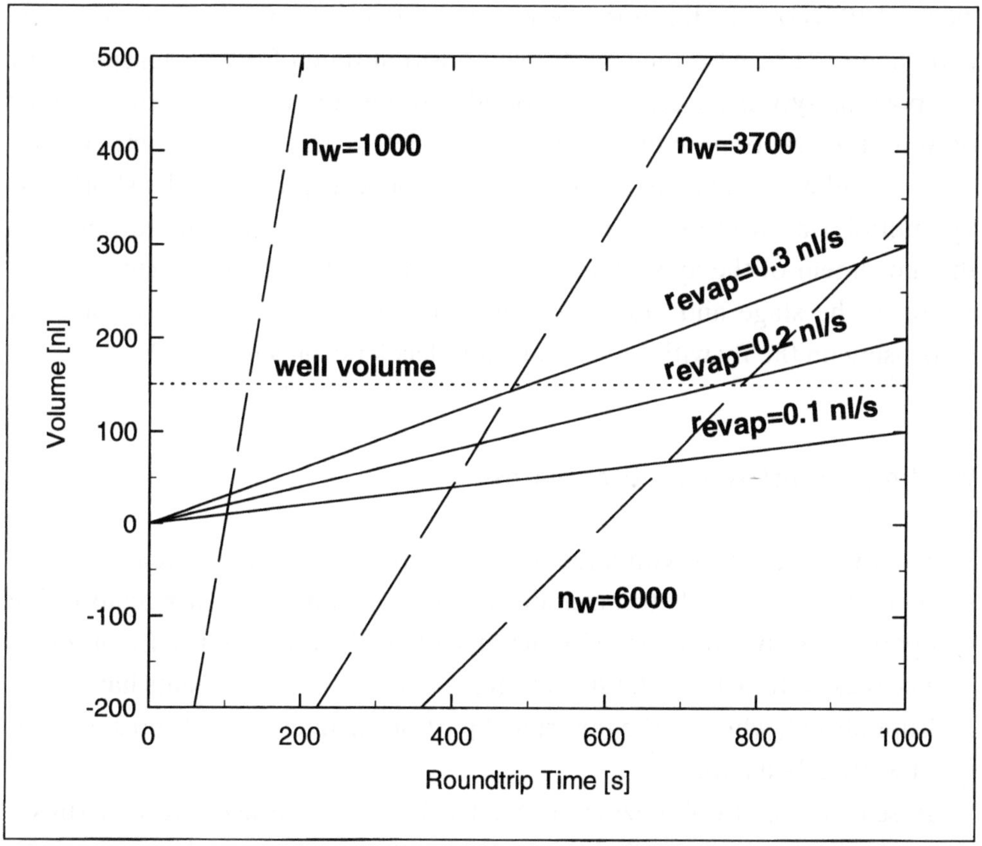

Figure 19 *Evaporation of solvent during refill operation for different evaporation rates (solid lines) and number of wells to refill (dashed lines). The dotted line represents the well volume of 150 nl. Refilling is possible where the solid lines are below the dashed ones. Parameters used are V_{drop} = 0.5 nl, f_{drop} = 10 KHz and τ_{ww} = 0.1 s*

trip. The lefthand side is the evaporated solvent per well and round-trip. It depends on the evaporation rate and the round-trip time, which is given by the time to position the dispenser and the actual dispensing time times the number of wells to be served.

These relations are visualized in Figure 19, where the evaporated volume (solid lines) and the volume to refill (dashed lines) are displayed as a function of the time needed for a full round-trip T_{rt}:

$$T_{rt} = \left(\frac{n_{drop}}{f_{drop}} + \tau_{ww} \right) n_w \tag{22}$$

Refilling is possible where the solid lines are below the "refill" lines and the line representing the well volume. As parameter for the dispenser frequency we used 10 KHz, which is possible with the microdrop system we used in our experiments [17]. Also the well-to-well position time of 0.1 s is feasible with fast position systems. Because the round-trip time increases with the number of wells, the evaporation in large nano titer plates can hardly be compensated by one refilling system alone. However, the round-trip time can be shortened by multiplexing with two or more dispensers. It is also possible to optimize the movement of the xy-stage, e.g., dispensing starts within the deceleration phase of the stage and also makes use of the acceleration phase, or even dispensing on the fly with no acceleration/deceleration steps at all.

Wetting control in nano titer plates

In case of experiments with aqueous solutions, the wetting of areas between the compartments leads to an undesired interaction between neighbouring compartments by diffusion of reactants through liquid films. In order to reduce these interactions the surface areas between compartments have to be hydrophobized, whereas the innermost part of the micro vessels must exhibit good wetting behaviour.

A selective hydrophobizing succeeded by a print procedure and subsequent surface reaction of alkyl thiols with gold-covered nano titer plates [72].

We applied a stamping technique with either a selective gold etching process [63] or a selective gold-thiol chemistry to modify gold sputtered silicon nano titer plates in such a way that the bars between the micro compartments

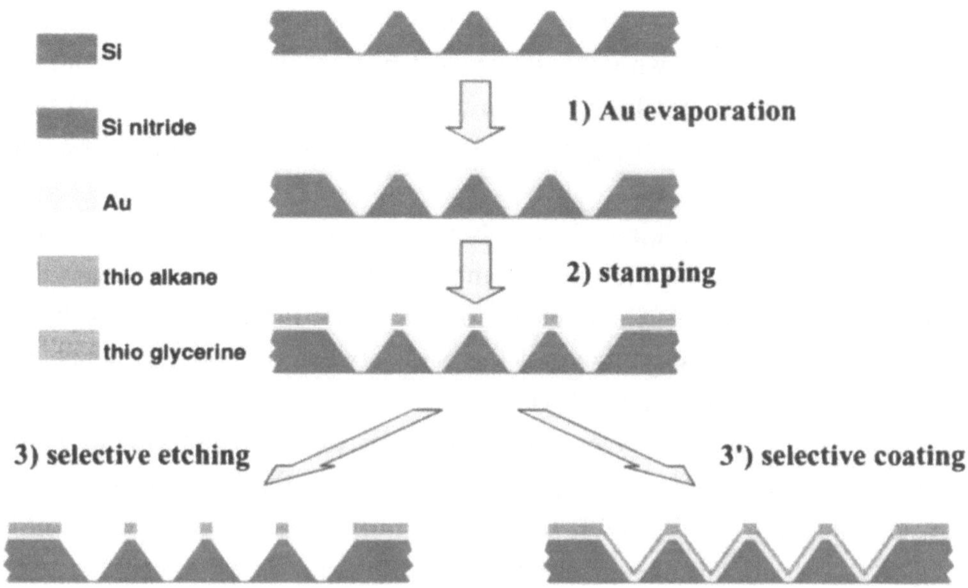

Figure 20 *Principal sketch of the chemical surface modification of micro compartments (see colour plate 3)*

became hydrophobic, whereas the surface of each compartment remained hydrophilic.

This treatment has an important practical aspect for liquid handling of micro compartments since it avoids any cross-talking of liquids between neighbouring micro compartments. The treatment was carried out according to the following procedure (see also the principal sketch in Fig. 20).

First a silicon nano titer plate is sputtered with a thin gold layer (50–100 nm).

The selective modification of any structured gold surface can be achieved by a combination of a stamping step and a selective gold etching or selective gold-thiol-chemistry step: liquid silicone (polydimethylsiloxane, PDMS) is spin-coated on an appropriate substrate like a silicon or glass wafer and baked for polymerization to form a thin silicone layer. This polymer layer is inked with an ethanolic solution of a thio-functional compound like hexadecane-thiol to be stamped on any gold-covered surface.

With putting an array of gold sputtered silicon compartments on such an activated polymer stamp, a covalent bond between the gold surface and the thiol-compound will be immediately formed (step 2 in Fig. 20). In case of a long alkyl group (12 to 18 carbon atoms length) of the thiol compound, the gold surface in contact with the stamp will become hydrophobic (contact angle to water about 100°), whereas in case of a hydroxyl- or carboxyl terminated thiol compound the gold surface will become hydrophilic (contact angle 25°).

To avoid any cross-talking of liquid between compartments it is necessary to make the walls inside the micro compartments hydrophilic, whereas the bars between them should remain hydrophobic. Since the compartments have a relief structure, after stamping only the bars between the compartments are covered with the hydrophobic alkyl thiol.

In a third step the gold layer of the still underivatized compartment walls is selectively etched without interfering with the gold-thiol bond of the compartment bars. For etching a 0.1 molar potassium cyanide solution with 1 mM $K_3Fe(CN)_6$ as oxidizing agent is used. The etching rate for a 50-nm-thick gold layer is about 1 min. After etching, the compartment surface is again the initial silicon which is moderately hydrophilic (contact angle 62°).

As an alternative to the etching procedure the compartment walls can be selectively modified without interfering with the gold-thiol bond of the compartment bars by a reaction with a hydroxyl or carboxyl functional groups like thioglycerine. The reaction is carried out simply by putting the compartments in an ethanolic solution of thio glycerine while stirring. After finishing the reaction, the compartment walls are strongly hydrophilic (contact angle to water 25°). The difference in wetting behaviour between compartment bars and walls after this treatment is about 70°, which is very favourable to avoid any cross-talking of aqueous solutions between neighbouring compartments (see also Fig. 21).

The excellent control of the wettability is demonstrated by the addition of water to well-filled chambers. The water almost does not wet the hydrophobized surface. This way, a ball-like drop can be piled onto a micro compartment by adding water through a microdrop dispenser.

The liquid drops on the compartments hold their shape also in case of very small distances between liquid surfaces between to two neighbouring compartments. The shape of the liquid drops does not collapse before the liquid surfaces touch each other with addition of more water into both compartments.

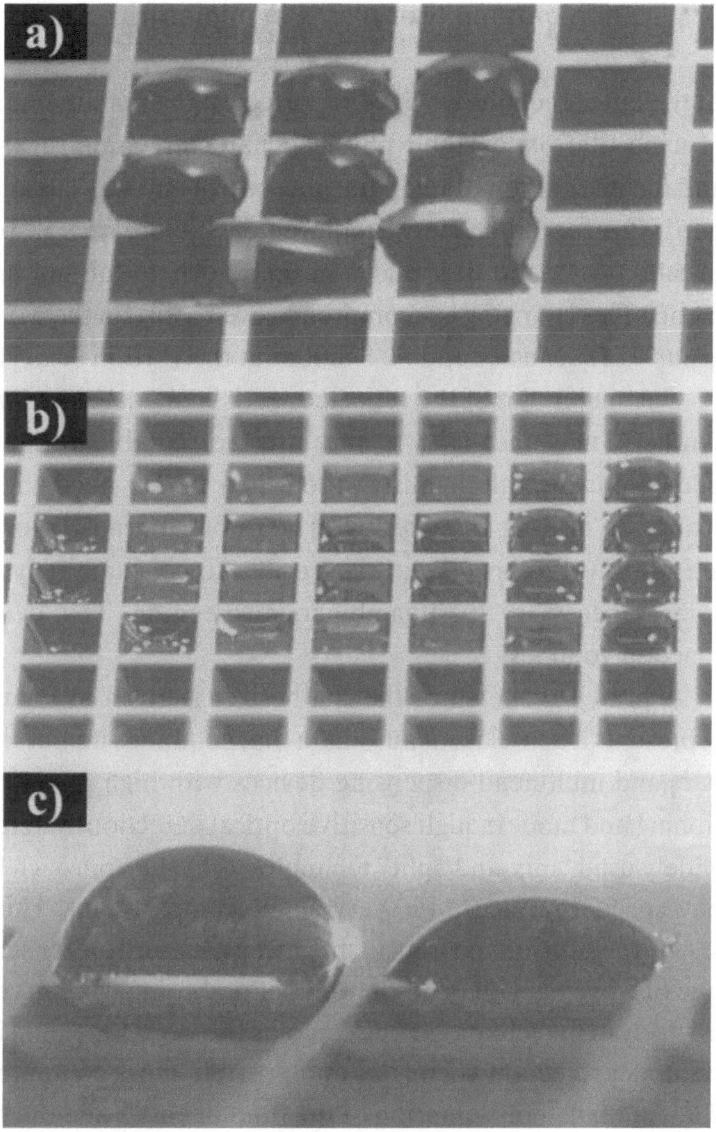

Figure 21 *a) Water filled in unmodified nano titer plates. The poor water-repellency of the rims lets the water creep out of the well and leads to a strong fluidic cross-talking of adjacent wells. b) Water filled in surface modified nano titer plates in different states of evaporation. Due to the strong water-repellency of the rims the water is confined to the wells. Note also the good wettability of the compartment walls. c) Close-up view of piled-up droplets of water in surface modified micro compartments. The estimated volume of the left water drop is 250 nl. The well volume is ~ 150 nl in all cases with a 800 × 800 µm² well opening*

Application of nano titer plates in miniaturized automata

In the latter sections we have reported on the manufacturing of various types of nano titer plates, their physical characterization regarding fluid handling and the integration of temperature control and ways to adapt the surface properties to special chemical requirements. Now we will briefly show the applicability of nano titer plates in real biochemical and biomolecular experiments. The following is a comprehensive report on the work carried out by the group of P. Schuster and A. Schober at the IMB in Jena [17].

One major aim of the reported work is the development of a miniaturized synthesis machine. Due to the effect of combinatorial explosion in fields like combinatorial chemistry there is a strong need for highly parallelized strategies in drug discovery. Furthermore, because biochemical substances are often very expensive, it is sensible to consume only the amount of substances necessary for unambiguous screening. This demands miniaturized reaction volumes.

A principal sketch of the synthesizer is given in Figure 22. The machine represents a new class of automata that combines miniaturized reaction chambers and multihead dispensing devices with high precision mechanical positioning and modern high sensitive optical detection systems. In contrast to the widely used "mix-and-split" technology [1], the approach here is to synthesize one specific sequence in every single compartment. This way the complete sequence information is available instantaneously after screening. The technique of working with beads in well defined compartments defined by means of micro system technologies also avoids the need for extensive and complicated photodeprotection chemistry and possible cross-contamination problems [3, 73]. A detailed presentation of the fundamentals and the components is given in [20] and [74], respectively.

Preparatory experiments for combinatorial synthesis of oligo peptides were performed using the miniaturized automaton and nano titer plates. The synthesis has been performed using polystyrole beads with a diameter of ~100 µm, one in each compartment. The substances were dispensed using Microdrops adapted to the amino acids. The chemical synthesis protocols have been carefully optimized with respect to the miniaturized scheme, resulting in coupling efficiencies exceeding 95%. This way 100 tripeptides and a small library of 27 different tripeptides have been synthesized and confirmed

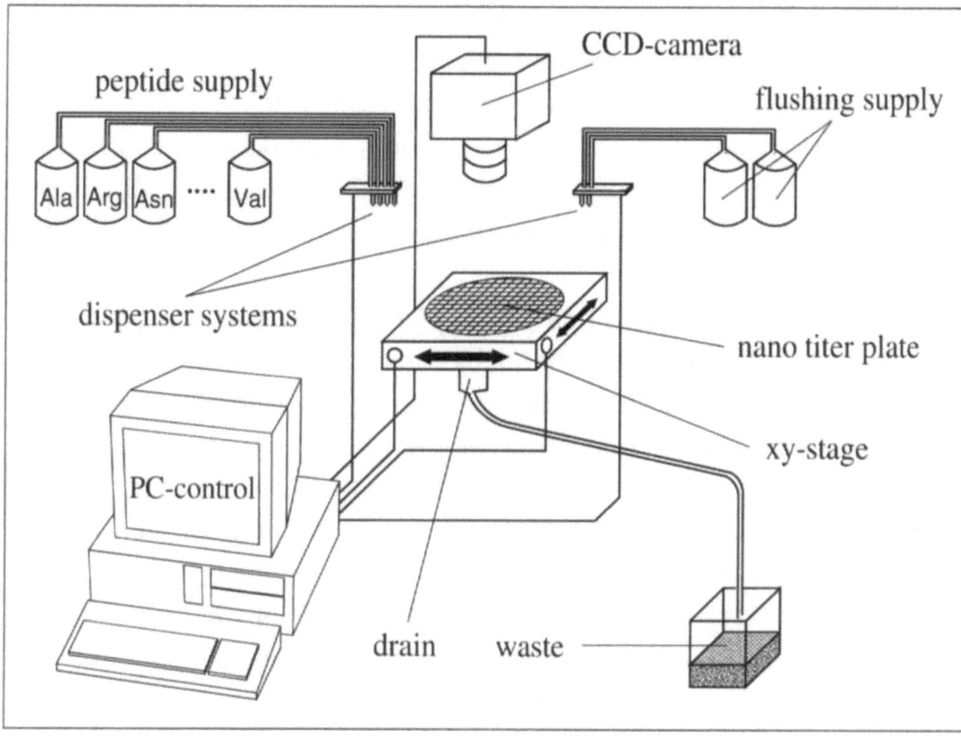

Figure 22 *Principal sketch of a miniaturized automaton for combinatorial synthesis with integrated screening. For an explanation see text*

by HPLC [17] and recently a library of 243 pentapetides has been generated [75]. The results may serve as a model to further experiments on a larger scale. Currently a tool for handling the beads is under development.

Another class of experiments deals with evolutive strategies, which is a completely different approach for the optimization of molecules in the search for novel agents [76–78]. As has already been shown by Beaudry and Joyce, the Darwinian evolution is suitable for engineering ribozymes, i.e. nucleic acids that exhibit catalytic properties [79].

The principle of the experiments is simple: take a self-replicating system with a certain mutation rate. After a while there will be a population with different species. Then we transfer a small sample to another compartment, which may have different stress factors, i.e. temperature, pH-value, nutritional medium, etc. Due to the differences in stress and starting conditions the second system will evolve in a different way. Repeat as often as necesary.

Finally, one will have a number of systems with different populations, from which the best suited can be chosen for further analysis.

The so-called serial transfer machine is quite similar to the synthesizer, but needs no washing and filtering. Instead it requires the possibility to transfer a small sample from one compartment to another, which is realized by a steel needle that can be sterilized after the transfer. More details are given in [20, 80].

To show the applicability of nano titer plates for serial transfer investigations and to test the compatibility for biochemical and biomolecular assays two kinds of experiments were performed. As an *in vivo* model system for cell based evolution the workhorse of cell biology, *E. coli* (strain MC1016), was used. The bacteria were dispensed with Microdrops and the encapsulated nano titer plate was incubated at 37°C. Growth of bacteria has been verified by taking aliquots at defined intervals and determining the optical density at $\lambda = 600$ nm. The serial transfer served as clarification of whether growth of bacteria can be incubated with a steel needle. Ten serial transfers were carried out in series and the growth examined after 2, 4 and 6 h of incubation as described.

As an *in vitro* amplification system the self-sustained sequence replication (3SR) reaction [81, 82] was chosen for testing the biomolecular compatibility of silicon compartments. The 3SR reaction allows the isothermal amplification of target DNA and RNA sequences *in vitro* using the HIV-1 reverse transcriptase and the T7 RNA polymerase. For the test experimental conditions concerning the evaporation, temperature of reaction and supply vessels and dispenser system had to be optimized. The test has been performed in 200 compartments and the growth has been detected online by fluorometric analysis. The specificity has been shown by gelelectrophoresis.

4.8 Conclusions

The crosslinked handling of information and substances by highly automated and parallelized liquid automata is in its infancy. It is to be expected that the development of nano titer plates will contribute decisively to this development. The potential of microfabrication for the geometric definition of small reaction vessels is much higher than the actual need of miniaturization of compartment arrays. The first types of nano titer plates are realized. Picoliter compartment arrays can be made using conventional microfabrication techniques. Arrays of femtoliter or attoliter compartments are imaginable, but probably not meaningful. Surface forces dominate the behaviour of liquids in the case of such small volumes. Therefore, close capillaries like those used in micro channel reactors, and spot technologies might be favoured over open titer plates in the liquid handling below the picoliter range.

All miniaturization cannot solve the problem of combinatorial explosion by itself. The imaginable number of substances is much higher than number of particles of the observable universe. The meaningful and efficient management of conversion of information into chemical structures and *vice versa* represents the key in the solution of molecular optimization. The interfaces between substance flows and bit flows can be best realized in miniaturized configurations, which offer the additional advantages to ensure short ways and short transport times, and can integrate a lot of interfaces in small spaces. This way, biomolecular studies and future biomolecular technologies will definitely profit from miniaturization and from the nano titer technology. Last but not least, the discovery of new molecular structures is possible, and they may act as nano tools in a future superminiaturized world.

Acknowledgement

We thank M. Sossna, G. Palitza, H. Roth, H. Porwol, F. Jahn, J. Tuchscheerer, E. Ermantraut and T. Kaiser for technical assistance and the team at the "Institut für molekulare Biotechnologie (IMB)", Jena, Germany: P. Schuster, D. Vetter and G. Schlingloff and also R. Günther from EVOTEC, Hamburg, Germany for helpful discussions. The financial support by the "Bundesministerium für Bildung, Wissenschaft, Forschung und Technologie (BMBF)" under contract no. 0310717 is gratefully acknowledged.

References

1 Gallop MA, Barrett RW, Dower WJ, Fodor SPA, Gordon EM (1994) Applications of combinatorial technologies to drug discovery: 1. Background and peptide combinatorial libraries. *J Med Chem* 37(9): 1233–1251

2 Gordon EM, Barrett RW, Dower WJ, Fodor SPA, Gallop MA (1994) Applications of combinatorial technologies to drug discovery: 2. Combinatorial organic synthesis, library screening strategies, and future directions. *J Med Chem* 37(10): 1385–1401

3 Jacobs JW, Fodor SPA (1994) Combinatorial chemistry – applications of light-directed chemical synthesis. *TibTech* 12: 19–26

4 Elson EL, Magde D (1974) *Biopolymers* 13: 1–27 and 29–61

5 Rigler R, Mets Ü, Widengren J, Kask P (1993) Fluorescence correlation spectroscopy with high count rate and low background: analysis of translation diffusion. *Eur Biophys J* 22: 169–175

6 Eigen M, Rigler R (1994) Sorting single molecules: Application to diagnostics and evolutionary biotechnology. *Proc Natl Acad Sci USA* 91: 5740–5747

7 Gradl G et al; *this volume*

8 Heuberger A, Löchel B (1996) Optical DUV-lithography for high microstructures. *Microsystem Technol* 3: 1–6

9 Kroy W, Seidel H, Dette E, Deimel P, Binder F, Hilpert R, Königer M. Mikromechanische Struktur. German patent: DE 3915920 A1 (16.5.1989/22.11.1990)

10 Becker EW, Ehrfeld W, Hagmann P, Maner A, Münchmeyer D (1986) Fabrication of microstructures with high aspect ratios and great structural heights by synchrotron radiation lithography, galvanoforming and plastic molding (LIGA process). *Microelectronic Engineering* 4: 35–56

11 Ehrfeld W, Hagemann P, Mohr J, Münchmeyer D. Verfahren zur Herstellung eines Filters (13.2.87). German patent: DE 3704546/European patent: EP 0278059 B1 vom 15.7.92)

12 Ruther P, Bacher W, Feit K, Heckele M, Weindel K (1994) LIGA- made pneumatically driven micro-actuator for use in a micro-testing system. *Proc of Micro System Technol* '94, Berlin, 1994, H Reichl, A Heuberger (eds), VDE-Verlag, Berlin-Offenbach, p 899

13 Weibezahn KF, Knedlitschek G, Bier W, Schaller Th (1994) Mechanically processed microstructures used to establish an *in vitro* tissue model. *Proc of Micro System Technol* '94, Berlin 1994, H Reichl, A Heuberger (eds), VDE-Verlag, Berlin-Offenbach, p 873

14 Köhler JM, Mayer G, Poser S, Schulz T, Schober A (1996) Chip elements for combinatorial chemistry, fluid processing and PCR. *Proc of Micro System Technologies* 96, Potsdam, Germany, 17./19.9.96, H Reichl, A Heuberger (eds), VDE-Verlag, p 693

15 Petersen KE (1982) Silicon as a mechanical material. *Proc IEEE* 70 (5): 420

16 Sato K, Kawamura Y, Tanaka S, Uchida K, Kohida H (1990) Individual and mass operation of biological cells using micromechanical silicon devices. *Sensors and Actuators* A 21–23: 948–953

17 Schober A, Schlingloff G, Thamm A, Vetter D, Tomandl D, Gebinoga M, Kiel HJ, Scheffler Ch, Döring M, Köhler JM, Mayer G (1996) Systemintegration of microsystems/chip elements in miniaturized automata for high-throughput synthesis and screening in biology, biochemistry and chemistry. *Proc of Micro System Technologies* 96 (MST), Potsdam, Germany, 17.–19.9.96,

H Reichl, A Heuberger (eds), VDE-Verlag, p 705–710

18 Jäckel K-P (1996) Microtechnology: Application opportunities in the chemical industry. In: W Ehrfeld (ed): *Microsystem technology for chemical and biological microreactors*. DECHEMA monographs vol. 132, p 29–50, VCH Weinheim, Basel, Cambridge, New York

19 Ehrfeld W, Hessel V, Möbius H, Richter Th, Russow K (1996) Potentials and realization of microreactors. In: W Ehrfeld (ed): *Microsystem technology for chemical and biological microreactors*. DECHEMA monographs vol. 132, p 1–28, VCH Weinheim, Basel, Cambridge, New York

20 Köhler JM, Schober A, Schwienhorst A (1994) Micromechanical elements for microchemical systems. *Exp Tech Phys* 40(1): 35–56

21 Atkins PW (1986) Physical chemistry. 3rd ed, p 674 ff, Oxford University Press

22 Döring M (1991) Flüssigkeiten mikrofein dosieren. *F & M* 99(11): 459–463 (*in German*)

23 Bean KE (1978) Anisotropic etching of silicon. *IEEE Trans Electr Dev* ED-25(10): 1185

24 Seidel H, Csepregi L, Heuberger A, Baumgärtel H (1990) Anisotropic etching of crystalline silicon in alkaline solutions: I. Orientation dependence and behavior of passivation layers. *J Electrochem Soc* 137(11): 3612; H Seidel, L Csepregi, A Heuberger, H Baumgärtel (1990) Anisotropic etching of crystalline silicon in alkaline solutions: II. Influence of dopants. *J Electrochem Soc* 137(11): 3626

25 Kendall DL (1979) Vertical etching of silicon at very high aspect ratios. RA Huggins (ed): *Ann Rev Mater Sci* 9: 373

26 Price JB (1973) In: Semiconductor silicon. HR Huff, RR Burgess (eds): *The Electrochemical Society Softbound Proceedings Series*. Princeton, NJ, p 339

27 Zangwill A (1988) Physics at surfaces. Cambridge University Press, Cambridge, MA, p 91

28 Palik ED, Bermudez VM, Glembocki OJ (1985) *J Electrochem Soc* 132: 871

29 Kendall DL, de Guel GR (1985) Orientation of the third kind: The coming age of (110) silicon. In: CD Fung, PW Cheung, WH Ko, DG Fleming (ed): *Micromachining and micropackaging of transducers*, Elsevier p 107–124

30 Völklein F (1990) Thermal conductivity and diffusivity of a thin film SiO_2-Si_3N_4 sandwich system. *Thin Solid Films* 188: 27–33

31 Meinel W, Müller J, Keßler E, Völklein F, Wiegand A (1988) Multijunction thin-film radiation thermopile sensors. *Measurements* 6: 2–4

32 Elbel Th, Müller JE, Völklein F (1985) Miniaturisierte thermische Strahlungssensoren: Die neue Thermosäule TS 50.1. *Feingerätetechnik* 34: 113–115 (*in German*)

33 Sarro L (1994) Thermal sensors. In: GCM Meijer, AW van Herwaarden (eds): *Institute of Physics Publishing*, p 154

34 Steinke SO, Schulz T, Köhler JM (1995) Mikromechanischer Modul für den Stoffaustausch zwischen zwei flüssigen Phasen. *Proc of Mikrosystemtechnik, Mikromechanik & Mikroelektronik*, Chemnitz, 16.-17.10.1995, p 121–129 (*in German*)

35 Fink H-W, Schmid H, Ermantraut E, Schulz, T (1997) Electron holography of individual DNA molecules, accepted for publication in *J Am Opt Soc A*

36 Wallis G, Pomerantz, DI (1969) Field assisted glass-metal sealing. *J Appl Phys* 40: 3946–3949

37 Kanda Y, Matsuda K, Murayama C, Sugaya J (1990) The mechanism of field-assisted silicon-glass bonding. *Sensors and Actuators* A 21–23: 939–943

38 Hanneborg A (1991) Silicon wafer bonding techniques for assembly of micromechanical elements. *Proc of MEMS, Nara, Jpn, IEEE,* p 92

39 Hachitani Y, Sagara H (1993) Glass substrates for silicon sensors. *Proc of Transducers* '93, 7.–10.6.93

40 Hilgendorf K, Krause P, Obermeier E (1996) Reduction of the influence of the anodic bonding process on the behavior of pressure sensors by using new glass substrates. *Proc of Micro System Technologies* (MST) 96, Potsdam, Germany, 17.–19.9.96, H Reichl, A Heuberger (eds), VDE-Verlag, p 331–336

41 Baier V, Schmidt K, Straube B, Horst H-J (1997) Anodic bonding at low temperatures using microstructurable Li-doped glass. *Proc of 192nd Meeting of The Electrochemical Society*, Paris, France, 31.8.–5.9.97

42 Hanneborg A, Nese M, Jakobsen H, Holm R (1992) Silicon-to-thin film anodic bonding. *L Micromech Microeng* 2: 117

43 Brooks AD, Donovan RP, Hardesty CA (1972) Low-temperature electrostatic silicon-to-silicon seals using sputtered borosilicate glass. *J Electrochem Soc* 119(4): 545

44 Esashi M, Nakano A, Shoji S, Hebiguchi H (1990) Low-temperature silicon-to-silicon anodic bonding with intermediate low melting point glass. *Sensors and Actuators* A 21–23: 931–934

45 Choi W-B, Ju B-K, Lee Y-H, Haskard MR, Sung M-Y, Oh M-H (1997) Anodic bonding technique under low temperature and low voltage using evaporated glass. *J Vac Sci Technol B* 15(2): 477–481

46 Wallis G (1975) Field assisted glass sealing. *Electrocomponent Science and Technology* 2(1): 45

47 Maas D, Fahrenberg J, Keller W, Mihalj G, Seidel D (1994) Applicability of anodic and adhesive bonding to join microstructures consisting of various materials. *Proc of Micro System Technol.* '94, Berlin, 1994, H Reichl, A Heuberger (eds), VDE-Verlag, Berlin-Offenbach, p 481

48 Oikawa M et al (1991) Miniature and micro-optics: Fabrication and system applications. *Proc SPIE* 1544: 226–236

49 Karthe W, Göring R, Kley EB (1994) Micro-optical elements: fabrication and application. *Proc of Micro System Technol* '94, Berlin, Germany, 19.–21.10.94: H Reichl, A Heuberger (eds), VDE-Verlag, Berlin-Offenbach, p 1047–1053

50 Göttert J, Fischer M, Müller A (1995) High-aperture surface relief microlenses fabricated by X-ray lithography and melting. In: D Daly (ed): *Microlens arrays.* EOS Topical Meeting Digest Series Vol. 5, Teddington, UK

51 Elderstig H, Arvidsson G, Forssén L, Henriksson P, Laurell F, Palmskog G, Tikkanen G (1994) Silicone as an optical material. *Proc of Micro System Technol* '94, Berlin, 1994, H Reichl, A Heuberger (eds), VDE-Verlag, Berlin-Offenbach, p 1055

52 Hülsenberg D (1992) Glas in der Mikrotechnik. In: Sitzungsberichte der Sächs. Akad Wiss, math-naturwiss Klasse, Band 123, Heft 6, p 3, Akademie Verlag (*in German*)

53 Dietrich TR, Ehrfeld W, Lacher M, Krämer M, Speit B (1996) Fabrication technologies for microsystems using photoetchable glass. *Micoelectronic Engineering* 30: 497–504

54 Hülsenberg D, Bruntsch R, Schmidt K, Reinhold F (1990) Mikromechanische Bearbeitung von fotoempfindlichem Glas. *Silikattechnik* 41(11): 364–366 (*in German*)

55 Legtenberg R, Tilmans HAC, Elders J, Elwenspoek M (1994) Stiction of surface micromachined structures after rinsing and drying: model and investigation of adhesion mechanisms. *Sensors and Actuators* A 43: 230–238

56 Bauer J, Drescher G, Illig M (1996) Surface tension, adhesion and wetting of materials for photolithographic process. *J Vac Sci Technol* B 14(4): 2485–2492

57 Chidsey CED, Loiacono DN (1990) Chemical functionality in self-assembled monolayers: Structural and electrochemical Properties. *Langmuir* 6: 682–691

58 Neinhuis C, Barthlott W (1997) Characterization and distribution of water-repellent, self-cleaning plate surfaces *Ann Bot* 79: 667–677

59 Shibuichi S, Onda T, Satoh N, Tsuji K. Super water-repellent surfaces resulting from fractal structure (1996) *J Phys Chem* 100: 19512–19517

60 Whitesides GM, Mathias, JP, Seto ChrT (1991) Molecular self-assembly and nanochemistry: A chemical stragtegy for the synthesis of nanostructures. *Science* 254: 1312

61 Mrksich M, Whitesides, GM (1995) Patterning self-assembled monolayers using microcontact printing: a new technology for biosensors? *TIBTech* 13(6): 228–235

62 Ulman A (1991) An introduction to ultrathin organic films: from Langmuir-Blodgett to self-assembly. Academic Press, San Diego, p 288 ff

63 Kumar A, Biebuyck HA, Whitesides GM (1994) Patterning self-assembled monolayers: Applications in material sciences. *Langmuir* 10: 1498–1511

64 Arzt E, Kraft O, Möckl UE (1996) Metalle unter Extrembedingungen. *Physikal Blätter* 52(3): 227–231 (*in German*)

65 Dijksman JF (1984) Hydrodynamics of small tubular pumps. *J Fluid Mech* 139: 173–191

66 Schober A, Günther R, Schwienhorst A, Döring M, Lindemann BF (1993) Accurate high speed liquid handling of very small biological samples. *Biotechniques* 15(2): 324

67 Schwesinger N (1993) Planarer Tintenstrahldruckkopf mit piezokeramischem Antrieb. *F & M* 101(11–12): 456–460 (*in German*)

68 Fiehn H, Howitz S, Wegener Th (1997) Eine neue Technologie für die Präzisionsdosierung von Flüssigkeiten im Submikroliterbereich. *BIOforum*(1–2): 22–25 (*in German*)

69 Krause P, Obermeier E, Wehl W (1995) Backshooter – a new smart micromachined single-chip inkjet printhead. *Proc of 8th Int Conf on Solid-State Sensors and Actuators/ Eurosensors IX*, Stockholm, Sweden, 25.–29.6.95, p 325–328

70 Howitz S. Components and systems for microliquid handling, *this volume*

71 Mayer G, Köhler JM (1997) Micromechanical compartments for biotechnological applications: Fabrication and investigation of liquid evaporation. *Sensors and Actuators* A 60: 202–207

72 Mayer G, Tuchscheerer J, Kaiser Th, Wohlfart K, Ermantraut E, Köhler JM (1997) Nano titer plates: Micro compartment arrays for biotechnological applications. *In:* W Ehrfeld (ed) Microreaction Technology – Proc. of the First Int. Conf. on Microreaction Technology, Springer-Verlag p 112–119

73 Fodor JPA, Read JL, Pirrung MC, Stryer L, Lu AT, Solas D (1991) Light-directed, spatially addressable parallel chemical synthesis. *Science* 251: 76

74 Schober A, Schwienhorst A, Köhler JM, Fuchs M, Günther R, Thürk M (1995) Microsystems for independent parallel chemical and biological processing. *Microsystem Technol* 1: 168

75 Schober A et al, *unpublished*

76 Eigen M, Gardener jr WC (1984) Evolutionary molecular engineering based on RNA replication. *Pure and Appl Chem* 56(8): 967

77 Eigen M (1986) The physics of macromolecular evolution. *Chemica Scripta* 26B: 13
78 Schuster P (1996) Evolutionary biotechnology – theory, facts and perspectives. *Acta Biotechnol* 16: 3–17
79 Beaudry AA, Joyce GF (1992) Directed evolution of an RNA enzyme. *Science* 257: 635–641
80 Köhler JM, Pechmann R, Schaper A, Schober A, Jovin ThM, Schwienhorst A (1995) Micromechanical elements for Detection of molecules and molecular design. *Microsystem Technol* 1: 202
81 Breaker RR, Joyce GF (1994) Emergence of a replicating species from an *in vitro* RNA evolution reaction. *Proc Natl Acad Sci USA* 91: 6093–6097
82 Gebinoga M, Oehlschläger F (1996) Comparison of self-sustained sequence replication reaction systems. *Eur J Biochem* 235: 256–261

5 Chip technology for micro-separation

Andrew J. de Mello and Andreas Manz

5.1 Introduction

The continuous monitoring of a chemical parameter, usually the concentration of a molecular species, is of central importance in analytical science. It also has commercial significance in chemical production, environmental analysis, and medical/clinical diagnostics [1]. Consequently, modern sensing technology must provide real-time measurements at low analyte concentrations.

Ideally, a molecular "sensor" is highly selective, i.e. it transduces chemical information of *one* species to electronic information, at the exclusion of all other components in a sample matrix (Fig. 1a). In addition, a molecular sensor should possess a large dynamic (concentration) range, be highly reproducible, have a short response time, and be able to operate *in situ*. Many different approaches to the molecular (or chemical) sensor have been investigated [2].

In the real world, chemical analysis methods are at best selective; few if any, are truly specific. Consequently, the discrimination of an analyte from potential interferences is more often than not the key step in an analytical procedure. Accordingly, an alternative approach to molecular recognition is to incorporate a separation step prior to transduction (Fig. 1b). This lessens selectivity requirements, and leads to improved sensitivity (due to an appreciable reduction in background signal). This general strategy can be amalgamated into the concept of the Total (chemical) Analysis System (TAS) [1]. A TAS is a device that periodically transforms chemical information into electronic information. Sampling, sample transport, sample pre-treatment, analyte separation, analyte detection and data storage are automatically performed in a flowing stream. Sample pre-treatment and analyte separation serve to discriminate molecular signals, and consequently lower the selectivity requirements of the detector. This approach to analysis allows for continuous

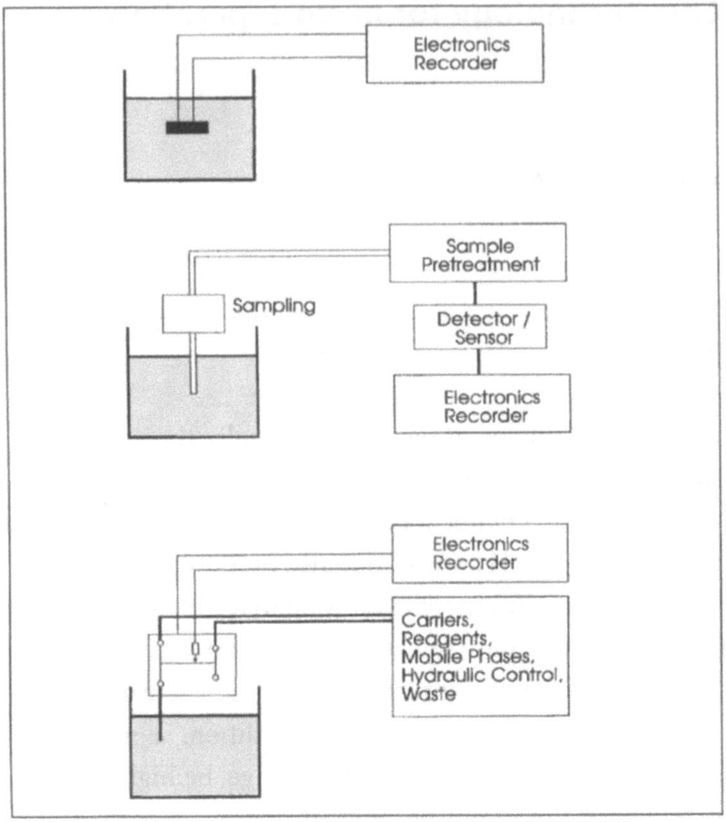

Figure 1 *Schematic diagram of: (a) an ideal chemical sensor, (b) a total chemical analysis system (TAS), and (c) a miniaturised total chemical analysis system (μ-TAS)*

or "real-time" monitoring of molecular systems with time-dependent compositions.

Without question, the most widely used methods for effecting analytical separations are chromatography and electrophoresis. Both have been used with considerable success in TAS research. Each technique exploits the fact that the physical properties of similar molecules are non-identical. In chromatographic separations, the sample matrix is dissolved in a *mobile phase*, which may be a gas, liquid or supercritical fluid. Subsequently, this mobile phase is forced through an immiscible *stationary phase*, which is immobilised on a surface or in a column. The two phases are chosen so that the components of the sample matrix distribute themselves between mobile and stationary phases to varying degrees. Components which interact strongly with the stationary phase move slowly with the flow of mobile phase, whereas weakly interacting species travel more rapidly. Accordingly, sample components separate into discrete bands that can be analysed quantitatively.

Electrophoresis describes the movement of electrically charged species (ions or molecules) in a conductive, liquid medium under the influence of an applied electric field [3]. Figure 2 illustrates the basic concept. A sample containing a mixture of anions, cations and neutral species is injected at the aniodic end of the system. If an electric field is then applied across the electrolyte, the ions in the sample will tend to migrate through the column at varying rates, and in different directions. The rates and directions of migration depend on the sizes of the ions, and the magnitude and polarity of their charges. For example, cations will migrate toward the negative electrode at a rate that is directly proportional to their charge-to-mass ratio. Separation can then be effected through inherent differences in electrophoretic mobilities. Even though anions will be attracted to the positive electrode (anode), net displacement under normal conditions is toward the cathode. This is due to an electrokinetic phenomenon termed electroosmosis. Electroosmosis occurs as a result of surface charge on capillary walls. This charge causes the formation of an electrical double layer. When an electric field is applied to the system, mobile positive charges within this layer migrate in the direction of the catho-

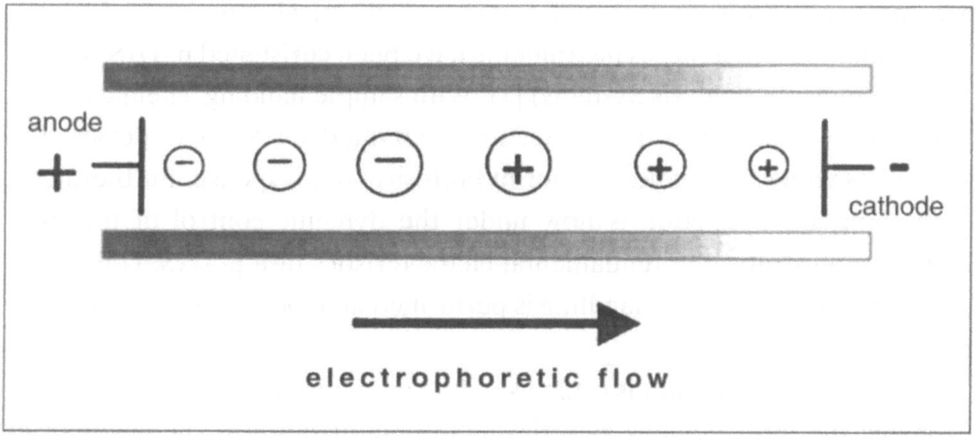

Figure 2 *Schematic representation of capillary electrophoresis. In this situation, sample is introduced at the anode. The circled +'s and –'s describe cationic and anionic species, respectively*

de and drag solvent (electrolyte) molecules along with them. This effect is transmitted throughout the diameter of the capillary, causing a net flow toward the cathode. Detailed descriptions of electroosmotic theory can be found elsewhere [3, 4].

TAS have been applied to a diversity of chemical problems. The main driving force for much research is the need for real-time monitoring and environmental control in industrial processing. For example, TAS able to monitor hazardous substances such as DMS (dimethylsulfate), ECH (epichlorohydrine) and BCME (bis(chloromethyl) ether) have proved extremely useful in the chemical industry [5].

Unfortunately, the TAS approach to chemical analysis is beset by significant problems which hinder its more widespread commercial application. These include slow transport of sample during analysis (especially in the liquid phase), poor separation speed (in the case of electrophoresis and liquid chromatography), excessive consumption of reagents and carrier solutions, and the necessity to fabricate efficient interfaces between the individual components of a TAS.

The concept of *miniaturisation* addresses all of the above problems. Miniaturisation of system dimensions reduces the transport distance between sampling and detection. Furthermore, and more importantly, diffusion and hydrodynamic theories predict higher separation speeds and improved resolution of sample components as channel dimensions diminish [2, 6]. If instrumental dimensions can be made small enough, then analysis times can become comparable to those of traditional chemical sensors (< 500 ms). These miniaturised analysis instruments have been christened μ-TAS (micro-Total (chemical) Analysis Systems) [1]. With sample handing, chemical reactions, sample separation and detection integrated on a *single* device, the μ-TAS resembles a chemical sensor in both size and response. Furthermore, each component function is now under the dynamic control of the user. Figure 1c illustrates the fundamental characteristics of a μ-TAS. Unlike the TAS (Fig. 1b), all sample handling is performed at a location extremely close to the initial sampling.

The following question is often asked: *Is miniaturisation really necessary*? Probably the most obvious gain through miniaturisation is the lowering of sample requirements. Diminished sample consumption permits greater numbers of analyses, which can be significant when working with small amounts of pharmaceutical and biological material. However, as already noted, a more significant benefit of miniaturisation is the dramatic reduction of analysis times. This reduction allows for the "real-time" monitoring of many molecular systems. Currently, the most obvious need for fast analyses lies in areas such as high-throughput drug screening [7] and genomics [8]. In particular, completion of the Human Genome Project is an enormous task of the highest priority [9]. Unfortunately, current opinion calls for a processing capability of at least 2 orders of magnitude greater than existing DNA analysis methods. In addition, combinatorial chemistry holds the promise of vastly accelerating the rate of drug discovery [10]. The combination of chemical building blocks using parallel, multiple reactions can generate huge numbers of compounds on very short timescales. Subsequently, these combinatorial libraries of compounds need to be rapidly screened to generate new drugs, or explore structure-activity correlations.

Finally, the most significant and exciting possibility afforded by miniaturisation is the creation of highly integrated μ-TAS. These systems will combine all aspects of chemical analysis on a single, integrated device (such as a microchip). Integration will permit automation, which in turn will allow unrivalled reliability, ease of use, and low cost. Yes, miniaturisation is a necessary step in the creation of the ultimate μ-TAS: the *laboratory-on-a-chip*.

The aim of this chapter is to provide a panoramic perspective of the basic philosophies, concepts and current advances in μ-TAS technology. The theoretical aspects of miniaturisation will be introduced, so as to develop a few simple rules for microstructure design. Subsequently, the basic techniques in microstructure fabrication will be outlined, and the use of μ-TAS as tools for microseparation will be discussed. Finally, some novel approaches to detection and component integration will be surveyed. It should be noted that some aspects (such as detection) will be discussed in greater detail elsewhere in this volume.

5.2 Theory of miniaturisation

As already stated, the primary objective of miniaturisation is to enhance analytical performance (i.e. speed of separation, component resolution and

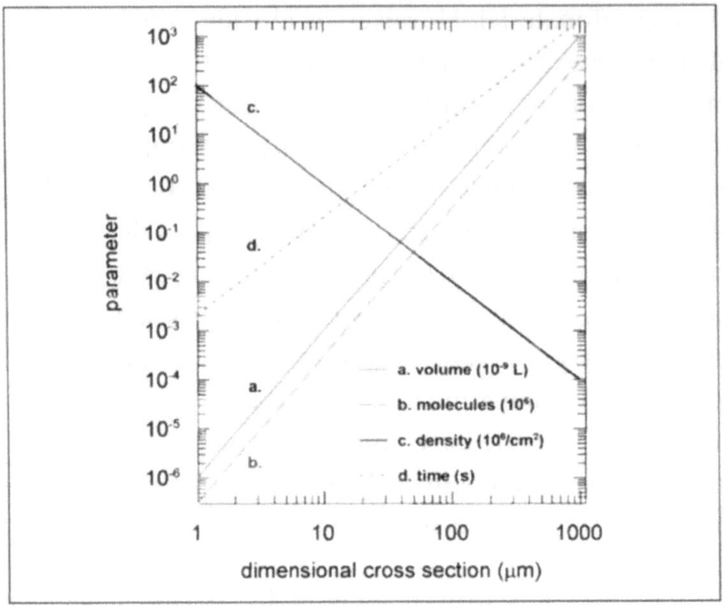

Figure 3 *Scaling of: (a) volume of solution, (b) number of analyte molecules, (c) spatial density, and (d) time to diffusively mix, with dimensional cross section Analysis is performed within a cube of dimension d. A bulk solution concentration of 0.5 nanomolar and a diffusion coefficient of 5 × 10^{-6} cm^{2}s^{-1} are assumed. See text for discussion*

throughput). According to a method described by Ramsey, the scaling of a chemical system can be illustrated simply through reference to Figure 3 [11]. Here, analysis is performed within a detection cube of dimensional cross-section, d (1–1000 μm). The variation of four relevant parameters (probe volume, molecular population of probe volume, spatial density and diffusion time) as a function of d are calculated.

Probe volume and molecular population

It is observed that at the 1 μm dimension our probe volume is approximately 1 fL. For a bulk solution concentration of 5 × 10^{-10} moldm^{-3} this would mean, on average, less than one target molecule existing at any instant within the volume element. Detection under these conditions is extremely difficult, although not impossible with modern technology [12]. At the 50 μm scale, the probe volume is now 125 pL and contains approximately 75000 molecules. Molecular recognition, using various protocols, is easily attainable in this regime. At the 1 mm scale, the probe volume is 1 μL and at least 600 million molecules exist in the volume element! According to this rough examination, detection issues imply a normal dimension range of between 10 and 100 μm.

Diffusion time

The ultimate μTAS will perform chemical reactions. On an ultra-small scale these reactions will be mediated by the diffusion of reagent molecules within the reaction chamber. The diffusion of any molecular species is characterised by a specific diffusion coefficient, D (cm^2/s) [13]. In the simplest case of a spherical (or globular) molecule,

$$D = \frac{kT}{6\pi\eta r} \tag{1}$$

where $k = 1.38 \times 10^{-23}$ JK^{-1}, T is the absolute temperature (K), η is the absolute viscosity (gcm^{-1}s^{-1}) and r is the hydrodynamic radius. A small molecule (molecular weight *ca.* 500–1000) would be expected to have a diffusion coefficient of approximately 5×10^{-6} cm^{2}s^{-1}. Under these conditions the molecule will diffuse across a dimension of 1 μm in approximately 2 ms. For a dimension of 50 μm, the molecule will take 5 s, and for a dimension of 1000 μm the diffusion time will increase to 2×10^3 s! This clearly demonstrates that purely diffusive mixing of reagents is only viable for dimensions well below 100 μm.

Scale of integration

It is also of commercial interest to consider the potential scale of integration (device density). This has obvious implications in terms of parallel processing and cost. As can be seen from Figure 3 device density varies as a function of 1/d^2. Thus for a dimension of 1 μm, 10^8 devices per cm^2 is a theoretical possibility. This decreases to 10^2 devices per cm^2 for a dimension of 1000 μm. Clearly, smaller dimensions are desirable, but can only be achieved using superior fabrication technologies.

The use of capillary electrophoresis (CE) within the μ-TAS format was realised and pioneered by Manz and Harrison in the early 1990s, and it has become the dominant liquid phase separation technique used in the microchip format [14]. Consequently, if we consider how the *parameters of interest* in CE (e.g., number of theoretical plates and Péclet number) vary as a function of the *variables to miniaturised* (space and time), we can see how miniaturisation affects the characteristics of the separation process.

Broadly, a CE system can be defined according to its column length L, inner diameter d, and analysis time t. Time is not directly related to L or d, resulting in one degree of freedom. Table 1 illustrates two possible systems where time is proportional to either L (length) or $d.L$ (cross-sectional area) with power per unit volume and power per unit length as a constant, respectively. Accordingly, it can be seen that all remaining parameters are solely related to L and d.

Table 1 Example of proportionality analysis for capillary electrophoresis. The miniaturisation factors are d and L. Two arbitrarily chosen time dependencies are shown here. The remaining parameters are calculated using the basic definition of d, L and time

Parameter	Symbol	L system	dL system
Capillary diameter	d	d	d
Capillary length	L	L	L
Time	t	L	$d.L$
Linear flow rate	$u = L/t$	constant	$1/d$
Péclet number	$v \propto u.d$	d	constant
Reduced plate height	$h = 2/v$	$1/d$	constant
Number of theoretical plates	$N = L/(d.h)$	L	L/d
Electric field	$E \propto u$	constant	$1/d$
Applied voltage	$U = E.L$	L	L/d
Electric current	$I \propto U.d^2/L$	d^2	d
Power/volume	$P_v \propto (I.U)/(d^2.L)$	constant	$1/d^2$
Power/length	$P_l \propto (I.U)/L$	d^2	constant
Temperature difference	$\Delta T \propto l^2$	d^4	d^2

The L system is characterised by a timescale that is proportional to L. Thus, if L is reduced by a factor of two, the retention time will also be reduced by a factor of two (assuming a constant electric field). A concomitant decrease in the number of theoretical plates is noted. In the $d.L$ system the timescale varies with the cross-sectional area. This means that the Péclet number, reduced plate height and power per unit length remain constant. Importantly, it can now be seen that an improvement in both analysis time and separation performance can be achieved by simply reducing the ratio of d to L. For example, if the diameter of a capillary is reduced from 100 to 10 µm, whilst the capillary length is reduced from 10 to 2 cm, the number of theoretical plates would increase by a factor of two, in 1/50th of the retention time! This system has been experimentally shown to exist [6].

A thorough analysis of miniaturisation theory by Manz and co-workers yields some simple rules for the design of electrophoretic microstructures [2, 15].

Separation efficiency

(a) The *absolute voltage* drop between the points of injection and detection defines the maximum number of theoretical plates (N), and the corresponding resolution of two neighbouring peaks ($\sqrt{N}$). The higher the applied electric field the higher N is.

(b) To minimise Joule heating and heat transfer across the capillary, a power of 1 W/m should not be exceeded.

(c) When separating species of different charge 10^2 theoretical plates is sufficient. When separating species of similar or identical charge 10^5 theoretical plates are necessary to give satisfactory resolution.

Consequently, the maximum number of theoretical plates (N_{max}) subsequent to miniaturisation is defined according to the following proportionality,

$$N_{max} \propto \frac{L}{d} \qquad (2)$$

In other words, if both L and d are reduced by the same fraction, there is no loss in analytical performance. Furthermore, N_{max} will be increased by either decreasing d or increasing L.

Analysis time

(a) The length of the capillary dictates the analysis time.
(b) The efficiency of the separation column will only be discernible if the contributions of injection and detection processes are small with respect to both the volume or length of the column.

Accordingly, the minimum analysis time is proportional to the product of the capillary length and internal diameter, i.e.

$$t_{min} \propto L \cdot d \tag{3}$$

Hence, a reduction in analysis time can be achieved by either reducing L or d. However, it must be noted that the only way to achieve a reduction in analysis time whilst maintaining the efficiency of the separation will be to reduce L and d by the same factor.

5.3 Fabrication and structure of µ-TAS

The stringent demands of µ-TAS have meant that conventional manufacturing methodologies are not applicable to miniaturised systems. The fabrication of flow manifolds, reactors, electrodes, sieves etc. on a micron scale is a non-trivial task. Consequently, the standard form of a µ-TAS is a planar, chip created using *micromachining* techniques. Micromachining can be defined simply as "the sculpturing of silicon and silicon compatible materials" to produce devices having no direct electrical function [16]. Micromachining methods were conceived as early as the 1850s through the work of Grove, Edison and Fleming [17]. Nevertheless, the real applications of micromachining technology were only realised a century later when the microelectronic revolution began. Over the next three decades a diversity of silicon-based sensors and actuators were developed, at an ever increasing rate and scale of integration. Today, techniques such as reactive ion etching [18] and LIGA [19] permit even more elaborate 3D microstructures to be made. The creation of sophisticated microelectromechanical systems (MEMS) such as micro-accelerometers, blood pressure sensors, infusion pumps and ultra-high resolution inkjet nozzles are now commonplace [20, 21, 22]. The μ-TAS can be considered a distinct subcategory of the MEMS concept, i.e. a microsystem for chemical analysis.

One of the primary strengths of micromachining is the reproducibility achieved in manufacturing. Typically, each silicon wafer will eventually yield hundreds or thousands of individual and identical "devices". More specifically, micromachining brings together the generic methods of photolithography, special etching, film deposition and bonding to create three-dimensional microstructures [23]. The fundamental idea is to sequentially superimpose two-dimensional patterns onto a substrate surface. Each pattern defines an area to be etched or an area onto which a new material can be deposited. The final, "three dimensional" structure is simply the result of these additive or subtractive steps.

A variety of substrate materials can be used in the manufacture of μ-TAS. These include silicon, glass, quartz, metals, plastics and ceramics. To date silicon, quartz and glass are the substrate materials of choice (primarily due to chemical, electrical and optical properties). However, it is noted that recent developments in polymer micromachining technology have enabled the construction of novel, moulded microstructures which offer high analytical performance at low cost [24, 25]. Since, detailed discussions of micromachining technologies are given elsewhere in this volume, only a cursory survey of these methods will be provided in the current chapter.

Photolithography

Photolithography is universally used to define the regions where subtractive and additive processes will act. The basic procedure starts with the deposition of a durable, photosensitive polymer (photoresist) onto the substrate surface. This is normally achieved by dropping a solution of the photoresist onto the surface, whilst spinning the substrate at many thousand rpm. This results in the formation of a uniform, photoresist layer a few microns thick. The coated surface is baked, to drive off any excess solvent, and then exposed to UV radiation though a chrome mask. Subsequent development in an organic solvent allows the removal of exposed portions of photoresist, leaving a polymerised resist pattern with high chemical resistance. This resist mask is then used to define one layer of the microstructure being fabricated.

Etching

The most simple subtractive process is etching. Etching allows the two-dimensional photoresist pattern to be transferred to the substrate material, through use of "wet" or "dry" protocols. Wet etching involves the use of aqueous etchants. These etchants are differentiated on the basis of whether they act in an isotropic or anisotropic manner. Examples include HF, HNO_3 (isotropic etchants), KOH and tetramethyl ammonium hydroxide (anisotropic etchants). Etchants act by oxidizing silicon to silicates [23].

Isotropic and anisotropic "dry" etch protocols exist for silicon, dielectric materials, metals and organics. These methods generally involve the use of partially or fully ionized gas plasmas to effect the etch. Extremely high degrees of anisotropy are possible using plasma-based techniques, allowing the fabrication high aspect-ratio microstructures.

Film deposition

This term encapsulates a wide variety of additive processes that enable microstructures to be built on the substrate surface. For silicon, the simplest additive process is thermal oxidation to generate a SiO_2 layer. More correctly this is a reactive process since temperatures in excess of 900°C are normally necessary. For subsequent deposition of secondary materials a variety of low-temperature processes can be used. These include spin-coating, physical vapour deposition (PVD), chemical vapour deposition (CVD), low pressure CVD, plasma enhanced CVD (used when low temperature processing is desired) and sputter deposition [23, 26, 27]. A huge number of materials can be uniformly deposited using these techniques. Table 2, lists some examples of these materials and the processes that enable their deposition. Depending on the material and the deposition process used, film thicknesses of a few Angstroms to many microns can be obtained.

Table 2 Common additive processes used in micromachining

	Thermal > 900°C	CVD 600–1000°C	PECVD 100–600°C	Sputter < 300°C	Electro-plating ambient	Spincasting/ screenprinting ambient
SiO_2	x	x	x	x		
Glass				x		x
Single crystal silicon		x				
Metals				x	x	x
Polysilicon		x				
Amorphous silicon			x	x		
Polymers		x	x			x

Bonding

This is the final stage in fabrication of a μ-TAS. Bonding simply refers to the assembly of the substrate materials, e.g., silicon-to-silicon, glass-to-silicon, silicon-to-oxide and glass-to-glass. Aniodic bonding is commonly used to bond glass and silicon substrates. This procedure utilises electrostatic attraction to form covalent bonds between the surface atoms of the glass and the silicon. For glass or quartz, thermal bonding (450°–900°C) provides the simplest way to assemble the substrate materials. Furthermore, the planarity of the substrate surfaces generally means that bonding is a simple and efficient final step in the manufacture of μ-TAS [2].

Figure 4 illustrates a simple one-mask micromachining procedure. The process begins with thermal oxidation of the silicon to generate a surface oxide layer. A photoresist film is then deposited on top of the oxide layer. Photolithography and subsequent chemical development define the region to be etched. The first etch step removes the oxide layer. Subsequent etching of the bulk substrate yields the desired channel. If further structural definition is required, additional processes would augment the sequence shown in Figure 4 (e.g., CVD of metal films). The final step in creation of the complete microstructure is bonding to a top plate. This allows encapsulation of the open channel. An example of a μ-TAS is shown in Figure 5.

The procedure illustrated in Figure 4 describes the manufacture of the simplest kind of μ-TAS; an enclosed channel on a microchip. The need for

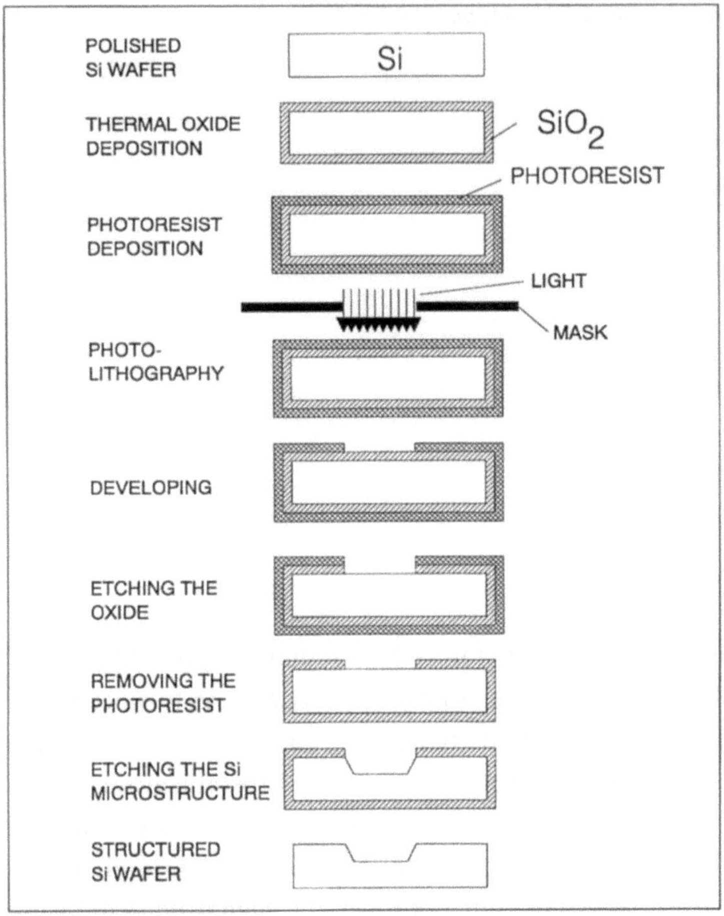

Figure 4 *Process steps of a standard one-mask micromachining procedure for etching a single channel into a substrate material*

additional components (for flow handling and detection) is obvious, but instantly adds an extra dimension to the level of micromachining technology necessary. The following paragraphs briefly outline some of these components.

Devices for handling liquids and gases are key elements in μ-TAS. Research on silicon micropumps was pioneered by Wallmark and Smits in the early 1980s [28]. Since then, a diversity of micropumps and microvalves have been developed. Most micropumps are of the reciprocating kind (i.e. diaphragm pumps) [29, 30, 31]. These pumps in their most basic form consist of a dis-

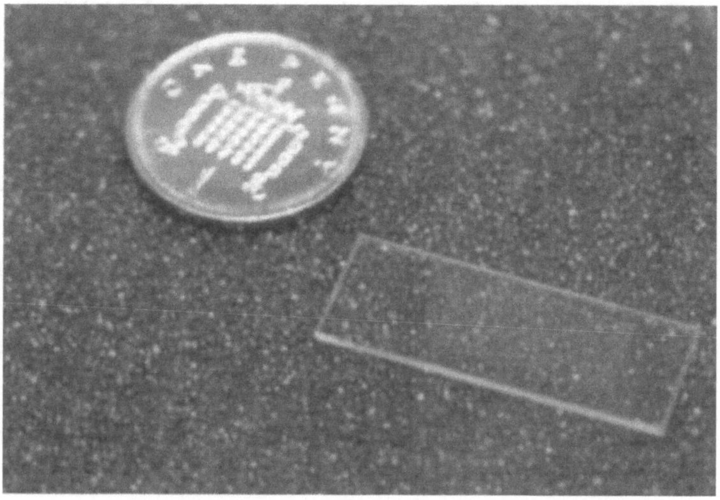

Figure 5 *Scanning electron micrograph of part of a continuous flow PcR microchip*

placement unit and two passive check valves. Almost all liquids can be manipulated using diaphragm pumps, with achievable flow rates of nanoliters per minute up to a few milliliters per minute. Several physical principles have been applied to the actuation of micro diaphragm pumps. These include piezoelectric actuation [32], electrostatic actuation [33], electromagnetic actuation [34], bubble actuation [35] and thermopneumatic actuation [35]. To date no commercial fluid systems incorporate micropumps. This is primarily due to problems in pump priming, particulate pollution and the fact that construction is both complex and expensive [36]. In addition, although flow-rate stabilities and lifetimes are reasonably good, most pumps are not suited to operate in high pressure situations. More recent research has led to the development of pumps (with high pressure check valves) with improved forward-to-reverse flow behaviour at pressures above 1 atm [37].

Valves are one of the most fundamental components of a μ-TAS and need to be designed with the operation of a particular chemical system in mind. Fortunately, under many circumstances (e.g., microelectrophoresis) the need for physical valves is obviated through the use of electrokinetic switching [38]. Nevertheless, in more sophisticated procedures efficient and robust microvalves are necessary.

To date, most microvalves have been developed for handling gases [33, 34, 39], although many can now handle both liquids and gases [32, 40]. These valves are actuated using a variety of mechanisms [41]. System aspects such as dead volume, leakage, size and simplicity play a dominant role in design. Microvalves can work discretely or in combination. By coupling two microvalves with a membrane, a simple micropump can be easily formed. When the membrane is contacted with the glass surface analyte cannot move between the inlet and outlet. However, if the membrane is pulled away from the surface, analyte can flow freely between channels. In this case actuation is based on pneumatics [42]. In addition Esashi and co-workers have reported several actuated, three-way valves for gases and liquids, as well as miniaturised mass flow controllers for gases [32].

Temperature sensors can be easily fabricated on most substrates. The most obvious form of a temperature sensor is a thin film thermocouple, where sensing is based on thermoelectric phenomena. Thermocouples of this kind can be made using evaporative deposition techniques [43].

The potentials of µ-TAS will only be realised if "real samples" can be effectively handled and analysed. Samples of this kind (e.g., blood, semen and saliva) will almost certainly contain particulates and gas bubbles, which pollute the microsystem both chemically and physically. Particulates cause the most serious problems. For example, any particulate matter in a microchannel may cause a local blockage or render delicate micropumps and microvalves useless. Consequently, the need for microsieves, filters and bubble separators is apparent. These can be made using fairly simple micromachining technologies [44, 45]. Pore diameters can be precisely defined to match particular species present in sample. Sieves will remove both particulates and bubbles.

The fabrication of mixers/reaction chambers is a non-trivial process. The downsizing of the fluidic process to the femtoliter-nanoliter range leads to Reynolds numbers which are significantly smaller than values characteristic for the onset of turbulence. Consequently, novel mixing protocols must be developed, an example of which has been presented by Miyake [46]. They propose a situation where a fluid is forced through a matrix of holes perpendicular to the flow direction of a second fluid. The fluid that is forced through the "sieve" emerges as micron-sized globules. This leads to an increase in the contact area of the fluids, yielding efficient mixing of the two fluids. More recently, Larsen and co-workers have demonstrated a novel static mixer for

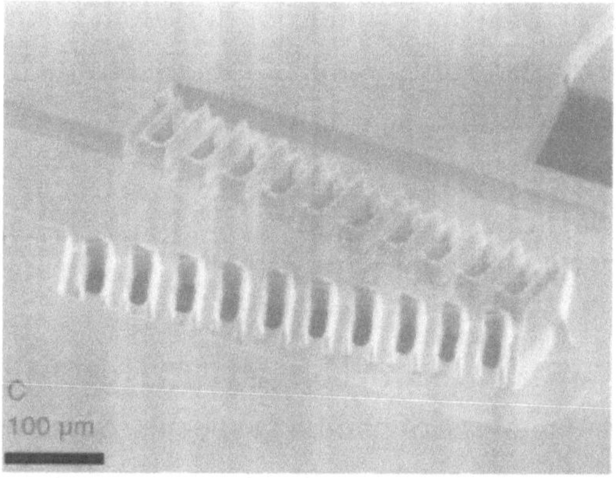

Figure 6 *Scanning electron micrograph of a novel mixer for multilayer lamination (as described in [47])*

chemical and biochemical reactions [47]. Rapid mixing is achieved by multilayer lamination of two liquids in a single microchannel. Figure 6 is an SEM image of the silicon mixer with laminated flows. The advantage of this design stems from the robustness of the mixer to both particulate and bubble pollution.

5.4 Chemical analysis using µ-TAS

The original application of microsystem technology to chemical analyses occurred nearly two decades ago, when Terry and co-workers created an integrated gas chromatograph on a 50 mm silicon wafer [48]. Unfortunately, this seminal piece of work received limited attention at the time; primarily due to performance issues. Nevertheless, in the mid to late 1980s several research groups realised the potentials of miniaturisation in chemical analysis, and applied the concepts of micromachining to liquid phase separation techniques. This was in large part due to the advent of capillary electrophoresis as a dominant analytical tool [3].

Early research was pioneered by Manz and colleagues initially at Hitachi Ltd. [49] and subsequently at Ciba-Geigy [50], where the term µ-TAS was christened. Since then, a number of laboratories world-wide have developed the µ-TAS concept and applied it to a variety of chemical systems. As stated previously, microfabricated devices have been demonstrated for a diversity of chromatographic and electrophoretic separation methods. These µ-TAS have analytical performances equivalent to, or in excess of their conventional counterparts. This section summarises much of this research, and serves to highlight the diversity of μ-TAS.

Capillary electrophoresis

Capillary electrophoresis (CE) is a relatively new technique for the separation and analysis of chemical species. Since its inception in the early 1980s, its use in the chemical and more significantly, the biological sciences has grown at an exponential rate [51]. CE complements the traditional separation techniques of high-performance liquid chromatography (HPLC), gas chromatography (GC) and slab gel electrophoresis. Many analyses currently utilising HPLC or slab gel electrophoresis will probably convert in time to CE, due to its high efficiency and high throughput. In addition, CE sample requirements are low and capillaries are relatively cheap and long-lasting. As noted previously, CE separations occur due to inherent differences in the electrophoretic mobilities of molecular species. Furthermore, since electroosmotic flow (EOF) is normally in excess of electrophoretic migration, all sample components move in the same direction (but at differing rates). Consequently, EOF can be used to manipulate sample within the capillary.

CE is perfectly suited for application within the μ-TAS framework. This statement can be demonstrated by reference to some fundamental equations. For an ideal CE system, the separation efficiency can be defined in terms of the number of theoretical plates N, i.e.

$$N = \frac{\mu \cdot V}{2D} \tag{4}$$

Here, D is the specific diffusion coefficient (cm^2/s), μ is the electrophoretic mobility (cm^2/Vs) and V the total applied voltage. N is proportional to the applied voltage, regardless of the capillary dimensions. In addition, the time for a zone to migrate a distance L is given by Eq. 5.

$$t = \frac{L^2}{\mu \cdot V} \tag{5}$$

Consequently, it can be seen that high voltages and short capillary lengths generate fast, high efficiency separations. For example, if the length of a capillary is reduced by a factor of two, and the applied electric field doubled, the efficiency of the separation will increase by a factor of two, in 1/8th of the retention time. It should be noted that high applied electric fields however introduce the additional problem of Joule heating. This effect must be offset by reducing the internal diameter of the capillary. Micromachining technology allows the creation of channels which are short in length and have small cross-section dimensions. Thus, the μ-TAS approach should theoretically yield ultra-fast, ultra-high efficiency electrophoretic separations.

Figure 7(a) depicts the layout of a glass chip for CE with integrated sample injection. The separation channel is 50 μm wide and approximately 12 μm deep. Figure 7(b) is an electron micrograph of the region where injection and separation channels intersect. As stated, the primary method of sample transport on a CE chip is electrokinetic pumping (a combination of electroosmosis and electrophoresis). Significantly, sample transport, sample injection and other handling steps can be performed and controlled by the selective application of electric potentials across channels. In contrast, approaches using micropumps and valves suffer from high back pressures, and are generally not suited to the delivery of very small dose volumes [52].

Application of an electric potential between reservoirs 1 (sample) and 4 (injection waste) causes the geometrically defined injection volume ("double-Tee" injection volume ~ 90 pL) to be filled by electrokinetic migration of sample. Once loading is complete, application of a high potential between reservoirs 2 (buffer) and 5 (waste) drives the injected sample plug into the separation channel, and effects the electrophoretic separation of sample components. Figure 8 illustrates an electropherogram obtained using this protocol. A mixture of 6 FITC-labelled amino acids; arginine (Arg), glutamine (Glu), phenylalanine (Phe), asparagine (Asp), serine (Ser) and glycine (Gly) are separated on a CE chip within 14 s. Plate numbers range from 5800 to 160000 for only 20 amol of injected sample. In addition, shorter analysis times were achieved by reducing the separation length (at the expense of resolution) [53]. Due to the simplicity of this generic approach, i.e. microfabricated, interconnected channels on a planar substrate, CE chips have been applied to a diversity of chemical systems. Inspection of the literature illustrates that the vast majority of microchip separation research has centered on electropho-

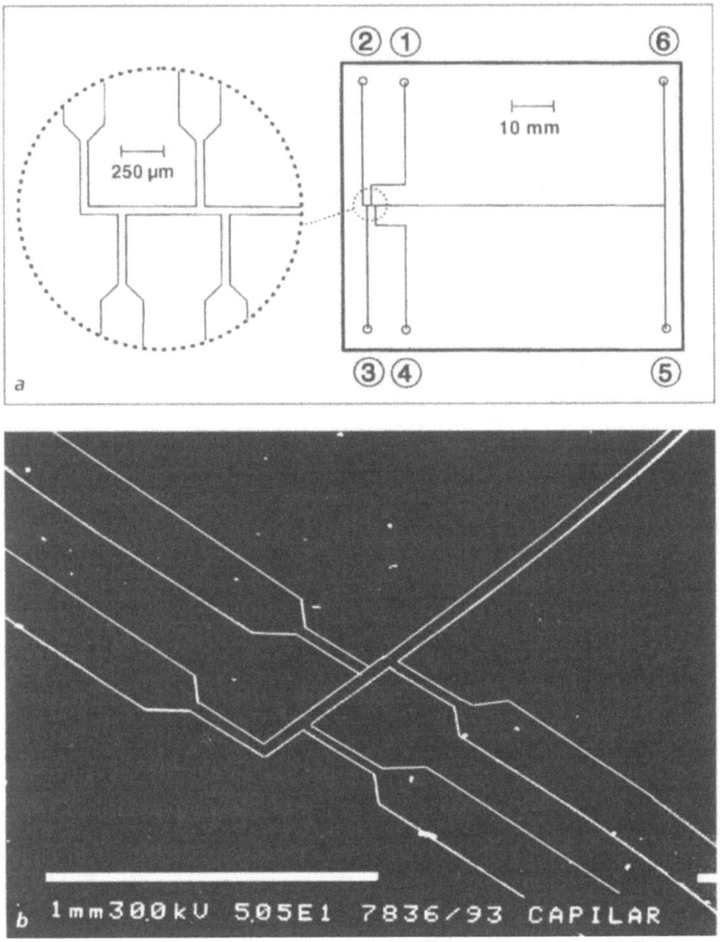

Figure 7 *(a) Layout of glass CE microchip with integrated sample injector. Channel cross-sections: 50 × 12 µm (tin channels) and 250 × 12 µm (wide channels). See text for description of sample manipulation. (b) Scanning electron micrograph of region where injection and separation channels intersect*

retic and electroosmotic mechanisms. The composition of this review will reflect that fact.

The feasibility of using electroosmotic pumping and electrophoretic separation within a planar, glass chip was first demonstrated by Manz and Harrison in late 1991 [54, 55]. Glass CE chips were used to separate a mixture of two fluorescent dyes (calcein and fluorescein), with efficiencies approaching those obtained from conventional fused silica capillaries. Reduction of channel

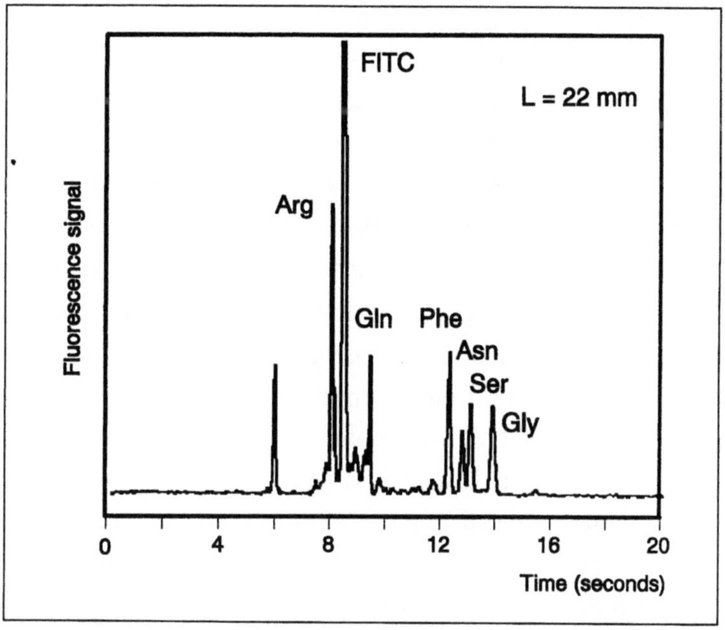

Figure 8 *Electropherogram of a mixture of six FITC-labelled amino acids recorded at a separation length of 22 mm, and an electric field strength of 1060 V/cm. Formal concentration of each amino acid is 10 μM. Solutions were made in a pH 9.0 buffer (20 mM boric acid – 100 mM Tris)*

lengths allowed ultra-high speed separations of fluorescein dye derivatives. For instance, fluorescein and fluorescein sulphonate dyes were separated in a channel 75 mm long within 4 s of injection [56]. Performance evaluation demonstrated that electric fields of over 10^3 V/cm could be applied to the glass substrates without apparent dielectric breakdown [57]. This work was swiftly followed by the application of CE chips to more interesting media. As noted, Figure 8 illustrates a fast and efficient separation of amino acids on a glass microchip. Analysis times for this system ranged from a few seconds to a few tens of seconds, and plate numbers of up to 160000 were realised. In addition, repetitive sample injection and separation cycles yielded highly reproducible migration times and peak areas. These results were significant in that they demonstrated the feasibility of automated, quasi-continuous, on-line monitoring of chemical species [53, 58]. Studies at about the same time indicated that the use of silicon as a possible substrate material was not desirable. Insulation characteristics placed limitations on the size of the applied electric fields, and thus impaired separation efficiencies [59].

In 1994, a number of additional research groups began to apply micromachining techniques to CE. Ramsey and co-workers at the *Oak Ridge National Laboratory* fabricated glass microchips similar in concept to those previously described. Serpentine column geometries allowed channels 165 mm in length to be confined within a 10 mm × 10 mm area. Fast, efficient separations of rhodamine dyes, with minimal bandbroadening effects, were demonstrated [60]. This work was soon followed by an elegant demonstration of ultra-fast CE on a glass microchip. Fluorescein and Rhodamine B mixtures were separated along a microfabricated channel using fluorescence detection (Fig. 9) [61]. At an electric field strength of 1.5 kV/cm, analytes are resolved in less than 1.6 s for an 11.1 mm separation length, in less than 260 ms for a 1.6 mm separation length, and in less than 150 ms for a 900 µm separation length! These results established the suitability of microchip CE devices in sensor applications.

A fundamental study of EOF in microfabricated columns by Seiler et al. deemed channels to behave in an analogous way to resistors (in terms of their electrical characteristics). Accordingly, the potential at any point within the channel network can be estimated using *Kirchoff's Rules* (assuming the solution resistivities are known). Use of this concept enables the direction and magnitude of sample flow to be precisely calculated, thus ensuring a totally valveless fluidic system [38, 62]. Conversely, Jacobsen et al. minimised EOF by covalently bonding polyacrylamide to the channel walls of fused silica CE chips [63]. The reduction of EOF leads to an enhancement in the separation efficiency.

We have seen that miniaturisation of the analytical process yields improvements in efficiency and speed. In addition, sample and reagent requirements are reduced. However, perhaps the most exciting possibility afforded by miniaturisation is the creation of highly integrated µ-TAS. Many laboratory-based procedures require sample manipulation prior to the measurement process, and some analyses are fully automated to avoid operator bias. Consequently, the evolution of µ-TAS must encompass the whole analytical process; not just the separation. The first attempt to address the idea of integration involved the construction of a *postcolumn* reactor on a glass microchip [64]. Two amino acids were separated in a microfabricated channel. Each component was then reacted with *o*-phthaldialdehyde, as it passed through a mixing tee, to form a fluorescent moiety. These are subsequently detected

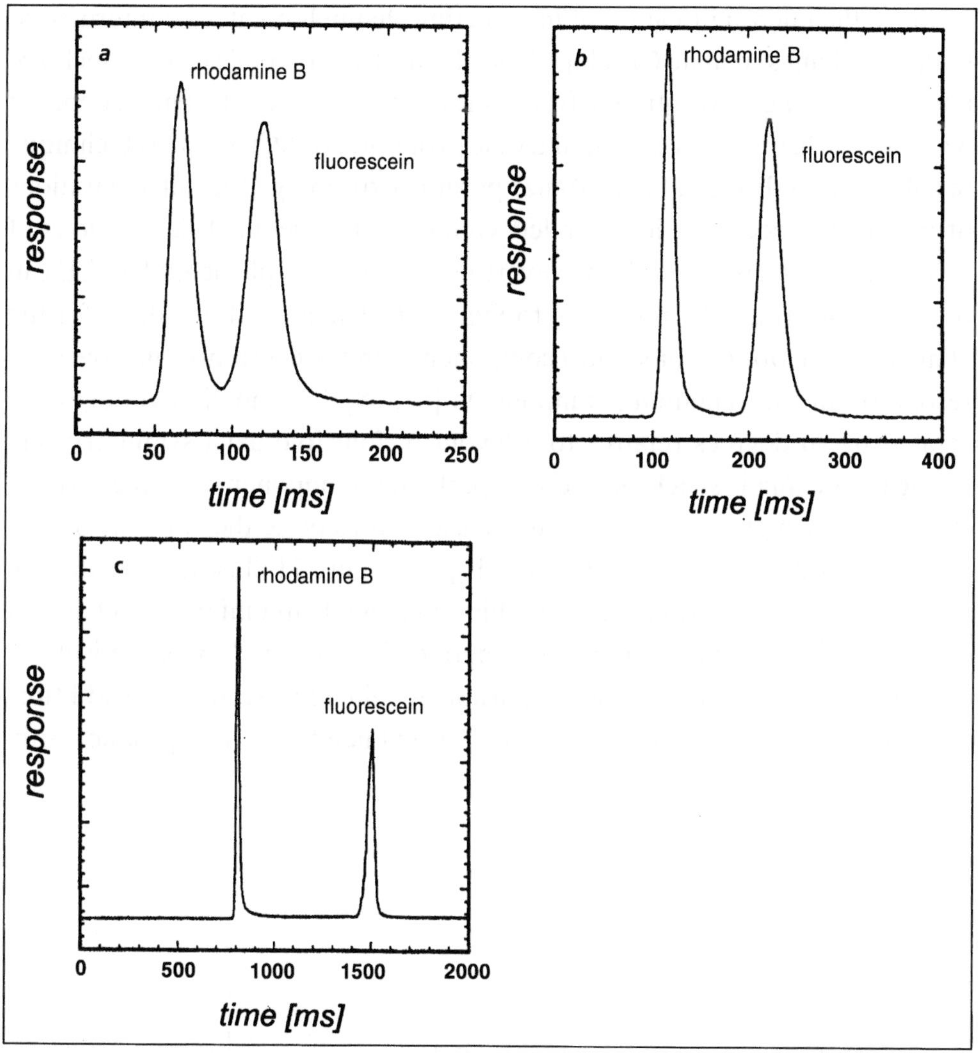

Figure 9 *Electropherograms of rhodamine B and fluorescein with a separation field strength of 1.5 kV/cm and a separation length of: (a) 0.9 mm, (b) 1.6 mm, and (c) 11.1 mm*

using laser induced fluorescence. This work was soon followed by the first example of an on-line *precolumn* reaction and CE on a glass microchip [65]. Amino acids (arginine and glycine) were reacted with *o*-phthaldialdehyde in a 1 nL reaction chamber (96 µm × 6 µm × 2000 µm) at room temperature, to yield fluorescent products. The product mixture is then electrokinetically injected into the separation channel, and electrophoresed. Species were detected using confocal laser induced fluorescence. Highly reproducible separations were obtained within a few seconds of injection.

Since the microfabrication of intersecting channels creates no intermediate dead volumes, planar CE chips lend themselves to applications which are impractical using conventional, fused silica capillaries. An elegant example of which is synchronised cyclic capillary electrophoresis (SCCE). This technique (established by Burggraf and Manz) permits extremely efficient separations of molecular species at low applied voltages [66]. Detailed descriptions of SCCE are given elsewhere [66]. Nevertheless, the principle behind SCCE can be clearly illustrated by reference to Figure 10. The procedure allows for the elimination of *too slow* or *too fast* components, and concurrently improves the resolution of the remaining components [67, 68]. The applied voltages are rotated by 90 degrees relative to a layout of four-fold axial symmetry, and voltage switching is synchronised to a particular component. During a single cycle, species migrate in a field that corresponds to twice the applied voltage, thus yielding higher plate numbers (*cf.* Eq. 4). Figure 11 illustrates an electropherogram obtained using a SCCE chip of channel circumference 80 mm. It can be seen that using a separation length of 20 mm, six fluorescein isothiocyanate labelled amino acids are separable within a few seconds. In addition, it is noted that any fully integrated μ-TAS will need to isolate products sub-

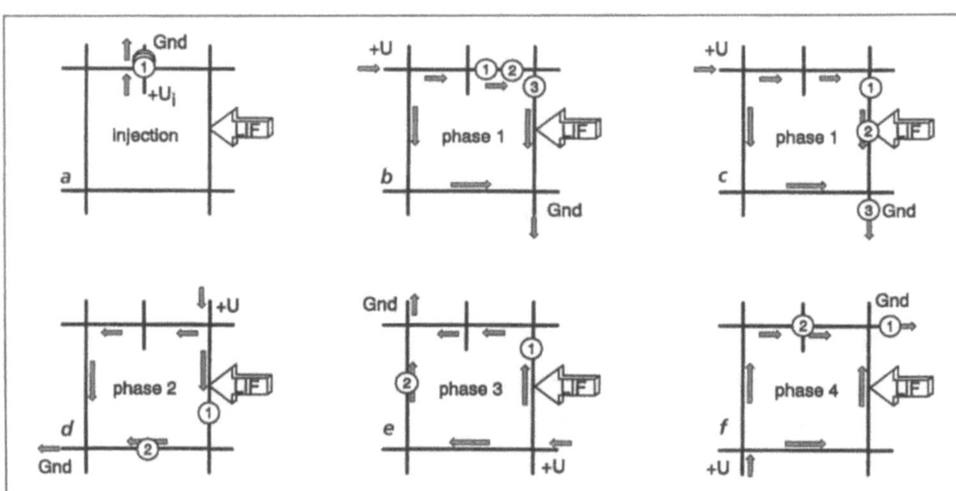

Figure 10 *Principle behind synchronized cyclic capillary electrophoresis (SCCE). Three sample components are represented by circled numbers (① ② ③). The voltage switching procedure is synchronized to component 2: (a) injection phase, (b) during phase 1, (c) at the end of phase 1, (d) phase 2, (e) phase 3, (f) end of cycle*

sequent to reaction and separation stages. SCCE provides a well defined mechanism for doing this, even for complex mixtures.

We have seen that miniaturisation of the electrophoretic process leads to huge gains in analysis time and efficiency [61]. In addition, the nature of the microfabrication process allows the manufacture of many identical devices in a single batch (hundreds to thousands). Consequently, the combination of microfabricated electrophoresis channels with the ability to analyse multiple channels in parallel leads to enormous gains in information density and sample throughput. As noted previously, this has the most significant implications in the areas of combinatorial chemistry [10], drug screening and genomics [9].

The human genetic blueprint – the human genome – consists of an estimated 100000 genes, which encode information in DNA for making and maintaining human life. These genes are distributed among 23 pairs of chromo-

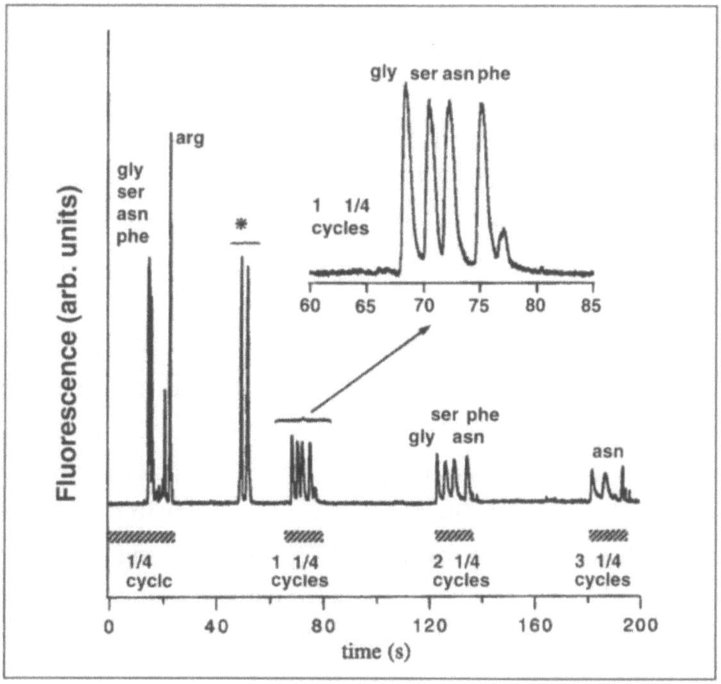

Figure 11 *Synchronised cyclic CE separation of FITC-labelled amino acids. Synchronization time interval is 14.4 s, with the hatched bars indicating times during which zones of interest are detected. The peak marked with an asterisk represents samples that are redetected due to channel switching*

somes, each containing a long DNA molecule. There are a total of 3 billion bases in the complete human genome. CE has proved to be a powerful technique for DNA analysis, and has been applied to restriction fragment sizing, PCR product analysis and DNA sequencing [51]. Although considerably faster than conventional slab gel separations, the current demands of the Human Genome Project require increased processing capabilities [8]. The ability to separate DNA on microfabricated CE arrays would clearly meet that need, and establish the feasibility of integrated DNA analysis devices. The first examples of DNA separations on a microfabricated CE devices were presented in 1994. Woolley and Mathies performed high resolution electrophoretic separations of ϕX174 *Hae III* DNA restriction fragments (72–1353 base pairs) in microfabricated channels filled with a hydroxyethylcellulose sieving matrix [69]. DNA fragments were labelled with dye in the running buffer, and detected using a laser-excited, confocal fluorescence system. Electrophoresis in 35-mm-long channels allowed complete separation of all fragments in less than 2 min. This work clearly established protocols for high-speed, high-throughput DNA analysis. Concurrently, Effenhauser et al. demonstrated fast separations of synthetic phosphorothioate oligonucleotides (10–25 base pairs) on similar glass microchips [70]. Application of electric fields of up to 2.3 kV/cm resulted in size separation of single-stranded oligonucleotides in less than 45 s. Both these studies suggested that with sensitivity and resolution improvements DNA sequencing on chips was possible.

Within a year, Woolley and Mathies published the first demonstration of DNA separations with single base pair resolution on microfabricated CE glass chips [71]. Electrophoresis was performed in 50×8 µm cross-section channels, using a denaturing 9% *T*, 0% *C* polyacrylamide sieving matrix. Fragment ladders were fluorescently labelled using novel energy transfer dye-labelled sequencing primers [72]. Sequencing extension fragments were separated to 433 bases within 600 s using a one-colour detection protocol and a separation distance of 35 mm. Using a four-colour chemistry, DNA sequencing (with 97% accuracy) to 150 base pairs was achieved in only 540 s. Although improvements in resolution and run length are necessary, the authors postulate that CE array chips could realistically yield a raw DNA sequencing rate of 4×10^4 bases/hour per chip! Very recent work by Woolley et al. has established the feasibility of high speed DNA genotyping using CE array chips. Both restriction fragment sizing and genotyping of the HLA-H gene have been performed in 12 parallel channels within 160 s [73]. In addition, Effenhauser and co-workers have developed moulded, silicone elastomer microchips for DNA fragment analysis. PDMS-microchip separations of ϕX174 *Hae III* DNA restriction fragments stained with an intercalating fluorescent dye are complete in a few minutes. Improvements in detection protocols have also allowed the detection of single λ-DNA molecules under electrophoretically controlled flow conditions [24]. It should be noted that the manufacturing cost of a silicone chip is approximately \$ 0.7. This low price should justify a single-use, disposable approach to μ-TAS devices in areas such as medical diagnostics and screening.

One of the eventual requirements of a fully integrated μ-TAS will be the ability to isolate a product subsequent to reaction and separation stages. In conventional CE, this is a non-trivial task [74, 75]. However, this assignment becomes far simpler when applied to microfabricated networks of intersecting channels. A refined study by Effenhauser et al. used a very simple channel network for selective isolation of fraction zones after fast sizing of phosphorothioate oligonucleotides [76]. Withdrawal of sample was effected by automated switching of applied potentials. Importantly, the authors demonstrated that a single sample component, originally injected into a 90 pL volume prior to separation, was still confined to a volume of approximately 300 pL after isolation. This degree of sample control has particular relevance for future developments the creation of μ-TAS for biochemical analysis.

To further develop these microsystems towards a complete μ-TAS, some degree of sample pre-treatment will be required. This may involve filtration or sieving (as discussed previously), isolation of components, or even complex chemical reactions. Raymond and co-workers have applied the technique of free-flow electrophoresis (FFE) to the problem of sample pre-treatment using a silicon microchip device [77]. The general principle of FFE is illustrated in Figure 12. A narrow sample stream is fed into a carrier solution which flows perpendicular to an applied electric field. Consequently, charged species are deflected from the direction of flow at an angle determined by a combination of the carrier flow velocity and the electrophoretic mobility of the component [77, 78]. After separation, the individual sample components can be collected as they leave the separation bed. This approach is advantageous since isolation and subsequent delivery to the analysis system can be done in a con-

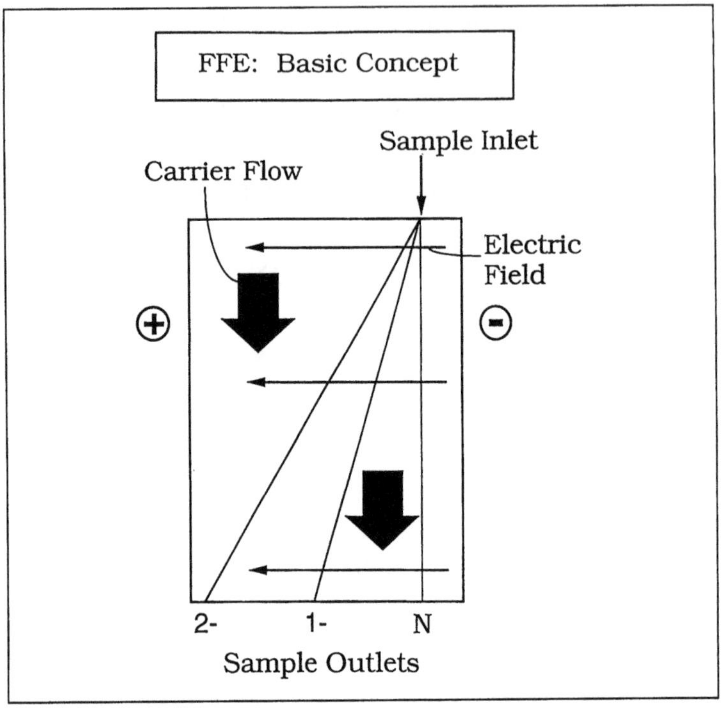

Figure 12 *Principle behind free-flow electrophoresis. The positions N, −1, −2 illustrate where species of formal charge 0, −1, −2 respectively exit the microstructure*

tinuous manner. Figure 13(a) shows a schematic of the silicon FFE chip. Carrier is introduced through two inlets on either side of the sample inlet. Once sample is transported into the separation bed (10 mm × 50 mm × 50 μm) electrophoresis occurs at right angles to the direction of carrier flow. Figure 13(b) is a three-dimensional view of the inlet region. Using this silicon chip it is possible to achieve a continuous separation of small ions according to their charge. Figure 14 illustrates the continuous separation of three labelled amino acids (lysine, glutamine and glutamic acid). An applied potential of 50 V gave baseline resolution of ions differing by one charge unit in 2–5 min. The application of μ-FFE to the continuous separation of high molecular weight compounds has also been demonstrated [79]. Using a silicon device with a separation bed volume of 25 μL, the continuous separation of human serum albumin (HSA), bradykinin and ribonuclease was achieved at low applied fields (25–100 V/m). Studies indicated that for maximum throughput, high field strengths and high flow rates are necessary. In addition, modifications to

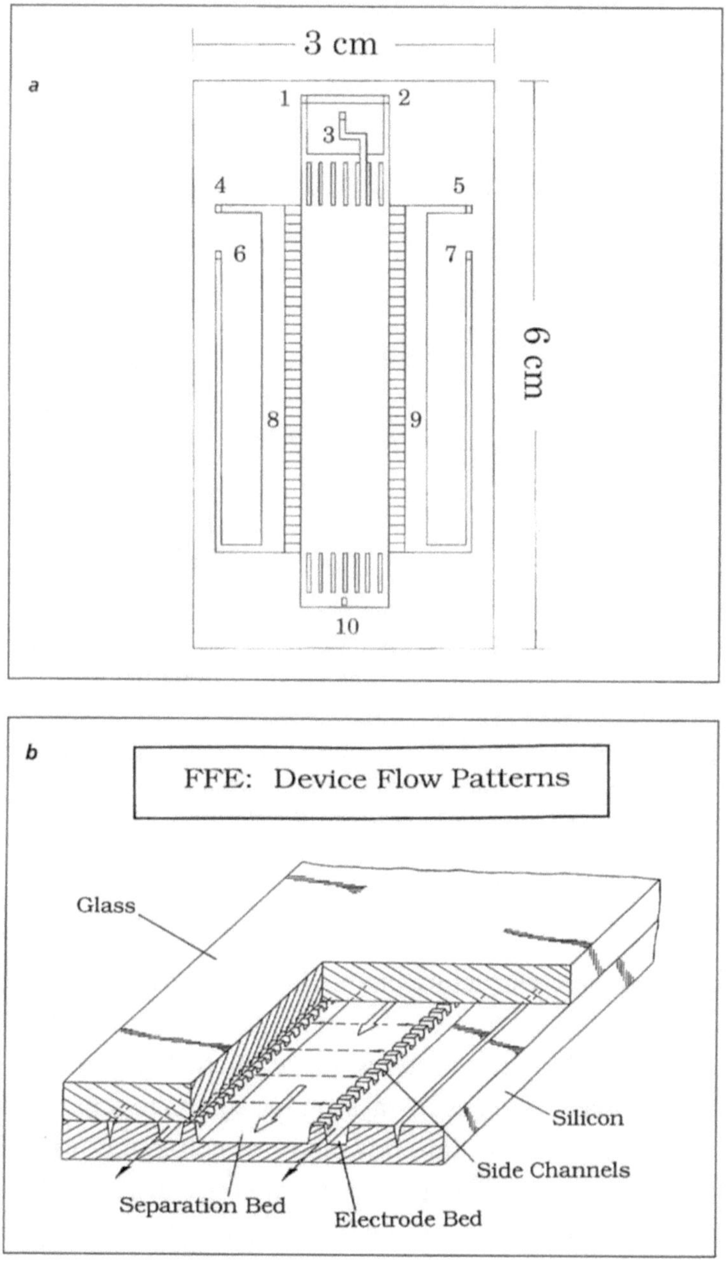

Figure 13 *(a) Schematic of a free-flow electrophoresis (FFE) microchip: 1,2 – carrier buffer inlets; 3 – sample inlet; 4,5 – side bed inlets; 6,7 – side bed outlets; 8,9 – side beds containing Pt electrodes; 10 – outlet. (b) Three-dimensional view of the inlet region on a FFE microchip*

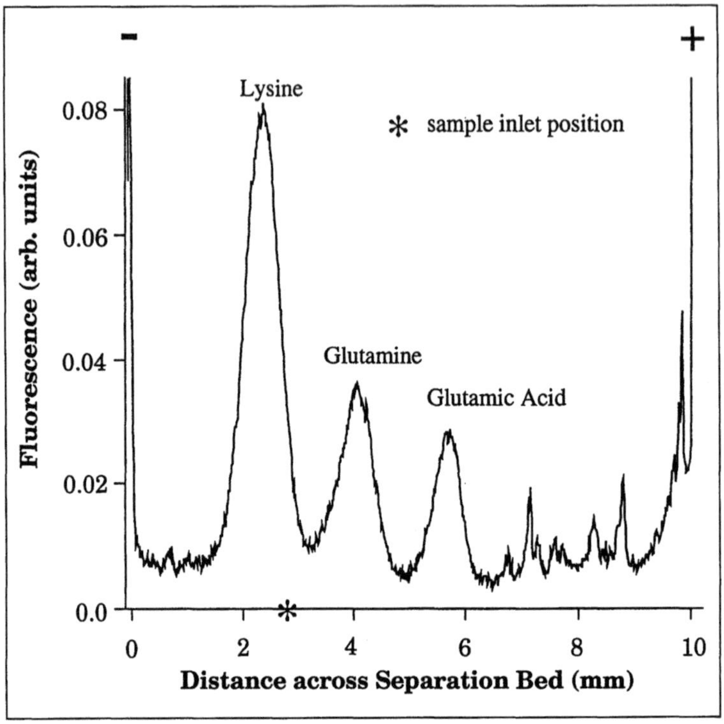

Figure 14 *Separation profile of rhodamine B isothiocyanate labelled amino acids (lysine, glutamine, glutamic acid), obtained 22 mm along the separation bed with an applied voltage of 50 V (current 7.5 mA). Sample and carrier flow rates were 0.2 and 5 μL/min respectively*

the silicon microstructure enabled efficient fraction collection subsequent to separation.

Much of the current interest in CE microchips has centred around their use in immunoassays. These assays are based on the separation of free and bound forms of antigen or antibody, and are commonly used for selective determination of molecular species at low concentration. The unique fluidic control system afforded by a chip format and reduced analysis times should allow for the development of fast, automated immunoassays for use "in-the-field". The first example of a microchip based immunoassay was presented by Koutny et al. in 1996 [80]. A competitive immunoassay for the analysis of cortisol in serum was performed in a fused-silica microchip. Fast, reproducible separations allowed the determination of cortisol in blood serum over the range of clinical interest. More recently, Harrison and co-workers have demonstrated on-chip, homogeneous, immunological analyses for proteins such as immunoglobulin G (IgG) and drugs such as theophylline [81]. The problem of protein adsorption to channel walls was minimised through careful choice of buffer systems.

Chromatography

Chromatography encompasses a diverse group of techniques that permit the separation of closely related chemical components. The most fundamental classification of chromatographic methods is based upon the type of mobile phase used in separation. Consequently, two generic methodologies can be defined: *gas chromatography* (GC) and *liquid chromatography* (LC). In GC, elution is effected by the flow of an inert gaseous mobile phase. Many volatile, stable chemicals can be separated using this method. For example, GC is the most common technique used in the screening of drugs and narcotics, and it is estimated that over 200000 gas-chromatographs are in current use worldwide [82]. LC, on the other hand, utilises a liquid mobile phase. It has wider application than GC due to the fact that 85% of all known compounds are not sufficiently stable or volatile for use in GC. The modern forms of LC are generically termed HPLC (high-performance liquid chromatography). HPLC is unquestionably the most widely used separation and quantitative analysis technique, with annual sales of HPLC equipment of approximately 1 billion dollars [82]! HPLC techniques have undergone much development recently.

Both techniques appear to be particularly amenable to a micromachined format, and indeed, the very first example of a microfabricated chemical separation device was a gas chromatograph. As noted previously, this GC analyser was fabricated by Terry and co-workers at Stanford University in 1975 (over 15 years before the manufacture of the first microfabricated CE device) [48, 83]. Integrated onto a single silicon wafer (50 mm diameter) were a sample injection valve, a 150-cm-long spiral separation channel (30 μm deep and 200 μm wide), and a thermal conductivity detector (based on a nickel film resistor). Connections for a gas inlet and the detector were etched through the 200 μm silicon wafer. The GC chip was used in conjunction with the carrier gas supply and data processing apparatus, as shown in Figure 15. Simple separations of gaseous hydrocarbons were effected in less than 5 s.

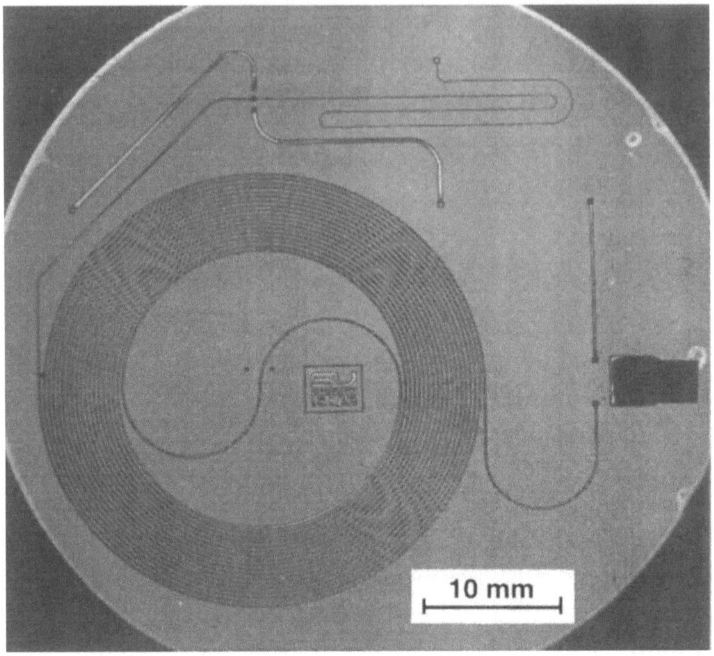

Figure 15 *Photograph of integrated Gas Chromatography silicon chip manufactured by Terry and co-workers [48]*

The microfabricated GC was small and compact, but performance and peak capacity were poor compared to conventional GC equipment used in the analysis of organic species. Consequently, the "analytical community" showed little interest in this new approach to high-speed, integrated chromatography [84]. The only other example of a microfabricated GC research in the literature is that of Reston and Kolesar [85–87]. Their approach utilises silicon micromachining and integrated circuit processing techniques. The GC system consists of a miniature sample injector, 90 cm rectangular column coated with a copper phthalocyanine (CuPc) stationary phase, and a dual detector (a CuPc coated chemiresistor and a thermal conductivity detector). Silicon micromachining is used to interface the injector and column. These miniaturised GC devices are able to separate NH_3 and NO_2 at ppm concentrations in less than 30 min [86]. More recently, Tjerkestra et al. have reported the fabrication of high performance GC columns [88]. These are made by isotropically etching half-circular channels in silicon substrates. These half-

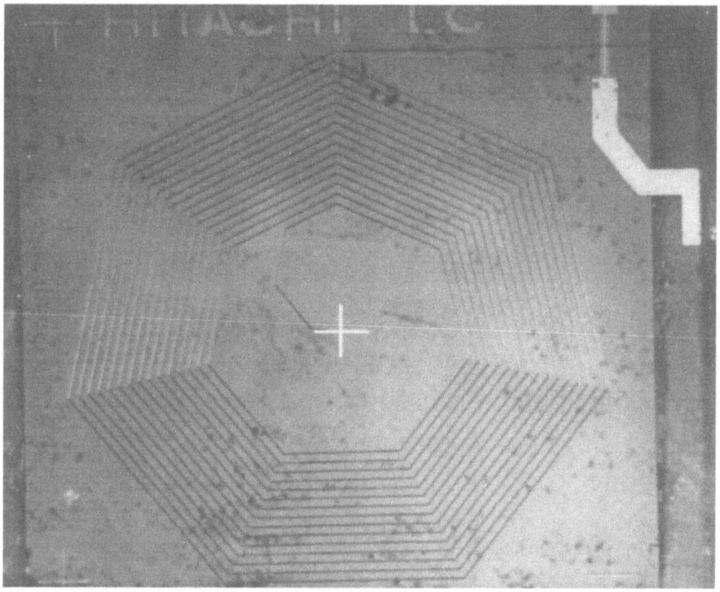

Figure 16 *Micrograph of Liquid Chromatograph microchip manufactured by Manz and co-workers [49]*

circular channels can be bonded on top of each other to yield channels with circular cross-section.

HPLC techniques have also undergone advances through the miniaturisation of component parts. For example, the use of packed and open microcolumns have increased separation efficiencies, but are difficult to implement, and generally result in long analysis times. Open-tubular column LC is an approach designed to reduce analysis times. According to theory, LC in open columns should allow for short analysis times with increased separation efficiencies compared to standard packed-column HPLC [89]. The first demonstration of a microfabricated, liquid-phase separation device was in 1990, by Manz and colleagues at Hitachi Ltd. [49]. They constructed a silicon LC microchip. Integrated on this chip was an open-tubular separation column and a conductometric detector. A micrograph of the LC microchip is shown in Figure 16. The open column has dimensions 6 µm × 2 µm × 15 cm, resulting in a total column volume of 1.5 nL, and the detection cell volume is 1.2 pL. Subsequent efforts at creating a workable HPLC chip demonstrated integration of a split injector, a packed small-bore column, a frit and an optical detector cell onto a silicon chip [90]. In this instance, the separation column volume was 500 nL, and the total dead volume less than 2.5 nL. Figure 17 illustrates a LC separation of fluorescein and acridine orange using this microchip. Cowen and Craston have also reported successful open-tubular LC separations on a silicon chip [91, 92].

At approximately the same time, Jacobson et al. reported open-channel electrochromatography on a glass microchip [93]. The internal surface of a microfabricated channel was chemically modified with octadecylsilane, producing an open-tubular reversed phase LC column. As for CE separations on microchips, EOF was used to load sample, and also to "pump" the mobile phase. Using fluorescence detection, a mixture of coumarin dyes was separated within 150 s at a linear flow rate of 650 µm/s. This study clearly demonstrated the potential of microchip devices in the separation of neutral molecules.

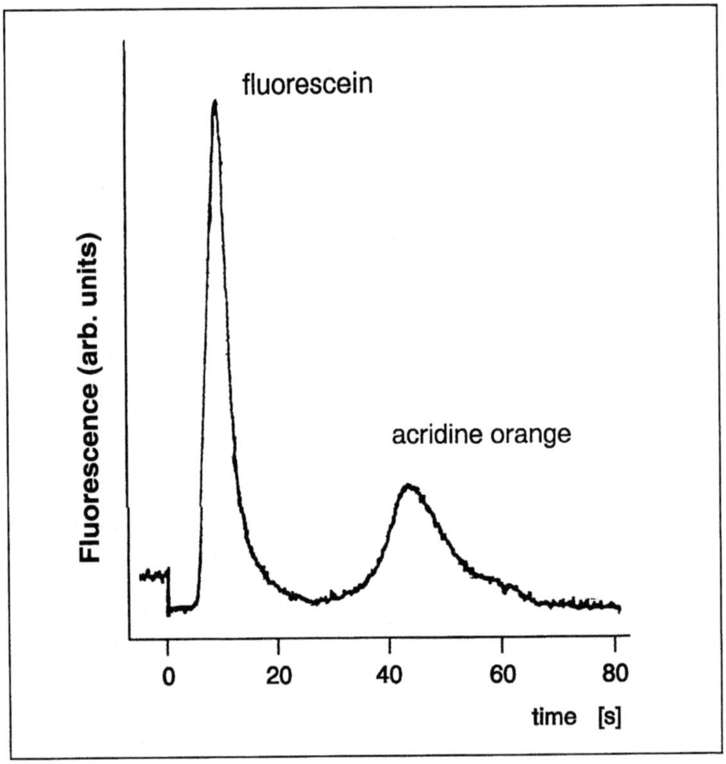

Figure 17 *LC separation of fluorescein and acridine orange using a packed separation channel. Sample: 100 µM fluorescein-Na, 100 µM acridine orange-HCl. Stationary phase: Nucleosil 100-5-C8. Injected volume before split: 60 nL.* F_{col}*: 3.2 µL min*$^{-1}$

Within the last 3 years micellar electrokinetic capillary chromatography (MECC) on planar, glass microchips has been reported by both von Heeren et al. [94] and Moore et al. [95]. In MECC, a surfactant above the critical micelle concentration is added to the running buffer. This allows the separation of uncharged solutes based upon differential partitioning [96]. The first demonstration of MECC on a microchip reported separations of three neutral coumarin dyes. Detection was performed using laser-induced fluorescence. At low applied electric fields highly reproducible separations were obtained. High electric field strengths, resulted in shorter analysis times but poorer separation efficiencies [95]. More recently, von Heeren and co-workers reported a comprehensive study of MECC on a glass microstructure. A cyclic channel geometry was used to effect the separation of six fluorescently labelled amino acids within a few seconds. More significantly, a chip-based MECC immunoassay for serum theophylline was presented. Compared to conventional MECC in fused silica capillaries, MECC microchip analyses were performed 1–2 orders of magnitude faster, with higher efficiencies [94]!

5.5 Detection

The scope of this chapter does not permit a comprehensive survey of all detection protocols used in microseparation science, and indeed detection is addressed in depth elsewhere in this volume. Nevertheless, the chapter would be incomplete without cursory reference to some of the detection modes used by researchers in the field of µ-TAS technology.

The adaptation of a number of detection protocols to measurement in small volumes (and therefore small numbers of species) has accompanied the development of µ-TAS. Detection volumes may be as low as a few hundred attoliters in some microfabricated systems, and the very nature of microseparations means that detection times must also be rapid.

Small volume optical detection is generally based around either absorption or fluorescence measurements. Absorbance is directly proportional to the optical pathlength as well as the sample concentration. Consequently, the major difficulty faced in small volume absorption measurements is the ability to probe small volume cells whilst maintaining a long enough pathlength.

With microfabricated CE chips, this pathlength problem is enhanced (due to reduced channel dimensions) and, to date, fluorescence detection has been the principal optical detection method (in more than 95% of all studies). Although fluorescence is inherently a far more sensitive technique [12], absorbance detection has a much wider applicability (i.e. not all species that absorb radiation fluoresce), and is therefore of importance as a detection protocol. In addition both absorption and fluorescence detection are generally highly specific, and less prone to interferences than their electrochemical counterparts.

Alternative approaches have been developed to enable on-chip absorption measurements. Verpoorte and Manz reported a micromachined absorbance Z-cell for use in liquid chromatography [97]. In this design, the crystal planes of silicon (111) are used to form "micro-mirrors" which effectively extend the optical pathlength. CE microchips, on the other hand, cannot be fabricated in silicon due to adverse electrical characteristics. Hence, an alternative strategy to launch and collect light is required. Liang and co-workers have recently demonstrated the use of a planar, optical U-cell in glass, for both absorption and fluorescence detection [98]. The cell provides a ten-fold improvement in absorbance detection limits by probing the channel in a longitudinal rather than transverse direction. The cell also allows for improvements in the S/N ratio of fluorescence measurements.

Other approaches to integrated optical detection have recently been reported. For example, Hoppe and co-workers have described a novel method of fabricating integrated optical waveguides for use in fluidic microsystem [99]. In addition, Bruno et al. have reported the construction of μ-optics specifically geared for use in μ-TAS. These components include light emitting diodes, gradient index optics and thermo-optical devices [100]. Weigl and co-workers have also developed a novel optical chemical detection protocol for use in silicon flow structures. This device allows for the continuous monitoring of analytes within particle laden solutions, using either fluorescence or absorption spectroscopies [101].

Other generic detection methods have been used in conjunction with μ-TAS. The original gas chromatograph developed by Terry and co-workers incorporated a microfabricated thermal conductivity detector [48, 83]. Furthermore, among the earliest detector cells for liquid-based analysis were devices that combined solid state chemical sensors with small-volume sample chambers defined in silicon [102]. Interestingly, Burggraf has also demonstrated refractive index detection for CE on glass microchips [103]. Although detection limits are poor compared to more conventional protocols, refractive index measurements afford the possibility of truly *universal detection.*

As stated, the primary consideration when choosing a detector for a µ-TAS is that it can probe small volumes. Electrochemical detection protocols have been used within a µ-TAS format for this very reason [104, 105]. Furthermore, the whole electrochemical sensor itself can be made very small. This can be of importance if portability is a system requirement.

A key aspect of chemical analysis is *structure elucidation.* Mass spectrometry (MS) is one of the most effective tools available for this purpose. Very recently electrophoresis microchips have been successfully combined with the technique of electrospray mass spectrometry (ESI-MS) [106, 107]. Karger and co-workers demonstrated high sensitivity (low nanomolar) microchip ESI-MS in the analysis of proteins, and soon after, Ramsey reported a method of generating electrospray from solutions emerging from microchannels etched on planar glass substrates. This approach to detection extends the applicability of µ-TAS to molecules which are non-fluorescent, and leads to the possibility of high-throughput MS analysis in screening and diagnostic applications.

5.6 The "laboratory-on-a-chip" – fact or fiction?

The ultimate µ-TAS will combine all aspects of the analytical process. Component processes such as sampling, sample pre-treatment, sample transport, chemical reaction, product separation, product detection and product isolation will all be performed in an automated manner on a single microchip device. The previous section has summarised the dramatic advances made in "microseparation" technology over the past 7 years. Classical separation methodologies (CE, LC and GC) can now be effectively converted to a microchip format, leading to improvements in analysis time and separation efficiency (a direct result of the miniaturisation process). In addition, other potential µ-TAS components such as reactors, micropumps, sieves, heaters, filters, microvalves and detector cells have all been fabricated and characterised on a micron scale. Consequently, the next step is the integration of these modules to create a complete, integrated analytical system (ideally on a microchip).

To date there have been few real attempts to fabricate a fully integrated μ-TAS (based on a "*react-separate-detect*" principle). As recognised in the previous section, Ramsey and co-workers reported both pre- and post-separation reactions on glass microchips [64, 65]. For the pre-separation reaction, a fast derivatisation was performed by mixing reagents in a dilated section of channel (1 nL volume). This reaction, although extremely simple (i.e. diffusional mixing two reagents at ambient temperature), established the feasibility of integrated reactions and separations on a single microchip. As Ramsey says, the system accomplishes the same things that a laboratory technician or an automated robotic system could do. Specifically, the microchip mixes two reagents, incubates them for a given time, injects the product mixture into an analytical system and presents the results to a digitising detector. All these processes are performed at an extremely high level of precision.

Many biological and clinical problems will benefit from the ability to perform rapid, automated analyses on minute quantities of material in a high-throughput fashion. As noted, fulfilment of the Human Genome Project will require an increase in processing speed of up to 500-fold by the year 2000 [9]. Consequently, miniaturised, integrated DNA analysis systems should provide a route to faster, cheaper analyses. Demonstration of an integrated device of this kind was first provided by Jacobson et al. in 1996 [108]. An integrated monolithic device that performs restriction digestion followed by an electrophoretic sizing of the products was fabricated in glass using standard photolithographic methods. The DNA plasmid *pBR322* was mixed with the *Hin*fI restriction enzyme in a 700 pL reaction chamber. Subsequent to digestion, the DNA fragments are injected into a 67 mm separation channel and sized. Digestion of the pBR322 plasmid by *Hin*fI, and fragment analysis were completed within 300 s. The success of this chip based DNA analysis indicated the possibility of miniaturising more sophisticated biochemical procedures.

The polymerase chain reaction (PCR) has revolutionised the biological and medical sciences since it was formally introduced at the Cold Spring Harbor 51st Symposium on Quantitative Biology [109]. The PCR process allows virtually any nucleic acid sequence to be generated *in vitro* in abundance. This is done by repeated cycles of heating and cooling of sample material and enzyme in a reaction chamber. Although PCR is simpler, faster and more flexible than traditional cloning techniques, it is limited by instrumental technology. Conventional thermal cycling systems require long cycling times due to their large thermal masses (approximately 2–3 h for a 30 cycle amplification). In addition, the physical size of thermal cyclers has forced PCR to remain a laboratory technique. A number of microfabricated PCR devices have been proposed in the last 4 years. PCR amplification has been performed in 4–12 µL silicon/glass reaction chambers placed in larger thermal cyclers [110], in 20–50 µL microfabricated silicon chambers with integrated heaters and temperature control [111], and on a 25 µL drop of solution above a microfabricated heater [112]. All these devices call for fast, portable PCR.

The integration of miniaturised PCR amplification and ultrafast DNA analysis (two fundamentally different processes) on a single microchip device would allow for vastly reduced analysis times, instrumental portability, reduced sample requirements and high-throughput. All these characteristics are critical for the completion of the Human Genome Project. In addition, a device of this kind could genuinely be termed a *laboratory-on-a-chip.*

A first generation device of this kind has recently been reported by Woolley et al. [113]. An integrated PCR-CE microdevice was fabricated and shown to perform ultra-fast amplification followed by electrophoretic DNA analysis, with a complete absence of manual sample transfer. Figure 18 shows a schematic of the integrated PCR-CE microdevice. The CE manifold is a glass microchip containing microfabricated electrophoresis channels (separation channel – 8 µm × 100 µm × 46 mm filled with a HEC sieving matrix). The microfabricated PCR reactor has an etched silicon reaction chamber, LPCVD doped polysilicon heaters, and gold/titanium heater contacts. Disposable polypropylene liners were inserted into the reactor to minimise contamination and adsorption effects. The PCR chamber and CE manifold were coupled at one of the sample reservoirs. Electrophoretic injection directly from the PCR chamber through the injection channel was used as an "electrophoretic valve". The functionality of the system was initially demonstrated by amplifying a 268 bp β-globin target cloned in M13. A 30 cycle PCR amplification followed by electrophoretic separation, and fluorescence detection took less than 20 min! In addition, a rapid assay for genomic *Salmonella* DNA was

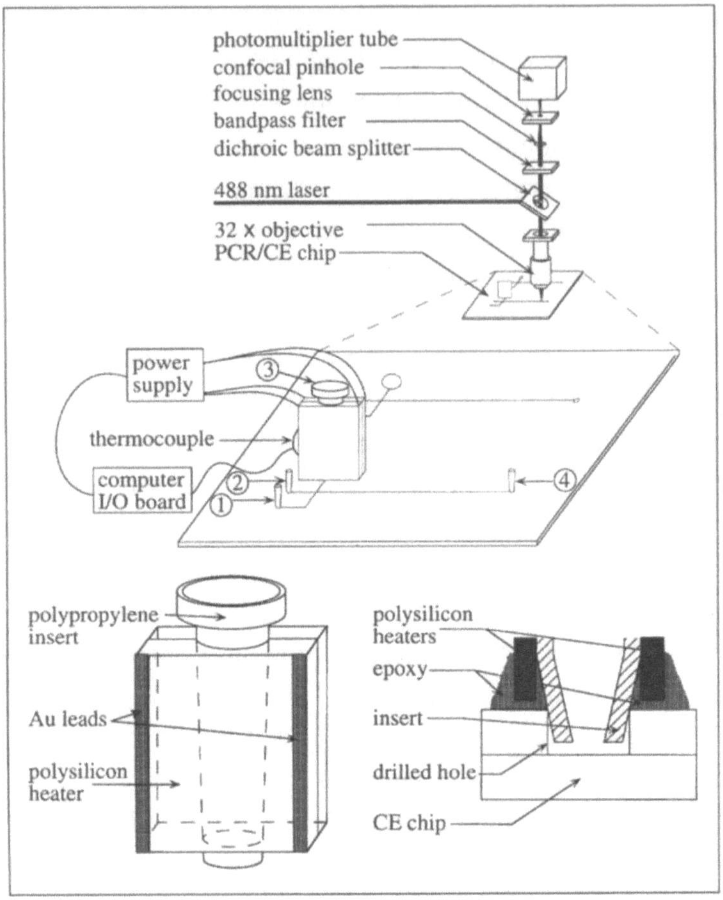

Figure 18 *Schematic of an integrated PCR-CE microdevice. (A) Laser-excited confocal fluorescence detection apparatus and an integrated PCR-CE microdevice. (B) Expanded view of the microfabricated PCR chamber. (C) Expanded cross-sectional view of the junction between the PCR and CE devices*

performed in under 45 min (Fig. 19). Finally, the authors established the feasibility of real-time monitoring of PCR target amplification. The demonstration of a functional integrated PCR-CE microdevice is a significant step toward the complete integration of DNA analysis, and more generally chemical analysis.

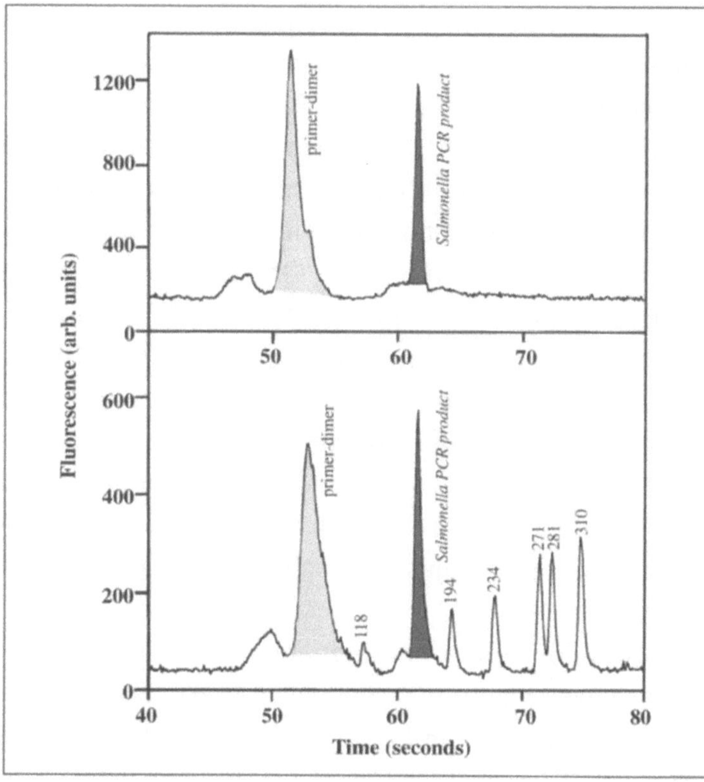

Figure 19 *High-speed integrated PCR-CE microdevice assay of genomic Salmonella DNA. (Top) Chip CE separation of the Salmonella PCR product was performed immediately following a 39-min PCR amplification in the integrated PCR-CE microdevice. The primer-dimer peak (light grey) appears at 51 s and the PCR product peak (dark grey) appears at 61 s. Total analysis time for the Salmonella sample using the integrated PCR-CE microdevice was under 45 min. (Bottom) Sizing of the Salmonella PCR product (1 : 100 dilution) using ϕX174 HaeIII DNA (1 ng/µL) in a separate CE chip*

5.7 Conclusions

The title of a review article over 6 years ago posed the following question: "Miniaturisation of Chemical Analysis Systems – A look into Next Century's Technology or Just a Fashionable Craze"? [1]. At the time, the answer was not an obvious one. Fundamental diffusional and hydrodynamic theories suggested that *faster* and *more efficient* separations would result, by miniaturising both chromatographic and electrophoretic processes [114]. In addition, miniaturisation of the analytical process reduced reagent requirements and afforded the possibility of fully integrated and automated systems. However, little experimental work had been undertaken to address these theoretical claims, and thus the authors were incapable of answering their own question. Six years on, and we can now begin to answer that question. Miniaturisation is surely not a fashionable craze? Or if it is, it is one which is standing the test of time! However, it is still difficult to foresee the precise impact that μ-TAS will have on the analytical sciences of the 21st century.

The ultimate challenges are clear: Can a complete analytical procedure be miniaturised? In addition, can the component processes be integrated on a single device (most likely a microchip)? And finally, can the operation of the complete process be automated? The research summarised in the preceding sections demonstrate technological advances in all of these three areas.

The development of microseparation technology has been the area of most obvious development over the past 6 years. The separation of both charged and neutral species can be now effected using chip-based formats of many techniques. These include CE, SCCE, FFE, micellar electrokinetic chromatography, open-channel electrochromatography and packed bed, liquid chromatography. Through direct experiment, the theoretical predictions about separation speed and efficiency have now been validated, and the use of μ-TAS as a highly efficient alternative to their conventional counterparts can be justified. In addition, the gains in separation speed and efficiency now permit CE microchips to be used in similar ways to chemical sensors for certain applications [61].

Integration of component processes will be key to the continued development of μ-TAS. In many ways the "separation" element of the analytical process is now well defined. For example, the construction and operation of a CE microchip is now a simple task for many research groups. The real challenge is to link in the sample pre-treatment, detection and data processing to the separation stage. Of these, sample pre-treatment appears to hold the most scope for development. Although sample manipulation through electrokinetic pumping mechanisms is well understood, the analysis of real samples will require sophisticated handling steps, such as filtration [44], mixing [47], heating [113]. Many of these components have been reported individually, but very few studies have approached analysis in a truly integrated manner.

Fortunately, more contemporary research is now beginning to address the issue of integration. As shown previously, microchip devices for immunoassays [81] and DNA restriction fragment analysis [108] have been reported. However, the recent demonstration of an integrated PCR-CE microchip is probably the most significant step towards the true *laboratory-on-a-chip* [113]. This integrated device performs a diagnostically relevant and sophisticated reaction followed by the separation and detection of reaction products, with the absence of manual transfer. The only point of human contact is the loading of the reactants into a reaction chamber at the very start. This is significant since the whole analytical process then becomes automated. The final steps in creating a fully integrated/automated device will be the integration of all sample handling and product detection on a single microdevice. This may not be too long away.

The generality of the miniaturised analysis system concept makes it applicable to many areas of science and technology. These include industrial process control, environmental analysis, medical and clinical diagnostics and forensics. At the start of the decade the clinical diagnostics market alone was over $ 9 billion US dollars annually, and the market for chemical sensors was estimated at around $ 5 billion US dollars [2]. These figures give an idea of the size of the potential markets that µ-TAS can tap into. Indeed considerable corporate research is currently focused on the development of µ-TAS technology. Large scale business activities include those of Caliper Technologies, Affymetrix, Ciba-Geigy, Soane Biosystems, Nanogen, Hitachi, Orchid, Shimadzu Scientific Instruments and PerSeptive Biosystems.

In conclusion, it is fair to say that the transition between the conventional, "bench-top" approach to chemical analysis and the µ-TAS described in this chapter will undoubtedly prove to be as significant an advancement as the microelectronic revolution was over three decades ago. However, to finish on a more futuristic note, imagine a hapless analytical chemist who was lost at sea in the year 1990 *anno domini*. After many adventures and trials he at last finds his way back to his native laboratory, 20 years later. The job of today's µ-TAS researcher will be to make him feel like Gulliver landing on the Island of Lilliput and exclaiming "what on earth are all these tiny, strange objects!"

References

1 Manz A, Harrison DJ, Verpoorte EMJ, Fettinger JC, Lüdi H, Widmer HM (1991) Miniaturization of chemical analysis systems – a look into next century's technology or just a fashionable craze? *Chimia* 45: 103–105

2 Manz A, Harrison DJ, Verpoorte EMJ, Widmer HM (1993) Planar chips technology for miniaturization of separation systems; a developing perspective in chemical monitoring. *Advances in Chromatography* 33: 1–66

3 Baker DR (1995) *Capillary electrophoresis*. Wiley Interscience, New York

4 Tsuda T, Nomura K, Nakagawa G (1982) Open tubular microcapillary liquid-chromatography with electroosmosis flow using a UV detector. *J Chromatog* 248: 241–247

5 Widmer HM (1996) A survey of the trends in analytical chemistry over the last twenty years, emphasising the development of TAS and µ-TAS. Analytical Methods and Instrumentation Special Issue µ-TAS '96, pp 3–8

6 Monnig CA, Jorgenson JW (1991) On-column sample gating for high-speed capillary zone electrophoresis. *Anal Chem* 63: 802–807.

7 Murphy MJ, Fushimi F, Parchment RE, Barberaguillem E (1995) Automated imaging and quantitation of tumor-cells and CFU-GM colonies in microcapillary cultures – toward therapeutic index-based drug screening. *Investigational New Drugs* 13: 303–314

8 Smith LM (1993) The future of DNA sequencing. *Science* 262: 530–531

9 Collins F, Galas D (1993) A new five-year Plan for the US human genome project. *Science* 262: 43–46

10 Floyd CD, Lewis CN, Whittaker M (1996) More leads in the haystack. *Chemistry in Britain* 32: 31–35

11 Ramsey JM, Jacobson SC, Knapp MR (1995) Microfabricated chemical measurement systems. *Nature Medicine* 1, pp 1093–1096

12 Barnes MD, Whitten WB, Ramsey JM (1995) Detecting single molecules in liquids. *Anal Chem* 67: 418A–423A

13 Elson EL, Magde D (1974) Fluorescence correlation spectroscopy. I. Conceptual basis and theory. *Biopolymers* 13: 1–27

14 Manz A, Verpoorte EMJ, Raymond DE, Effenhauser CS, Burggraf N, Widmer HM (1995) µ-TAS: Miniaturised chemical analysis systems. In: A van den Berg, P Bergveld P. *Micro total analysis systems*. Kluwre Dordrecht, pp 5–27

15 Manz A (1996) The secret behind electrophoresis microstructure design. Analytical methods and instrumentation special issue µTAS '96, pp 28–30

16 Wolffenbuttel RF (1992) Silicon micromachining for integrated radiant sensors. *Sensors & Actuators A* 30: 109–115

17 Holland L (1965) Thin film microelectronics. Chapman and Hall London, pp xi–xii

18 Moreau D, Baptist R, Peccoud L (1996) Silicon microtip fabrication techniques. *Vide-Science Technique et Applications* 52: 463

19 Menz W (1996) LIGA and related technologies for industrial application. *Sensors and Actuators A* 54: 785–789

20 Ko WH (1996) The future of sensor and actuator systems. *Sensors and Actuators A* 56: 193–197

21 Sen M, Wajerski D, Gadelhak M (1996) A novel pump for MEMS applications. *J Fluid Eng* 118: 624–627

22 Lang W (1996) Silicon microstructuring technology. *Materials Science & Engineering Reports* 17: 1–55

23 Kovacs GTA, Petersen K, Albin M (1996) Silicon micromachining sensors to systems. *Anal Chem* 68: 407A–412A
24 Effenhauser CS, Bruin GJM, Paulus A, Ehrat M (1996) Detection of single DNA molecules and DNA fragment analysis in molded silicone elastomer microchips. Analytical Methods & Instrumentation Special Issue µ-TAS '96, pp 124–125
25 Kim E, Xia Y, Whitesides GM (1995) Polymer microstructures formed by moulding in capillaries. *Nature* 376: 581–584
26 Doyle LE (1985) Manufacturing processes and materials for engineers. Prentice Hall, Englewood Cliffs, NJ
27 Runyan WR, Bean KE (1990) Semiconductor integrated circuit processing technology. Addison-Wesley Reading, MA
28 Smits JG (1990) Piezoelectric micropump with three valves working peristaltically. *Sensors and Actuators A* 21: 203–206
29 Van Lintel HTG, Van de Pol FCM, Bouwstra S (1988) A piezoelectric micropump based on micromachining of silicon. *Sensors and Actuators* 15: 153–167
30 Shoji S, Nakagawa S, Esashi M (1990) Micropump and sample injector for integrated chemical analyzing systems. *Sensors and Actuators A* 21–A23: 189–192
31 Zengerle R, Ulrich J, Kluge S, Richter M, Richter A (1995) A bi-directional silicon micropump. *Sensors and Actuators A* 50: 81–86
32 Esashi M (1990) Integrated micro flow control systems. *Sensors and Actuators A* 21–23: 161–167
33 Ohnstein T, Fukiura T, Ridley R, Bone U (1990) Micromachined silicon microvalve. Proc IEEE Micro Electro Mechanical Systems Napa Valley, pp 95–98
34 Bosch D, Heimhofer B, Muck G, Seidel H, Thumser U, Welser W (1993) A silicon microvalve with combined electromagnetic/electrostatic actuation. *Sensors and Actuators A* 37–A38: 684–692
35 Richter M, Prak A, Naundorf J, Eberl M, Leeuwis H, Woias P, Steckenborn A (1996) Development of a micro-fluid system as a demonstrator for a µ-TAS. Analytical Methods & Instrumentation Special Issue µTAS '96, pp 129–137
36 Zengerle R, Stehr M, Freygang M, Haffner H, Messner S, Rossberg R, Sandmaier H (1996) Microfabricated devices and systems for handling liquids and gases. Analytical Methods & Instrumentation Special Issue µTAS '96, pp 91–94
37 Shoji S, Esashi M, van der Schoot B, de Rooij N (1992) A study of a high-pressure micropump for integrated chemical analyzing systems. *Sensors and Actuators A* 1 –A3: 335–339
38 Seiler K, Zhonghui HF, Fluri K, Harrison DJ (1994) Electroosmotic pumping and valveless control of fluid flow within a manifold of capillaries on a glass chip. *Anal Chem* 66: 3485–3491
39 Yanagisawa K, Kuwano H, Tago A (1993) An electromagnetically driven microvalve. Proc Transducers '93, pp 102–105
40 Freygang M, Haffner H, Messner S, Schmidt B (1995) A new concept of a bimetallically actuated, normally-closed microvalve. Proc. Transducers '95, pp 73–34
41 Minami K, Sim DY, Kurabayashi T, Esashi M (1996) Bakeable pneumatic microvalve for advanced semiconductor processing. Analytical Methods & Instrumentation Special Issue µTAS '96, pp 65–71
42 Branebjerg J, Fabius B, Gravesen P (1995) Application of miniature analyzers: from microfluidic components to µTAS in micro total analysis systems (van den Berg, A, Bergveld, P.) Kluwre Dordrecht, pp 141–151
43 de Mello AJ (1996) Unpublished results
44 Larsen UD, Blankenstein G, Branebjerg J (1996) A novel design for chemical and bio-

chemical liquid analysis system. Analytical Methods & Instrumentation Special Issue µ-TAS '96, pp 113–115
45 Fluitman JH, van den Berg A, Lammerink TS (1995) Micromechanical components for µTAS. In micro total analysis systems (van den Berg, A, Bergveld, P.) Kluwre Dordrecht, pp 73–83
46 Miyake R, Lemmerink TSJ, Elwenspoek M, Fluitman JHJ (1993) Micro mixer with fast diffusion. Proc. MEMS-Worksop 1993, pp 248–253
47 Larsen UD, Branebjerg J, Blankenstein G (1996) Fast mixing by parallel multilayer lamination. Analytical Methods & Instrumentation Special Issue µTAS '96, pp 228–230
48 Terry SC, Jerman JH, Angell JB (1979) A gas chromatographic air analyser fabricated on a silicon wafer. *IEEE Trans Electron Devices* 26: 1880–1886
49 Manz A, Miyahara Y, Miura J, Watanabe Y, Miyagi H, Sato K (1990) Design of an opentubular column liquid chromatograph using silicon chip technology. *Sensors and Actuators B*1: 249–255
50 Manz A, Graber N, Widmer HM (1990) Miniaturised total analysis systems: A novel concept for chemical sensors. *Sensors and Actuators B*1: 244–248
51 Huang XHC, Quesada MA, Mathies RA (1992) DNA Sequencing using capillary array electrophoresis. *Anal Chem* 64: 2149-–2154
52 Manz A, Fettinger JC, Verpoorte E, Lüdi H, Widmer HM, Harrison DJ (1991) Micromachining of monocrystalline silicon and glass for chemical analysis systems – a look into next century's technology or just a fashionable craze? *Trends in Anal Chem* 10: 144–149
53 Effenhauser CS, Manz A, Widmer HM (1993) Glass chips for high-speed capillary electrophoresis separations with submicrometer plate heights. *Anal Chem* 65: 2637–2642
54 Manz A, Harrison DJ, Verpoorte EMJ, Fettinger JC, Paulus A, Lüdi H, Widmer HM (1992) Planar chips technology for miniaturization and integration of separation techniques into monitoring systems. *J Chromatography* 593: 253–258
55 Harrison DJ, Manz A, Fan Z, Lüdi H, Widmer HM (1992) Capillary electrophoresis and sample injection systems integrated on a planar glass chip. *Anal Chem* 64: 1926–1932
56 Harrison DJ, Fan Z, Seiler K, Manz A, Widmer HM (1993) Rapid separation of fluorescein derivatives using a micromachined capillary electrophoresis system. *Analytica Chimica Acta* 283: 361–366
57 Seiler K, Harrison DJ, Manz A (1993) Planar glass chips for capillary electrophoresis: Repetitive sample injection, quantitation and separation efficiency. *Anal Chem* 65: 1481–1488
58 Harrison DJ, Fluri K, Seiler K, Fan Z, Effenhauser CS, Manz A (1993) Micromachining a miniaturised capillary electrophoresis-based chemical analysis system on a chip. *Science* 261: 895–897
59 Harrison DJ, Glavina PG, Manz A (1993) Towards miniaturised electrophoresis and chemical analysis systems on silicon: An alternative to chemical sensors. *Sensors and Actuators B*10: 107–116
60 Jacobson SC, Hergenröder R, Koutny LB, Warmack RJ, Ramsey JM (1994) Effects of injection schemes and column geometry on the performance of microchip electrophoresis devices. *Anal Chem* 66: 1107–1113
61 Jacobson SC, Hergenröder R, Koutny LB, Ramsey JM (1994) High-speed separations on a microchip. *Anal Chem* 66: 1114–1118
62 Manz A, Effenhauser CS, Burggraf N, Harrison DJ, Seiler K, Fluri K (1994) Electroosmotic pumping and electrophoretic

separations for miniaturised chemical analysis systems. *J Micromech Micro Eng* 4: 257–265

63 Jacobson SC, Moore AW, Ramsey JM (1995) Fused quartz substrates for microchip electrophoresis. *Anal Chem* 67: 2059–2063

64 Jacobson SC, Koutny LB, Hergenröder R, Moore AW, Ramsey JM (1994) Microchip capillary electrophoresis with an integrated postcolumn reactor. *Anal Chem* 66: 3472–3476

65 Jacobson SC, Hergenröder R, Moore AW, Ramsey JM (1994) Precolumn reactions with electrophoretic analysis integrated on a microchip. *Anal Chem* 66: 4127–4132

66 Burggraf N, Manz A, Verpoorte E, Effenhauser CS, Widmer HM, de Rooij NF (1994) A novel approach to ion separations in solution: Synchronized cyclic capillary electrophoresis (SCCE). *Sensors and Actuators B*20: 103–110

67 Manz A, Verpoorte E, Effenhauser CS, Burggraf N, Raymond DE, Widmer HM (1994) Planar chip technology for capillary electrophoresis. *Fresnius J Anal Chem* 348: 567–571

68 Burggraf N, Manz A, Effenhauser CS, Verpoorte E, de Rooij NF, Widmer HM (1993) Synchronized cyclic capillary electrophoresis – a novel approach to ion separations in solution. *J High Res Chromatography* 16: 594–596

69 Woolley AT, Mathies RA (1994) Ultra-high-speed DNA fragment separations using microfabricated capillary array electrophoresis chips. *Proc Natl Acad Sci* 91: 11348–11352

70 Effenhauser CS, Paulus A, Manz A, Widmer HM (1994) High-speed separation of antisense oligonucleotides on a micromachined capillary electrophoresis device. *Anal Chem* 66: 249–253

71 Woolley AT, Mathies RA (1995) Ultra-high-speed DNA sequencing using capillary array electrophoresis chips. *Anal Chem* 67: 3676–3680

72 Ju JY, Kheterpal I, Scherer JR, Ruan CC, Fuller CW, Glazer AN, Mathies RA (1995) Design and synthesis of fluorescence energy-transfer dye-labeled primers and their application for DNA sequencing and analysis. *Anal Biochem* 231: 131–140

73 Woolley AT, Sensabaugh G, Mathies RA (1997) High-speed DNA genotyping using microfabricated capillary array electrophoresis chips. *Anal Chem*; 69: 2181–2186

74 Camilleri P, Okafo GN, Southan C (1991) Separation by capillary electrophoresis followed by dynamic elution. *Anal Biochem* 196: 178–182

75 Eriksson KO, Palm A, Hjerten S (1992) Preparative capillary electrophoresis based on adsorption of the solutes (proteins) onto a moving blotting membrane as they migrate out of the capillary. *Anal Biochem* 201: 211–215

76 Effenhauser CS, Manz A, Widmer HM (1995) Manipulation of sample fractions on a capillary electrophoresis chip. *Anal Chem* 67: 2284–2287

77 Raymond DE, Manz A, Widmer HM (1994) Continuous sample pre-treatment using a free-flow electrophoresis device integrated onto a silicon chip. *Anal Chem* 66: 2858–2865

78 Mesaros JM, Luo G, Roeraade J, Ewing AG (1993) Continuous electrophoretic separations in narrow channels coupled to small-bore capillaries. *Anal Chem* 65: 3313–3319

79 Raymond DE, Manz A, Widmer HM (1996) Continuous separation of high molecular weight compounds using a microliter volume free-flow electrophoresis microstructure. *Anal Chem* 68: 2515–2522

80 Koutny LB, Schmalzing D, Taylor TA, Fuchs M (1996) Microchip electrophoresis

immunoassay for serum cortisol. *Anal Chem* 68: 18–22
81 Chiem N, Harrison DJ, (1997) Micro-chip based capillary electrophoresis for immunoassays: Analysis of monoclonal antibodies and theophylline. *Anal Chem* 69: 373–378
82 Skoog DA, Leary JJ (1992) Principles of instrumental analysis 4th ed. Saunders 605–669
83 Terry SC (1975) A gas chromatographic air analyser fabricated on a silicon wafer using integrated circuit technology. Ph.D. Dissertation Stanford University
84 Some commercial products do exist. For example, *Chrompack (NL)* manufacture microfabricated GC components
85 Reston RR, Kolesar ES (1994) Silicon-micromachined gas-chromatography system used to separate and detect ammonia and nitrogen dioxide. 1. Design, fabrication and integration of the gas-chromatography system. *J. Microelectromechanical Systems* 3: 134–146
86 Kolesar ES, Reston RR (1994) Silicon-micromachined gas-chromatography system used to separate and detect ammonia and nitrogen dioxide. 1. Evaluation, analysis and theoretical modeling of the gas-chromatography system. *J Microelectromechanical Systems* 3: 147–154
87 Kolesar ES, Reston RR (1994) Silicon-micromachined gas-chromatography system with a thin-film copper phthalocyanine stationary-phase and its resulting performance. *Surface and Coatings Tech* 68: 679–685
88 Tjerkstra RW, Gardeniers JGE, van den Berg A, Elwenspoek MC (1996) Isotropically etched channels for gas chromatography. Analytical Methods & Instrumentation Special Issue µ-TAS '96 247–249
89 Knox JH, Gilbert MT (1979) Kinetic optimisation of straight open tubular liquid chromatography. *J Chromatography* 186: 405–418
90 Ocvirk G, Verpoorte E, Manz A, Grasserbauer M, Widmer HM (1995) High performance liquid chromatography partially integrated onto a silicon chip. *Anal Methods & Instrum* 2: 74–82
91 Cowen S, Craston DH (1996) Initial stages in the development of a miniature integrated liquid analysis system. Analytical Methods & Instrumentation Special Issue µ-TAS '96 196
92 Cowen S. Craston DH, Wasson E (1994) An on-chip miniature liquid-chromatography system with on-line electrochemical detection. Deauville Conference, Montreux. Oral presentation
93 Jacobson SC, Hergenröder R, Koutny LB, Ramsey JM (1994) Open channel electrochromatography on a microchip. *Anal Chem* 66: 2369–2373
94 von Heeren F, Verpoorte E, Manz A, Thormann W (1996) Micellar electrokinetic chromatography separations and analyses of biological samples on a cyclic planar microstructure. *Anal Chem* 68: 2044–2053
95 Moore AW, Jacobson SC, Ramsey JM (1995) Microchip separations of neutral species via micellar electrokinetic capillary chromatography. *Anal Chem* 67: 4184–4189
96 Terabe S, Otsuka K, Ando T (1985) Electrokinetic chromatography with micellar solution and open-tubular capillary. *Anal Chem* 57: 834–841
97 Verpoorte E, Manz A, Lüdi H, Bruno AE, Maystre F, Krattiger B, Widmer HM, van der Schoot BH, de Rooij NF (1992) A silicon flow cell for optical detection in miniaturised total chemical analysis systems. *Sensors and Actuators B*6: 66–70
98 Liang Z, Chiem N, Ocvirk G, Tang T, Fluri K, Harrison DJ (1996) Microfabrication of a planar absorbance and fluorescence cell for integrated capillary electrophoresis devices. *Anal Chem* 68: 1040–1046

99 Hoppe K, Svalgaard M, Kristensen M (1996) Integrated optical waveguides in fluidic microsystems. Analytical Methods & Instrumentation Special Issue µ-TAS '96 164–166
100 Bruno AE, Barnard S, Krattiger B, Ehrat M, Völkel R, Nussbaum P, Herzig HP, Dändliker R (1996) µ-Optics for µ-TAS. Analytical Methods and Instrumentation Special Issue µ-TAS '96 163
101 Weigl BH, Holl MR, Schutte D, Brody JP, Yager P (1996) Diffusion based optical chemical detection in silicon flow structures. Analytical Methods & Instrumentation Special Issue µ-TAS '96 174–184
102 van de Schoot B, Bergveld P (1985) An ISFET-based microliter titrator – integration of a chemical sensor/actuator system. *Sensors and Actuators* 8: 11–22
103 Burggraf N PhD Thesis (1995) Synchronized cyclic capillary electrophoresis. Université de Neuchâtel
104 van der Schoot B, Jeanneret S, van den Berg A, de Rooij NF (1993) A modular miniaturised chemical analysis system. *Sensors and Actuators B*13–B14: 333–335
105 Verpoorte EM, van der Schoot B, Jeanneret S, Manz A, Widmer HM, de Rooij NF (1994) Three-dimensional micro flow manifolds for miniaturised chemical analysis systems. *J Micromech Microeng* 4: 246–256
106 Xue Q, Foret F, Dunayevskiy YM, Zavracky PM, McGruer NE, Karger BL (1997) Multichannel microchip electrospray mass spectrometry. *Anal Chem* 69: 426–430
107 Ramsey RS, Ramsey JM (1997) Generating electrospray from microchip devices using electroosmotic pumping. *Anal Chem* 69: 1174–1178
108 Jacobsen SC, Ramsey JM (1996) Integrated microdevice for DNA restriction fragment analysis. *Anal Chem* 68: 720–723
109 Mullis K, Faloona F, Scharf S, Snikl R, Horn G, Erlich H (1986) Specific amplifications of DNA *in vitro*: The polymerase chain reaction. *Cold Spring Harbor Symp Quant Biol* 51: 260
110 Cheng J, Shoffner MA, Hvichia GE, Kricka LJ, Wilding P (1996) Chip PCR. 2. Investigation of different PCR amplification systems in microfabricated silicon-glass chips. *Nucleic Acids Research* 24: 380–385
111 Northrup MA, Gonzalez C, Hadley D, Hills RF, Landre P, Lehew S, Saiki R, Sninsky JJ, Watson R, Watson R Jr (1995) In Digest of Technical Papers: Transducers 1995: IEEE New York 1 764–767
112 Burns MA, Mastrangelo CH, Sammarco TS, Man FP, Webster JR, Johnson BN, Foerster B, Jones D, Fields Y, Kaiser AR, Burke DT (1996) Microfabricated structures for integrated DNA analysis. *Proc Natl Acad Sci USA* 93: 5556–5561
113 Woolley AT, Hadley D, Landre P, DeMello AJ, Mathies RA, Northrup MA (1996) Functional integration of PCR amplification and capillary electrophoresis in a microfabricated DNA analysis device. *Anal Chem* 68: 4081–4086
114 Manz A, Graber N, Widmer HM (1990) Miniaturised total chemical analysis systems: A novel concept for chemical sensing. *Sensors and Actuators B*1: 244–248

6 Rapid multisample PCR in miniaturized ultrathin-walled microwell plates

Alexander N. Tretyakov, Rimma A. Pantina, Oleg K. Kaboev and Hans Peter Saluz

From its first-published account, the polymerase chain reaction (PCR) has become a standard research tool in a wide range of laboratories (Mullis and Faloona, 1987; Saiki et al., 1988). Initially described as a tool to "find a needle in a haystack and to subsequently produce a pile of needles from the hay," PCR has been transformed into a myriad array of methods and diagnostic assays. Its impact is felt in basic molecular biology, clinical research, evolutionary studies and the various genome sequencing projects. Therefore, it is not surprising that the demands for PCR are still increasing and that the need of rapid, inexpensive and efficient high-throughput thermocycling systems that can be coupled with rapid post-PCR processing of multiple samples becomes more and more apparent.

At present various systems based on temperature cycling of small samples in glass capillary tubes (Wittwer et al., 1990) or silicon-glass microfabricated reactors (Northrup et al., 1993) are used as rapid alternatives to conventional heat block temperature cycling procedures in plastic tubes or microplates (Wilding et al., 1994; Burns et al., 1996; Cheng et al., 1996; Shoffner et al. 1996; Kalinina et al., 1997; Taylor et al., 1997; Wittwer et al., 1997; Kopp et al., 1998). The surface to volume ratio of the above reactors is 5- to 13-fold increased compared with conventional plastic PCR tubes (Shoffner et al., 1996). In addition, the thermal conductivity of silicon and glass containers is better than the one of plastic materials. Therefore, some of these systems can be used to perform up to 30 amplification cycles in less than 15 min. This is several-fold faster than conventional heat block cycling in plastic tubes or microplates. However, these rapid DNA amplification techniques are also connected with various disadvantages. For example, in these techniques the samples are handled and sealed one-by-one, sometimes in a relatively cumbersome manner due to the special features of the microreactors. The delivering and recovering of the small samples is more complicated compared to conventional plastic tubes or microplates. Therefore, the experimental throughput using the above systems is limited. In addition, the increased glass or silicon surface of the reactors actively adsorbs components of the reaction mixture and can inactivate the reaction. Therefore, in order to minimize this effect, the silicon surfaces have to be treated (Shoffner et al., 1996) and in addition, bovine serum albumin (BSA) has to be added to the reaction mixture (Wittwer and Garling, 1991). The PCR products are accumulated at local parts of the highly expanded reactors (i.e. capillaries, chips) and saturate the reaction locally (Kalinina et al., 1997). This may explain the reduction of sensitivity of the techniques compared to conventional heat block temperature cycling (Chapin and Lauderdale, 1997). Furthermore, the price of such reactors is relatively high. Therefore, we tested alternative sample containers and could show that cheep plastic reactors, such as pipette tips, could be used for rapid heat block temperature cycling (Tretyakov et al., 1994). By this means DNA could be efficiently amplified in 15 min or less. However, the handling and sealing of such sample reactors was still too complicated for routine use. Here we describe a new technique for rapid temperature cycling of small samples in miniaturized ultrathin-walled microwell plates vacuum-formed directly on the heat block of rapid thermal cycler. The main advantage of this technique is the use of arrayed microreactors, i.e. microwell plates that allow an easy access to deliver and recover small samples. In addition, it is suitable for parallel multisample processing. Two unrelated technologies are integrated in this system: first, vacuum forming of three-dimensional structures of cheep plastic sheets and second, rapid heat block temperature cycling.

The rapid heat block thermal cycler (Fig. 1) operates as a compact vacuum forming machine and has been constructed according to an earlier concept of ours in which the block is heated by intense light and cooled by circulating water (Tretyakov et al., 1994). Rapid, uniform heating of the block is achieved by means of a transilluminator-like "optical heater." The cooling is obtained by intense, turbulent circulation of water through tiny channels in the body of the block using a high-speed water pump. Five standard double-ended halogen lamps (500 W) have been arranged in a specially designed aluminium-mirror, air-cooled reflector. The power of the "optical heater" is 2500 W by this means. An efficient water-to-air heat exchanger has been used to draw off the heat load into the surrounding air. The channels have been arranged in two balanced fluid counter-circuits (not shown) in order to guarantee a

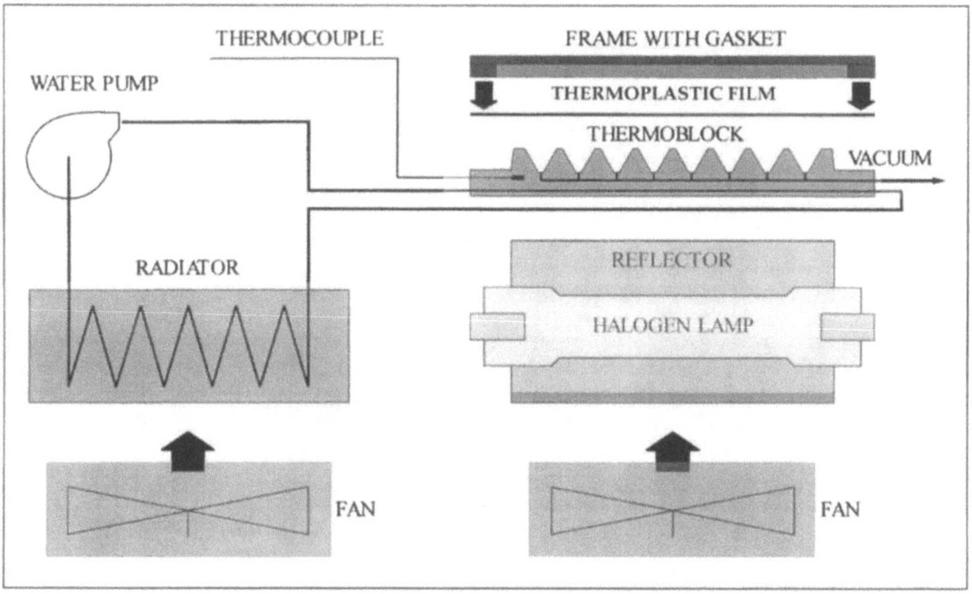

Figure 1 *Diagram of the rapid cycling vacuum-forming system. This rapid heat block thermal cycler operates as a compact vacuum-forming machine and has been constructed according to an earlier concept of ours in which the block is heated by intense light and cooled by circulating water (Tretyakov et al., 1994. Rapid, uniform heating of the block is achieved by means of a transilluminator-like "optical heater." The cooling is obtained by intense, turbulent circulation of water through tiny channels in the body of the block using a high-speed water pump. Five standard double-ended halogen lamps (500 W) are arranged in a specially designed aluminium-mirror, air-cooled reflector. The maximum heating rate of the block is 15°C/s and the cooling rate is 15°C/s. An efficient water-to-air heat exchanger is used to draw off the heat load into the surrounding air. The channels are arranged in two balanced fluid counter-circuits (not shown) in order to guarantee a uniform cooling, i.e. to exclude overcooling of the samples nearby the water entry and an undercooling nearby the water exit. The cycler is equipped with a miniaturized "wine press" heated lid and exchangeable 48-well and 96-well heat blocks with a distance of 4.5 mm between the wells. The free volume of the wells is 10 µl. A system of tiny canals connects the holes in the wells to a vacuum pump*

uniform cooling, i.e. to exclude overcooling of the samples nearby the water entry and an undercooling nearby the water exit. The cycler has been equipped with a miniaturized "wine press" heated lid and small exchangeable 48-well and 96-well heat blocks with a distance of 4.5 mm between the wells. The free volume of the wells is 10 µl. The maximum ramping rate of the cycler is 15 °C/s. The cycler has been transformed into a compact vacuum forming apparatus by introducing a system of transformed tiny channels that introduced to connect the holes in the wells to a vacuum pump.

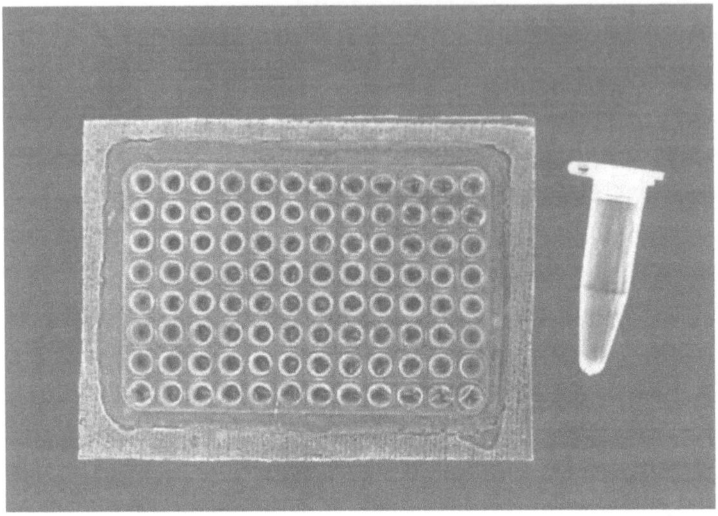

__Figure 2__ Small ultrathin-walled microwell plastic plate. The plate for sample volumes of 0.5 – 4 µl is vacuum formed directly on the heat block of the cycler (see Fig. 1). The main advantages of such microwell plates are the excellent heat-transfer parameters and the easy access to deliver and recover small samples. It is suitable for parallel multisample processing

The microplates are vacuum formed directly on the heat block of the cycler. For this, a precut sheet of a thin thermoplastic film is clamped to the edges of the block using a rectangular stainless steel frame of appropriate size and a rubber gasket. The vacuum is switched on and the block is heated to the forming temperature. By this means the thermoplastic film is formed into a microwell plate. The plate tightly fits with the structured top surface of the heating block. Upon cooling of the block to room temperature, the plate is ready for use. The thickness of such miniature plates is drastically reduced. A comparison with the standard "96V" PCR plates revealed 5–10 fold thinner walls at the top of the wells and 50–200 fold thinner walls at their bottom. The mass of the plate is 0.13 g only, i.e. 60–150 fold less compared with commercial 96-well PCR plates. The ultrathin-walled plates match perfectly with the surface of the block of the cycler thus guaranteeing an optimum heat transfer to the samples. The plate is pressed onto the surface of the block by "vacuum" and can be compared with a thin plastic "skin" which covers the entire top surface of the block (including the holes). The microsamples are transferred into the wells by conventional pipetting. This procedure usually takes several minutes. All the samples are tightly and quickly sealed by a standard sealing film and a heated lid.

The heated-lid technology for thermal cyclers was introduced in 1991, and was developed to allow oil-free thermal cycling of aqueous solutions in the bottoms of plastic vessels placed in a cycler's block. By this means no overlayer of oil is needed and no water vapor condenses on the upper parts of the vessels which rise above the block. Condensation on the upper parts changes the concentration of reactants. This can cause an inhibition of the reaction. A chief problem is that the vapor pressure of water at the DNA-denaturation temperature (92°–94 °C) is sufficiently high to pop the tops of some vessels. This allows water to escape and spoils the reactions. On the other hand, applying high pressure to the top of the tubes or plates to seal efficiently is tricky, because different brands of tubes differ in height, and excessive pressure can craze or crack the vessels, leading to leakage. In order to address this problem some of the manufactures supply the cyclers with adjustable heatable lids. In our case this problem does not arise because the relatively high pressure caused by the conventional "wine press" lid is practically applied through a silicon mat or sealing film to the top surface of the plate. The thin walls of the wells cannot be destroyed because there are no parts of the plate rising above the top surface of the block (Fig. 1). Therefore, compared to commercially available systems, the wells can be tightly closed without any danger of evaporation of the samples and without any danger of braking the side walls of the plates.

Despite the extremely thin walls of the plates, i.e. 10–20 micron at the bottom of the wells, no loss of the small sample volumes (0.5–2 µl) can be measured upon amplification in this rapid cycling regime using the heated-lid technology. The control of the actual temperature of the microsamples is not required because the samples are "trapped" in the relatively massive body of the copper block which is covered by the heated-lid. In addition, the temperature transfer through the ultrathin walls of the plate is almost instantaneous. The temperature equilibrium in samples of a volume of 0.5–4 µl is reached after 1–3 s, respectively. Using the above technology it was possible to amplify DNA in volumes as small as 0.5 µl. What is even more, it was possible to perform 30 cycles in less than 10 min without any detectable loss of the product yield. So far several thousands of different DNA targets have been amplified by this means and the amplification products have been very clean

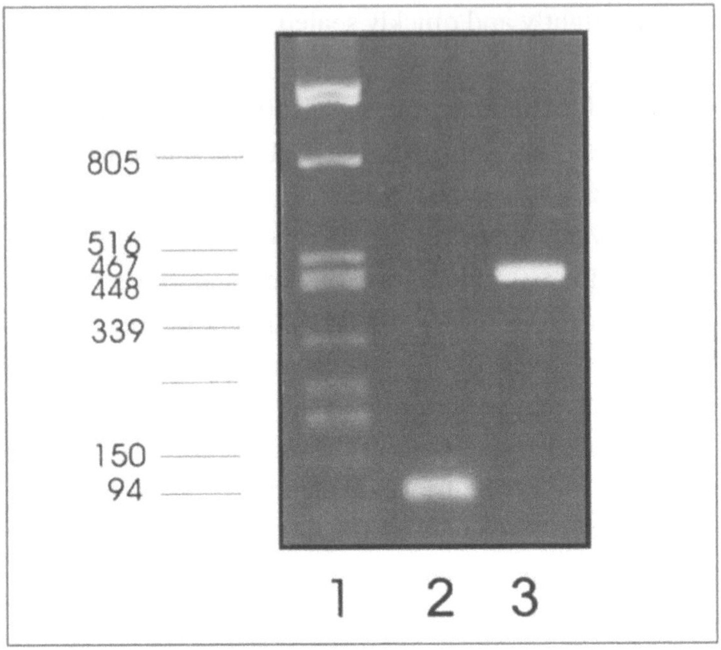

Figure 3 *Example of DNA amplification products using the miniaturized ultrathin-walled microplates. The amplified DNA fragments of HIV-1 (2) and HPV-18 (3) have been analyzed on a agarose gel. PstI digested λDNA was used as size-marker (1)*

(Fig. 3) due to the increased specificity of rapid cycle DNA amplification (Wittwer and Garling, 1991).

The technology of rapid multisample PCR in ultrathin-walled microwell plastic plates has several advantages compared with rapid cycling of small samples in glass capillaries or microfabricated silicon-glass chips. First, a relatively high throughput is possible due to rapid and easy loading and post-PCR recovering of multiple microsamples for electrophoresis or hybridization on solid supports. In addition, all samples can be rapidly sealed in parallel by using conventional means. Second, the plates are formed of PCR- friendly material. This material does not adsorb components of the reaction mixture. Therefore, a high PCR yield can be obtained without adding vast amounts of bovine serum albumin (BSA) to the reaction mixture. Third, the thermo- and vacuumformed microwell plates are very cheap because both the thermoplastic films and the technology for their production, i.e. vacuumforming, are inexpensive.

The great and unique advantage of this technology for rapid cycle DNA amplification of multiple microsamples is that it can be easily integrated into fully automated systems for large scale genome sequencing or high-throughput screening approaches. The ultrathin-walled high-sample density plates (e.g.1536-well plate) can be directly formed on the heat block of the cycler. In addition, the robotic dispensing system can load the micro samples directly into the wells. Furthermore, the rapid cycling guarantees high-quality sequencing reactions in 10 to 15 minutes (Swerdlow et al., 1993). However, this technique is too complicated to serve as a base for the development of rapid, miniaturized and automated DNA-diagnostic assays.

Recently we have developed a technology for rapid DNA amplification using specially designed, preformed, ultrathin-walled microplates supported by a rigid frame (Tretyakov and Saluz, 1998). Preliminary experiments showed that even at a relatively low ramping rate of the miniaturized, custom-made single-module Peltier instrument, i.e. 4 °C/s, small samples (approx. 0.5–3 ml) can be easiliy amplified in 15–20 minutes. This new development fulfills the general requirements for miniaturized, inexpensive DNA diagnostics (Abramowitz, 1996). The sample volume is reduced by a factor greater than 10 but remains still within the range of modern pipetting systems. In contrast to all previously described rapid-cycling systems our system uses the advantage of extremely thin walls of the reactors for a rapid heat transfer. By this means we have solved the well-known problem of both, the enzyme inactivation and the adsorption of the reaction components on the large surface of the microreactor walls. Finally, this system perfectly suites for the requirements for parallel processing of multiple microsamples. An additional great advantage of this technology is that all the procedure can be performed according to the well-established PCR-protocols with a guaranteed performance of the standard heat block thermocycling.

In its present state, the system can be used for rapid post-PCR detection of products by means of fluorogenic probes (Livak et al., 1996; Tyagi and Kramer, 1996; Nazarenko et al., 1997) and a simple fluorescent microplate reader. In addition, the thermocycler can be transformed into a rapid miniaturized analytical cycler for real time detection of multiple microsamples (i.e. 36–100) (Heid et al 1996; Wittwer et al 1997; Taylor et al., 1997).

Much effort has been expended in developing alternative thermocycling technologies, but little research has been conducted on improving the basic performance of conventional heat-block thermocycling in sample size and speed. Our results show that the use of vacuum forming of three dimensional structures of thin thermoplastic films opens a very attractive and novel way to develop new types of reactors for the performance of enzymatic reactions, such as DNA-amplification reactions and high throughput PCR-based DNA-diagnostic assays.

References

Abramowitz S (1996) Towords inexpensive diagnostics. *Trends Biotechnol* 14, 397–441

Burns MA, Mastrangelo CH, Sammarco TS, Man FP, Webster JR, Johnsons BN, Foerster B, Jones D, Fields Y, Kaiser AR, Burke DT (1996) Microfabricated structures for integrated DNA analysis. *Proc Natl Acad Sci USA*, 93: 5556–5561

Chapin K, Lauderdale TL (1997) Evaluation of a rapid air thermal cycler for detection of Mycobacterium tuberculosis. *J Clin Microbiol* 35: 2157–2159

Cheng J, Shoffner MA, Hvichia GE, Kricka LJ, Wilding P. (1996) Chip PCR. II. Investigation of different PCR amplification systems in microbabricated silicon-glass chips. *Nucleic Acids Res* 24: 380–385

Heid CA, Stevens J, Livak KJ, Williams PM (1996) Real time quantitative PCR. *Genome research* 6: 986–994

Kalinina O, Lebedeva I, Brown J, Silver J (1997) Nanoliter scale PCR with TaqMan detection. *Nucleic Acids Res* 25: 1999–2004

Kopp UK, deMello AJ, Manz A (1998) Chemical aplication: Continuous flow PCR on a chip. *Science* 280, 1046–1048

Livak KJ, Flood SJA, Marmaro J, Giusti W, Deetz K (1995) Oligonucleotides with fluorescent dyes at opposite ends provide a quenched probe system useful for detecting PCR product and nucleic acid hybridization. *PCR Methods Appl* 4: 375–362

Mullis KB, Faloona FA (1987) Specific synthesis of DNA *in vitro* via a polymerase-catalyzed chain reaction. *Methods Enzymol* 155: 335–350

Nazarenko IA, Bhatnagar SK, Hohman RJ (1997) A closed tube format for amplification and detection of DNA based on energy transfer. *Nucleic Acids Res* 25: 2516–2521

Northrup MA, Ching MT, White RM, Watson RT (1993) DNA amplification with a microfabricated reaction chamber. Proceeding of the 7th international Conference on solide-state sensors and actuators, Yokohama, Japan, pp 924–927

Saiki RK, Gelfand DH, Stoffel S, Scharf SJ, Higuchi R, Horn GT, Mullis KB, Erlich HA (1988) Primer-directed enzymatic amplification of DNA with a thermostable DNA polymerase. *Science* 239: 487–491

Shoffner MA, Cheng J, Hvichia GE, Kricka LJ, Wilding P (1996) Chip PCR. 1. Surface passivation of microfabricated silicon-glass chips for PCR. *Nucleic Acids Res* 24: 375–379

Swerdlow H, Dew-Jage K, Gestland RF (1993) Rapid cycle sequencing in an air thermal cycles. *Biotechniques* 15: 512–519

Taylor TB, Winn-Deen ES, Picozza E, Woudenberg TM, Albin M (1997) Optimization of the performance of the polymerase chain reaction in silicon-based microstructures. *Nucleic Acids Res* 25: 3164–3168

Tretyakov AN, Gelfand VM, Pantina RA, Shevtsov SP, Bulat SA (1994) Rapid DNA amplification in pipettor tips. *Molecular Biology* (Englisch Translation), 28: 441–443

Tretyakov AN and Saluz HP (1998) Ultrathin-walled multiwell plate for heat block thermocycling. EP 98120187

Tyagi S and Kramer FR (1996) Molecular beacons. *Nature Biotechnology* 14: 303–308

Wilding P, Shoffner MA, Kricka LJ (1994) PCR in a silicon microstructure. *Clin Chem* 40: 1815–1818

Wittwer CT, Fillmore GC, Garling DJ (1990) Minimizing the time required for DNA amplification by efficient heat transfer to small samples. *Analytical Biochem* 186: 328–331

Wittwer CT, Garling DJ (1991) Rapid cycle DNA amplification: time and temperature optimization. *BioTechniques* 10: 76–83

Wittwer CT, Herrmann MG, Moss AA, Rasmussen RP (1997) Continuous fluorescence monitoring of rapid cycle DNA amplification. *Biotechniques* 22: 130–131

7 Generation of large libraries

Carola Burgtorf and Hans Lehrach

7.1 Introduction

DNA is one of the most efficient and fascinating "information storage systems" known to mankind. A few pg of DNA encodes the information in the form of genes necessary for development of an oocyte into a complex organism, for example a human. It is estimated that there are about 80000–100000 genes in a vertebrate genome. However, only a small proportion of the DNA present in the cell encodes genes, which are transcribed to mRNA and then translated into proteins or other non-translated RNAs such as rRNAs and tRNAs. This 1–5% of the genome is usually referred to as coding DNA.

In order to analyze a DNA region that contains information of interest, for example a specific gene or regulatory elements, it is necessary to fragment the genome into easy manipulatable pieces. Libraries constructed from these pieces, derived from the cell's DNA, are called genomic libraries.

Most genes of higher eucaryotes are interrupted by non coding sequences on the DNA level, which are removed only after transcribing the gene to RNA (Fig. 1). Although it is often possible to deduce from the genomic sequence which parts of a gene are coding, it is often easier and more conclusive to analyze only the coding parts. Since it is not possible to directly make libraries from single stranded mRNAs, the RNAs are first copied into cDNA (complementary DNA). Libraries derived from a cell's RNA are called cDNA libraries. Analysis of cDNA libraries gives insight into the concert of genes active in the particular cell type the library was constructed from.

As different as the two libraries physically are, they still share many biological, technical, and logistical problems. In both library types foreign DNA/cDNA has to be ligated to a vector and transferred into a bacterial or yeast host, where it is then maintained as one "volume" of the library. Vectors

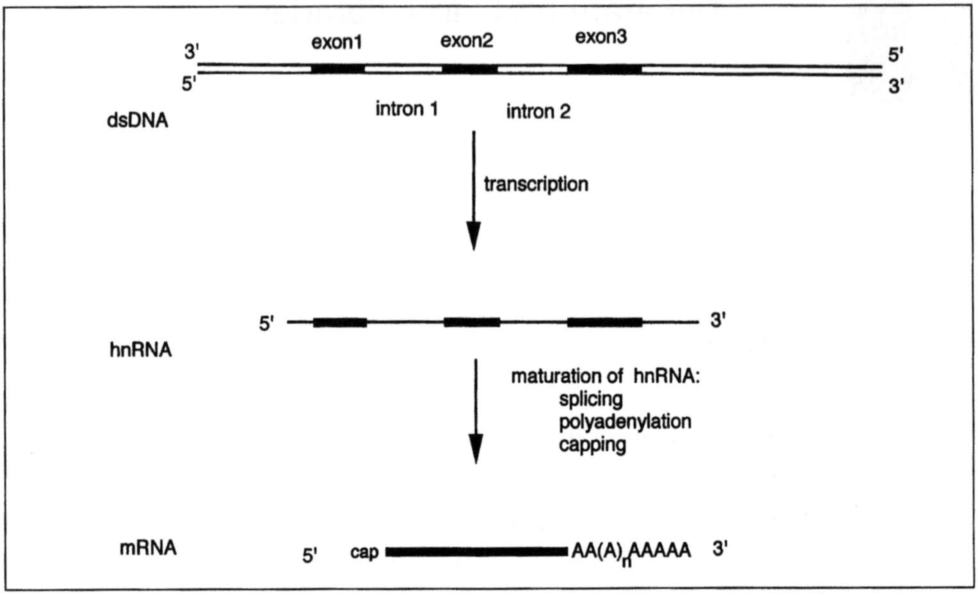

Figure 1 *Schematic of mRNA transcription*

are pieces of DNA which carry genes necessary for their own maintenance in a host and that of the foreign DNA fragment which is physically ligated to the vector. The host has a selective advantage to maintain the vector as it is designed to confer resistance to an antibiotic.

Both cDNA and genomic libraries can potentially be valuable: It can be very hard to get sufficient material of certain developmental stages for making special cDNA libraries, where a certain set of genes is expressed at a sufficiently high level to have a realistic chance to be able to isolate it from the library. Handling large pieces of genomic DNA is not very easy and the efficiency by which DNA molecules can be introduced to the host (transformation efficiency) drops with increase of their size. Therefore it requires experience and patience to make genomic libraries which are of a high quality, both in the number of clones and the sizes of the DNA fragments inserted (inserts) in the vector to have a realistic chance of finding all available sequences.

As a consequence of this, once a library has been constructed it represents a valuable resource and should be stored in a manner which makes it available for many experiments and as a common research resource for many laboratories world wide.

Not all clones of a library grow at exactly the same speed. Bacteria carrying a plasmid with a small insert might grow faster than those having to replicate a large insert. Therefore keeping the libraries as a mixed pool of clones carries the danger of some fast growing clones outgrowing the slowly growing ones. This means that only after a short period the library's content is distorted, no longer reflecting the original distribution of fragments. This problem can be circumvented by growing the primary library on nutrient agar plates, allowing colonies to form, each colony representing one single transformation event. Each clone is then picked using a robotic device and transferred into an individual microtitre plate well. The library can then be stored at –80°C and maintained indefinitely. Our laboratory has worked out the basic logistical problems of storing and retrieving of specific clones on a large scale. If many laboratories use the same set of libraries a database which integrates all the information obtained from a specific clone, like its sequence, assignment to a gene family or to a chromosomal location can be developed.

In the following section, we describe how genomic and cDNA libraries are made, stored and handled in our laboratory and how they are used in modern molecular genetics. We also outline some of their uses as the basic tools of modern molecular genetics.

Tables 1 and 2 summarize the commonly used cloning systems for genomic cloning and for making cDNA libraries, respectively. The terminology used in the tables will be explained in detail in the following paragraphs.

Genomic libraries

The aim of a genomic library is it to provide a tool which gives the ability to study specific genomic regions. At times these regions will span a greater physical distance than the length of an individual insert. Therefore, it is necessary to build up a contig, i.e. a series of overlapping clones spanning the entire region of interest. To fulfill this criterion the fragments must be derived randomly and the library must be large enough that it contains many overlapping clones.

What is large enough is defined by the physical size of the source genome (i.e. the total number of base pairs in the genome) and the average size of the fragments which are cloned. Assuming that the generation of DNA fragments

Table 1 Cloning systems most commonly used for genomic cloning

Cloning system	Common vectors	Cloning capacity	Cloning procedure	Advantages/disadvantages
Plasmid	pBR322 pUC19 pBluescript	0.1–10 kb	Fragments are ligated transformed to *E. colic* by electroporation or heatshock.	Easy, but low cloning capacity.
λ replacement vector	EMBL3	12–20 kb	Size-selected fragments are ligated to phagearms, packaged *in vitro* into λ-heads and transfected into *E. coli*.	High cloning efficiency, but still low cloning capacity.
Cosmid	Lawrist Supercos	40 kb	Fragments are ligated to cosmid arms, packaged *in vitro* and transfected into *E. coli*.	Relatively high cloning efficiency, relatively large inserts, but delete and rearange frequently.
PI	pAd*sacB*II	70–100 kb	Size-selected fragments are ligated to phagearms, packaged *in vitro* into PI heads and transfected into *E. coli*.	Large insert size, stable propagation of inserts, but fair cloning efficiency, and relatively large vector.
PAC	pCYPACII	100–300 kb	Size-selected fragments are ligated to the vector and electroporated into *E. coli*.	Large insert size, stable propagation of inserts, but fair cloning efficiency, and relatively large vector.
BAC	pBeloBACII	100–300 kb	Size-selected fragments are ligated to the vector and electroporated into *E. coli*.	Large insert size, stable propagation of inserts, relatively smaller vector but fair cloning efficiency.
YAC	pYAC4	0–2000 kB	Size-selected fragments are ligated to YAC arms, and Yeast-spheroplasten are transformed.	Large inserts, but low cloning efficiency, more difficult to work with than bacterial plasmids,and large YACs are often chimeric or have deletions.

Table 2 Most commonly used cloning systems for cDNA cloning

Cloning system	Common vectors	Cloning capacity	Cloning procedure	Advantages/disadvantages
λ insertion vector	λ gt-10	up to 7.6 kb	cDNA synthesis, protection of internal EcoRI sites with EcoRI methylase, ligation of EcoRI linkers, cutting with EcoRI, ligation to lambda arms, *in vitro* packaging.	High cloning efficiency, easy screening with nucleic acid probes, but not very versatile vector, not suitable as reference library.
λ insertion vector	λ gt-11	up to 7.2 kb	Same as for λ gt-10; with λ gt-11 Sfi-Not vector first strand synthesis with oligo dT/NotI primer.	High cloning efficiency, expression possible as fusion protein with lacZ, easy screening with nucleic acid probes and antibodies possible, not suitable as reference library, with λ gt-11 Sfi-Not vector directional cloning possible.
λ insertion vector	λ-GEM	up to 7.1 kb	Similar as fro λ gt vectors; first strand synthesis with oligo dT/XbaI primer.	High cloning efficiency, easy screening with nucleic acid probes, *in vitro* transcription of the inserts from the flanking Sp6 and T7 promoters possible, not suitable as reference library.
Plasmid	pSport	no lower or upper size limit	First strand cDNA synthesis with oligodT/NotI primer, ligation of adapters after second strand synthesis, NotI digest, ligation to double cut plasmid and electroporation.	Relatively high cloning efficiencies, expression of insert possible, screening with antibodies and nucleic acid probes possible, suitable as reference library.

and their propagation in cloning experiments is independent of their DNA sequence, then the probability of a given sequence being present in a library can be calculated by the following equation:

$$P = 1 - (1 - f)^N$$

where f is the fraction of the genome that the average insert DNA represents and N is the number of clones in the library.

For most small scale mapping projects a three times genome coverage (i.e. the sum of the insert lengths of all clones equals three times the length of the source genome), which gives > 95% chance of finding a specific sequence is sufficient. However, for covering a large region, which would take successive isolation of several clones, the overall probability covering the region drops rapidly. This means, for large scale mapping projects of a complex genome either the average insert size has to be large or a very large library is needed.

Table 3 gives an overview over genome sizes from some of the most common model organisms, together with the number of clones necessary to give a 95% and 99% probability of finding a specific sequence. For this example we have used the cosmid cloning system, which has an average insert size of 40 kb. During some large scale mapping projects it has become apparent that the use of the different types of genomic libraries complement each other well (Francis et al., 1994). Therefore we describe here all major genomic cloning systems currently in use.

Plasmid cloning

In the beginning of genetic engineering small circular DNAs called plasmids were used as vectors. Plasmids occur naturally in bacteria. They often carry an antibiotic resistance gene which allows the host to grow on media containing an antibiotic which would otherwise kill it. Researchers found that plasmids could be linearized with restriction enzymes, which cleave DNA at a specific, short sequence motif, often creating “sticky ends”. This linearized DNA can then be joined (ligated) back together but now with a fragment of foreign DNA with compatible ends inserted. The result is a circular hybrid molecule of the plasmid and the insert. These hybrid molecules can then be transformed into *E. coli* (Dagert and Ehrlich, 1979). Currently, plasmid-cloning is

Table 3 Genome sizes of some common research organisms and number of clones necessary for a representative cosmid library

Species	Genome size (bp)	Number of cosmid clones for p = 0.95	Number of cosmid clones for p = 0.99
Arabidopsis thaliana	$7 \cdot 10^7$	5241	8058
Saccharomyces cerevisiae (yeast)	$1.6 \cdot 10^7$	1200	1842
Caenorhabditis elegans	$8.8 \cdot 10^7$	6590	10131
Drosophila melanogaster (fruitfly)	$1.4 \cdot 10^8$	10482	16118
Fugu rubipes (puffer fish)	$4.8 \cdot 10^8$	35940	55262
Danio rerio (zebra fish)	$1.7 \cdot 10^9$	127286	195718
Xenopus laevis (frog)	$3.6 \cdot 10^9$	269547	414462
Gallus domesticus (chicken)	$1.2 \cdot 10^9$	89850	138154
Mus musculus (mouse)	$2.7 \cdot 10^9$	202160	310850
Homo sapiens (human)	$3 \cdot 10^9$	224623	345385

still the preferred system for small DNA fragments. It has recently found much further development in the PAC and BAC cloning systems (see below for more detail).

Lambda phage cloning

Another cloning system that was developed at this time is based on the bacteriophage lambda (Murray and Murray, 1974). Bacteriophages are viruses that infect bacteria and duplicate at the expense of their host. The genome of bacteriophage lambda is 48.5 kbp long and carries genes for the two manifestations of lambda's lifecycle, i.e. its lysogenic and lytic phases. In the lytic phase the phage uses the transcription and translation machinery of the bacterium to duplicate its own genome and to produce the proteins that form its capsule also called phage heads and tails. It then packages its genome into the phage heads and induces the lysis of the bacterium to release several hundred infectious phage particles. If this occurs on a lawn of bacteria, clear circles, called plaques, of lysed bacteria form. Each circle being derived from a single infected bacterium, which has infected other surrounding bacteria. The plaques in a bacteria lawn contain phage particles which can easily be disrupted to release essentially pure DNA.

The genes for the lysogenic cycle are not essential for the use of lambda as a cloning vector and are removed in "replacement vectors" like the EMBL series (Frischauf et al., 1983). Since the phage heads can accept only DNA fragments of a certain size (38–52 kbp) the vector alone is too small to be packaged. This means there is a positive selection for recombinant DNA fragments where the lysogenic genes are replaced by a genomic fragment of 10–18 kb. The hybrid DNA molecules are then packaged *in vitro* into phage heads using packaging extracts. These packaging extracts are commercially available and consist of a mixture of empty phage heads, tails, and enzymes which stuff the DNA into them. These phage particles are then used to infect bacteria, the amplified phage are washed from the plaques and the library is then stored as a phage suspension. Every time the library is used, bacteria are infected with an aliquot of this phage suspension mixed with top agarose and plated. For screening the library a nylon membrane which adsorbs some of the phage particles is carefully placed on top of the agarose surface. The DNA is then released and bound to the membrane. The membranes can be used for screening with a radioactively labeled probe as described below. Since *in vitro* packaging of DNA into infectious phage particles is a very efficient way to introduce relatively large DNA fragments into *E. coli* and it is possible to plate the phage at high densities (20000–100000 clones on a 22 × 22 cm plate) it is relatively easy to construct and screen large phage libraries. However, because of the large number of clones that have to be handled and the difficulties of storing large numbers of individual phage clones separately lambda libraries are not commonly used as reference libraries (i.e. a collection of individual clones picked into microtitre plates).

Cosmid cloning

Cosmid cloning combines many of the advantageous features of plasmid and of lambda cloning (Collins and Bruning, 1978). A cosmid vector contains an origin of replication, an antibiotic resistance gene and one or two cos sites.

The cos sites are only a few nucleotides long and are the recognition sequence for the lambda packaging machinery. All DNA between two cos sites, which must be separated by approximately 40–50 kbp can be packaged *in vitro* into lambda heads. Since cosmid vectors are usually only a few kbp in size, this leaves more then 40 kbp for foreign DNA fragments. When the cosmid phage particles are used to infect bacteria, the cosmid DNA forms a circular molecule, which is propagated similarly to plasmids, conferring a selective antibiotic resistance on the host. In Figure 2 all steps of cosmid cloning are summarized schematically. Cosmid libraries are fairly simple and efficient to construct. It is possible to construct cosmid libraries from very small amounts of DNA (100 ng), such as flow sorted chromosomes (Nizetic et al., 1994) or even from microdissected chromosome regions (Tohma et al., 1993). Current cosmid vectors carry two cos sites which simplifies the cloning procedure considerably compared with the initial protocols, which required either size fractionating of the insert molecules or preparation of the two cosmid arms separately (Collins and Bruning, 1978; Ish-Horowicz and Burke, 1981). They have a cloning site that carries single restriction sites for several different enzymes and is flanked by RNA polymerase promoters which facilitate simple probe generation from the ends of the insert. A detailed protocol for making a cosmid library is given below.

YAC cloning

The YAC (yeast artificial chromosome) cloning system was developed after the three cis acting DNA sequences had been identified which are essential for yeast chromosome function. These are: ARS (autonomous replicating sequence) which allows the molecule to be maintained in yeast, CEN (centromere) which is essential for accurate mitotic segregation, and TEL (telomere) which confers structural integrity on linear DNA molecules. Since yeast is a eukaryote it was hoped that many of the problems encountered in bacterial cloning, like under-representation, deletion and rearrangement of some sequences, would be overcome with YAC cloning (Burke et al., 1987). The insert size which can be propagated in yeast can be anything, from just the empty vector to even bigger than the largest yeast chromosome (2.5 Mbp). With such a large insert size YACs are still the preferred tool to quickly cover very large genomic regions. However, a number of problems have been

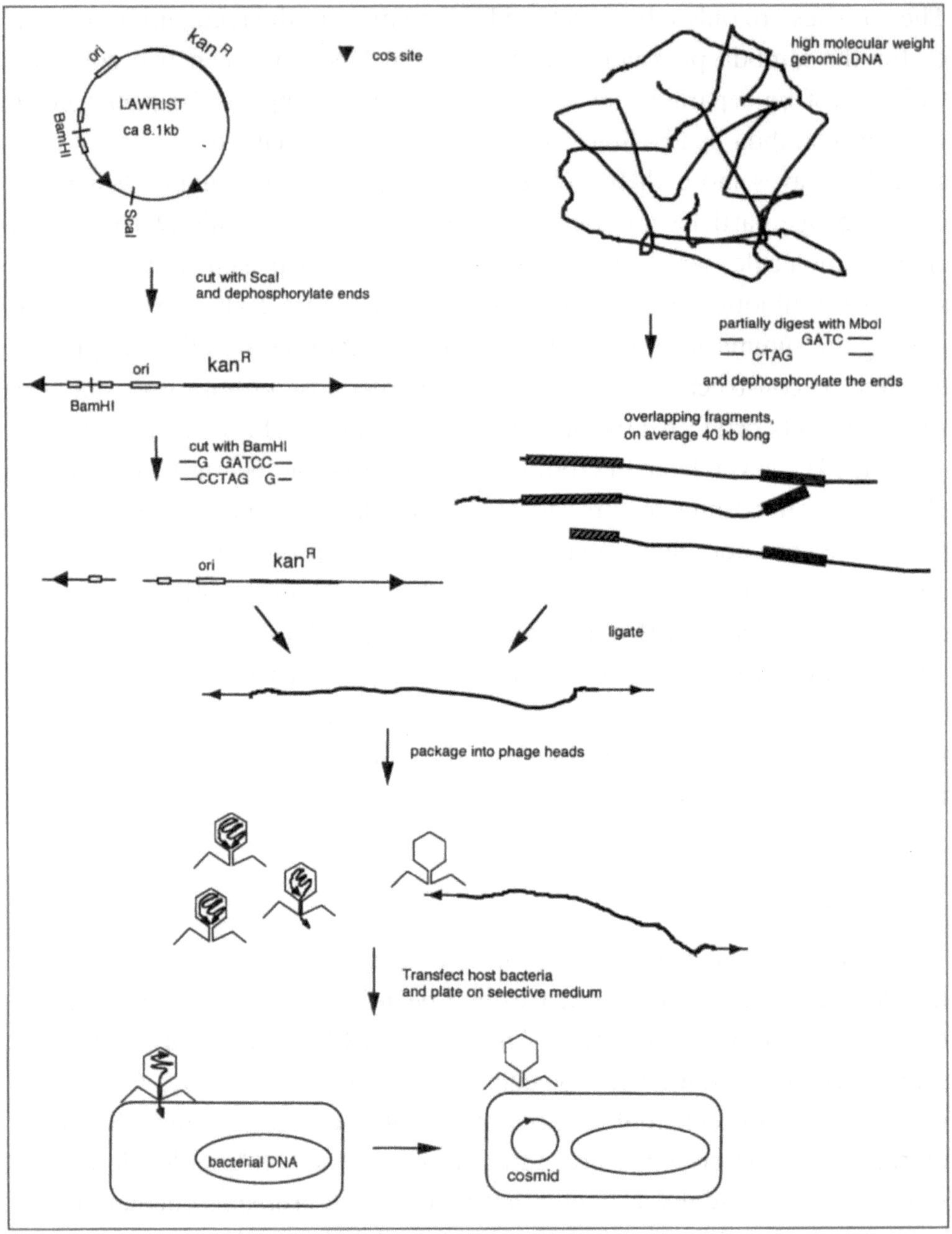

Figure 2 *Schematic of cosmid cloning*

encountered with YAC cloning. These include: i) co-ligation of two unrelated inserts, due to the high DNA concentrations required during YAC library construction and thus creating chimeric YACs, ii) co-transformation of two YACs into one host cell, and iii) recombination *in vivo* between repeated sequences within the cloned inserts (Schalkwyk et al., 1995).

It is difficult to obtain more than a few micrograms of YAC DNA for analysis. This is due to the structural similarity of the YAC and the yeast chromosomes which makes it difficult to separate the YAC from the 16 Mbp of yeast genome. This problem can be partially overcome by using only the subset of insert sequences which can be amplified by IRS-PCR (Cox et al., 1991; Nelson et al., 1989). IRS-PCR takes advantage of interspersed repeat sequences (IRS) which occur in the donor genome, but not in the yeast genome. DNA between two IRS elements can be amplified specifically in a Polymerase Chain Reaction (PCR), provided they are close enough (200 bp–3 kbp) together. Due to the problems encountered in YAC cloning much effort was put into further development of bacterial cloning systems.

P1 cloning

The first step towards increasing the insert size of *E. coli* based libraries above the 50 kbp limit of the bacteriophage lambda was the use of vectors based on the Bacteriophage P1. This phage has a packaging capacity of 105 kbp of DNA (Sternberg et al., 1990; Sternberg, 1992, 1994). The vector carries a short DNA sequence, called the *pac* site, which is the recognition sequence for the P1 packaging system. The two LoxP sites enable recircularisation within the bacterial host. P1 clones are propagated with only very few copies per bacterium. This means that one of the main disadvantages of cosmids, namely recombination and deletion of insert DNA within the host is observed in P1 clones at much lower frequencies. Furthermore, a positive selection system is introduced into the P1 cloning vector pAd*sacB*II, by which the insert is cloned in the *sacB* gene. The gene product of the *sacB* gene converts sucrose into the bacteriotoxic levan, which means that only recombinant clones where the gene is inactived by an insert can survive on a sucrose containing medium. Although *in vitro* packaging of DNA into phage particles is a very efficient way of transforming bacteria, the size limitations which the phage heads impose mean that only fragments up to 105 kb can be packaged. Furthermore the P1 packaging machinery does not exclude relatively small fragments from being packaged. Therefore it is necessary to remove all fragments smaller than the desired size range (70–100 kbp), e.g. by a special type of agarose gel electrophoresis (PFGE).

BAC and PAC cloning

A method for transforming large DNA molecules with high efficiencies into *E. coli* using electroporation was recently described (Calvin and Hanawalt, 1988; Sheng et al., 1995). The bacteria are mixed with the recombinant DNA and a high voltage pulse (2 kV/2 mm) is applied. This method made genomic library construction with large inserts in plasmid-like cloning systems possible. The BAC and PAC cloning system are very similar and differ only in the vectors used. The BAC cloning system is based on the F' episome of *E. coli* and can discriminate between insert carrying and "empty" clones with a color selection. The PAC cloning system is based on the P1 vector and has also the positive selection system, by which only clones containing an insert can survive (Ioannou et al., 1994). Both vectors are maintained at a low copy number. Cloning procedures in both systems are almost identical: High molecular weight DNA is partially cut with a restriction enzyme to result fragments which are around 100–300 kb. All fragments smaller than the desired insert size are removed by electrophoresis on a pulsed field gel (PFGE). The area containing the fragments with the size of interest (100–300 kb) is excised, the agarose is removed by enzymatic digestion and the DNA fragments are ligated to the cloning vector. The BACs or PACs are then transformed into the host by electroporation. One electroporation can yield up to 1000 clones. Efficiencies per microgram of vector DNA are comparable to the P1 system and at least one order of magnitude higher than for YAC cloning. DNA can be isolated from BACs and PACs with the same simple protocols as from small plasmids. Sets of good mouse and human libraries already exist in an arrayed format and high density colony membranes (see below) or PCR pools are available for screening.

cDNA libraries

A cDNA library is a population of bacteria or bacteriophage which contain a cDNA insertion in a plasmid or their genome, respectively. The starting material is mRNA, of which a faithful DNA copy (complementary DNA/cDNA) is made, since the single stranded RNA is not cloneable. The first strand of the cDNA is synthesized by reverse transcriptase which requires a free 3′-hydroxyl group of a short DNA primer and dNTPs. The two different strategies of either using random primer of oligo(dT) primer are discussed later on in this chapter. The RNA is partially hydrolyzed and the second strand which resembles the primary sequence of the mRNA, is synthesized by nicktranslational replacement with *E. coli* PolymeraseI. Figure 3 summarizes one possible route for cDNA cloning. If no enrichment procedures for certain sequences are carried out, the proportion of a specific cDNA in the library reflects the expression level of the gene in the source tissue. However, only around 1% of a cell's RNA population is protein coding mRNA and it is estimated that of the 80–100000 vertebrate genes only a set of 10–20000 is expressed in any specific tissue. The vast majority are highly abundant rRNAs and tRNAs. Since only mRNAs have a 3′ poly-(A) tail of 50–200 nucleotides this can be used to specifically enrich this species, for example by binding to an oligo(dT)-column. This means that the rRNAs and tRNAs can largely be excluded from the future library.

Since not all genes are expressed at the same level in every tissue or developmental stage, the number of clones of the library must usually be much larger than the number of different genes expressed in the donor tissue. Table 4 correlates the probability of finding a specific transcript with its relative

Table 4 Number of cDNA clones necessary for finding transcripts of a certain abundance

Number of clones	Probability of finding a clone representing a proportion of:			
in the cDNA library	1%	0.01%	0.0001%	0.000001%
10000	1	0.999955	0.632121	0.00995
20000	1	1	0.864678	0.019801
50000	1	1	0.993264	0.048771
100000	1	1	0.999955	0.095163
500000	1	1	1	0.393469
1000000	1	1	1	0.632121

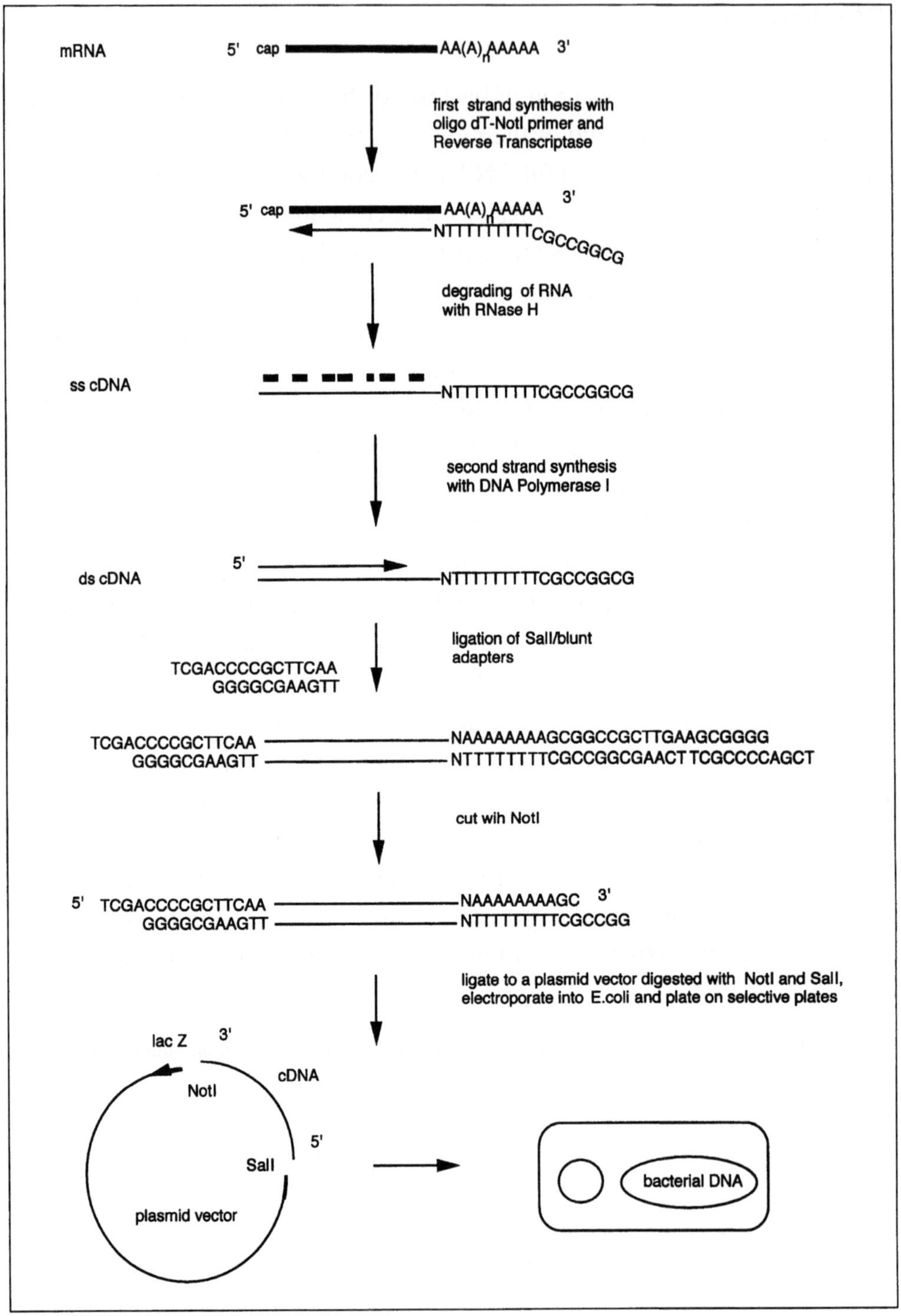

Figure 3 *Schematic of cDNA cloning*

expression level and the number of clones in a library, and demonstrates that very large libraries are necessary to find rare transcripts.

Apart from the non-translated regions at the 3′ and 5′ ends and the poly(A)-tail at the 3′ end mRNAs are free from non-coding intronic sequences, which on the genomic DNA level interrupt with a few exceptions eukaryotic genes (Breathnach and Chambon, 1981) (Fig. 1). Since bacteria do not possess the machinery for processing intron-containing RNAs, the complex eukaryotic genes can only be expressed in bacteria if the introns, which contain stop codons or cause frame shifts, are removed before cloning. Therefore cDNA libraries cannot only be screened with nucleic acid probes, but if the cDNA is expressed and translated in the bacterial host, also with antibodies. This is of particular interest, if one is interested in finding a particular gene in different species. Because of the redundance of the genetic code genes can differ substantially on the DNA level, but can still be very similar or identical on protein level. In some vectors the cDNA can be expressed as a fusion protein with another small protein or a few additional aminoacids. This acts as a tag. The use of a tag facilitates protein purification via affinity chromatography.

Furthermore, the comparison of cDNA libraries of different tissues, disease stages, or developmental stages offers the possibility of studying expression levels of genes and to identify genes involved in these processes. Sequencing effort for a gene of interest can easily be reduced by a factor 10–50 compared to genomic sequencing and functional studies can be carried out: *In vitro* RNA transcripts of the cDNAs can be injected into mutant animals to rescue the mutant phenotype or if the antisense strand is used, a mutation can be induced in a genetically wildtype background.

7.2 Protocols

Generating a cosmid library

Isolation of high quality genomic DNA

DNA can be isolated from almost any tissue using a number of different protocols. For cloning of fragments derived from a partial digest with a restriction enzyme (see p. 206), the starting DNA must obviously be larger than the fragments to be cloned, ideally multiple times. This means that for cosmid cloning, which uses fragments of around 40 kbp, the starting DNA should be at least 200 kpb in length. Furthermore the DNA should be free from contaminants which inhibit the subsequent enzyme reactions. Sometimes different protocols have to be tested to obtain intact and sufficiently pure DNA. We present here two protocols, one that yields intact chromosomal DNA embedded in agarose blocks and one which gives DNA in solution.

For embedding DNA in agarose (Birren et al., 1988) a tissue is dispersed into single cells or 5×10^8 cells of a cell line are harvested. The cells are washed twice in cold PBS by collecting them at 1000 g for 10 min at 4°C, and resuspended by tapping the tube. After the last wash the cells are resuspended to double the desired final concentration. As a guideline it is convenient to work with DNA concentrations of around 20–40 µg/100 µL agarose block, which equals $5–10 \times 10^6$ for example cells per agarose block. The cell suspension is then mixed with an equal volume of 1.5% low melting point agarose in 125 mM EDTA and pipetted into block molds for setting. The set agarose blocks are then transferred to a NDS solution (500 mM EDTA, pH 8.0; 1% N-lauroylsarcosinate) with 1–2 mg/mL ProteinaseK and incubated for 20 h at 50°C. If the blocks do not appear to be clear, the supernatant gets replaced by fresh NDS with ProteinaseK and incubation at 50°C continues for another 12 h. Before the DNA can be used, the blocks have to be washed extensively in TE (10 mM Tris-HCl, 1 mM EDTA, pH 8.0) to remove the ProteinaseK and excess of EDTA. (The blocks can be stored in NDS or 0.5 M EDTA at 4°C until needed for at least 12 month.)

To obtain liquid DNA (Herrman and Frischauf, 1987) around 5×10^8 cells are harvested, washed, and resuspended in 20 mL of TEN9 (50 mM Tris-HCl, pH 9.0, 100 mM EDTA, 200 mM NaCl). If the DNA is to be prepared from whole tissue, the tissue is cut into small pieces, quickly frozen by dropping into liquid nitrogen, and ground to powder using a precooled mortar and pestle. The powder is then carefully spread on the surface of 20 ml of TEN9 buffer in a 500-ml beaker and the beaker is swirled to obtain a homogeneous suspension. SDS and ProteinaseK to final concentrations of 1% and 0.5 mg/mL, respectively, are added, and the cells are incubated for 20 h at 55°C. The cell lysate is extracted twice with TE saturated phenol/chloroform/isoamylalcohol (25:24:1), once with chloroform/isoamylalcohol (24:1) and the DNA is precipitated by addition of two volumes ethanol. The precipitate can usually be removed with a plastic loop, and is then washed in 70% ethanol and dissolved in TE to a concentration of 0.5 mg/mL.

Both methods usually lead to DNA preparations which are of sufficient quality for cosmid cloning. The advantage of DNA in agarose blocks is that the DNA is not exposed to shearing forces during pipetting and therefore remains more intact. Also it is easier to handle an agarose block, e.g. by simply cutting it into equal pieces, than it is to precisely pipette small amounts of a very viscous high quality DNA solution. However, some people find the handling of small, almost clear agarose blocks difficult. Also using agarose blocks two additional steps in the protocol are necessary, which are pre-incubation with restriction enzyme of the blocks for the partial digest (see p. 206) and subsequent removal of the agarose.

Vector preparation

It is inevitable that the first task in constructing a library is to choose the vector. Choice depends largely on the intended purpose of the library. The vector should be small to leave as much as possible space for the fragments to clone; but on the other hand it should contain all features that make it a useful and convenient tool to work with. For cosmid libraries we usually use a vector from the lawrist series. Lawrist vectors are double cos vectors, and confer kanamycin resistance on the bacteria. In comparison with ampicillin, kanamycin is superior, because it is not metabolized and is more stable. Sp6 and T7 promoters are flanking the cloning site, which facilitate generation of end probes of the insert in genomic walking experiments. The cloning site is *Bam*HI which is compatible to the frequentyl cutting restriction enzyme *Mbo*I. Furthermore it has recognition sites for rare cutting enzymes flanking the cloning site, which can be used for excising the insert.

From a few single colonies of bacteria containing the cosmid vector inoculate 3 ml LB medium with 80 mg/mL ampicillin and 25 mg/ml kanamycin. Use 2 mL to perform a small scale plasmidprep (Sambrook et al., 1989) and check the integrity of the plasmid. Use the remaining mL of the miniculture to inoculate 1 L medium and grow for 12 h at 37°C. For purifying the plasmid DNA follow either the Qiagen Maxiprep instructions (Qiagen) or other standard Maxiprep-protocols (Sambrook et al., 1989). Digest 20 μg with 100 U ScaI, which linearizes the vector within the stuffer fragment between the two cos-sites, in a total volume of 200 µL under the conditions recommended by the supplier. Extract with phenol:chloroform:isoamylalcohol (25:24:1) and precipitate the DNA with ethanol. Disolve the DNA in 179 µl H_2O, add 20 µL 10 × CIP-buffer (10 mM $ZnCl_2$, 10 mM $MgCl_2$, 100 mM Tris-HCl, pH 8.3), add 1 U calf intestinal phosphatase (Boehringer-Mannheim), and incubate at 37°C for 30 min. This removes the phosphate groups at the ends of the DNA molecule, which prevents these ends from ligating with each other in the subsequent ligation reaction. Inactivate the phosphatase by adding 20 µL EGTA (100 mM, pH 8.0) and heating to 65°C for 10 min followed by a phenol:chloroform:isoamylalcohol (25:24:1) extraction and precipitate the DNA by adding NaCl to a final concentration of 0.5 M and 2 vol. ethanol; recover by centrifugation. Disolve the DNA in 175 µl H_2O, add 20 µl 10 × BamHI buffer and 5 µl *Bam*HI (20 U/µL) and incubate at 37°C for 30–40 min. Remove the enzyme by phenol:chloroform:isoamylalcohol (25:24:1) extraction and precipitate the DNA as before; recover by centrifugation and disolve in 30 µL TE.

It is essential to check all steps of the vector preparation for completeness, i.e. the restriction reactions, by electrophoresis of an aliquot on an agarose gel, and the dephosphorylation reaction by setting up a ligation reaction with a small aliquot, which should not yield any higher molecular weight products.

Partially digesting the genomic DNA

The most critical step in the construction of a genomic library is to generate random fragments of the genomic DNA of the size necessary for the chosen cloning system. For cosmid cloning the majority of the fragments should be around 40 kb. This can be achieved by mechanically shearing the DNA by passing it through a needle. This could be followed by ligating adapters compatible to the cloning site to the ends of this material. The only advantage of this method is that it ensures a maximum of sequence independence. The more common approach is to enzymatically hydrolyze the DNA with a relatively frequently cutting enzyme. A restriction enzyme with a recognition site of 4 bp cuts the DNA on average every 256 pb (4^4 bp). For genomic DNA which should be hydrolyzed to 40 kb fragments a partial digest with such an enzyme is regarded as sufficiently random. In a pilot experiment the amount of enzyme which yields most of the fragments in the desired size range is determined. Often *Mbo*I is used in a competition reaction with another enzyme (dam methylase) (Nizetic et al., 1994) that "hides" the recognition site by methylating it. This is especially advisable if only very limited amount of material is available. We present here a titration series for liquid DNA and for chromosomal DNA embedded in agarose blocks. Partial digestions for cosmid and PAC cloning are carried out by incubating the DNA with a limited amount of enzyme for a defined period of time.

The ensure even digestion, for DNA in agarose blocks it is necessary to pre-incubate the DNA with the enzyme and then start the reaction by adding of Mg Acetate.

Equilibrate the blocks in 1 × A^--buffer (33 mM Tris-acetate, pH 7.8, 66 mM potassium acetate), then 25 µl of a block (1/4 block) in an Eppendorf tube with: 17.5 µl 10 × A^-, 4.0 µl BSA (10 mg/ml), 5.0 µl Spermidine (250 mM), 5.0 µl DTT (100 mM), and 140.5 µ H_2O. After 30 min on ice add 1 µl of a MboI dilution in EDB (0.5 U/µl, 0.25 U/µl, 0.125 U/µl, 0.0625 U/µl, 0.03125 U/µl) and incubate with the enzyme for 2 h on ice. (EDB: 50% Glycerol, 50 mM KCl, 10 mM Tris/HCl pH 7.5, 1 mM DTT, 200 µg/ml BSA.)

To start the reaction add 2 µl 1 M magnesium acetate, and incubate for 4 h at 37°C. To stop the reaction exchange the reaction buffer with TE, chill on ice, and check the digests by PFGE. Choose the enzyme concentration which gives the appearance to result in slightly underdigested DNA to scale up the reaction, or if necessary, repeat the titration in a narrower range of different enzyme concentrations.

For partially digesting liquid DNA preparations a preincubation with the enzyme is not necessary. Prepare eight Eppendorf tubes with 1 µg of genomic DNA, 5 µl 10 × A^+ (100 mM magnesium acetate, 660 mM potassium acetate, 330 mM Tris acetate, pH 7.8) and 2.5 µl 2% Gelatin in a total volume of 45 µl. On ice prepare a range of seven different *Mbo*I dilutions in EDB ranging from 1 U/µL to 0.0016 U/µL and one without any enzyme, and add 5 µL of the dilution to the corresponding tube. Incubate for 30 min at 37°C and stop the reaction by adding 10 µL 0.5 M EDTA and heating to 65°C for 15 min. Check the digests either by PFGE (Fig. 4) or by electrophoresis through a 0.3% agarose gel at 1 V/cm. Special care must be taken when pipetting high molecular weight DNA. To prevent shearing pipette slowly and use wide bore tips.

Once the optimal conditions for the partial digest are determined, scale up the digest with two blocks or with 5–10 µg DNA, exactly mimicking the con-

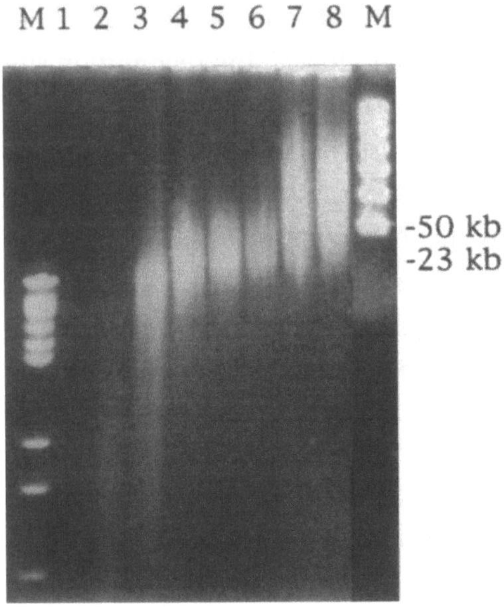

Figure 4 *PFGE of a restriction digest of liquid DNA with a serial dilution of MboI (lanes 1–7). Although the DNA preparation is far from optimal (some fragments in the undigested DNA (lane 8) are already smaller than 50 kbp) a partial digest with an intermediate enzyme concentration of those used in lanes 6 and 7 was successfully converted into a cosmid library*

ditions of the pilot experiment. If DNA in agarose-blocks was used, the agarose has now to be removed by digestion with gelase (Epicentre, Cambridge), carefully extracted with phenol:chloroform:isoamylalcohol (25:24:1), and the DNA is recovered by precipitation with ammonium acetate (final concentration 2.5 M) and 2 vol. ethanol.

To prevent self-ligation of smaller inserts, the fragments are then treated with alkaline phosphatase. Resuspend the DNA (ca. 5 µg) in 89 µL H_2O and add 10 µL 10 × CIP-buffer. Add 0.5 U alkaline phosphatase and incubate for 30 min at 37°C. The successful removal of the 5′ phosphate groups should be checked by setting up a ligation reaction with a small aliquot of the DNA, which should not yield any higher molecular weight products.

Ligation and *in vitro* packaging

Ligation of the genomic fragments to the vector arms is done with a molar excess of vector fragments. In a total volume of 10 µL mix in an Eppendorf tube: 1–2 µg genomic fragments (from above), 1 µg vector arms (see p. 206), 1 µl 10 × ligase buffer (500 mM Tris-HCl, pH 7.8, 100 mM $MgCl_2$, 10 mM DTT, 10 mM ATP, 0.5 mg/ml BSA) and 10 U ligase and incubate the reaction at 16°C for 12–16 h. Control ligations with the vector arms alone and genomic fragments alone should also be performed in the same manner.

The cosmids are now ready for *in vitro* packaging. Packaging extracts can be prepared from mutant *E. coli* lysogens (Sambrook et al., 1989), but are also available from a number of commercial sources. Routinely, we use the Gigapack III Gold packaging extracts (Stratagene) which are convenient single tube reactions. The extracts are removed from – 80°C and rapidly thawed. Five µL of the ligation reaction are immediately added to the tube and the reaction is allowed to proceed for 2 h at room temperature. Then 500 µL PDB (100 mM NaCl, 8 mM $MgSO_4$, 50 mM Tris-HCl, pH 7.5, 0.01% gelatine) and one drop chloroform are added.

Titrating and plating the library

From a single colony of DH5α (F^-*mcr*A, Δ(*mrr-hsd* RMS-*mcr* BC), φ80d*lac* ZΔM15, Δ(*lac* ZYA-*arg* F), U169, *deo* R, *rec* A1, *end* A1, *sup* E44, λ^-, *thi* – 1, *gyr* A96, *rel* A1.) a 50 mL culture LB (10 g/L BactoTryptone, 5 g/L Yeast extract, 10 g/L NaCl), with 0.2% Maltose and 10 mM $MgSO_4$, is inoculated and grown at 37°C for 6–7 h. The cells are spun down and resuspended in 25 ml of 10 mM $MgSO_4$. To titrate the library 25 µL of a 1:10 and 1:50 dilution of the phage-suspension in PDB are incubated with 25 µL of plating cells for 30 min at 25°C. Then 200 µL LB are added and incubation continues for 1 h at 37°C. The cells are then plated on LB-agar (LB with 15 g/L agar) plates with 25–30 µg/mL kanamycin. As a control the plating cells alone should also be plated on selective medium. Those and the packaged ligations of vector arms and genomic fragments from above should not give any colonies, where the 1:10 dilution of the cosmids can easily yield more than 1000 colonies.

Generating a cDNA library

There is a vast variety of different strategies and different protocols for construction of cDNA libraries (Kimmel and Berger, 1987). Although the easiest and most common method is to buy a cDNA construction kit (e.g. from Gibco BRL) and to exactly follow the instructions of the supplier, we present still a brief outline of a protocol, added with some comments.

mRNA isolation

The quality of cDNA library depends on the integrity of the mRNA. Although there are now direct mRNA isolation methods available (e.g. Oligotex from Qiagen), for the amount and quality required the most commonly used protocols still start with the isolation of total RNA and then extracting the mRNA (poly(A)$^+$ RNA) subsequently. Because of the omnipresence of RNA degrading enzymes a successful RNA isolation requires immediate inactivation of all RNases upon lysis of the cells. This can be achieved by guanidinium isothiocyanate. In the original protocol the RNA is isolated by centrifugation through a CsCl step gradient with the density of 1.76 g/mL (Chirgwin et al., 1979), which leaves all proteins behind. The easier and less time consuming way is to extract the lysate with phenol, which can be done simultaneously with the lysis in a single step procedure (Chomczynski and Sacchi, 1987). Analysis of the RNA by gel-electrophoreses shows a smear of mRNAs of different sizes and a few prominent bands which result from the rRNAs (Fig. 5). The poly(A)$^+$ RNA is then isolated by binding it to an oligo(dT) cellulose column (Aviv and Leder, 1972) or to oligo(dT) beads (Promega, Dynal). By washing all RNAs (rRNA, tRNA, and partially degraded mRNA) which are not bound to the oligo(dT) matrix are removed and then the mRNA is eluted.

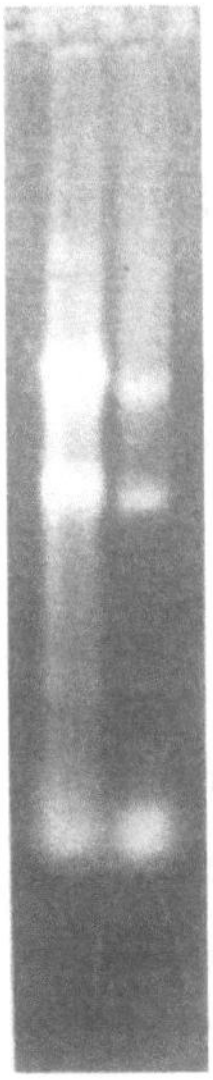

Figure 5 *Gelelectrophoresis of an RNA preparation*

Vector preparation

For cDNA libraries the choice of vector determines not only the ease with which the resulting clones can be grown and manipulated but also with what methods the library can be screened. If the library is to be screened with an antibody an expression vector has to be used. In choosing between phage and plasmid cloning systems several points should be considered. Formerly the cloning efficiencies of a factor 10–50 higher in the lambda phage system compared to plasmid cloning definitely favored the phage systems. However, with electroporation (see p. 215) similar efficiencies can be achieved using plasmid cloning. Plasmid vectors offer a wider variety of cloning sites, the possibility for vector-primer cloning, and different tailing strategies as discussed in (Kimmel and Berger, 1987). In addition, in most experiments plasmid clones are much easier to handle than phage clones. If there is no intention to array the library (see p. 215) and if the library is to be screened for transcripts of low abundance, phage are the system of choice, because they can be efficiently screened at high plaque densities with either nucleic acid or antibody probes. In our laboratory cDNA libraries are mainly constructed for fingerprinting with oligo-nucleotide probes (Meier-Ewert et al., 1993, 1994), which requires arraying of the library to high density filters. This is considerably easier using the plasmid cloning system, e.g. pSport (Gibco/BRL). This vector contains a multiple cloning site with unique restriction sites for 19 enzymes (arranged for easy generation of nested deletions), which is flanked by Sp6 and T7 RNA polymerase promoters. Cloned genes can be expressed by inducing the *lac*P promoter, which is otherwise efficiently repressed by the *lac*I gene product, also contained on the plasmid. This means clones coding for proteins interfering with the growth of *E. coli* are not lost.

The vector is commercially available already precut with *Sal*I and *Not*I, which allows directional cloning and ensures low vector background. If other primers and/or adapters are used, prepare a large batch vector according to standard procedures (Sambrook et al., 1989) and digest the DNA with the appropriate enzymes, to generate the compatible ends.

First strand synthesis

First strand synthesis is done with an RNA dependent DNA polymerase (reverse transcriptase) (Baltimore, 1970; Temin and Mizutani, 1970) which requires a primer with a 3′ hydroxyl group. One can either use an oligo(dT)-primer, which binds to the 3′ poly(A) tail of the mRNA or a mixture of all possible 6 mers (random primer). There are good reasons for either strategy: Oligo(dT) primer theoretically gives rise to full length copies. The yield however depends on the quality of the mRNA to be copied and the degree to which secondary structure of the mRNA interferes with extension of the primer. Random primers ensure equal representation of all mRNA sequences in the cDNA library, however at the expence of full-length cDNA copies. This strategy is often chosen, when the mRNA of interest is expected to be of such a large size that it seems unlikely that a complete oligo(dT) primed cDNA can be obtained. The sequence of the original mRNA can then be deduced from overlapping clones. Oligo (dT) priming is variation on a theme with a $(dT)_{12-14}$ primer. Anchored oligo(dT) primers include on or two bases other than Thymidine at the 3′ end of the primer, which "anchors" the priming at the start of the poly(A) tail, and prevent long stretches of non informative poly(A) from being cloned. A few specific extra bases at the 5′ end of the oligo (dT) primer result in a restriction enzyme site (e.g. of the rare cutter *Not*I) after second strand synthesis and facilitate easy cloning.

In an Eppendorf tube combine in a total volume of 13 µl: 1 µg mRNA and 1 µg *Not*I oligo (dT)-primer. To remove secondary structure of the mRNA heat up to 70°C for 10 min and chill on ice. Then add the following: 4 µL 5 × first strand buffer (250 mM Tris-HCl, pH 8.3, 375 mM KCl, 15 mM $MgCl_2$, 50 mM DTT), 1 µL dNTP mix (10 mM each), and to facilitate easier monitoring of all following reactions: 1 µL [α-^{32}P]dCTP (1 µCi/µL) and after brief equilibration at 37°C 1 µL ™RTII (Gibco BRL). Incubate for 1 h at 37°C; an aliquot of 2 µL is removed at this stage to evaluate the yield of first strand synthesis by determining the fraction of TCA insoluble (incorporated) radioactivity. With the remaining 18 µl immediately proceed to the following Section.

Second strand synthesis

During the second strand synthesis the mRNA strand is replaced with a DNA copy which resembles the original sequence of the message. In early protocols a small loop at the 3′ end of the first strand was used as "primer" for the second strand synthesis, after the RNA strand had been hydrolyzed with NaOH. The loop was cut open with S1 nuclease and the now double-stranded

DNA could be cloned. However, this route is no longer recommended, because S1 nuclease is a rather aggressive enzyme which is difficult to control. Today the RNA is partially degraded with RNaseH, and the RNA primers are utilised for replacement synthesis with DNA PolymeraseI. Its 5′–3′ exonuclease activity removes the RNA as it synthesizes the DNA strand.

To the first strand reaction (from p. 212) add 93 µl H_2O, 30 µL 5 × second strand buffer (250 mM Tris-Hcl, pH 7.5, 500 mM KCl, 25 mM $MgCl_2$, 50 mM $(NH_4)_2SO_4$, 250 µg/mL BSA, 7.5 mM DTT), 3 µL dNTP mix (10 mM), 4 µL *E. coli* DNA PolymeraseI (10 U/µL), 1 µL RNaseH (2 U/µL), and 1 µL *E. coli* ligase (10 U/µL), which especially enhances the cloning of larger cDNAs. The reaction is incubated at 16°C for 2 h. To ensure completely blunt ends of the DNA molecules 2 µL T4 DNA polymerase (5 U/µL) are added and incubation is continued for an additional 5 min. The reaction is stopped with 10 µL 0.5 M EDTA, 5 µg of yeast tRNA are added, and the mixture is extracted with an equal volume phenol:chloroform:isoamylalcohol (25:24:1). The cDNA is precipitated by addition of ammonium acetate to a final concentration of 2.5 M and 2 vol. ethanol and recovered by centrifugation.

SalI/blunt adapter addition, *NotI* digestion, and size fractionation

Ligations between blunt ended DNA molecules are not very efficient. To increase the efficiency of the ligation between insert and vector, *SalI*/blunt-adapters are ligated to the cDNA (of course any other adapter for which the vector has a cloning site can be chosen, too). Adapters are short double-stranded DNA molecules which allow ligation of DNA molecules with otherwise not compatible ends, in this case between SalI fragments and blunt ended fragments. To prevent forming of long adapter concatemers, only one strand is phosphorylated.

To the cDNA from above add in a total volume of 50 µL: 10 µL 5 × PEG-ligase buffer (250 mM Tris-HCl, pH 7.8, 50 mM $MgCl_2$, 25% w/v PEG8000, 5 mM dTT, 5 mM ATP, 0.25 mg/mL BSA), 10 µg *SalI*/blunt adapters, and 5–10 U T4 DNA ligase, and incubate the mixture at 16°C for 12–16 h. The mixture is extracted with an equal volume phenol:chloroform:isoamylalcohol (25:24:1). The cDNA is precipitated by addition of ammonium acetate to a final concentration of 2.5 M and 2 vol. ethanol and recovered by centrifugation.

To release the *Not*I site, which generates an asymmetry to the cDNA ends and thereby allows directional cloning, resuspend the DNA in 40 µl H_2O, add 5 µl 10 × *Not*I reaction buffer (500 mM Tris-HCl, pH 8.0, 100 mM $MgCl_2$, 1 M NaCl, 1 µg/µL BSA) and 5 µL *Not*I (10 U/µL) and incubate for 1–2 h at 37°C. Phenol extraction and ethanol precipitation of the DNA follow as described above. The DNA is resuspended in 100 µL TEN buffer (10 mM Tris-HCl, pH 7.5, 0.1 mM EDTA, 25 mM NaCl).

Gel-chromatography on a Sepharose CL-4B or Sephacryl S-500 HR column efficiently removes not only excess of adapter, but also allows size fractionation of the cDNA molecules. Load the cDNA on top of a mini-column (ca. 1 mL bed volume) which was been equilibrated with TEN buffer. Discard the non-radioactive effluent and collect the radioactive effluent in one-drop fractions (ca. 30–40 µL), wash the column with TEN buffer until all radioactivity is removed from the column. The longest fragments will be in the first fractions. To determine the actual size range a small aliquot of each fraction can be run on an agarose gel together with radioactively labeled size markers, which is then dried and exposed to x-ray film. To evaluate the amount of cDNA, the amount of radioactivity in each fraction is determined by Cerenkoff counting.

Vector ligation

To ligate the cDNA fragment to the vector combine in a 20 µl reaction 4 µl 5 × PEG-ligase buffer (see p. 213) 50 ng of *Sal*I/*Not*I cut vector (e.g. pSport1 from Gibco/BRL) 10 ng size fractionated cDNA (from above), and 1–5 U T4 DNA ligase. Incubate for several hours at room temperature, and then precipitate the DNA by addition of ammonium acetate and ethanol as above. Collect the DNA by centrifugation and carefully wash the pellet with 70% ethanol to remove as much as possible of the salts as they will interfere with the electroporation. Dissolve the DNA in 1 mM Tris-HCl, pH 8.0.

Transformation and titrating the library

Electro-competent cells are commercially available, but they can easily be prepared by washing *E. coli* (e.g. DH10B), which are growing in the logarithmic phase, three times in ice-cold 10% v/v glycerol in water. After the last wash the cells are resuspended in as small as possible a volume. Aliquots are frozen in a dry ice-ethanol bath and kept frozen at –80°C until needed.

Twenty-five µL of cells are thawed and 1 µL of the ligation reaction is added and mixed. All 26 µL are transferred to an electroporation cuvette with a 1 mm gap (Biorad or Euro Gentech) and pulsed at 1.68 kV, 25 µF and 200 Ω in the Biorad genepulser. The displayed time constant (τ) should be around 4.8–5.3 ms. Immediately add 1 mL SOC medium (20 g/L BactoTryptone, 5 g/L Bacto yeast extract, 0.58 g/L NaCl, 0.186 g/L KCl, 3.6 g/L glucose, 10 mM $MgCl_2$) and incubate with vigorous shaking at 37°C for 1 h. Make a serial dilution and plate the equivalent of 1, 0.1, and 0.01 µL on LB plates with 80–100 µg/mL ampicillin. The remaining cell suspension is stored overnight at 4°C, or if it is not possible to screen or array the library immediately mixed with 100 µL glycerol and stored frozen in several aliquots at – 80°C, which will however result in the loss of around 50% of the clones.

Arraying of libraries

As discussed in the introduction the only efficient way to preserve the original distribution of clones in a library is to array it. This means that each primary clone is transferred to an individual well of a microtitre plate. Growing the clones separately slows down fast growing clones as the media in the well gets depleted of nutrients and prevents them from competing with the more slowly growing clones.

For robotic picking the library is plated to a density of ca. 8000 colonies on a 22 × 22 cm LB plate containing the appropriate antibiotic selection and incubated at 37°C for 18–20 h. Using an XYZ computer controlled picking head with a CCD camera and a 96-pin picking gadget (Fig. 6), individual colonies are identified, picked, and transferred to wells of quadruple density (384 well) microtitre plates, filled with 60 µL 2YT medium per well, containing the selective antibiotic and 1 × HMFM as cryoprotectant. (10 × HMFM: 360 mM K_2HPO_4, 132 mM KH_2PO_4, 17 mM sodium citrate, 68 mM $(NH_4)_2SO_4$, 36 mM $MgSO_4$ (autoclaved separately and added after cooling to room temperature) 44% w/v glycerol). The robot can pick up to 4000 clones per hour. After inoculation the microtitre plates are incubated at 37°C for 20 h and at

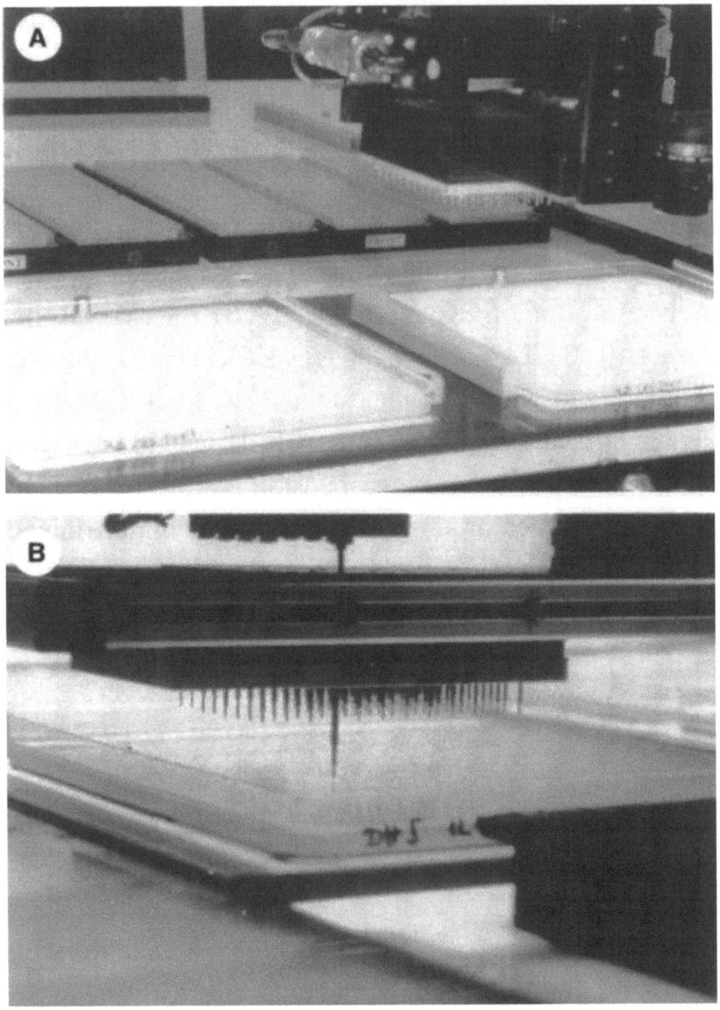

Figure 6 *Picking robot:* (A) Picking head with CCD camera and 96-pin picking gadget, two 22 × 22 cm plates with randomly plated primary clones of a cosmid library. (B) One pin of the picking gadget extended to pick a single colony

least one further duplicate is made of each plate, which is used as spotting copy. The original plate can be stored indefinitely at – 70°C.

To generate high density hybridisation grids, the clones are spotted onto 22 × 22 cm HybondN$^+$ membranes (Amersham), which are placed on blotting paper pre-wetted with 2YT media (16 g/L BactoTryptone, 10 g/L Yeast extract, 5 g/L NaCl), using a 384 pin spring loaded spotting gadget, which can be mounted on the same robot used for picking. The computer controlled XYZ drives can place the gadget within a tolerance of 5 µm on a ca. 1.2 × 1.2 m working surface, which allows simultaneous production of up to 15 identical hybridization membranes with colony densities of up to 112 000 colonies on one membrane (7 × 7 spotting pattern). Routinely, we spot the filters in a 5 × 5 spotting pattern, which means that 12 different 384-well microtitre plates are spotted in duplicate on the same 8 × 12-cm-sized membrane surface (field), each shifted by ca. 0.8 mm against each other. To facilitate easy identification the distance and relative orientation of the duplicate spots to each other are different for each microtitre plate of a field. This way 27 648 different clones are spotted on a 22 × 22 cm membrane. At the end of the robot run the membranes are placed on 2YT agar plates containing the appropriate antibiotic selection and incubated for 16–20 h at 37°C. The colonies are lysed *in situ* and the DNA is bound to the membranes using a modified protocol of Grunstein and Hogness (Grunstein and Hogness, 1975; Nizetic et al., 1991):

The membrane is placed on a blotting paper prewetted with denaturant (0.5 M NaOH, 1.5 M NaCl) for 3 min. Then the membrane is transferred together with the blotting paper to a glass plate above the water level in a 95°C water bath and is steamed for 3 min. The membrane is then put on a fresh blotting paper prewetted with neutralising buffer (1 M Tris-HCl, pH 7.5, 1.5 M NaCl) and then transferred into Pronase buffer (50 mM Tris-HCl, pH 8.5, 50 mM EDTA, 100 mM NaCl, 1% N-Lauroylsarcosinate) with 150 µg/mL Pronase and incubated at 37°C for 20–40 min. After drying the DNA is crosslinked to the membrane with UV light, e.g. in the Stratalinker (Stratagene) in the autocrosslink mode.

Hybridization of high density filters

High density colony membranes can be hybridized with different nucleic acid probes, e.g. RNA, DNA, or oligonucleotides which are labeled either radioactively, which allows detection by autoradiography, or another small group like biotin which can be detected via enzyme linked antibodies. Although during recent years the nonradioactive detection systems have improved much, both in sensitivity and ease of handling, radioactively labeled probes offer still highest sensitivity with least hands on time. The most commonly used probes, PCR products or other DNA fragments are labeled to high specific activity by random primed labeling (Feinberg and Vogelstein, 1983). The high density colony membranes are prehybridized for at least 1 h in the hybridization buffer, and the buffer is replaced with fresh hybridization buffer, especially if the filters are used for the first time. The denatured (and in case of repetitive elements containing probes with competitor preannealed) probes are then added. We commonly use Church buffer (Church and Gilbert, 1984) (7% w/v SDS, 0.5 M sodium phosphate, pH 7.4) or formamide hybridization mix (30–50% v/v formamide, 8% w/v dextran sulphate, 4 × SSC, 50 mM sodium phospate pH 7.2, 1 mM EDTA, 10 × Denhardt's, 25 µg/ml sonicated salmon sperm DNA, 1% SDS). The latter with only 30% formamide is especially used for cross species hybridizations at 42°C. Church buffer is routinely used at 65°C, but to increase the specificity the temperature can be raised to 72°C. Hybridization is allowed to proceed for 16 h, and the membrane is washed to remove all probe not specifically hybridized to the DNA on the membrane. The stringency of the washing steps depends on a number of characteristics of the probe and has to be determined empirically. It is possible to raise the stringency gradually and expose the membrane to x-ray film between the washing steps, providing the membrane was never allowed to dry completely. As a rule of thumb, good single copy probes can be washed at a very high stringency (0.2–0.1 × SSC, 0.1% SDS at 65°C, 20 × SSC: 3 M NaCl, 300 mM sodium citrate) for 1–1.5 h with several changes of the washing buffer. Probes intended to detect new members of a gene family or cross species hybridizations can be washed at 2–5 × SSC, 0.1% SDS. After the last wash, the membrane is wrapped between two layers of Saran wrap and exposed to x-ray film in a light tight exposure cassette with intensifying screens at – 70°C.

7.3 Use of reference libraries and building up a database

On a high density array of colonies each coordinate corresponds to a specific clone of the library and determines row and column as well as number of the microtitre plate. Therefore the result of any hybridization experiment is more or less black spots on an x-ray film (Fig. 7) which can be converted to a list of potentially positive clones. If many hybridizations are performed, and this list

Figure 7 *Autoradiography of a 22×22 cm colony membrane with 18400 Fugu-cosmid clones spotted in a duplicate pattern (representing a 1.5 times genome coverage), hybridized with a single cDNA fragment. The two positives (double spots) are marked with their coordinates*

together with information on the probe is entered into a searchable database several things can be achieved: i) The library is thoroughly characterized, and ii) if the position of the probe on a genetic map was known this links the positive clones to this position, which especially in the case of large insert libraries eventually will result in a physical map. iii) If cDNA libraries are screened one will eventually characterize the expression level of each gene of the source tissue. This is especially interesting if different tissues or development stages are compared this way with each other. iv) cDNAs used as probe on genomic libraries will eventually result in a transcript map.

High density colony filters are available for screening from a number of genome centers *via* the worldwide web (e.g.: http://www.hgmp.mrc.ac.uk/ or http://www.rzpd.de) funded by national governments and from commercial sources (e.g.: http://genomesystems.com/). The German RZPD (Ressourcen-

ZentrumPrimärDatenbank) for example offers a total of 52 publicly available gridded libraries, and holds copies of ca. 80 further not publicly available libraries which are available for collaborative efforts. By March 1997, almost 1000 researchers requested hybridization membranes and almost 10 000 membranes have been sent out. More than 5000 probes have been used to screen one or more of the libraries, and 82 300 potentially positive clones have been sent out by the RZPD and its predecessor the RLDB (ReferenceLibrary-DataBase) which was established in 1989 at the ICRF in London (Günther Zehetner, personal communication). It is obvious that this represents a major combined effort, which – to avoid unnecessary duplication of work – should be made publicly available (Zehetner and Lehrach, 1994). Therefore each clone request to the RZPD from a hybridization experiment has to be accompanied with a probe information sheet which characterizes the origin, type, and chromosomal location of each probe. And this information, together with the hybridization data gets entered into a database and becomes public after 6 months.

7.4 Conclusions

In many different large scale and small scale genome mapping projects high quality arrayed libraries have played an important role. It is easier and faster to screen a membrane with a clone array of an existing library than it is to construct the library and screen it using colony lifts from randomly plated clones. Isolating clones from randomly plated libraries usually requires several rounds of rescreening, whereas a clone from an arrayed library is in most cases already pure.

Although the libraries available already represent an invaluable tool, more and better libraries using better and more versatile vectors are being constructed. In spite of the vast variety of vectors available already (http://www.atcg.com/vectordb/), almost weekly more vectors with additional features are being advertised. Especially for PAC and BAC vectors, which often allow cloning of complete genes, eukaryotic selectable markers, allowing functional studies *in vivo*, would be very useful. But for sequencing of clones from these libraries the vectors should remain as small as possible. A good alternative might be "modular" vectors which are easily modified, e.g. using transposon based systems, the Cre/lox system, or rare cutting restriction enzymes (Chatterjee and Sternberg, 1996; Mejía and Monaco, 1997).

The development of data bases will allow immense amounts of information to be rapidly accessed. This will lead to the development of so-called "*in silica*" experiments and will lead a veritable explosion of knowledge as this information is used.

Acknowledgement

We thank Michael V. Wiles and Leo C. Schalkwyk for comments and suggestions on the manuscript.

References

Aviv H, Leder P (1972) Purification of biological active globin mRNA by chromatography on oligothymidylic acid cellulose. *Proc Natl Acad Sci USA* 69: 1409–1412

Baltimore D (1970) RNA-dependent DNA polymerase in virions of RNA tumour viruses. *Nature* 226: 1209–1211

Birren BW, Lai E, Clark SM, Hood L, Simon MI (1988) Optimized conditions for pulsed field gel-electrophoretic separations of DNA. *Nucleic Acids Research* 16: 7563–7582

Breathnach R, Chambon P (1981) Organization and expression of eucaryotic split genes coding for proteins. *Ann Review Biochem* 50: 349–383

Burke DT, Carle GF, Olson MV (1987) Cloning of large segments of exogenous DNA into yeast by means of artificial chromosome vectors. *Science* 236: 806–812

Calvin NM, Hanawalt PC (1988) High-efficiency transformation of bacterial cells by electroporation. *J Bacteriology* 170: 2796–2801

Chatterjee PK, Sternberg NL (1996) Retrofitting high molecular weight DNA cloned in P1: introduction of reporter genes, markers selectable in mammalian cells and generation of nested deletions. Genetic Analysis: *Biomolecular Engineering* 13: 33–42

Chirgwin JM, Przybyla AE, MacDonald RJ, Rutter WJ (1979) Isolation of biological active ribonucleic acid from sources enriched in ribonuclease. *Biochem* 18: 5294–5299

Chomczynski P, Sacchi N (1987) Single-step method of RNA isolation by aid guanidinium thiocyanate-phenol-chloroform extraction. *Anal Biochem* 162: 156–159

Church GM, Gilbert W (1984) Genomic sequencing. *Proc Natl Acad Sci USA* 81: 1991–1995

Collins J, Bruning HJ (1978) Plasmids useable as gene-cloning vectors in an *in vitro* packaging by coliphage lambda: "cosmids". *Gene* 4: 85–107

Cox RD, Stubbs L, Evans T, Lehrach H (1991) A mouse specific polymerase chain reaction (PCR) primer: probe generation from somatic cell hybrids. *Nucleic Acids Research* 19: 2503

Dagert M, Ehrlich SD (1979) Prolonged incubation in calcium chloride improves the competence of *Escherichia coli* cells. *Gene* 6: 23–28

Feinberg AP, Vogelstein B (1983) A Technique for Radiolabeling DNA Restriction Endonuclease Fragments to High Specific Activity. *Anal Biochem* 132: 6–13

Francis F, Zehetner G, Höglund M, Lehrach H (1994) Construction and Preliminary Analysis of the ICRF Human P1 Library. *Genetic analysis: Techniques and Applications* 11: 148–157

Frischauf AM, Lehrach H, Poustka A, Murray N (1983) Lambda-replacement vectors carrying polylinker sequences. *J Molec Bio* 170: 827–842

Grunstein M, Hogness DS (1975) Colonie Hybridisation: A Method for the Isolation of Cloned DNAs That Contain a Specific Gene. *Proc Nat Acad Sci USA* 72: 3961–3965

Herrman BG, Frischauf AM (1987) Isolation of genomic DNA. *Methods in Enzymology* 152: 180–183

Hoheisel JD, Lennon GG, Zehetner G, Lehrach H (1991) Use of high coverage reference libraries of *Drosophila melanogaster* for relational data analysis. A step towards mapping and sequencing of the genome. *J Molec Bio* 220: 903–914

Ioannou PA, Amemiya CT, Garnes J, Kroisel PM, Shizuya H, Chen C, Batzer MA, Dejong PJ (1994) A new bacteriophage P1-derived vector for the propagation of large human DNA fragments. *Nature Genet* 6: 84–89

Ish-Horowicz D, Burke JF (1981) Rapid and efficient cosmid cloning. *Nucleic Acids Research* 9: 2989–2998

Kimmel AR, Berger SL (1987) Preparation of cDNA and the Generation of cDNA Libraries: Overview. *Methods of Enzymology* 152: 307–337

Meier-Ewert S, Maier E, Ahmadi A, Curtis J, Lehrach H (1993) An automated approach to generating expressed sequence catalogs. *Nature* 361: 375–376

Meier-Ewert S, Rothe J, Mott R, Lehrach H (1994) Establishing catalogues of expressed sequences by olionucleotide fingerprinting of cDNA libraries. *In*: U Hochgeschwender, K Gardiner (eds): *Identification of transcribed sequences.* Plenum Press, New York

Mejía JE, Monaco AP (1997) Retrofitting Vectors for *Escherichia coli*-based Artificial Chromosomes (PACs and BACs) with Markers for Transfection Studies. *Genome research* 7: 179–186

Murray NE, Murray K (1974) Manipulation of restriction targets in phage lambda to form receptor chromosomes for DNA fragments. *Nature* 251: 476–481

Nelson DL, Ledbetter SA, Corbo L, Victoria MF, Ramirez-Solis R, Webster TD, Ledbetter DH, Caskey CT (1989) Alu polymerase chain reaction: a method for rapid isolation of human-specific sequences from complex DNA sources. *Proc Nat Acad Sci USA* 86: 6686–6690

Nizetic D, Monard S, Young B, Cotter F, Zehetner G, Lehrach H (1994) Construction of cosmid libraries from flow-sorted human chromosomes 1, 6, 7, 11, 13, and 18 for reference library resources. *Mammalian Genome* 5: 801–802

Nizetic D, Drmanac R, Lehrach H (1991). An improved bacterial colony lysis procedure enables direct DNA hybridisation using short (10, 11 bases) oligonucleotides to cosmids. *Nucleic Acids Research* 19: 182

Sambrook J, Fritsch EF, Maniatis T (1989) Molecular Cloning: A Laboratory Manual, Second ed. Cold Spring Harbor Laboratory, Cold Spring Harbor, New York

Schalkwyk LC, Francis F, Lehrach H (1995) Techniques in mammalian genome mapping. *Curr Opin Biotechnol* 6: 37–43

Sheng Y, Mancino V, Birren B (1995) Transformation of *Escherichia coli* with large DNA molecules by electroporation. *Nucleic Acids Research* 23: 1990–1996

Sternberg N (1994) The P1 cloning system – past and future. *Mamm Genome* 5: 397–404

Sternberg N, Ruether J, deRiel K (1990) Generation of a 50000-member human DNA library with an average DNA insert size of 75–100 kbp in a bacteriophage P1 cloning vector. *New Biol* 2: 151–162

Sternberg NL (1992) Cloning high-molecular-weight DNA fragments by the bacteriophage-P1 system. *Trends In Genetics* 8: 11–16

Temin HM, Mizutani S (1970) RNA-dependent DNA polymerase in virions of Rous sarcoma virus. *Nature* 226: 1211–1213

Tohma T, Tamura T, Ohta T, Soejima H, Kubota T, Jinno Y, Tsukamoto K, Nakamura Y, Naritomi K, Niikawa N (1993) Cosmid clones from microdissected human chromosomal region 15q11–q13. *Jap J Human Gen* 38: 267–275

Zehetner G, Lehrach H (1994) The reference library-system – sharing biological-material and experimental-data. *Nature* 367: 489–491

8 Generation and screening of solution-phase synthetic peptide combinatorial libraries

Andrew Wallace, Kenneth S. Koblan, Riccardo Cortese, Jackson B. Gibbs and Antonello Pessi

8.1 Introduction

Following the initial use of libraries of peptides displayed on filamentous bacteriophage [1] it was inevitable that the potential of solid-phase peptide synthesis should be harnessed to produce libraries of peptides by synthetic chemical methods. These libraries made their first appearance in the mid-1980s following the development of multiple peptide synthesis by the groups of Geysen [2], Furka [3, 4] and Houghten [5], but it was not until the early 1990s that synthetic peptide combinatorial libraries (SPCLs) were described in the literature [6–8]. This was followed by a series of papers detailing the use of peptide libraries for the discovery of novel antibacterial and antinoistic compounds [9–12] (for a review, see [13]).

Classification of peptide libraries

Peptide libraries can usefully be classified into three types:

1. Bacteriophage surface-displayed libraries
2. Support-bound libraries (solid-phase libraries)
3. Non-support-bound or solution-phase libraries

It is this latter class of library which will be discussed in this chapter. There are, however, some similarities and distinctions between the various classes which should be pointed out first. Library class 1 makes use of molecular biological methods to generate libraries of peptides attached as fusion proteins to the filamentous bacteriophage coat proteins (Ff type phage are the most commonly used) [14]. Diversity in these libraries is introduced by genetic and recombinant DNA methods on a suitable bacteriophage vector. The vector is then propagated in an appropriate bacterial host where biosynthesis of the peptide-coat protein fusion product and subsequent phage assembly takes place. Mature phages displaying the peptide library are secreted by the bacterial host, recovered and screened by a variety of methods [15].

Library classes 2 and 3 are, in contrast to the above, both produced by chemical methods such as solid-phase peptide synthesis but are screened for activity by different methods. Library class 2 is somewhat similar to biological class 1 libraries in that peptides are screened whilst "displayed" on a solid-phase support [8, 16]. Library class 3 is, on the other hand, usually synthesised by solid support-bound methods but then the support bound peptides are released into solution for screening [7, 9, 11]. This leads to considerable differences used to identify active peptides from these latter two types of libraries. Class 2 is usually screened by a "one bead, one compound" approach where a variety of methods such as automated Edman degradation [17, 18], chemical tagging [19–24], radiofrequency encoding of library-microchip assemblies [25] or mass spectrometry [26–29] are used to detect the active species. Class 3 libraries can be screened by deductive approaches such as iterative deconvolution [7] or positional scanning [9].

An example of the screening of a class 3 library for inhibitors of farnesyl protein transferase by a positional scanning deductive method will now be given [30].

8.2 Inhibitors of Farnesyl-protein transferase selected from a positional scanning tetrapeptide combinatorial library

Farnesyl-protein transferase (FPTase) catalyses the transfer of a farnesyl (C_{15}) isoprenoid moiety from Farnesyl pyrophosphate (FPP) to a cysteine thiol in protein substrates such as the *ras* oncogene protein Ras. Farnesylation is an important post-translational modification of Ras which anchors the protein to the inner surface of the cell membrane. This membrane-anchoring is essential to enable Ras to transform cells, thus the discovery of FPTase inhibitors is at the focus of intense research to develop novel antiproliferative agents.

The cysteine thiol recognised by FPTase is located within a specific C-terminal tetrapeptide based on the CAAX sequence, where C is cysteine, A is an aliphatic amino-acid and X is serine or methionine. Peptidomimetic inhibitors based on the CAAX tetrapeptide have been developed which show nanomolar potency towards FPTase [31–34]. These peptidomimetic inhibitors typically contain a free thiol with the C-terminal carboxylate esterified to create a prodrug, both of which are features that may be metabolic liabilities to the activity of FPTase inhibitors in animals. We wished to discover new FPTase inhibitors which did not contain these undesirable features and therefore synthesised and screened a peptide combinatorial library composed of 2×10^7 C-terminally amidated tetrapeptides where no free thiol-containing amino acid was used amongst the library building blocks.

Synthesis of the tetrapeptide library

In an attempt to compensate for the loss of the known FPTase binding features we used an expanded repertoire of library building blocks which included both L- and D-isomers of the genetically encoded amino acids as well as many non-coded amino acids of both L and D chirality, giving 67 building blocks in all (Tab. 1). These are commercially available and were purchased from Novabiochem (Läufelfingen, Switzerland), Bachem (Bubendorf, Switzerland) or Neosystem (Strasbourg, France) in the form of Fmoc (*N*-(9-fluorenyl)methoxycarbonyl)/*tert*-butyl protected derivatives. Peptides were assembled on batches (100 mg; 0.03 meq) of Polyhipe SU 500 resin to which a Rink amide linker had been coupled (NovaSyn PR 500 resin; Novabiochem) using a Zinsser Analytic SMPS 350 multiple peptide synthesiser (Frankfurt, Germany) with TBTU/HOBt activation (single coupling protocol, five-fold excess of activated species). For the tetrapeptide positional scanning combinatorial library we synthesised four different sublibraries, $H\text{-}O_1XXX\text{-}NH_2$, $H\text{-}XO_2XX\text{-}NH_2$, $H\text{-}XXO_3X\text{-}NH_2$ and $H\text{-}XXXO_4\text{-}NH_2$. Each sublibrary was composed of 67 peptide mixtures, with the defined amino acid (O_n) being one of those indicated in Table 1. The X or "mixed" positions were incorporated using a kinetically-balanced mixture of Fmoc amino acids whose relative concentration was adjusted to obtain equimolar incorporation (± 25%) [35] of each residue. The peptides were cleaved from the resin by 1.5 h

Table 1 Amino acids used in the kinetically-adjusted mixture coupling

Name (abbreviation)	Relative ratio
L-α-Aminobutyric (Abu)	1
γ-Aminobutyric (γAbu)	1
5-Aminovaleric (Ava)	1
6-Aminohexanoic (εAhx)	1
8-Aminooctanoic (Aoc)	1
Aminoisobutyric (Aib)	2
β-Alanine (βAla)	1
β-Cyclohexyl-L-alanine (CHA)	1
β-Cyclohexyl-D-alanine (cha)	1
3,4-Dehydro-L-proline (dhP)	1
γ-Carboxyglutamic (Gla)	1
Hydroxy-L-proline (Hyp)	1
L-Norleucine (NLE)	1
D-Norleucine (nle)	1
L-Norvaline (NVA)	1
D-Norvaline (nva)	1
L-Ornithine (ORN)	1
Homo-L-phenylalanine (Hph)	1.4
L-Phenylglycine (PHG)	1.4
D-Phenylglycine (phg)	1.4
p-Chloro-L-phenylalanine (FCL)	1.4
p-Chloro-D-phenylalanine (fcl)	1.4
p-Nitro-L-phenylalanine (FNO)	1.4
Sarcosine (Sar)	1
2,3-Diamino-a-L-propionic (αDP, either α or β-NH_2 free for coupling)	1
(3S,4S)-4-amino-3-hydroxy-5-phenylpentanoic (AHPPA)	1
(3S,4S)-4-amino-3-hydroxy-5-cyclohexylpentanoic (ACHPA)	1
(3S,4S)-4-amino-3-hydroxy-6-methylheptanoic (Statine)	1
1,2,3,4-Tetrahydroisoquinoline-3-L-carboxylic (TIC)	1.4
1,2,3,4-Tetrahydroisoquinoline-3-D-carboxylic (tic)	1.4
L/D-Ala	1
L/D-Arg	2
L/D-Asn	1
L/D-Asp	1
L/D-Gln	1
L/D-Glu	1
Gly	1
L/D-His	1

Table 1 (continued)

Name (abbreviation)	Relative ratio
L/D-Ile	2
L/D-Leu	1
L/D-Lys	1
L/D-Met	1
L/D-Phe	1.4
L/D-Pro	1
L/D-Ser	1
L/D-Thr	1
L/D-Trp	1.4
L/D-Tyr	1.4
L/D-Val	2

treatment with Reagent B [36], precipitated with cold methyl-*tert*-butyl ether, dissolved in 5% acetic acid/50% acetonitrile and lyophilized twice in pre-weighed microcentrifuge tubes. The lyophilized material was then dissolved at the appropriate concentration in dimethyl sulphoxide. The concentrations of the resulting sublibrary pools were determined by amino acid analysis. They were all of comparable concentration (± 15%) and all the samples showed the expected amino acid ratios when assesed by quantitative amino acid analysis using two standard amino acid mixtures (natural and uncoded) for calibration. Trp, labile to the acid hydrolysis conditions used (tube sealed under vacuum, azeotropic HCl, 110°C, 18 and 24 h), was not determined. Asn/Asp and Gln/Glu each eluted as a single peak for Asx and Glx, respectively, and for a few of the 41 chromatographic peaks we could not achieve baseline separation; within these limits, we estimated the maximum deviation from equimolarity to be ± 30%.

Screening of the tetrapeptide library

FPTase inhibition assay

The sublibrary pools were screened for inhibition of recombinant FPTase activity using a published assay [37]. Briefly, a typical reaction mixture in a final volume of 50 µl contained 100 nM (^{3}H)FPP and 100 nM Ras-CVIM substrate peptide in 50 mM HEPES buffer, pH 7.5, containing 5 mM $MgCl_2$, 5 mM dithiothreitol, 1% polyethylene glycol 15 000, 100 mM $ZnCl_2$ and 1 nM FPTase. The individual sublibrary mixtures were prepared as 40 mM stock solutions in 100% dimethyl sulphoxide. The solutions were diluted in the assay buffer to the appropriate final concentration, yielding a final dimethyl sulphoxide concentration of 5%. Reactions were initiated at 30°C and stopped after 15 min by the addition of 10% HCl in ethanol (2 ml). The quenched reactions were vacuum-filtered through Whatman GF/C filters. Filters were washed four times with 3 ml-aliquots of 100% ethanol, mixed with scintillation fluid (10 ml) and counted in a Beckman LS3801 scintillation counter. IC_{50} values were determined using a five-point titration where the concentration of inhibitor was varied from 0.1 to 1000 nM.

Library screening

The peptide mixtures of the H-O_1XXX-NH_2 sublibrary were initially screened for FPTase inhibition at a final concentration of 2 mM (Fig. 1; grey bars) where each peptide is at a final individual concentration of around 6 nM. Since several pools gave >50% inhibition, the most active mixtures were rescreened at 0.4 mM Fig. 1; black bars). Under these conditions, D-Trp was the most active residue, followed by L-Hph. The other sublibraries H-XO_2XX-NH_2, H-XXO_3X-NH_2 and H-XXXO_4-NH_2 were all screened at 0.4 mM (Figs 2–4).

Identification of novel FPTase inhibitors

The consensus sequences derived from the screening are shown in Table 2. Large aromatic residues were preferred in the first position, D-amino acids in the second and third positions, but the equivalent L-residues were also quite active. The dicarboxylic acid Gla was by far the preferred residue in the fourth position; however, neither L-/D-Asp nor Glu was active in this position. All the possible combinations of the selected residues would include 60 sequences, therefore we selected a more restricted consensus sequence (residues in boldface in Tab. 2) and prepared 16 peptides derived from it. When screened for inhibition of FPTase activity, all the peptides were active, with IC_{50} values

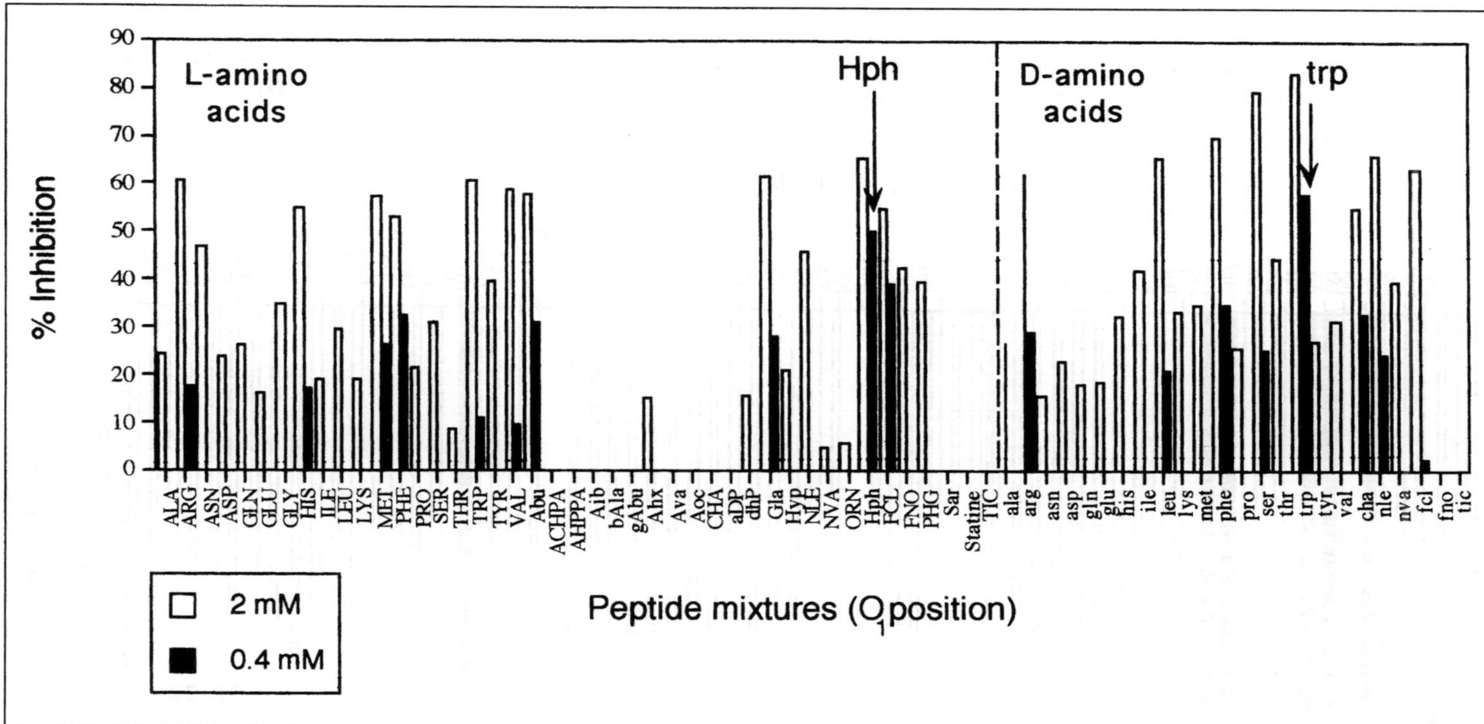

Figure 1 *Inhibition of FPTase activity by sublibrary H-O_1XXX-NH_2. The graph shows percent inhibition of FPTase activity plotted against the identity of the O_1 amino acid for each of the sublibrary pools. Grey bars, final concentration of the peptide mixtures is 2 mM (about 6 nM for each individual peptide); black bars, final concentration is 0.4 mM. When both the L- and D-forms of an amino acid are present, the L-isomer is shown in upper case three letter code and the D-isomer in lower case; abbreviations are defined in Table 1. Only the most active pools at 2 mM were retested at 0.4 mM and the most reactive residues at the lower concentration are indicated by arrows*

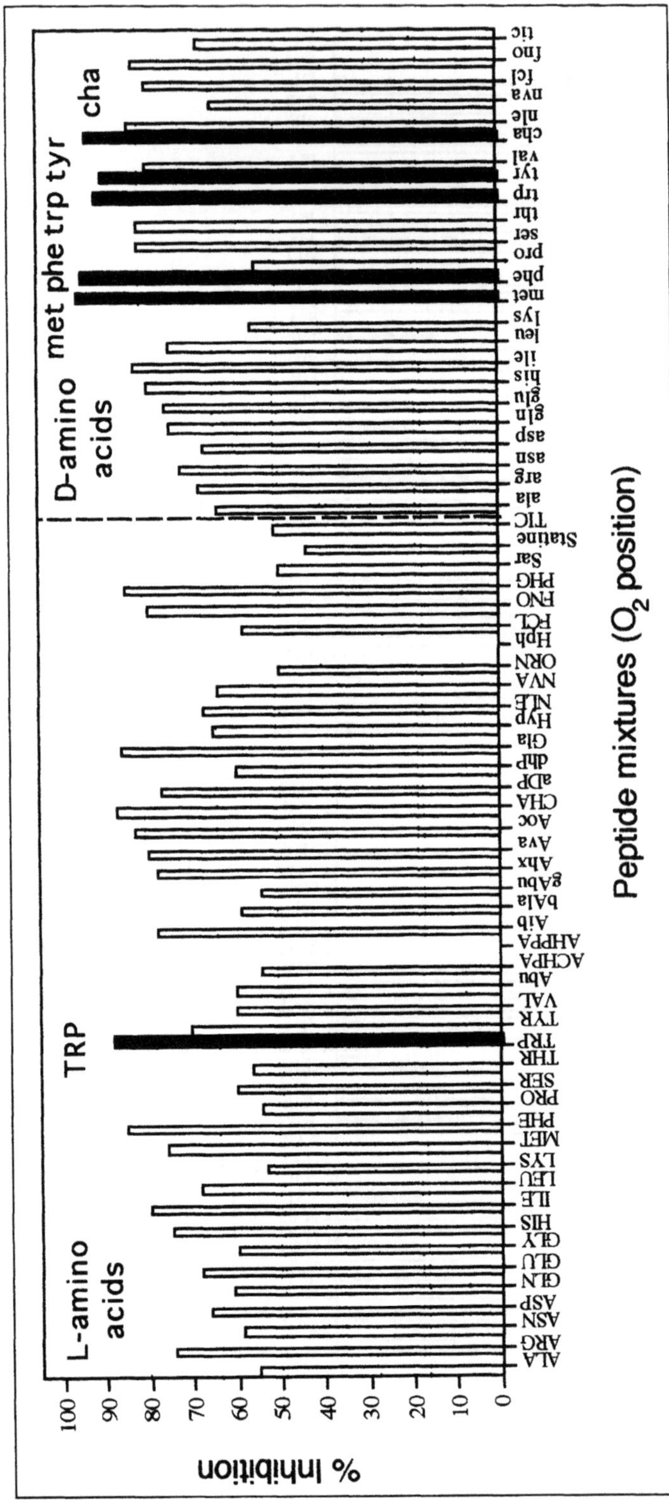

Figure 2 *Inhibition of FPTase activity by sublibrary H-XO_2XX-NH_2. The final concentration of each pool in the defined (O_2) position was 0.4 mM (about 1 nM for each peptide). The most active residues are shown as dark grey bars with the corresponding amino acid indicated above each bar. Other details are as for Figure 1*

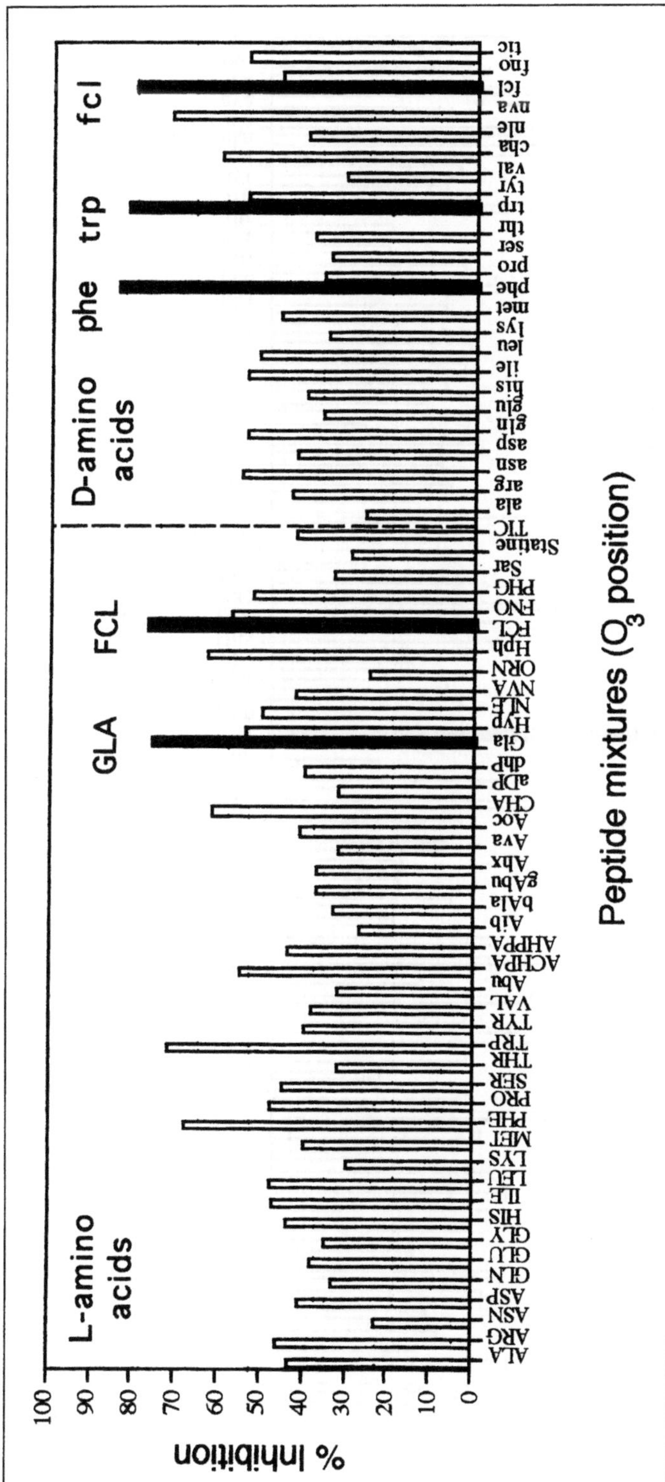

Figure 3 *Inhibition of FPTase activity by sublibrary H-XXO$_3$X-NH$_2$. The final concentration of each pool in the defined (O$_3$) position was 0.4 mM (about 1 nM for each peptide). Other details are as for Figures 1 and 2*

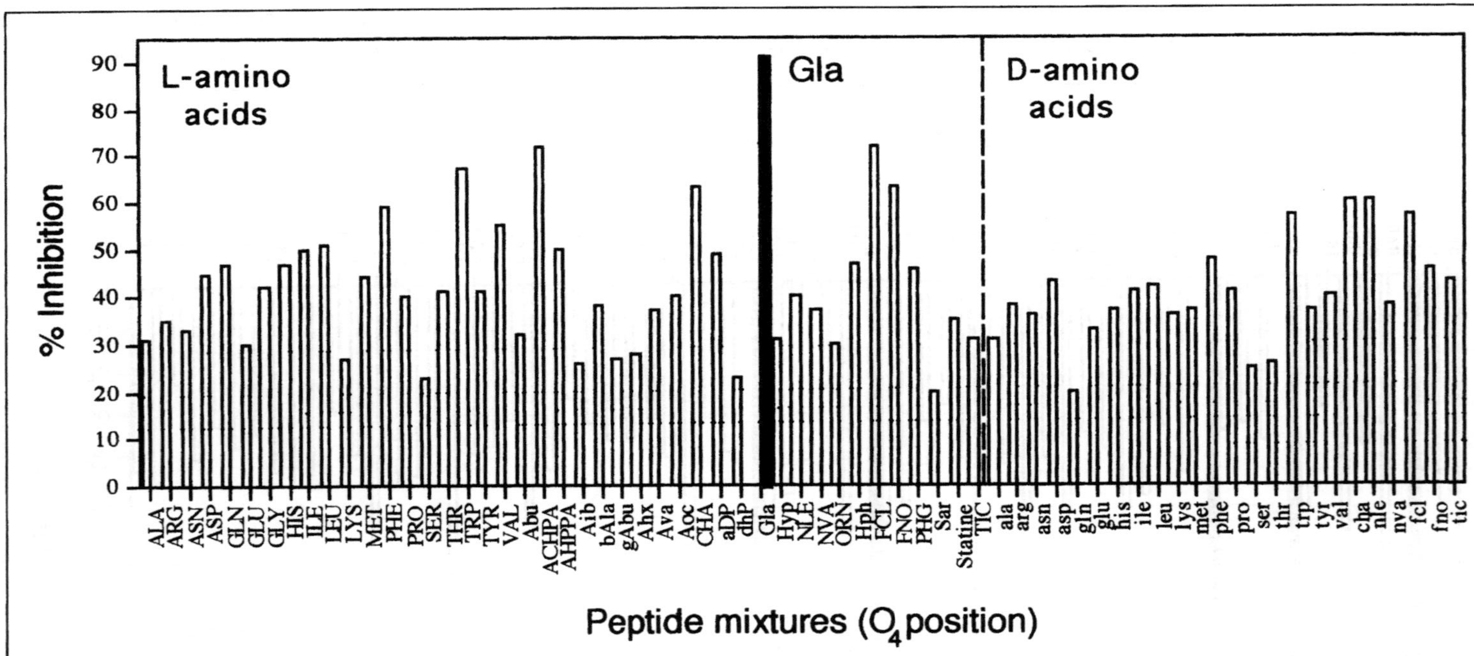

Figure 4 *Inhibition of FPTase activity by sublibrary H-XXXO_4-NH_2. The final concentration of each pool in the defined (O_4) position was 0.4 mM (about 1 nM for each peptide). Other details are as for Figures 1 and 2*

Table 2 Amino acids identified in each position of the active consensus sequence

Position 1 (O_1)	Position 2 (O_2)	Position 3 (O_3)	Position 4 (O_4)
D-Trp	**D-Met**	**D-Phe**	L-Gla
L-Hph	**D-Phe**	**D-Trp**	
	D-Trp	**D-FCl**	
	L-Trp	**L-FCl**	
	D-Cha	L-Gla	
	D-Tyr		

See Table 1 for a definition of amino acid abbreviations. Amino acids chosen for the more restricted consensus are highlighted in **boldface** type.

Table 3 Inhibitory activity of individual tetrapeptides identified from screening of the library

Peptide	IC_{50} against FPTase (nM)
1. D-Trp-D-Met-D-FCl-L-Gla-NH_2	42
2. D-Trp-D-Met-D-Phe-L-Gla-NH_2	62
3. D-Trp-D-Trp-D-FCl-L-Gla-NH_2	50
4. D-Trp-D-Phe-D-FCl-L-Gla-NH_2	90
5. D-Trp-D-Met-D-Trp-L-Gla-NH_2	100
6. D-Trp-D-Phe-D-Trp-L-Gla-NH_2	110
7. D-Trp-D-Phe-D-Phe-L-Gla-NH_2	160
8. D-Trp-D-Trp-D-Trp-L-Gla-NH_2	160
9. D-Trp-D-Trp-D-Phe-L-Gla-NH_2	230
10. D-Trp-D-Trp-L-FCl-L-Gla-NH_2	300
11. D-Trp-D-Met-L-FCl-L-Gla-NH_2	500
12. D-Trp-D-Phe-L-FCl-L-Gla-NH_2	900
13. D-Trp-L-Trp-L-FCl-L-Gla-NH_2	2300
14. D-Trp-L-Trp-D-Phe-L-Gla-NH_2	4300
15. D-Trp-L-Trp-D-Trp-L-Gla-NH_2	7400
16. D-Trp-L-Trp-D-FCl-L-Gla-NH_2	8800

ranging from 9 µM to 40 nM (Tab. 3). As predicted from the screening results, the D chirality was critical for positions 2 and 3 in the sequence. A change from D to L chirality gave a 20- to 150-fold increase in IC_{50} for position 2 (compare peptides 9 with 14 and 3 with 16) and a 6- to 10-fold increase for position 3 (compare peptides 1 with 11, 4 with 12 and 3 with 10). The effect was not strictly additive, with an L residue in both positions yielding a 50-fold IC_{50} increase (compare peptide 3 with 13). Overall, these results validate the initial choice of including D and non-coded residues in the library, as these make up all the residues in the nine most active sequences resulting from the reduced consensus set. In more general terms, they also underline one of the key advantages of chemically synthesised peptide libraries over those produced by genetic methods [13].

8.3 Conclusions

While several microbial FPTase inhibitors have been identified that act as FPP competitors [38], none of these is peptidic in nature. The compounds selected from the tetrapeptide library, such as D-Trp-D-Met-D-FCl-L-Gla-NH_2 thus represent a new class of inhibitors, opening the way to the synthesis of FPP peptidomimetics. The dicarboxylate of Gla most likely serves as a surrogate for the diphosphate of FPP, as has been suggested for the FPTase inhibitor chaetomellic acid that is a dicarboxylate-containing natural product [39]. It is possible that these dicarboxylate moeities mimic the FPP diphosphate by co-ordinating to the active site Zn^{2+} ion in a manner analogous to FPP, as suggested by the x-ray crystal structure of the FPTase-FPP complex [40].

Initial analyses of these compounds in cells suggest that they are not cell-active as assayed by inhibition of Ras processing (N.E. Kohl and J.B. Gibbs, unpublished observation). This result suggests that additional modifications will be necessary to render them biologically active. At this stage, the work presented here has already demonstrated that FPTase is amenable to screening of class 3 combinatorial libraries, which can be used to identify highly potent and novel compounds.

The above example is an illustration of the potential utility of solution-phase peptide libraries in drug discovery research. Attention is now turning away from simple peptides towards modified peptides (for example, to create libraries of chromogenic or fluorogenic proteinase substrates [41–43]) or even libraries of organic compounds such as benzodiazepines [44–46]. This creates an opportunity to search for active compounds from amongst classes of compounds which are more traditionally associated with valuable pharmacological properties. The pharmaceutical industry in particular is turning towards the latter, especially with the increasing number of organic chemical reactions which are being modified for use on solid phase. While this topic is beyond the scope of the present chapter it remains to be noted that in the absence of a traditional chemical lead, peptide libraries can provide a rapid and practical starting point in the drug discovery process. Further refinements and new applications of peptide libraries can be expected with the development of methods to synthesise arrays of peptides on solid supports at the nanometer scale level. When combined with novel methods of monitoring the interaction of such peptides with biomolecular ligands by, for example, surface plasmon resonance, real-time interaction analysis of highly diverse peptide arrays with large biomolecular repertoires (library-against-library screening) should be possible if sufficient sensitivity of interaction measurement can be obtained in these miniaturised systems. With such considerations in mind, it can be seen that peptide libraries have yet to realise their full potential and the prospects for the years ahead look interesting indeed.

References

1 Scott JK, Smith GP (1990) Searching for peptide ligands with an epitope library. *Science* 249: 386–390

2 Geysen HM, Meloen RH, Barteling SJ (1984) Use of peptide-synthesis to probe viral-antigens for epitopes to a resolution of a single amino-acid. *Proc Natl Acad Sci USA* 81: 3998–4002

3 Furka A, Sebestyen F, Asgedom M, Dibo G (1988) in: 14th international congress of biochemistry, vol. 5, pp 47, Prague, Czechoslovakia

4 Furka A, Sebestyen F, Asgedom M, Dibo G (1988) in: 10th international symposium of medicinal chemistry, pp 288, Budapest, Hungary

5 Houghten RA (1985) General-method for the rapid solid-phase synthesis of large numbers of peptides – specificity of antigen-antibody interaction at the level of individual amino-acids. *Proc Natl Acad Sci USA* 82: 5131–5135

6 Furka A, Sebestyen F, Asgedom M, Dibo G (1991) General method for rapid synthesis of multicomponent peptide mixtures. *Int J Pept Prot Res* 37: 487–493

7 Houghten RA, Pinilla C, Blondelle SE, Appel JR, Dooley CT, Cuervo JH (1991) Generation and use of synthetic peptide combinatorial libraries for basic research and drug discovery. *Nature* 354: 84–86

8 Lam KS, Salmon SE, Hersh EM, Hruby VJ, Kazmierski WM, Knapp RJ (1991) A new type of synthetic peptide library for identifying ligand-binding activity. *Nature* 354: 82–84

9 Pinilla C, Appel JR, Blanc P, Houghten RA (1992) Rapid identification of high-affinity peptide ligands using positional scanning synthetic peptide combinatorial libraries. *Biotechniques* 13: 901

10 Houghten RA, Appel JR, Blondelle SE, Cuervo JH, Dooley CT, Pinilla C (1992) The use of synthetic peptide combinatorial libraries for the identification of bioactive peptides. *Biotechniques* 13: 412–421

11 Salmon SE, Lam KS, Lebl M, Kandola A, Khattri PS, Wade S, Patek M, Kocis P, Krchnak V, Thorpe D, Felder S (1993) Discovery of biologically-active peptides in random libraries – solution-phase testing after staged orthogonal release from resin beads. *Proc Natl Acad Sci USA* 90: 11708–11712

12 Nikolaiev V, Stierandova A, Krchnak V, Seligmann B, Lam KS, Salmon SE, Lebl M (1993) Peptide-encoding for structure determination of nonsequenceable polymers within libraries synthesized and tested on solid-phase supports. *Peptide Res* 6: 161–170

13 Gallop MA, Barrett RW, Dower WJ, Fodor SPA, Gordon EM (1994) Applications of combinatorial technologies to drug discovery. 1. Background and peptide combinatorial libraries. *J Med Chem* 37: 1233–1251

14 Smith GP (1985) Filamentous fusion phage – novel expression vectors that display cloned antigens on the virion surface. *Science* 228: 1315–1317

15 Smith GP, Scott JK (1993) Libraries of peptides and proteins displayed on filamentous phage. *Meths Enzymol* 217: 228–257

16 Frank R (1995) Simultaneous and combinatorial chemical synthesis techniques for the generation and screening of molecular diversity. *J Biotech* 41: 259–272

17 Lam KS, Hruby VJ, Lebl M, Knapp RJ, Kazmierski WM, Hersh EM, Salmon SE (1993) The chemical synthesis of large random peptide libraries and their use for the discovery of ligands for macromolecular acceptors. *Bioorg Med Chem Letts* 3: 419–424

18 Chen CL, Strop P, Lebl M, Lam KS (1996) One bead-one compound combinatorial peptide library – different types of screening. *Meths Enzymol* 267: 211–219

19 Needels MC, Jones DG, Tate EH, Heinkel GL, Kochersperger LM, Dower WJ, Barrett RW, Gallop MA (1993) Generation and screening of an oligonucleotide-encoded synthetic peptide library. *Proc Natl Acad Sci USA* 90: 10700–10704

20 Nestler HP, Bartlett PA, Still WC (1994) A general-method for molecular tagging of encoded combinatorial chemistry libraries. *J Org Chem* 59: 4723–4724

21 Lam KS, Wade S, Abdullatif F, Lebl M (1995) Application of a dual-color detection scheme in the screening of a random combinatorial peptide library. *J Immun Meths* 180: 219–223

22 Baldwin JJ, Burbaum JJ, Chelsky D, Dillard LW, Henderson I, Li G, Ohlmeyer MHJ, Randle TL, Reader JC (1995) Combinatorial libraries encoded with electrophoric tags. *Eur J Med Chem* 30: S 349–S 358

23 Ni ZJ, Maclean D, Holmes CP, Murphy MM, Ruhland B, Jacobs JW, Gordon EM, Gallop MA (1996) Versatile approach to encoding combinatorial organic syntheses using chemically robust secondary amine tags. *J Med Chem* 39: 1601–1608

24 Ni ZJ, Maclean D, Holmes CP, Gallop MA (1996) Encoded combinatorial chemistry – binary coding using chemically robust secondary amine tags. *Meths Enzymol* 267: 261–272

25 Moran EJ, Sarshar S, Cargill JF, Shahbaz MM, Lio A, Mjalli AMM, Armstrong RW (1995) Radio-frequency tag encoded combinatorial library method for the discovery of tripeptide-substituted cinnamic acid inhibitors of the protein-tyrosine-phosphat-

ase PTP1b. *J Am Chem Soc* 117: 10787–10788

26 Chu YH, Dunayevskiy YM, Kirby DP, Vouros P, Karger BL (1996) Affinity capillary electrophoresis mass-spectrometry for screening combinatorial libraries. *J Am Chem Soc* 118: 7827–7835

27 Brummel CL, Lee INW, Zhou Y, Benkovic SJ, Winograd N (1994) A mass-spectrometric solution to the address problem of combinatorial libraries. *Science* 264: 399–402

28 Brown BB, Wagner DS, Geysen HM (1995) A single-bead decode strategy using electrospray-ionization mass- spectrometry and a new photolabile linker – 3-amino-3-(2-nitrophenyl)propionic acid. *Mol Diversity* 1: 4–12

29 Berlin K, Jain RK, Tetzlaff C, Steinbeck C, Richert C (1997) Spectrometrically monitored selection experiments: Quantitative laser desorption mass spectrometry of small chemical libraries. *Chem Biol* 4: 63–77

30 Wallace A, Koblan KS, Hamilton K, Marquisomer DJ, Miller PJ, Mosser SD, Omer CA, Schaber MD, Cortese R, Oliff A, Gibbs JB, Pessi A (1996) Selection of potent inhibitors of farnesyl-protein transferase from a synthetic tetrapeptide combinatorial library. *J Biol Chem* 271: 31306–31311

31 Garcia AM, Rowell C, Ackermann K, Kowalczyk JJ, Lewis MD (1993) Peptidomimetic inhibitors of ras farnesylation and function in whole cells. *J Biol Chem* 268: 18415–18418

32 Kohl NE, Mosser SD, Desolms SJ, Giuliani EA, Pompliano DL, Graham SL, Smith RL, Scolnick EM, Oliff A, Gibbs JB (1993) Selective-inhibition of ras-dependent transformation by a farnesyltransferase inhibitor. *Science* 260: 1934–1937

33 James GL, Goldstein JL, Brown MS, Rawson TE, Somers TC, McDowell RS, Crowley CW, Lucas BK, Levinson AD, Marsters JC (1993) Benzodiazepine peptidomimetics – potent inhibitors of ras farnesylation in animal-cells. *Science* 260: 1937–1942

34 Nigam M, Seong CM, Qian YM, Hamilton AD, Sebti SM (1993) Potent inhibition of human tumor P21(ras) farnesyltransferase by A1a2-lacking P21(ras)CA1a2X peptidomimetics. *J Biol Chem* 268: 20695–20698

35 Wallace A, Altamura S, Toniatti C, Vitelli A, Bianchi E, Delmastro P, Ciliberto G, Pessi A (1994) A multimeric synthetic peptide combinatorial library. *Peptide Res* 7: 27–31

36 Sole NA, Barany G (1992) Optimization of solid-phase synthesis of [Ala8]-Dynorphin-A. *J Org Chem* 57: 5399–5403

37 Omer CA, Diehl RE, Kral AM (1995) Bacterial expression and purification of human protein prenyltransferases using epitope-tagged, translationally coupled systems. *Meths Enzymol* 250: 3–12

38 Tamanoi F (1993) Inhibitors of ras farnesyltransferases. *Trends Biochem Sci* 18: 349–353

39 Singh SB, Zink DL, Liesch JM, Goetz MA, Jenkins RG, Nallinomstead M, Silverman KC, Bills GF, Mosley RT, Gibbs JB, Albersschonberg G, Lingham RB (1993) Isolation and structure of chaetomellic acids A and B from *Chaetomella acutiseta* – farnesyl pyrophosphate mimic inhibitors of ras farnesylprotein transferase. *Tetrahedron* 49: 5917–5926

40 Park H-W, Boduluri SR, Moomaw JF, Casey PJ, Beese LS (1997) Crystal structure of protein farnesyltransferase at 2.25 angstrom resolution. *Science* 275: 1800–1804

41 Meldal M, Svendsen I, Breddam K, Auzanneau FI (1994) Portion-mixing peptide libraries of quenched fluorogenic substrates for complete subsite mapping of endoprotease specificity. *Proc Natl Acad Sci USA* 91: 3314–3318

42 Meldal M, Svendsen I (1995) Direct visualization of enzyme-inhibitors using a portion mixing inhibitor library containing a quenched fluorogenic peptide substrate. 1. Inhibi-

tors for subtilisin Carlsberg. *J Chem Soc Perkin Trans* 1: 1591–1596

43 Olesen K, Meldal M, Breddam K (1996) Extended subsite characterization of carboxypeptidase-Y using substrates based on intramolecularly quenched fluorescence. *Prot Pept Letts* 3: 67–74

44 Bunin BA, Plunkett MJ, Ellman JA (1994) The combinatorial synthesis and chemical and biological evaluation of a 1,4-benzodiazepine library. *Proc Natl Acad Sci USA* 91: 4708–4712

45 Thompson LA, Ellman JA (1996) Synthesis and applications of small-molecule libraries. *Chem Revs* 96: 555–600

46 Bunin BA, Plunkett MJ, Ellman JA, Bray AM (1997) The synthesis of a 1680 member 1,4-benzodiazepine library. *New Journal of Chemistry* 21: 125–130

9 Using oligonucleotide probe arrays to access genetic diversity

R.J. Lipshutz, D. Morris, M. Chee, E. Hubbell, M.J. Kozal, N. Shah, N. Shen, R. Yang and S.P.A. Fodor

9.1 Introduction

The goal of the Human Genome Project is to sequence all 3 billion base pairs of the human genome. Progress in this has been rapid; GenBank® finished April 1997 with 843 million bases of sequence. The challenge to the scientific community is to understand the biological relevance of this genetic information. In most cases the sequence being generated for any single region of the genome represents the genotype of a single individual. A complete understanding of the function of specific genes and other regions of the genome and their role in human disease and development will only become apparent when the sequence of many more individuals is known.

Access to genetic information is ultimately limited by the ability to screen DNA sequence. Although the pioneering sequencing methods of Sanger et al. [15] and Maxam and Gilbert [11] have become standard in virtually all molecular biology laboratories, the basic protocols remain largely unchanged. The throughput of this sequencing technology is now becoming the rate-limiting step in both large-scale sequencing projects such as the Human Genome Project and the subsequent efforts to understand genetic diversity. This has inspired the development of advanced DNA sequencing technologies [9]. Incremental improvements to Sanger sequencing have been made in DNA labeling and detection. High-speed electrophoresis methods using ultrathin gels or capillary arrays are now being more widely employed. However, these methods are throughput-limited by their sequential nature and the speed and resolution of separations. This limitation will become more pronounced as the need to rapidly screen newly discovered genes for biologically relevant polymorphisms increases.

An alternative to gel-based sequencing is to use high-density oligonucleotide probe arrays. Oligonucleotide probe arrays display specific oligonucleotide probes at precise locations in a high-density, information-rich format [5, 4, 12]. The hybridization pattern of a fluorescently labeled nucleic acid target is used to gain primary structure information of the target. This format can be applied to a broad range of nucleic acid sequence analysis problems including pathogen identification, polymorphism detection, human identification, mRNA expression monitoring and de novo sequencing.

In this review, we briefly describe the method of light-directed chemical synthesis to create high-density arrays of oligonucleotide probes, the method of fluorescently labeling target nucleic acids for hybridization to the probe arrays, the detection of hybridized targets by epi-fluorescence confocal scanning and the data analysis procedures used to interpret the hybridization signals. To illustrate the use of specific high-density oligonucleotide probe arrays, we describe their application to screening the reverse transcriptase (*rt*) and protease (*pro*) genes of HIV-1 for polymorphisms and drug-resistance conferring mutations.

9.2 Light-directed chemical synthesis

High-density oligonucleotide arrays are created using light-directed chemical synthesis. Light-directed chemical synthesis combines semiconductor-based photolithography and solid-phase chemical synthesis [3]. To begin the process (Fig. 1A), linkers modified with photochemically removable protecting groups [12] are attached to a solid substrate. Light is directed through a photolithographic mask to specific areas of the synthesis surface, activating those areas for chemical coupling (Fig. 1B). The first of a series of nucleosides (MeNPOC-dT in this instance) harboring a photo-labile protecting group at the 5′ end [12] is incubated with the array, and chemical coupling occurs at those sites that have been illuminated in the preceding step (Fig. 1C). Next, light is directed to a different region of the substrate through a new mask (Fig. 1D), and the chemical cycle (with MeNPOC-dC in this instance) is repeated (Fig. 1E). The process is repeated (Fig. 1F). Using the proper sequence of masks and chemical steps, a defined collection of oligonucleotides can be constructed, each in a predefined position on the surface of the array.

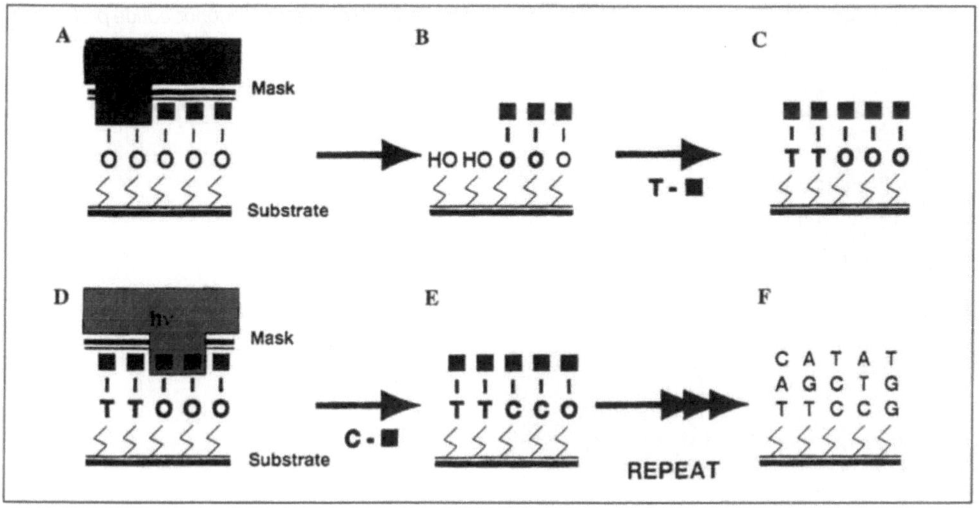

Figure 1 *High-density oligonucleotide probe array synthesis*

9.3 High-density probe arrays

Since the synthesis areas are defined by the photolithographic process, high-density arrays can be formed. Table 1 shows the relationship between the size of each synthesis site and the density of synthesis sites (a synthesis site being the region in which a homogeneous set of probes is synthesized). Productionscale instruments are currently synthesizing arrays with 24 μm synthesis sites. This corresponds to more than 250,000 synthesis sites in a standard 6.64 cm^2 synthesis area. Developmental instrumentation has demonstrated synthesis at a 10 μm resolution.

Table 1 ***Photolithographic resolution and synthesis site density***

Resolution	Synthesis site density
500 μm	400 sites/cm^2
200 μm	2500 sites/cm^2
100 μm	10000 sites/cm^2
50 μm	40000 sites/cm^2
20 μm	250000 sites/cm^2
10 μm	1000000 sites/cm^2

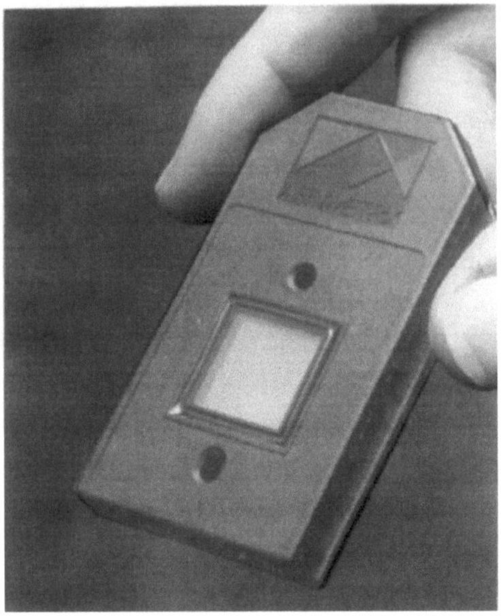

Figure 2 *High-density oligonucleotide probe array package*

To support the production of increased numbers of arrays, instrumentation has been developed to synthesize multiple arrays on a single large substrate (wafer). After synthesis, wafers are diced to yield individual arrays. These arrays are packaged in individual injection-molded flow cells, making them easier to handle (see Fig. 2). In addition, arrays synthesized by these methods have improved uniformity and quality and use significantly fewer resources to synthesize each array.

9.4 Combinatorial synthesis of a probe matrix

Light-directed synthesis provides a powerful and efficient method for generating random access molecular diversity in a spatially defined format. The location and composition of products depends on the pattern of illumination and the order of chemical coupling reagents (see Reference 5 for a complete description). In Figure 3 we illustrate the synthesis of all 256 4-mers in 16 chemical steps. Using the combinatorial synthesis strategy described in Figure 3, the set of all 4^k length k oligonucleotides (k-mers) can be generated in 4k syn-

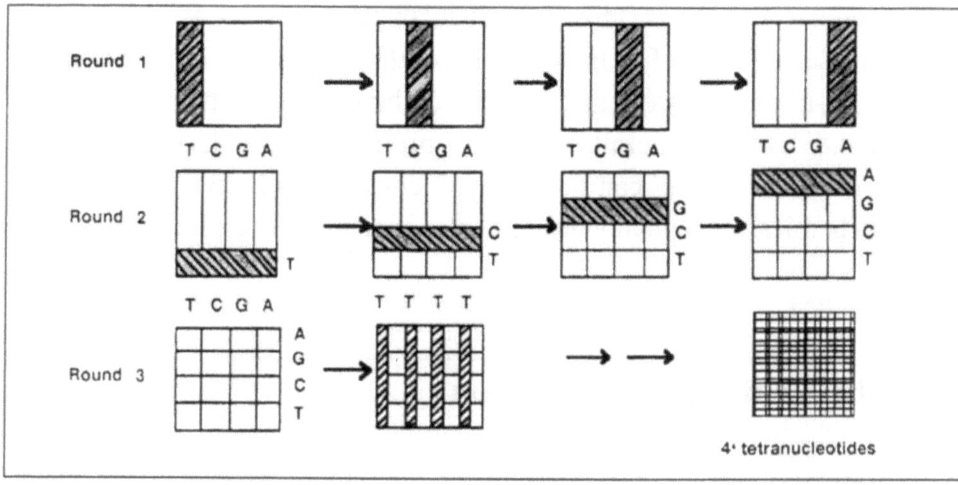

Figure 3 *Combinatorial synthesis of all 256 4-mers. In round 1, mask 1 activates one-fourth of the substrate surface for coupling with the first of four nucleosides in the first round of synthesis (MeNPOC-dT). In cycle 2 of round 1, mask 2 activates a different quarter of the substrate for coupling with the second nucleoside (MeNPOC-dC). The process is continued to build for regions of mononucleotides. The masks of round 2 are perpendicular to those of round 1, and each synthesis cycle generates four new dinucleotides. The process continues through round 2 to form 16 dinucleotides. The masks of round 3 further subdivide the synthesis regions so that each coupling cycle generates 16 trimers. The subdividion of the substrate is continued through round 4 to form the tetranucleotides (256 possibilities).*

Table 2 Combinatorial synthesis

Probe length	Chemical steps	Number of possible probes
4	16	256
8	32	65 536
10	40	1 048 576
15	60	1 073 741 824

thesis cycles. For example, the set of all 15-mers can be synthesized in 60 cycles. Using the current process, this synthesis can be completed in less than 10 h. Additional examples of the relationship between probe length, number of chemical synthesis steps and number of synthesis sites are given in Table 2.

The same relationship between probe length and number of synthesis cycles applies to any set of probes. For example, since all 15-mers can be synthesized in 60 chemical steps, any random subset of 15-mers can therefore be synthesized in 60 or fewer steps. In particular, given any set of 15-mers, it is possible to define a set of at most 60 photolithographic masks to synthesize the set of 15-mers. The strategy for defining the masking patterns is given in Fodor et al. [5]. More generally, any set of probes of length k or less (probes in different synthesis sites can be different lengths) can be synthesized in, at most, 4 k chemical steps in any desired arrangement on the chip surface.

9.5 Hybridization and detection

Hybridization of target nucleic acids to an oligonucleotide array yields sequence information. The nucleic acid to be interrogated, the target, is labeled with a fluorescent reporter group and incubated with the array. If the target nucleic acid has regions complementary to probes on the array, then the target will hybridize with those probes. Under a fixed set of hybridization conditions, e.g., target concentration, temperature, buffer and salt concentration, and so on, the fraction of probes bound to targets will vary with the base composition of the probe and the extent of the target-probe match. In general, for a given length, probes with high GC content will hybridize more strongly than those with high AT content. Probes matching the target will hybridize more strongly than probes with mismatches, insertions and deletions.

Hybridization of the target to the array is detected by epi-fluorescence confocal scanning (Fig. 4) [4]. The array is inverted in a temperature-controlled flow cell. The target solution is introduced to the flow cell and allowed to hybridize with the probes on the surface of the array. The confocal system then scans the array, measuring fluorescent signal from target nucleic acid bound at the surface. This system provides strong background rejection from the unbound target, the glass and other parts of the system, a large dynamic range (1:10,000), high resolution over a broad area and the ability to collect both kinetic and equilibrium data. The scanning resolution is chosen so that the synthesis sites of the array are oversampled, i.e., between 25 and 100 data points are collected from each synthesis site, corresponding to a 5×5 to 10×10 array of pixels. The collection of data forms an "image" of the array. Image processing software, GeneChip™ software (Affymetrix, Santa Clara, CA, USA), is used to segment the image into synthesis sites and integrate the data

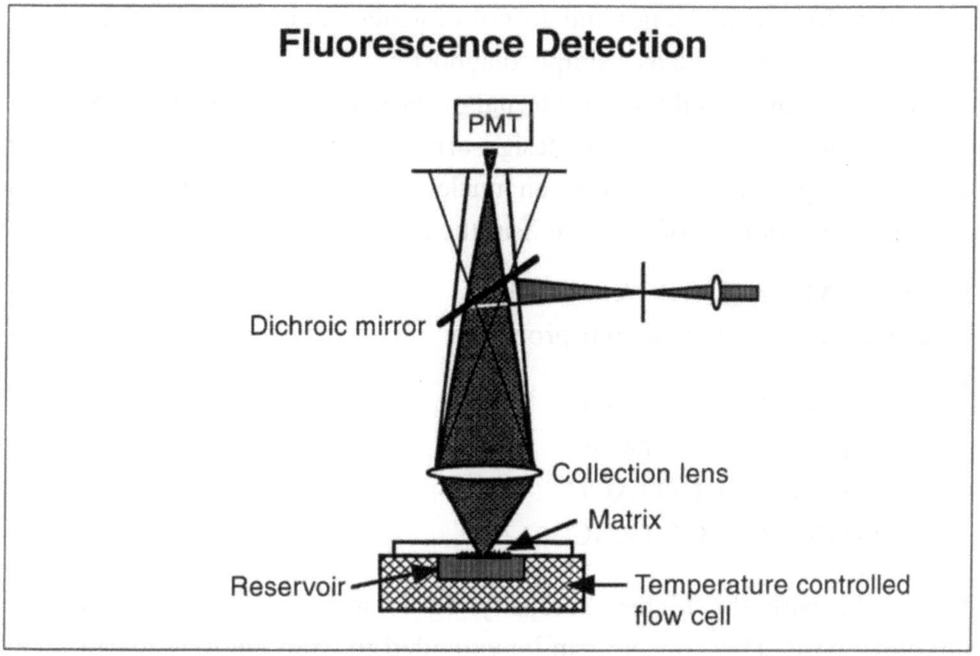

Figure 4 *Detection format. Probe-target hybridization is detected by epi-fluorescence confocal scanning*

Figure 5 *GeneChip scanner detection instrument*

over each synthesis site. A robust benchtop version of this detection instrument, the GeneChip scanner (Affymetrix), has been developed, (Fig. 5).

9.6 Polymorphism screening

The oligonucleotide arrays can be designed and used to rapidly and efficiently screen characterized genes for polymorphisms. The genetic diversity of a population can be explored, and the relationships between genotype and phenotype for specific genes and groups of genes can be discovered. In addition, probes can be used as a simple diagnostic.

The significant instability of internal probe-target mismatches, relative to perfect matches [3], is used to design arrays of probes capable of rapidly discriminating differences between nucleic acid targets. For example, to determine the identity of the base X in the target:

5′-GTATCAGCATXGCCATCGTGC
we could use the following four probes:

3′-AGTCGTAACGGTAGC
3′-AGTCGTACCGGTAGC
3′-AGTCGTAGCGGTAGC
3′-AGTCGTATCGGTAGC

The probe with the highest intensity would indicate the identity of the unknown base. This concept can be extended to examine long nucleic acid targets and detect polymorphisms/mutations relative to a characterized consensus sequence. Given a consensus sequence, a set of four probes can be defined for each nucleotide in the target as described above. Thus, to screen 1000 nucleotides for polymorphisms/mutations would require 4000 probes. A 1.28-cm^2 array designed with 100-μm synthesis sites will have about 16000 probes and could screen 4 kb of sequence. This approach has been applied to survey the *rt* and *pro* genes of HIV-1 for drug-resistance mutations and is described below [17].

9.7 HIV probe array

Currently, all of the approved HIV-1 therapeutics are targeted against the RT enzyme. These include AZT (Zidovudine), ddI, ddC and d4T. In addition, many of the drugs under development by the pharmaceutical industry target either RT or the protease (PRO) enzyme. Significant debate has arisen concerning the efficacy of the approved and investigation drugs in inhibiting HIV-1 proliferation and the value of drug therapy to patients. A significant portion of efficacy issues can be attributed to the acquisition of drug resistance by

HIV-1, governed by its ability to rapidly mutate while retaining pathogenicity.

A number of HIV-1 mutations have been cataloged that confer PRO or RT drug resistance to the approved antivirals. Depending upon the stage of disease when therapy is initiated, AZT, the first approved HIV-1 antiviral, may become ineffective after 3–12 months. Data from clinical specimens indicate that during this time period, a number of mutations arise in the *rt* gene that mediate this drug resistance [6–8]. The first four characterized mutations were Asp 67 → Asn, Lys 70 → Arg, Thr 215 → Phe/Tyr and Lys 219 → Gln. When all four mutations occur together, the viral isolate is more than 100 times less sensitive to AZT [14]. Most recently, several of the mutations associated with *pro* multi-drug resistance have been identified in the general population, specifically among patients who have not experienced any antiviral therapeutics [13].

Given the large number of possible mutation that have been associated with HIV-1 antiviral drug resistance, the rapid emergence of new mutations associated with new therapeutics and the distribution of the mutations throughout the *rt* and *pro* genes, there is significant value in being able to rapidly and cost effectively screen these genes for mutations.

9.8 HIV array design

The array design described above was applied to the HIV-1 *rt* and *pro* genes. The sequences analyzed consisted of 1040 bases of the Clade B consensus sequence, including the last 18 bases of *gag*, all 297 bases of *pro* and the first 723 bases of *rt* (codons 1–241). The 1040 nucleotides in the array survey the positions at which all known resistance-conferring mutations are located. The array contains 18495 93 × 95 µm synthesis sites (Fig. 7).

The assay consists of amplification of clonal DNA, in vitro transcription and label incorporation, hybridization and scanning, (Fig. 6). Nucleotide determinations were based on comparing the intensities of each set of four oligonucleotide probes interrogating each of the 1040 bases. Analysis tools are provided to view the intensity data in graphic and numeric formats, (Fig. 8). Even with this simple first generation array design and base-calling algorithm,

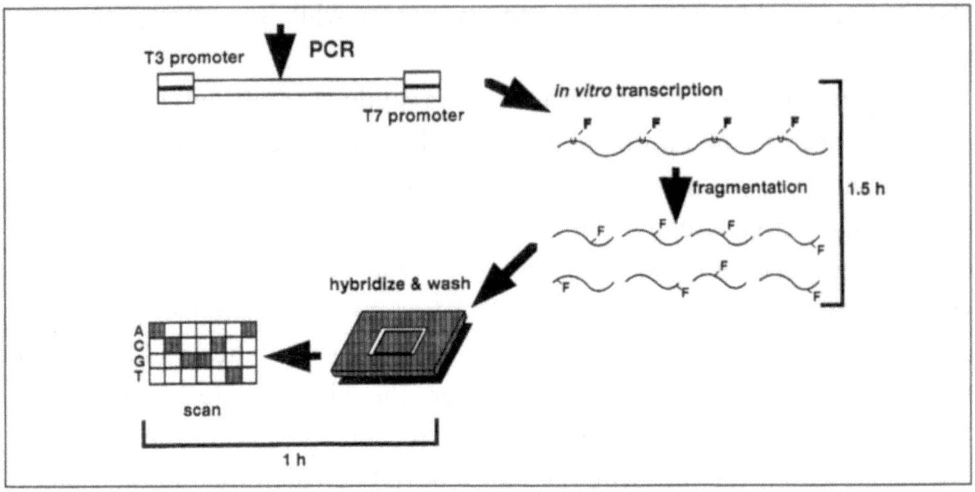

Figure 6 *Sample preparation consists of amplifications, labeling, fragmentation, hybridization and scanning*

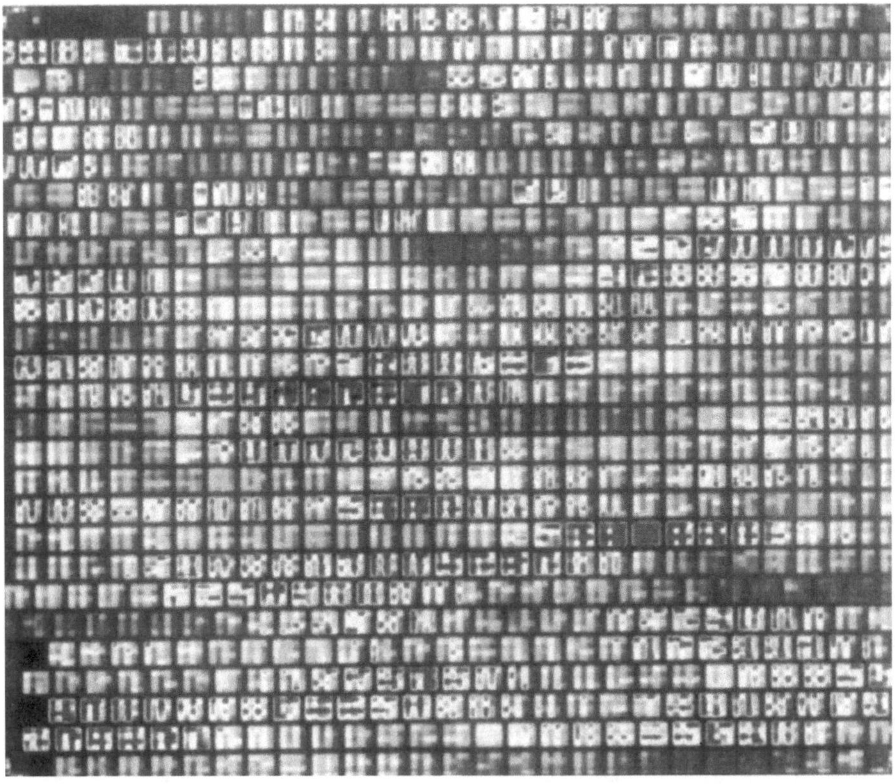

Figure 7 *HIV-1 oligonucleotide array image (see colour plate 4)*

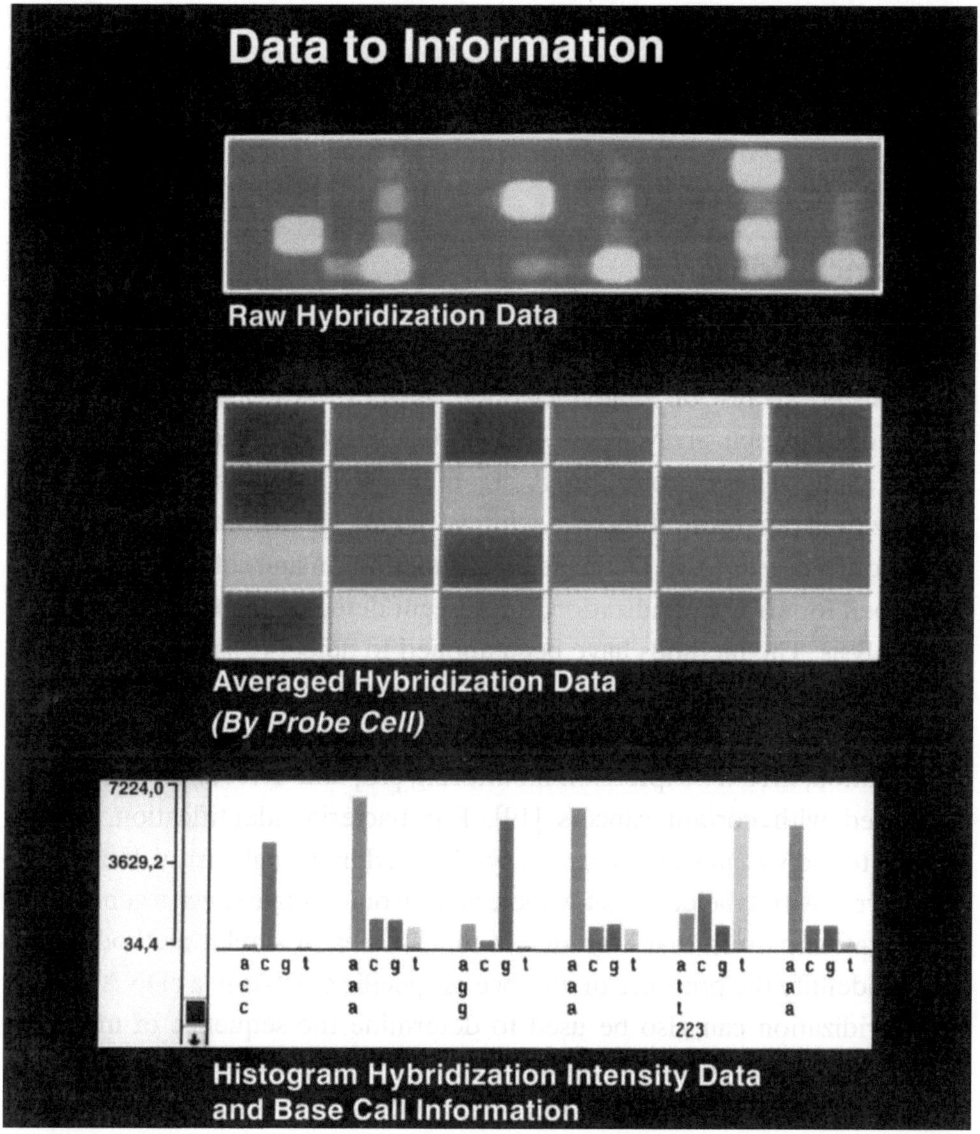

Figure 8 *GeneChip software. Separate windows are available to display the initial image, the integrated intensities and the resulting base calls (see colour plate 5)*

the method has proved to generate correct base calls with high confidence (unpublished). Future developments will focus on improving the design of the chip and the base-calling algorithm.

9.9 Conclusions

Light-directed chemical synthesis has been used to generate miniaturized, high-density arrays of oligonucleotide probes. Application-specifc oligonucleotide probe array designs have been developed to rapidly screen known genes. These probe arrays are then used for parallel nucleic acid hybridization analysis, directly yielding polymorphism information from genomic DNA sequence. Dedicated instrumentation and software have been developed for array hybridization, fluorescent detection and data acquisition and analysis. The methods have been applied to detect resistance-conferring mutations in the *rt* and *pro* genes of HIV-1 demonstrating their effectiveness.

Other applications of oligonucleotide probes arrays include, bacterial classification, mRNA expression monitoring [18] and detection of mutations correlated with cortain cancers [19]. For bacterial identification, probes specific to individual species can be synthesized in a single array. Incubation of the array with labeled target nucleic acids from a single species generates a hybridization pattern that is unique to that species. A similar method can be used to identify the presence or absence of specific cDNAs in a cDNA library.

Hybridization can also be used to determine the sequence of unknown DNA (de novo sequencing, sequencing by hybridization [SBH]) [1, 2]. In SBH, the sequence of an unknown target nucleic acid is reconstructed from the hybridization data. SBH on probe arrays has been demonstrated in small-scale experiments [3, 16]. These experiments, in combination with numerous simulation studies, have shown that de novo sequencing is an application with significant potential. There are, however, important challenges to be met before it can be broadly implemented. These include generating signal from GC-rich and AT-rich probes in the same experiment, being able to effectively distinguish perfect matches from highly stable mismatches and resolving ambiguities due to repeated sequences. High-density oligonucleotide probe array technology shows significant promise in enabling de novo sequencing [3].

The Human Genome Project and related efforts have undertaken the formidable task of identifying and determining the sequence of all of the human genes, and it is only through more efficient access to genetic information that the true benefit of the Human Genome Project will be realized. High-density oligonucleotide probes arrays should provide a basic platform for analyzing the genetic variation in the human genome.

Acknowledgments

The authors wish to acknowledge the support provided by the National Institutes of Health (National Center for Human Genome Research and National Institute for Allergy and Infectious Disease – Phase I SBIR), the Department of Energy (Office of Health Effects Research) and the National Institute of Standards and Technology (Advanced Technology Program).

References

1 Bains W, Smith, GC (1988) A novel method for nucleic acid sequence determination. *J Theor Biol* 135: 303–307

2 Drmanac R, Labat I, Brukner I, Crkvenjakov R (1989) Sequencing of megabase plus DNA by hybridization: theory of the method. *Genomics* 4: 114–128

3 Fodor SPA, Lipshutz RJ, Huang X (1993) DNA sequencing by hybridization, p 3–9. Proceedings of The Robert A. Welch Foundation 37th Conference on Chemical Research – 40 Years of the DNA Double Helix. Robert A. Welch Foundation, Houston

4 Fodor SPA, Rava R, Huang XC, Pease AC, Holmes CP, Adams CL (1993) Multiplexed biochemical assays with biological chips. *Nature* 364: 555–556

5 Fodor SPA, Read JL, Pirrung MC, Stryer L, Lu AT, Solas D (1991) Light-directed, spatially addressable parallel chemical synthesis. *Science* 251: 767–773

6 Kozal MJ, Merigan TC (1993) HIV resistance to dideoxynucleoside inhibitors. *Infect Dis Clin Practice* 2: 247–253

7 Land S, McGavin C, Lucas R, Birch C (1992) Incidence of zidovudine-resistant human immunodeficiency virus isolated from patients before, during, and after therapy. *J Infect Dis* 166: 1139–1142

8 Larder BA, Kemp SD (1989) Multiple mutations in HIV-1 reverse transcriptase confer high-level resistance to zidovudine (AZT). *Science* 246: 1155–1157

9 Lipshutz R, Fodor SPA (1994) Advanced DNA sequencing technologies. *Curr Opin Struct Biol* 4: 376–380

10 Lysov YP, Florentiev VL, Khorlin AA, Khrapko KV, Shik VV, Mirzabekov AD

(1988) DNA sequencing by hybridization with oligonucleotides. A novel method. *Dokl Acad Sci USSR* 303: 1508–1511

11 Maxam AM, Gilbert W (1977) A new method for sequencing DNA. *Proc Natl Acad Sci USA* 74: 560–664

12 Pease AC, Solas D, Sullivan EJ, Cronin MT, Holmes CP, Fodor SPA (1993) Light-generated oligonucleotide arrays for rapid DNA sequence analysis. *Proc Natl Acad Sci USA* 91: 5022–5026

13 Richman DD (1995) Protease uninhibited. *Nature* 374: 494–495

14 Richman DD, Shih CK, Lowy I, Rose J, Prodanovich P, Goff S, Griffin J (1991) Human immunodeficiency virus type 1 mutants resistant to nonnucleoside inhibitors of reverse transcriptase arise in tissue culture. *Proc Natl Acad Sci USA* 88: 11241–11245

15 Sanger F, Nicklen S, Coulson R (1977) DNA sequencing with chain-terminating inhibitors. *Proc Natl Acad Sci USA* 74: 5463–5467

16 Southern E, Maskos U, Elder R (1992) Hybridization with oligonucleotide arrays. *Genomics* 13: 1008–1017

17 Kozal MJ, Shah N, Shen N, Yang R, Fucini R, Merrigan T, Richman D, Morris D, Hubbell E, Chee MS, Gingeras TG (1996) Extensive polymorphisms observed in HIV-1 clade B protease gene using high-density oligonucleotide arrays. *Nature Medicine* 7: 753–759

18 Lockhart DJ, Dong H, Byrne MC, Follettie M, Gallo MV, Chee MS, Mittmann M, Wang C, Kobayashi M, Horton H, Brown EL (1996) Expression monitoring by hybridization to high-density oligonucleotide arrays. *Nature Biotech* 14: 1675–1680

19 Hacia JG, Brody LC, Chee MS, Fodor SPA, Collins FS (1996) Detection of heterozygous mutations in BRCA1 using high-density oligonucleotide arrays and two-colour fluorescence analysis. *Nature Genetics* 14: 441–447

10 Generation of libraries by print technologies

Eugen Ermantraut, Sefan Wölfl and Hans Peter Saluz

10.1 Introduction

An efficient production of nucleic acid arrays on chips and their applications will open a great potential for biological key indications, such as early diagnostics, diagnostics and therapy monitoring of infections diseases or cancer, diagnostics of genetic diseases, predisposition diagnostics, monitoring of therapeutical principles (gene therapy), etc. However, for a production of thousands of such arrays, appropriate means of oligonucleotide synthesis on various solid supports will be required. Unfortunately, only a few successful ways have been described so far.

Since large numbers of various molecules have to be assayed at the same time it is most comfortable to set up arrays of libraries, which is accomplished either in defined vessels, such as micro- or nanotiterplates, for libraries in solution or immobilized on beads [1], or at defined spots on the surface of a suitable carrier material (e.g. nylon membranes, modified glass or Si- wafers, etc.) for probe molecule libraries being immobilized directly onto the substrate [2]. Arrays offer the advantage to handle the entire array as one single assay, solvents for washing, test samples as well as modifying reagents are applied simultaneously to all of the immobilization spots, however every single spot remains a two-dimensional reactor, the character of the reaction being determined by the composition of the immobilized probe.

10.2 Methods

Microdispensing techniques bear the opportunity to become the method of choice to generate libraries with a low degree of integration. Because immobilization techniques usually require accurate pipetting of buffers, reagents and solutions with volumes in the microliter range, robotic pipetting workstations have been introduced.

Since piston operated pumps are commonly used in pipetting devices they cannot meet the demands for extremely accurate pipetting of very small volumes at the speed that is usually required in such applicationes as are high troughput sequencing projects. Therefore microdispensing with the aid of piezoelectric transducers is becoming a popular technique. Drops sizes of 5 picoliters and speeds up to 10000 drops per second have been achieved [3]. Various materials have been dispensed in such a way, including liquid metals, photoresists, organic solvents such as ethanol, acetone and acetonitril, various buffers, DNA and antibody containing solutions, thermoplastics, latex beads and metal particles [4].

With commercially availiable microdispensers it is already possible to produce spheres of liquid with diameters of 25–100 μm and rates of up to 8000 per second. In analogy to ink-jet printers it is possible to print multiple fluids. It is principally feasible to deliver various kinds of molecules in a variaty of buffers to a wide range of surfaces. The disadvantage of microdispensing is the limited resolution and especially crosstalking of already served areas on the immobilization matrix with spots to be served with the molceules to be immobilized.

Dedicated immobilization supports with predefined immobilization spots with optimized wetting characteristics for the dispensed liquid and non wetting bars inbetween the spots (Figs. 1 and 2) are extending the opportunities of producing library bearing microchips by microdispensing.

Another well known procedure was published already in 1994 by Jacobs and colleagues from Affymetrix [5]. They used masks – as commonly applied in photolithography – for the preparation of arrays. Molecules positioned on defined islands on the solid support were deprotected and coupled with the next activated monomer. By this means entire polymers were constructed. At that time, they were already capable to synthesize all possible variations of an oligonucleotide octamer on one and the same support, i.e. 65536 different molecular species [6]. Also peptide libraries were synthesized using the above approach [7].

Also in 1994, an alternative approach was described by Southern and Maskos [8]. It successfully allowed synthesis of large numbers of different

Figure 1 *Fluorescence image of a 10 × 10 µ UNIPAD immobilization matrix with FITC immobilized at predifened positions, the spots are clearly separated, Chip 1 cm × 1 cm*

Figure 2 *10 × 10 µ UNIPAD immobilization matrix with drops of water on hydrophilic spots, Chip 1 cm × 1 cm*

oligonucleotides on solid supports. They applied conventional solid phase synthesis chemistry for preparation of the oligonucleotides.

Recently, an approach combining standard phosporamidite chemistry and standard photolithographic resist technology has been reported [9]. In this approach, in analogy to techniques common to semiconductor manufacturing, the support is covered with a photosensitive resist. This resist is patterned by light exposure through a suitable photolithographic mask and serves as a physical barrier to protect selected areas from exposure to chemical reagents in the course of the synthesis cycle. Due to their nonlinear response to illumination intensity photoresists allow the generation of patterned surfaces with sub-micron features.

However, all these strategies have some drawbacks. The technique exploiting light sensitive deprotection involves a special cost intense chemistry, the application of which is still quite problematic. For example, the length of the oligonucleotides is limited to rather short molecules. It further involves photolithographic techniques requiring special clean rooms. In addition, the physical limitations do not allow high resolution due to the linear exposure characteristics of the light sensitive protecting groups of the deprotection chemistry applied.

The application of standard synthesis chemistry is therefore still a desirable feature. Because the application of photoresists as a wet masking technique is compatible with reagents common to standard phosphoramidite oligonucleotide synthesis chemistry and the resolution being near the physical limit, this approach seems to be a promising one. However, the adhesive interaction between the resist and the growing layer of synthesized material limits the synthesis yield. To increase compatibility between the resist and the chemistry would require a substantial re-engineering of the resist material. Since DNA is sensitive to UV-light, only resists which are imageable with near-UV or visible light are suitable for oligonucleotide synthesis applications.

The advantage of the procedure of Southern and Maskos concerns the use of conventional synthesis reagents. However, it is not very economic because a capillary system is involved, where all capillaries have to be filled with the appropriate reagent. Miniaturisation of a similar approach for immobilization of proteins has been reported [10] but remains difficult, since small capillaries are very sensitive to blockage by particles.

Therefore alternative developments of more efficient and economic procedures are required.

A strategy was worked out in our department [11], a procedure that can be applied in all laboratories. This technique allows a highly economic and efficient synthesis of all variations of a given polymer by conventional synthesis approaches. It can be used for the production of matrix-bound miniaturized combinatorial oligo- and polymer libraries of nucleic acids (in both directions), peptides, sugars and others that have to be synthesized in multistep approaches. It involves the use of chemically inert stamps [12], the active surface of which is microstructured according to defined synthesis algorithms. It allows the selective synthesis on defined loci on the surface of the substrate, an inert solid body, the surface of which is coated with a monolayer of a linker molecule covalently coupled to a terminal protecting group. The stamp is dipped into a deprotecting agent and positioned on the solid body in order to deprotect the linker areas of interest. Upon washing in an inert solvent the substrate is covered by the first activated monomer. These steps are repeated until the polymer of interest is synthesized (Fig. 3). The arrangement of the polymers on the surface of the solid body is given by the synthesis algorithm of choice. The polymers have a distinct length given by the algorithm and can be identified on the surface.

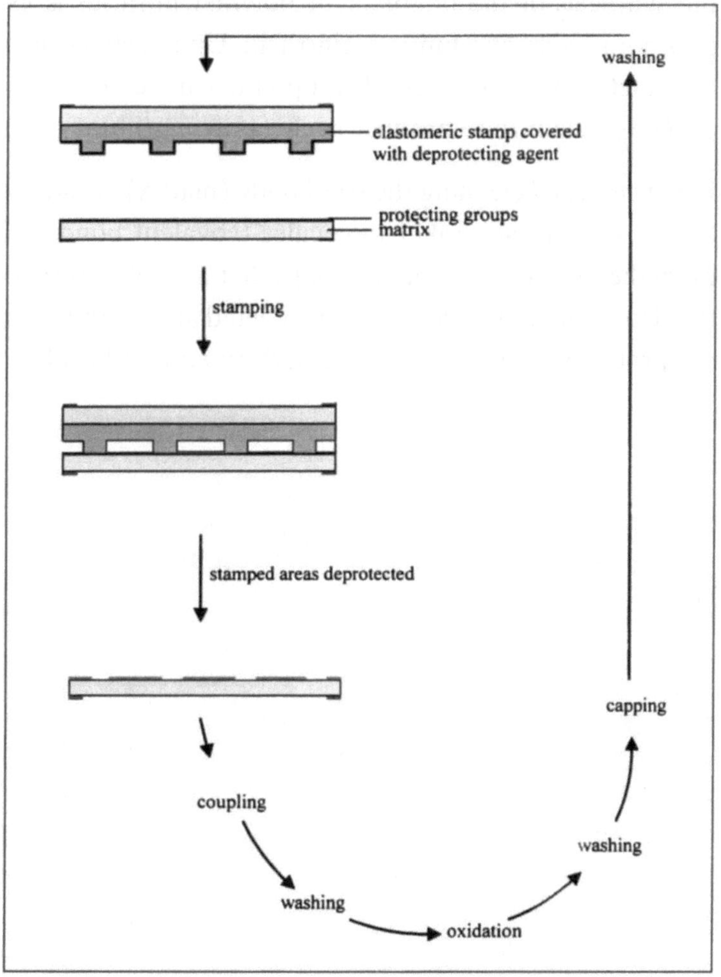

Figure 3 *Schematic representation of a synthesis cycle*

The combinatorial library is – as previously mentioned – synthesized on the surface of a solid body (matrix) that has to be inert against the solvents used for the synthesis. Such conditions are fulfilled by any kind of wafer used in microstructure technology, such as silicon/siliconoxide. However, also polished quartz glass, such as Suprasil, can be applied.

The linker is a multi-functional molecule that forms covalent bonds between the surface of the solid body and the first monomer of the molecule. It is chosen according to the prospective application. Excellent linkers for the synthesis of oligonucleotide arrays are 3-glycidoxypropyl-trimethoxysilan or 3-aminopropyl-triethoxysilan [13]. The first nucleic acid monomer used for the synthesis of the oligomer or polymer libraries is usually CE-nucleotide phosporamidite or a Fmoc-amino acid. Upon activation of the first compound the reaction with the second compound can be performed.

The procedure to produce such a polymer library is as follows:

Upon carefully cleaning the solid body (matrix), its surface is entirely covered by the appropriate linker molecules (covalent bonds). The linker molecules carry the protected terminal groups that are split off before they are bound to the first monomer which is also protected at its terminal end. The activation of the protected molecules can be performed in a locally specific manner using

Figure 4 *PDMS – stamp with 8 µm features*

special stamps (Fig. 4). The stamps consist of a chemically inert material, such as Poly-Di-Methyl-Siloxan (PDMS) capable to absorb the organic solvents sufficient for the specific deprotection reaction. By this means the coupling reactions on the areas of interest can be guaranteed. The preparation of the stamps containing a micro-structured surface is performed as follows:

First, a negative of the stamp is microstructured in a layer of photoresist that covers the wafer. The microstructuration is performed such that the synthesis algorithm can be realized. The PDMS mixed with an appropriate hardener is added to the negative and covered with a glass plate in order to avoid contamination. The stamp is fixed to the glass plate. However, it can be pulled off from the photoresist layer without any problem. This procedure is repeated until all positive stamp patterns required for the synthesis are produced. The number of stamps depends on the synthesis algorithm that has been chosen. The correct positioning of the stamp on the matrix is guaranteed by a special positioning device, similar to mask aligners commonly used in semiconductor fabrication. To activate a specific area on the matrix, the corresponding microstructured stamp is dipped into the deprotecting agent and put onto the surface of the matrix to take off the protection molecules within the areas of contact. After each cycle the matrix is washed thoroughly using an inert solvent. The steps are repeated until the length of the polymer is reached. The number of loci on the surface corresponds to the number of possible variations of the polymer ($Z = ML$; Z: number of separate loci; M: number of monomers; L: length of the polymer). The number of stamps (S) required for a synthesis is proportional to the product of M and L ($S = M \times L$). The potential to miniaturize the library appears to be identical with the limits approached in micro-systems technology [14]. The maximum length of a polymer is dependent on the type of synthesis technique. According to our own experience the synthesis of oligonucleotides of 20 nucleotides length was possible. The efficiency of the synthesis is measured by analyzing the number of the protection groups split-off after each cycle.

Example:

Preparation of a 64-species trimer-oligonucleotide array:

- Clean a quartz solid body in an ultrasound water bath.
- Dip the clean solid body into a closable glass containing 3-glycidoxypropyl-trimethoxysilan (55% in Xylole)

- and N-ethyldiisopropylamin (catalytic).
- Close the container tightly and keep it for 8 h at 80°C.
- Wash the solid body in xylol and ethyleneglycol.
- Put the body into ethyleneglycol and adjust to pH 5 with sulfuric acid.
- Close the container tightly and keep it for 8 h at 80°C.
- Wash the solid body thoroughly with methanol followed by diethylether.
- Dry the body in an oven for 4 h at 80°C.
- Perform now the reaction between the monolayer of hydroxyethyloxypropyl-siloxan on the glass surface with tetrazolyl-nucleotidephosphoramidite in an Atmosbag (Sigma) filled with argon gas as follows:

A FOD-phosphoramidite, such as FOD-dG-CE-phosphoamidite, carrying a protection group (dimethoxytrityl; DMT) at its 5′-OH-group, is dissolved in acetonitril: tetrazole = 1:1 (0.1M activated nucleotide phosphoamidite) and added to the container containing the solid body.

Incubate for 15 min at room temperature. Wash with acetonitril and add 0.02 M iodine-solution.

Wash with acetonitril (resulting in Di-methoxy-trityl layer on the surface of the solid body).

Cycles:

Each cycle needs a stamping step and includes the following steps (the solvents and reagents are commonly used in oligonucleotide synthesis; Users Manual, ABI, 1992):

- Wash with dichloromethane for 0.5 min.
- Deprotect – using the stamp – the 5′-hydroxyl group with 2% TCA in dichloromethane.
- Wash with acetonitril 2.5 min.
- Repeat this washing step several times.
- Add tetrazolyl-nucleotide-phosphoamidite and incubate for 5 min (condensation step).
- Wash with acetonitril for 2.5 min.
- Incubate for 5 min in 0.2M iodine in tetrahydrofuran, pyridine and water.
- Wash with Acetonitril for 2 min.

General: The protection groups of the exocyclic amino groups of the bases and at the phosphate are cleaved with concentrated ammonia solution (1 h, 55°C or 8 h at room temperature).

Conclusion

Microdispensing techniques are cheap and reliable, they are availiable and readily used for printing of low density chip libraries. Since dispensing is up to now mainly a serial process, other approaches are of interest to generate highly integrated libraries on chips.

Light directed synthesis has been proved to be the most powerful and developed method to set up complex libraries on chips. The method still remains expensive and bears a number of disadvantages, which might be overcome by other methods, of which the most promising one will be light directed synthesis using semiconductor photoresists. Novel photoresists dedicated to synthesis application will help overcome the existing problems. In the meantime, low-tech applications might remain of interest.

The application of micro contact printing to combinatorial synthesis bears a number of advantages:

- reusable stamps are applied in order to deprotect or to deliver the reagent of choice to the surface,
- the stamps are easily produced from a suitable resist master,
- no light damage occurs,
- no drying and baking steps are necessary, therefore the interaction between the immobilized growing molecules and the stamp material remains weak,
- standard phosphoramidite reaction chemistry may be applyed.

However, a number of substantial problems remain to be solved in order to implement the micro contact printing approach to synthesis. The swelling of the stamp material in organic solvents has to be taken into account when several synthesis steps are required. Positioning of the stamp needs to be carried out up to now in specially designed high precission machines, since a single misstamp leads to activation of the touched area.

Further work has to be run on homogenous stamping wafersized substrates.

References

1 Erb E, Jander KD, Brenner S (1994) Recursive deconvolution of combinatorial chemical libraries. *Proc Natl Acad Sci USA* 91: 11422–11426

2 Fodor SPA, Rava RP, Huang XC, Pease AC, Holmes CP, Adams CL (1993) Multiplexed biochemical assays with biochemical chips. *Nature* 364: 555–556

3 Schober A, Gunther R, Schwienhorst A, Döring M, Lindemann BF (1993) Accurate high-speed liquid handling of very small biological samples. *Biotechniques* 15: 324–329

4 Wallace DB. Ink-Jet based fluid microdispensing in Biochemical Applications, Laboratory Automation News, vol. 1, No. 5, 6–9 (Nov. 1996)

5 Jacobs JW, Fodor SPA (1994) Combinatorial Chemistry – applications of light directed chemical synthesis. *Trends in Biotechnology* 12(1): 19–26

6 Pease AC, Solas D, Sullivan EJ, Cronin MT, Holm CT, Fodor SPA (1994) Light generated oligonucleotide arrays for rapid DNA sequence analysis. *Proc Natl Acad Sci USA* 91(11): 5022–5026

7 Gruber SM, Yu-Pang P, Fodor SPA (1992) Light directed combinatorial peptide synthesis. *Proc Am Pep Symp* 12th

8 Southern EM, Maskos U (1994) Parallel synthesis and analysis of large numbers of related chemical compounds: applications to oligonucleotides. *J Biotechnology* 35(2–3): 217–227)

9 McGall G, Labadie J, Brock P, Wallraff G, Nguen T, Hinsberg W (1996) Light directed synthesis of high density oligonucleotide arrays using semiconductor photoresists. *Proc Natl Acad Sci USA* 93: 13555–13560

10 Delamarche E, Bernard A, Schmid H, Michel B, Biebuyck H (1997) Patterned delivery of immunoglobulins to surfaces using microfluidic networks. *Science* 276: 779–781

11 Ermantraut E, Wölfl S, Saluz HP. Generation of matrix – bound miniaturized combinatorial poly- and oligomere libraries. Patent No DE 195.43.232.A1

12 Kumar A, Biebuyck HA, Whitesides GM (1994) Patterning self assembled monolayers: Applications in materials science. *Langmuir* (10): 1498–1511

13 Engelhardt H, Mathes D (1977) Chemically bonded stationary phases for aqueous high performance exclusion chromatography. *Journal of Chromatography* 142: 311–320

14 Wilbur JL, Kumar A, Kim E, Whitesides GM (1992) Microcontact printing of self-assembled monolayers: A flexible new technique for microfabrication. *Journal of the American Chemical Society* 114: 9188–9189

11 Synthesis and screening of bead-based libraries

Dirk Vetter

11.1 Introduction

Heterogeneous solid phase assembly schemes start from the immobilization of one reaction partner on a support suited for treatment with solvent. The process offers a number of advantages which are most obvious for multi-step syntheses. Product remains bound to support during synthesis, isolation and purification of intermediates become obsolete. Simple washing steps remove educt and catalysts as well as additives, acid or base. The elegance of the approach was demonstrated by Merrifield's total synthesis of Ribonuclease A [1]. Other advantages of solid phase synthesis are less byproduct [2] and improved chemoselectivity [3]. A large excess of the solubilized reaction partner can be employed, in extreme cases up to a millionfold, driving reactions towards efficient conversion. On the other hand the excess of solubilized or liquid reagent is a necessity and effectively renders conversions more expensive. After completion of the assembly protocol and optional deprotection of side chains or functionalities the product either remains bound to the support for further evaluation such as orthogonal cleavage into aqueous media or is submitted to a simultaneous deprotection/cleavage workup. The latter case affords precipitation and/or solvent evaporation and purification by chromatography.

Common supports are particles (polymeric or glass [4, 5] or cellulose [6]), polymeric filter frits or pins [7], glass slides [8, 9], cellulose membranes [10] or polymer sheets. Methods in this area have been reviewed extensively [11].

11.2 Beads as synthesis support

Solid phase synthesis was strongly advanced by the pioneering work of Merrifield. In his original work on polypeptide synthesis he employed a beaded resin which is still highly popular in the field. The material is a commercially available divinylbenzene crosslinked polystyrene. Particle size distribution is polydisperse in the range of 50–200 µm. In his work he found a degree of crosslinking of 2% to give the best results in synthesis quality. This support offered an ideal compromise between mechanical stability during manipulation and solvent access. Intuitively, one considers a rigid material as less susceptible to damage during manipulation such as weighing and prolonged agitation in the presence of elevated temperature, while a soft and gel-like support is expected to exhibit a better performance during swelling, reagent uptake and exchange.

Merrifield resin is crosslinked with no permanent porosity, although it appears that macropores in Merrifield resin are memorized, even after swelling in a compatible solvent followed by vacuum evaporation [12]. Voids only unfold upon swelling in solvents such as dichloromethane, dimethylformamide, tetrahydrofurane or toluene, when polymer chains are separated from each other by solvation [13]. Solvent uptake occurs in a stepwise manner by filling of the fixed pores, expansion of fixed pores and swelling of the polymeric nuclei [14]. Various resins were compared for swelling behaviour, uniformity of particle size and degree of substitution in the context of synthesis quality control [15]. Diffusion of small organic molecules in preswollen polystyrene beads was also studied by NMR pulsed-gradient spin-echo [16], and rate constants of exchange of solvent in and out of solvated crosslinked polystyrene beads were detected by a ^{13}C-NMR magnetization transfer method [17]. Microviscous effects might be responsible for a rather slow solvent exchange in network microcavities. It remains a challenge in resin preparation to ensure uniformity of crosslinking as the monomers styrene and divinylbenzene exhibit varying reactivities depending on the polymerization process parameters [18]. A review on bead preparation methods is given by Arshady [19].

An important parameter to characterize synthesis supports is reaction site density, shown by Merrifield using autoradiography of labeled synthesis products to be homogeneous throughout the individual beads [20]. A more recent study investigated the distribution of functional groups in polymeric beads by laser-flash photolysis on benzoylated Merrifield resin [21]. The reaction site density is macroscopically described by "loading", the amount of functionalities per gram of resin. A high loading in polymer beads may be compared to a low loading on controlled pore glass, as CPG lacks any internal reaction sites. Therefore, the "surface" of a bead accounts for a mere 1% of its number of sites and most of a bead's functionalities are accessible only via diffusion through the polymer network.

As mentioned above, a heterogeneous solid phase reaction requires stoichiometric excess of the solubilized agent. This is due to a steric dilution or rather restricted mobility effect: polymer-bound functionalities cannot diffuse freely but rather have to sit and wait for a passing reaction partner. Therefore not only a molar excess but also a high concentration of the liquid phase reagent is desirable. It also helps to suppress intermolecular reactions between the immobilized reaction sites resulting in side reactions such as dimerization and chain aggregation. These occur as pseudo concentration effects on high-loading resins: if sites are sterically close, they will remain so for a much longer time than in solution, and although homogeneous reaction systems offer possibly more encounters, they will be less effective due to the short contact time. The steric concentration effect was studied on a model compound by means of electron paramagnetic resonance spectroscopy as a function of solvent and resin loading [22]. The feasibility of intermolecular solid-phase reactions in beads was demonstrated by intermolecular peptide cyclization in Merrifield resin [23] and investigations into site-site interactions by anhydride formation [24].

To circumvent many of the problems of the original Merrifield resin, a polyethylene glycol (PEG) -grafted polystyrene resin was developed. The resin exhibits improved solvent compatibility with a homogeneous swelling behavior over a great range of solvents, it is especially useful in protic polar liquids. The gel-like material carries a lower loading and reaction sites are exclusively located at the free termini of the polyethylene glycol chains and are highly mobile. The enhanced mobility is quite visible in nuclear magnetic resonance (NMR), as studied in a ^{13}C-NMR T1-relaxation time experiment of bead-bound peptides in deuterated DMSO: the immobilized probes behave similar to the components in free solution [25]. The resin structure was also investigated in a study of trypsin penetration of beads by confocal laser microscopy and NMR spectroscopy [26]. The on-resin pseudo dilution leads to less side reactions and more solution-like reactivity. Protocols established for solution phase systems are more easily adaptable to the solid phase format. Disadvantage of the PEG-polystyrene conjugate is a soft, gel-like appearance, which makes it more vulnerable to physical damage. Commonly employed gel resin is rather monodisperse with particle size distributions around 100 µm. Solid-phase synthesis has also been demonstrated with much smaller (5 µm-) [27] as well as much larger (750 µm-) [28] beads.

Other supports more recently developed for solid phase synthesis are higher loading PEG-polystyrene, cross-linked acrylamides [29–31], and cross-linked ethoxylate acrylate [32]. A trend goes towards macroporous resins, which offer the advantage of eliminating the cumbersome gel network, providing high mobility for the solvent and reagent, and little mobility and low steric density for the immobilized partner.

11.3 Libraries on beads

With the advent of combinatorial chemistry a dogma of organic synthesis, the sequential preparation of individual compounds, was broken. For rationalization purposes it was found advantageous to pursue the synthesis of as many compounds in parallel as possible. Methods and philosophies used in combinatorial approaches [33, 34] are not limited to solid phase synthesis on beads, but Furka's concept of split synthesis [35] opened the door to vast synthetic libraries of compounds. He realized that very little reaction steps can give a mixture of a great number of different chemical products. To illustrate the practice of this approach the simplest apparatus for this purpose can be envisioned: a 96-well microplate equipped with bottom filters is loaded with resin. After synthesis of 96 different intermediates and washing, the plate is covered with a single well microplate, bottom up. The plates are turned upside down, resin is mixed thoroughly and redistributed into the 96-well plate. The process is repeated until completion of the synthesis steps. Automated peptide split synthesis instruments have been constructed [36]. The advantage of split synthesis over coupling of solubilized mixtures of reagents to one resin sample is that in the final mixture products are more evenly distributed [37]. Mixture syntheses have to deal with individually differing coupling rates

[38]. Nevertheless it has to be taken into account that split synthesis libraries have to be redundant in terms of number of beads per compound [39].

When Furka carried out cleavage of the product mixture, he actually did not take advantage of the fact that the occurrence of one compound per bead allowed for more detailed library investigation. This was realized by Lam et al. [40] when investigating this process. They pointed out that in a mixture of beads prepared by Furka's split-and-pool methodology each bead carried exactly one compound. Or put more precisely, each bead experienced one well-defined protocol which in principle should result in the formation of only one chemical product: the one-bead-one-compound library [41].

The reason this became of great interest was the development of solid phase bead binding assays, where a biomacromolecule like a target protein was contacted with a number of library-derived beads carrying different chemical products. Binding events could be detected in mixtures of such beads by precipitative color formation, fluorescence staining, interaction with streptavidin-magnetic beads to pull out affinity-motif beads [42, 43] or incubation with radiolabled acceptor and subsequent immobilization in agarose followed by autoradiography [44, 45]. Beads associated with high assay signal can be picked out manually with the aid of forceps and microscope and analyzed for their bioactive chemical structure. Using this methodology, one-bead-one-compound libraries have proven effective in a number of applications [46–48].

Due to the sensitivity of modern analytical chemistry, the loading of a single bead, roughly 100 pmol, is sufficient for automated peptide sequencing by Edman degradation, high pressure liquid chromatography or gas chromatography. The latter methods are highly effective if coupled to mass spectrometry as in GC-MS: A recent example of a solid-phase split-mix organic synthesis of β-thioketones demonstrated that compound identities can be established by GC-MS analysis on a per bead basis [49]. Mass spectrometry in itself carries a great potential for library analysis [50]. Matrix assisted laser desorption ionization mass spectrometry can directly follow the course of multistep solid phase reactions in a single measurement if capping reagents created a mass-"ladder" during synthesis [51, 52]. Even spatially resolved spectra from the surface area of single beads were obtained by imaging-time of flight-secondary ion mass spectrometry [53, 54].

The disadvantage of the above-mentioned analytical tools is that they are destructive and use up the bead-borne amount of product. Non-destructive alternatives are equally sensitive and can also be used on individual beads, comprising methods such as infrared spectroscopy (IR) [54–57] and nuclear magnetic resonance (NMR) [58]. IR-microscopy is very well suited to follow reactions such as carbonyl formation on single beads [59–62]. These investigational tools are not primarily aimed at revealing the structure of an active compound but rather for quality control purposes of combinatorial liberaries [63]. Furthermore, solid phase chemistry often requires extensive adaptation and optimization of solution phase protocols. As this task is performed by ^{1}H-NMR in conventional synthetic work, the demonstration that the method is applicable even to individual beads is worthy of mention [64].

For a number of reasons named below, the full amount of product from one bead is rarely available for chemical analysis. It was realized that vast compound libraries on common 100 µm-beads require very large reaction vessels, kg-quantities of resin and a very long evaluation process. Therefore smaller beads were considered for library construction. Affymax' encoded synthetic library technology is a highly developed automated [65] scheme based on 10 µm beads. Screening was performed by bead-binding assays with fluorescently labeled proteins and binding event detection by means of a fluorescence activated cell sorter instrument. Common analytical tools are not able to handle these very small compound quantities and methods had to be found to "encode" the synthesis protocol for the product on its carrier, the bead, in a molecular fashion. In this approach, parallel peptide and oligonucleotide synthesis [66] were performed, with PCR and cloning as amplification tools to decode smallest traces of bead-bound structures [67, 68].

It was realized though that bead-binding assays are of limited use for medicinal chemistry. One has to fight problems arising from high local concentration effects [69] in the bead matrix as well as passive nonspecific adsorption of proteins, resulting in false positives and intense bead staining due to avidity rather than affinity. One way to address these issues could be bead-to-bead compound transfer, where a synthesized peptide is cleaved from the polystyrene support and bound to a more biocompatible hydrophilic material [70]. Alternatively, compounds may be cleaved from beads to be assayed under homogeneous conditions. Split synthesis libraries were divided into sublibraries and products were cleaved into small mixtures for screening. Pools of regular 100 µm sized beads were distributed in the 96 or 384 or more recently 1536 well format, cleaved and assayed for bioactivity. Often, orthogonal photocleavage with photolabile linker [71] was employed to directly release the products into the assay medium [72]. None or only fractions of that material remained on the bead, leaving traces that were insufficient for analysis.

For this reason, a fraction of the bead matrix was spared for a parallel encoding synthesis. The encoding chemistry initially focussed on microsequencing of peptides [73, 74] but then turned towards chemically more inert tagging structures [75, 76]. Binary encoding [77] uses mixtures of compounds which can be separated under optimized conditions into a coding digital "fingerprint". This way, a 23,540,625 member library was encoded by a mere 25 identifiers [78]. The encoded split synthesis [79] was customized to highly sensitive analytical methods such as electron capture detection GC [80], GC-MS, fluorescent confocal microscopy [81], tagging with NMR-readable isotopes [82], tagging with MS-readable isotopes [83], fluorescence detection HPLC [84, 85] or ICP-MS [86].

In this context the chemical differentiation of "surface" volume and internal volume of a single bead was developed. Spectrophotometric quantitation of enzyme-penetratable surface areas were performed [87]. This protein diffusion restriction was used to differentiate surface area and internal volume, and the approach was termed "shaved beads" [88, 89]. Alternatively, chemical derivatization of the polystyrene core and subsequent conjugation of biocompatible PEG-chains was used to create chemically different external and internal sites [90]. If products were to be tested in solution, the small surface fraction was spared for encoding chemistry and the large internal volume remained for bioactivity tests. Or otherwise, the binding assay was performed on the bead surface, protein target was stripped and promising beads were cleaved, microeluted and the mother bead preserved for sequencing [91]. Certain linker technologies even allowed for the controlled subsequent release of compound aliquots for multiple tests from a single bead [92, 93], termed "staged orthogonal release" [93]. This method was found useful if screening of mixtures preceded screening of individual beads [95].

As bioactivity testing of mixtures can result in wrong positives or deconvolution of the active member can prove difficult, approaches were taken to immobilize beads in a gel matrix in order to spatially separate them and release products after embedding of the beads. In suitable assays diffusion of active compounds created an aurora of cytotoxicity [96, 97]. This concept was further developed by the creation of regular arrays of miniaturized wells with one bead per well. The whole array was filled with agarose gel, immobilizing the beads and allowing for parallel washing. Due to the miniaturized well size, the compound diffusion was restricted to its home well [98].

Nevertheless, as beads were created by encoded split synthesis, the actual chemical structure of the compounds was not known at the bioassay stage and had to be learned by bead removal and subsequent decoding analysis. This disadvantage severely limits the use of encoded split synthesis libraries for structure activity relationship studies. The decoding process is of sequential nature and rather slow, only fractions of the library can be decoded to relate bioactivity to chemical structure.

To address this issue, miniaturized parallel single bead synthesis is currently being developed [99]. It is expected to create a greater wealth of bioactivity relationships. The advantage of this approach is that the location of a single bead is the same for synthesis and screening, hence full structural or protocol information is retained for every single member of the library.

References

1 Gutte B, Merrifield RB (1969) The total synthesis of an enzyme with ribonuclease A activity. *J Am Chem Soc* 91: 501–502

2 Routledge A, Abell C, Balasubramanian S (1997) An investigation into solid-phase radical chemistry. Synthesis of furan rings. *Synlett* (1): 61–62

3 Spitzer JL, Kurth MJ, Schore NE, Najdi SD (1997) Polymer-supported synthesis as a tool for improving chemoselectivity: Pauson Khand reaction. *Tetrahedron* 53 (20): 6791– 6808

4 Chong P, Sydor M, Wu E, Klein M (1992) Resin-bound synthetic peptides: use as antigens and immunogens. *Mol Immunol* 29 (3): 443–464

5 Parr W, Grohmann K (1971) New solid support for polypeptide synthesis. *Tetrahedron Lett* (28): 2633–2636

6 Englebretsen Darren R, Harding DRK (1994) Fmoc SPPC using Perloza beaded cellulose. *Int J Pept Protein Res* 43 (6): 546–554

7 Arendt A, McDowell JH, Hargrave PA (1993) Optimization of peptide synthesis on polyethylene rods. *Pept Res* 6 (6): 346–352

8 Fodor SPA, Read JL, Pirrung MC, Stryer L, Lu AT, Solas D (1991) Light-directed, spatially addressable parallel chemical synthesis. *Science* 767–773

9 Southern EM, Maskos U (1990) Support-bound oligonucleotides. *PCT Int Appl* 24 pp WO 9003382

10 Frank R (1992) Spot-synthesis: an easy technique for the positionally addressable, parallel chemical synthesis on a membrane support. *Tetrahedron* 48 (42): 9217–9232

11 Jung G, Beck-Sickinger AG (1992) Multiple peptide synthesis methods and their applications. *Angew Chem* 3 (4): 367–486

12 Yamamizu T, Akiyama M, Takeda K, Kobunshi R (1989) Synthesis and character-

ization of crosslinked polymers. II. Retention of pore structure in porous crosslinked polystyrenes. 46(1): 29–35

13 Svec F, Frechet MJ (1996) New designs of macroporous polymers and supports: from separation to biocatalysis. *Science* 273(5272): 205–211

14 Rabelo D, Coutinho FMB (1994) Porous structure formation and swelling properties of styrene-divinylbenzene copolymers. *Eur Polym J* 30(6): 675–682

15 Pugh KC, York EJ, Stewart JM (1992) Effects of resin swelling and substitution on solid phase synthesis. *Int J Pept Protein Res* 40(3–4): 208–213

16 Pickup S, Blum FD, Ford WT (1990) Self-diffusion coefficients of Boc-amino acid anhydrides under conditions of solid-phase peptide synthesis. *J Polym Sci, Part A: Polym Chem* 28(4): 931–934

17 Periyasamy M, Ford WT (1985) Rates of exchange of solvent in and out of crosslinked polystyrene beads. *React Polym, Ion Exch, Sorbents* 3(4): 351–355

18 Ding ZY, Ma S, Kriz D, Aklonis JJ, Salovey R (1992) Model filled polymers. IX. Synthesis of uniformly crosslinked polystyrene microbeads. *J Polym Sci, Part B: Polym Phys* 30(11): 1189–1194

19 Arshady R (1991) Beaded polymer supports and gels. I. Manufacturing techniques. *J Chromatogr* 586(2): 181–197

20 Merrifield RB, Littau VC (1968) Solid-phase peptide synthesis. Distribution of peptide chains on the solid support. *Colloq Int Centre Nat Rech Sci* No. 175: 179–182

21 Bourdelande JL, Font J, Wilkinson F, Willsher C (1994) Insoluble polymeric benzophenone: surface photophysical studies using diffuse-reflectance laser-flash photolysis (DRLFP). *J Photochem Photobiol, A* 84(3): 279–282

22 Cilli EM, Marchetto R, Schreier S, Nakaie CR (1997) Use of spin label EPR spectra to monitor peptide chain aggregation inside resin beads. *Tetrahedron Lett* 38(4): 517–520

23 Bhargava KK, Sarin VK, Nguyen LT, Cerami A, Merrifield RB (1983) Synthesis of a cyclic analog of oxidized glutathione by an intersite reaction in a swollen polymer network. *J Am Chem Soc* 105(10): 3247–3251

24 Scott LT, Rebek J, Ovsyanko L, Sims CL (1977) Organic chemistry on the solid phase. Site-site interactions on functionalized polystyrene. *J Am Chem Soc* 99(2): 625–626

25 Auzanneau F-I, Meldal M, Bock K (1995) Synthesis, characterization and biocompatibility of PEGA resins. *J Pept Sci* 1(Launch Issue): 31–44

26 Quarrell R, Claridge TDW, Weaver GW, Lowe G (1996) Structure and properties of TentaGel resin beads: implications for combinatorial library chemistry. *Mol Diversity* 1(4): 223–232

27 Bayer E, Rapp W (1988) Application of monodisperse polymer beads in polymer supported peptide synthesis. Synthesis of polymer peptides for immunological studies and affinity chromatography. *Pept Chem* Volume Date 1987 263

28 Pursch M, Schlotterbeck G, Tseng L-H, Albert K, Rapp W (1997) Monitoring the reaction progress in combinatorial chemistry: 1H MAS NMR investigations on single macro beads in the suspended state. *Angew Chem*, Int Ed Engl Volume Date 1996, 35(23/24): 2867–2868

29 Sparrow JT, Knieb-Cordonier NG, Obeyseskere NU, McMurray JS (1996) Large-pore polydimethylacrylamide resin for solid-phase peptide synthesis: applications in Fmoc chemistry. *Pept Res* 9(6): 297–304

30 Renil M, Meldal M (1995) Synthesis and application of a PEGA polymeric support for high capacity continuous flow solid-phase peptide synthesis. *Tetrahedron Lett* 36(26): 4647–4650

31 Alfred J-C, Daunis J, Jacquier R (1996) Facile synthesis of new amine high-loaded poly(meth)acrylamide-based resins for solid phase peptide synthesis. *Macromol Chem Phys* 197(1): 389–401

32 Kempe M, Barany G (1996) CLEAR: A Novel Family of Highly Cross-Linked Polymeric Supports for Solid-Phase Peptide Synthesis. *J Am Chem Soc* 118(30): 7083–7093

33 Gallop MA, Barrett RW, Dower WJ, Fodor SPA, Gordon EM (1994) Applications of combinatorial technologies to drug discovery. 1. Background and peptide combinatorial libraries. *J Med Chem* 37(9): 1233–1251

34 Gordon EM, Barrett RW, Dower WJ, Fodor SPA, Gallop MA (1994) Applications of combinatorial technologies to drug discovery. 2. Combinatorial organic synthesis, library screening strategies, and future directions. *J Med Chem* 37(9): 1233–1251

35 Furka Á, Sebestyén F, Asgedom M, Dibó G (1991) General method for rapid synthesis of multicomponent peptide mixtures. *Int J Peptide Protein Res* 37: 487–493

36 Bartak Z, Bolf J, Kalousek J, Mudra P, Pavlik M, Pokorny V, Rinnova M, Voburka Z, Zenisek K et al (1994) Design and construction of the automatic peptide library synthesizer. *Methods* (San Diego) 6(4): 432–437

37 Boutin JA, Fauchere AL (1996) Combinatorial peptide synthesis: statistical evaluation of peptide distribution. *Trends Pharmacol Sci* 17(1): 8–12

38 Ostresh JM, Winkle JH, Hamashin VT, Houghten RA (1994) Peptide libraries: determination of relative reaction rates of protected amino acids in competitive couplings. *Biopolymers* 34(12): 1681–1689

39 Burgess K, Liaw AI, Wang N (1994) Combinatorial Technologies Involving Reiterative Division/Coupling/Recombination: Statistical Considerations. *J Med Chem* 37(19): 2985–2987

40 Lam KS, Salmon SE, Hersh EM, Hruby VJ, Kazmierski WM, Knapp RJ (1991) A new type of synthetic peptide library for identifying ligand-binding activity. *Nature* (London) 354(6348): 82–84

41 Lebl M, Krchnak V, Sepetov NF, Seligmann B, Strop P, Felder S, Lam KS (1995) One-bead-one-structure combinatorial libraries. *Biopolymers* 37(3): 177–198

42 Samson I, Rozenski J, van Aerschot A, Samyn B, van Beeumen J, Herdewijn P (1995) Screening of a synthetic pentapeptide library composed of D-amino acids against fructose-1,6-biphosphate aldolase. *Lett Pept Sci* 2(3/4): 259–260

43 Fassina G, Belliti M, Cassani G (1994) Screening synthetic peptide libraries with targets bound to magnetic beads. *Protein Pept Lett* 1(1): 15–18

44 Turck CW (1994) Radioactive screening of synthetic peptide libraries. *Methods* (San Diego) 6(4): 396–400

45 Kassarjian A, Schellenberger V, Turck CW (1993) Screening of synthetic peptide libraries with radiolabeled acceptor molecules. *Pept Res* 6(3): 129–133

46 Pennington ME, Lam KS, Cress AE (1996) The use of a combinatorial library method to isolate human tumor cell adhesion peptides. *Mol Diversity* 2(1/2): 19–28

47 Lam KS, Lake D, Salmon SE, Smith J, Chen M-L, Wade S, Abdul-Latif F, Knapp RJ, Leblova Z et al. (1996) A one-bead one-peptide combinatorial library method for B-cell epitope mapping. *Methods* (San Diego) 9(3): 482–493

48 McBride JD, Freeman N, Domingo GJ, Leatherbarrow RJ (1996) Selection of chymotrypsin inhibitors from a conformationally-constrained combinatorial peptide library. *J Mol Biol* 259(4): 819–827

49 Chen C, Ahlberg Randall LA, Miller RB, Jones AD, Kurth MJ (1997) The solid-phase combinatorial synthesis of .beta.-

thioketones. *Tetrahedron* 53(19): 6595–6609

50 Loo JA, DeJohn DE, Ogorzalek Loo RR, Andrews PC (1996) Application of mass spectrometry for characterizing and identifying ligands from combinatorial libraries. *Annu Rep Med Chem* 31: 319–325

51 Youngquist RS, Fuentes GR, Lacey MP, Keough T (1995) Generation and screening of combinatorial peptide libraries designed for rapid sequencing by mass spectrometry. *J Am Chem Soc* 117(14): 3900–3906

52 Youngquist RS, Fuentes GR, Lacey MP, Keough T (1994) Matrix-assisted laser desorption ionization for rapid determination of the sequences of biologically active peptides isolated from support-bound combinatorial peptide libraries. *Rapid Commun Mass Spectrom* 8(1): 77–81

53 Brummel CL, Lee INW, Zhou Y, Benkovic SJ, Winograd N (1994) A mass spectrometric solution to the address problem of combinatorial libraries. *Science* 264: 399–402

54 Brummel CL, Vickerman JC, Carr SA, Hemling ME, Roberts GD, Johnson W, Weinstock J, Gaitanopoulos D, Benkovic SJ, Winograd N (1996) Evaluation of Mass Spectrometric Methods Applicable to the Direct Analysis of Non-Peptide Bead-Bound Combinatorial Libraries. *Anal Chem* 68(2): 237–242

55 Pivonka DE, Russell K, Gero T (1996) Tools for combinatorial chemistry: *in situ* infrared analysis of solid-phase organic reactions. *Appl Spectrosc* 50(12): 1471–1478

56 Pivonka DE, Russell K (1997) Monitoring reactions in solid phase peptide synthesis. *Brit UK Pat Appl,* pp 43 GB 2304410

57 Chan TY, Chen R, Sofia MJ, Smith BC, Glennon D (1997) High throughput on-bead monitoring of solid phase reactions by Diffuse Reflectance Infrared Fourier Transform Spectroscopy (DRIFTS). *Tetrahedron Lett* 38(16): 2821–2824

58 Sarkar SK, Garigipati RS, Adams JL, Keifer PA (1996) An NMR Method To Identify Nondestructively Chemical Compounds Bound to a Single Solid-Phase-Synthesis Bead for Combinatorial Chemistry Applications. *J Am Chem Soc* 118(9): 2305–2306

59 Deusen C (1997) Direct identification by IR microscopy. *Labor Praxis* 21(3): 32

60 Yan B, Sun Q, Wareing JR, Jewell CF (1996) Real-Time Monitoring of the Catalytic Oxidation of Alcohols to Aldehydes and Ketones on Resin Support by Single-Bead Fourier Transform Infrared Microspectroscopy. *J Org Chem* 61(25): 8765–8770

61 Yan B, Gstach H (1996) An indazole synthesis on solid support monitored by single bead FTIR microspectroscopy. *Tetrahedron Lett* 37(46): 8325–8328

62 Yan B, Kumaravel G, Anjaria H, Wu A, Petter RC, Jewell CF Jr, Wareing JR (1995) Infrared spectrum of a single resin bead for real-time monitoring of solid-phase reactions. *J Org Chem* 60(17): 5736–5738

63 Fitch WL (1997) Analytical methods for the quality control of combinatorial libraries. *Annu Rep Comb Chem Mol Diversity* 1: 59–68

64 Fitch WL, Detre G, Holmes CP, Shoolery JN, Keifer PA (1994) High-resolution 1H-NMR in solid-phase organic synthesis. *J Org Chem* 59(26): 7955–7956

65 Sugarman JH, Rava RP, Kedar H, Dower WJ, Barrett RW, Gallop MA, Needels MC (1995) Synthesizing and screening molecular diversity. PCT Int Appl, 201 pp. WO 9512608

66 Jones DG (1994) Applications of encoded synthetic libraries in ligand discovery. *Polym Prepr* (Am Chem Soc, Div Polym Chem) 35(2): 981–982

67 Lerner R, Brenner S (1992) Encoded combinatorial chemistry. *Proc Natl Acad Sci USA* 89: 5381–5383

68 Needels MC, Jones DG, Tate EH, Heinkel GL, Kochersperger LM, Dower WJ, Barrett RW, Gallop MA (1993) Generation and screening of an oligonucleotide-encoded synthetic peptide library. *Proc Natl Acad Sci USA* 90(22): 10700–10704
69 Erickson J, Goldstein B, Holowka D, Baird B (1987) The effect of receptor density on the forward rate constant for binding of ligands to cell surface receptors. *Biophys J* 52: 657–662
70 Canne LE, Winston RL, Kent SBH (1997) Synthesis of a versatile purification handle for use with Boc chemistry solid phase peptide synthesis. *Tetrahedron Lett* 38(19): 3361–3364
71 Holmes CP, Jones DG (1995) Reagents for combinatorial organic synthesis: development of a new o-nitrobenzyl photolabile linker for solid phase synthesis. *J Org Chem* 60: 2318–2319
72 Appell KC, Chung TDY, Ohlmeyer MJH, Sigal NH, Baldwin JJ, Chelsky D (1996) Biological screening of a large combinatorial library. *J Biomol Screening* 1(1): 27–31
73 Nikolaiev V, Stierandová A, Krchnák V, Seligmann B, Lam KS, Salmon SE, Lebl M (1993) Peptide-encoding for structure determination of nonsequenceable polymers within libraries synthesized and tested on solid-phase supports. *Pept Res* 6(3): 161–170
74 Kerr JM, Banville SC, Zuckermann RN (1993) Encoded combinatorial peptide libraries containing non-natural amino acids. *J Am Chem Soc* 115: 2529–2531
75 Borchardt A, Still WC (1994) Synthetic receptor binding elucidated with an encoded combinatorial library. *J Am Chem Soc* 116(1): 373–374
76 Ohlmeyer MHJ, Swanson RN, Dillard L, Reader JC, Asouline G, Kobayashi R, Wigler M, Still WC (1993) Complex synthetic chemical libraries indexed with molecular tags. *Proc Natl Acad Sci USA* 90(23): 10922–10926
77 Baldwin JJ, Burbaum JJ, Henderson I, Ohlmeyer MHJ (1995) Synthesis of a small molecule combinatorial library encoded with molecular tags. *J Am Chem Soc* 117(20): 5588–5589
78 Still WC, Wigler MH, Ohlmeyer MHJ, Dillard LW, Reader JC (1996) Complex combinatorial chemical libraries encoded with tags. U.S, 42 pp. Cont.-in-part of US Ser No 159,861. US 5565324
79 Baldwin JJ (1996) Design, synthesis and use of binary encoded synthetic chemical libraries. *Mol Diversity* 2(1/2): 81–88
80 Nestler HP, Bartlett PA, Still WC (1994) A General Method for Molecular Tagging of Encoded Combinatorial Chemistry Libraries. *J Org Chem* 59(17): 4723–4724
81 Egner BJ, Rana S, Smith H, Bouloc N, Frey JG, Brocklesby WS, Bradley M (1997) Tagging in combinatorial chemistry: the use of colored and fluorescent beads. *Chem Commun* (Cambridge) (8): 735–736
82 Garigipati RS, Sarkar SK (1997) A binary coding method for use in combinatorial chemistry. *PCT Int Appl*, 18 pp. WO 9714814
83 Geysen HM, Wagner CD, Bodnar WM, Markworth CJ, Parke GJ, Schoenen FJ, Wagner DS, Kinder DS (1996) Isotope or mass encoding of combinatorial libraries. *Chem Biol* 3(8): 679–688
84 Ni Z-J, MacLean D, Holmes CP, Murphy MM, Ruhland B, Jacobs JW, Gordon EM, Gallop MA (1996) Versatile approach to encoding combinatorial organic syntheses using chemically robust secondary amine tags. *J Med Chem* 39: 1601–1608
85 Maclean D, Schullek JR, Murphy MM, Ni Z-J, Gordon EM, Gallop MA (1997) Encoded combinatorial chemistry: synthesis and screening of a library of highly functionalized pyrrolidines. *Proc Natl Acad Sci USA* 94(7): 2805–2810
86 Rink H, Vetter D, Gercken B, Felder E (1996) Preparation of combinatorial com-

pound libraries coded with element atom tags. *PCT Int Appl*, 49 pp WO 9630392

87 Shin JS, Kim BG, Kim DH, Lee YS (1996) Spectrophotometric assay using streptavidin-alkaline phosphatase conjugate for studies on the proteolysis of polymer bead-bound peptides. *Anal Biochem* 236(1): 9–13

88 Vagner J, Barany G, Lam KS, Krchnak V, Sepetov NF, Ostrem JA, Strop P, Lebl M (1996) Enzyme-mediated spatial segregation on individual polymeric support beads: application to generation and screening of encoded combinatorial libraries. *Proc Natl Acad Sci USA* 93(16): 8194–8199

89 Lebl M, Lam KS, Salmon SE, Krchnak V, Sepetov N, Kocis P (1994) Preparation of solid phase libraries of test compounds and their topologically separated coding molecules. *PCT Int Appl*, 301 pp WO 9428028

90 Fleckenstein B, Wiesmueller K-H, Brich M, Jung G (1994) Novel heterobifunctionalized polystyrene-polyethylene glycol resin for simultaneous preparation of free and immobilized peptides and biological activity detected by confocal microscopy. *Lett Pept Sci* 1(3): 117–126

91 Felder ER, Heizmann G, Matthews IT, Rink H, Spieser E (1996) A new combination of protecting groups and links for encoded synthetic libraries suited for consecutive tests on the solid phase and in solution. *Mol Diversity* 1(2): 109–112

92 Lebl M, Pátek M, Kocis Krchnak V, Hruby VJ, Salmon SE, Lam KS (1993) Multiple release of equimolar amounts of peptides from a polymeric carrier using orthogonal linkage-cleavage chemistry. *Int J Peptide Protein Res* 41: 201–203

93 Kocis P, Krchnak V, Lebl M (1993) Symmetrical structure allowing the selective multiple release of a defined quantity of peptide from a single bead of polymeric support. *Tetrahedron Lett* 34(45): 7251–7252

94 Salmon SE, Lam KS, Lebl M, Kandola A, Khattri PS, Wade S, Patek M, Kocis P, Krchnak V et al. (1993) Discovery of biologically active peptides in random libraries: solution-phase testing after staged orthogonal release from resin beads. *Proc Natl Acad Sci USA* 90(24): 11708–11712

95 Lebl M, Krchnak V, Salmon SE, Lam KS (1994) Screening of completely random one-bead one-peptide libraries for activities in solution. *Methods* (San Diego) 6(4): 381–387

96 Salmon SE, Liu-Steven RH, Zhao Y, Lebl M, Krchnak V, Wertman K, Sepetov N, Lam KS (1996) High-volume cellular screening for anticancer agents with combinatorial chemical libraries: a new methodology. *Mol Diversity* 2(1/2): 57–63

97 Quillan JM, Jayawickreme CK, Lerner MR (1995) Combinatorial diffusion assay used to identify topically active melanocyte-stimulating hormone receptor antagonists. *Proc Natl Acad Sci USA* 92: 2894–2898

98 Schullek JR, Butler JH, Ni Z-J, Chen D, Yuan Z (1997) A high-density screening format for encoded combinatorial libraries: assay miniaturization and its application to enzymic reactions. *Anal Biochem* 246(1): 20–29

99 Vetter D, Thamm A, Schlingloff G, Schober A (1999) Single bead parallel synthesis and screening, *submitted*

12 Sensors for biomolecular studies

Jan Rickert, Thomas Wessa and Wolfgang Göpel

12.1 Introduction

Chemical sensing is a process by which selected information about chemical composition is obtained in real time. This usually takes the form of an amplified electrical signal related to the concentration of one or more chemical species present in the system. The signal can subsequently be manipulated in many ways, with varying degrees of sophistication, depending on the user's requirements.

A biosensor is a particular type of chemical sensor which uses the recognition properties of biological components . These are held in intimate contact with a suitable transducer, such as an electrode or an optical device. The transducer converts the (bio)chemical change into a signal which is then usually translated into a digital electronic result. Miniaturization, reduced cost and the improved processing power of modern microelectronics have further increased the analytical capabilities of such devices, and given them access to a wider range of applications.

The majority of current biosensors may be classified by four different principles. A distinction is usually made between bio-affinity, catalytic, transmembrane, and cell sensors (Fig. 1). The majority of transmembrane and cell sensors is based on affinity or catalytic principles as well. Current research and development aims at improved stability, selectivity and sensitivity of biosensing devices. On the other hand, there is an increasing trend towards miniaturization of those spectrometers, chromatographs and detectors, which are traditionally used in bioanalytical chemistry, by applying micro- and nanofabrication techniques. As a result, a clear distinction between miniaturized instruments and "real" biosensors becomes increasingly difficult and, in certain cases, meaningless. This statement holds in particular for recent trends in the development of optical and electroanalytical biosensors. In addition, the

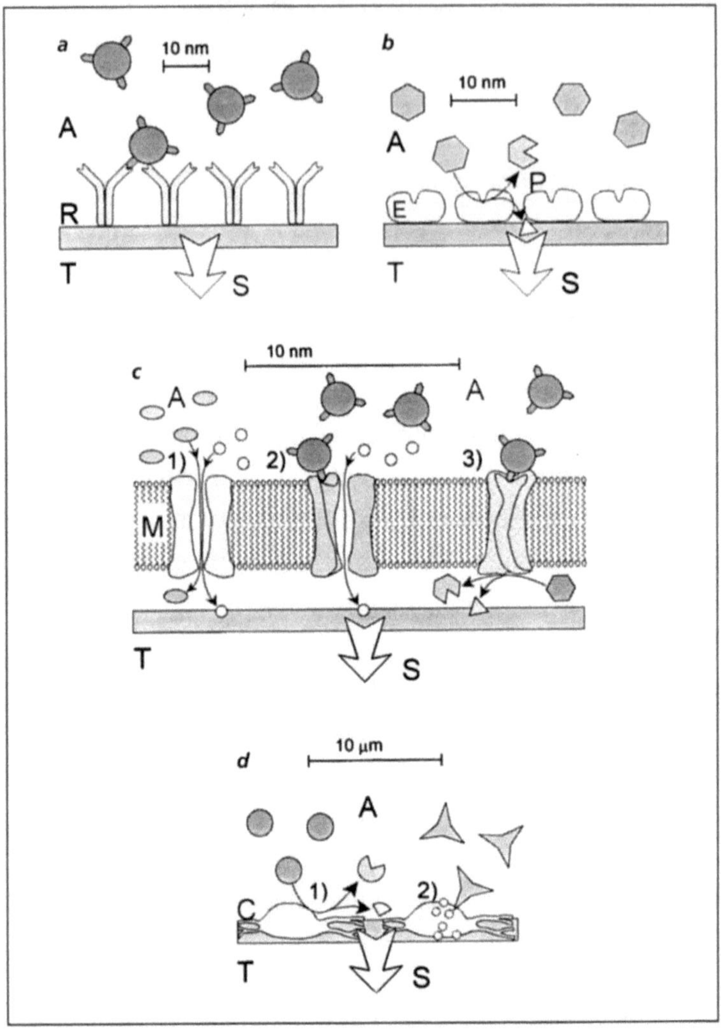

Figure 1 *Different principles of biosensing and typical interfaces: The analyte is denoted by **A**, whereas **T** and the arrow denote the transducer, **P** the reaction product, **E** the enzyme, **C** the cell, and **S** the recorded signal, respectively. (a) Bioaffinity sensors: The analyte (antibody, antigen,...) is recognized by immobilized recognition units **R** (antigen, antibody, DNA,...); (b) biocatalytic sensors: The analyte is converted by immobilized enzymes **E** to products **P**; (c) transmembrane sensors: Transport or channel proteins (1) or receptor proteins (2) are incorporated into a membrane **M**. These structures either move the analyte **A** through the membrane (1), bind the analyte and open a channel for another species (2), or activate subsequently a separate enzymatic cascade (3); (d) cell sensors (note the change in scale): Immobilized living cells **C** convert (1) or bind the analyte (2)*

practical application of any sensor or analytical instrument requires the use of complete analysis systems and hence the development and optimization of the same components for sample handling, pumps, filters, membranes, and sample conditioning (enrichment, conversion by catalysts or enzymes etc.).

For "real" biosensors, the detection occurs by direct conversion of biochemical information to electronic information via suitable transducers. The biochemical composition of complex mixtures can be characterized in selected cases by using arrays of biosensors. Here, the data pretreatment of results from different individual sensors determines suitable feature vectors, and by means of subsequent pattern recognition a biochemical analysis is performed.

There is a *common bottleneck* in the development of all next generation biosensors which is the *insufficient stability and reproducibility* of their interface properties in the different environments of their practical applications [1, 2].

12.2 Stable interfaces: The key in designing reliable biosensors

The selectivity of biosensors is usually obtained by utilizing natural or artificial biomolecular function units such as antibodies, enzymes, transmembrane proteins, etc. [3]. Because of the huge amount of electrical and optical transducers available in thin film technology, most developments of new biosensors aim at the preparation of controlled thin film structures in which these biomolecular function units may be arranged and addressed in reproducible and controlled geometric surroundings [2]. This requires the preparation of arrays of biomolecules on planar surfaces, and hence it is necessary to scale down the processes of covalent immobilization of complete molecules, the *in situ* synthesis of molecules on surfaces (e.g. by self-assembly) and the physical entrapment of molecules in defined areas to micro- and nanometer dimensions [4]. Various concepts to achieve their controlled geometric order include entrapment in polymeric matrices, physisorption, direct covalent attachment, ionic attachment, embedding in lipid membranes, coupling to ordered Langmuir-Blodgett or self-assembled monolayers with covalent or affinity-like linkers, etc. (Fig. 2, see also [5]). Another recent example is the formation of

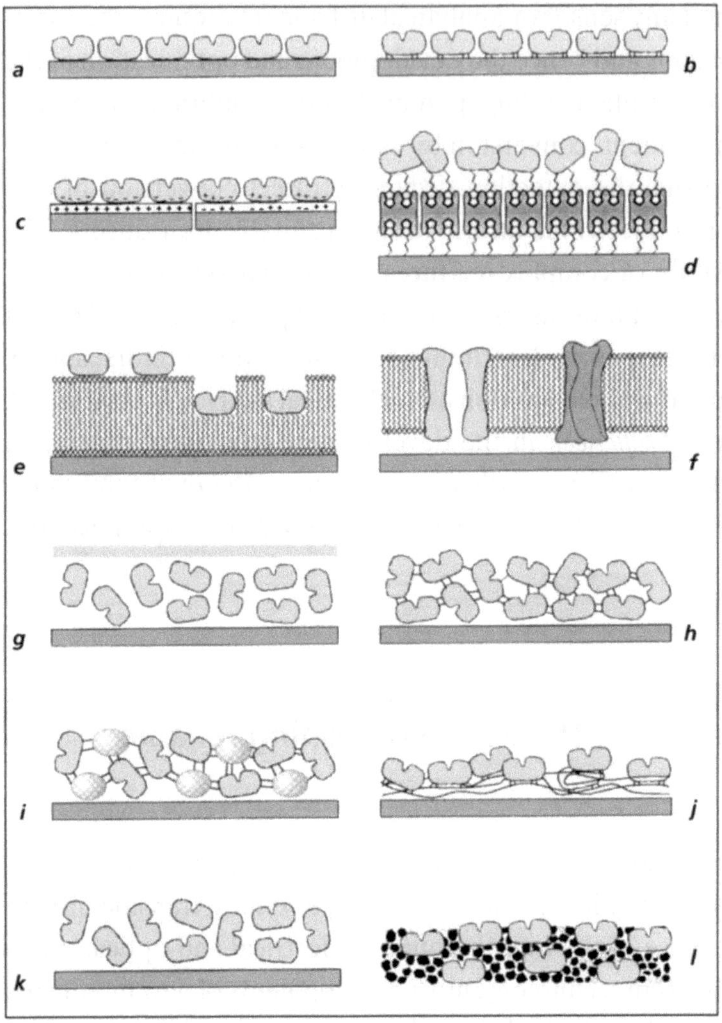

Figure 2 *Very schematic drawing of typical interfaces in biosensing with examples for the immobilization of biological recognition units on transducer surfaces: (a) direct adsorption, (b) covalent linking to the substrate, (c) adsorption by electrostatic forces, (d) coupling via biotin-avidin linkage, (e) adsorption at mono- or bilayers, (f) embedding in lipid bilayers, (g) entrapment behind a membrane, (h) cross-linking of the biological molecules, (i) cross-linking by other large molecules (e.g. BSA), (j) covalent linking to polymer chains, (k) entrapment within a gel of a polymer (non-conducting), (l) entrapment by mixing with carbon paste, conducting polymers or (conducting) organic salts*

nanocomposite films of polyelectrolytes or biological molecules [6]. In this context, a review on the immobilization of enzymes in polymers has recently been given by Geckeler and Müller [7]. A promising method for the controlled immobilization of various enzymes is electrochemical polymerization. In this case, the simultaneous entrapment or covalent coupling to monomer units occur either before or after the polymerization. For a recent review see Bartlett and Cooper [8]. Direct covalent coupling of enzymes to the substrate is another widespread technique. Examples have been given, for example, by Moser et al. [9]. Patterning of molecules at surfaces using photolithography or laser techniques becomes of increasing interest [10]. During recent years, applications of the high binding specificity of the biotin-avidin system became increasingly popular. As the first step, avidin or streptavidin may be coupled to the surface of interest, either directly or via biotin bound covalently to the surface [11]. In this concept, any species which can be biotinylated may be coupled to the surface. Examples include enzymes [12], antibody fragments [13], labeled antibodies [14], or even multilayers [15].

Four substrates are of key importance for most practical thin film devices, i.e.

- platinum,
- carbon,
- modified silicon surfaces or glasses (onto which different metals or metal oxides are deposited or different silanes are attached covalently), and
- gold (onto which mono- and disulphides, thioethers, thiocarbamates, nitriles or cyanides are attached covalently).

For further details, see for example, [11]. Perfect order in monolayer systems is usually not the only important property to be optimized in biosensors. Another criterion is that the biomolecular function units often have to be immobilized such that they keep their biological activity (catalytic or affinity properties). Their large size and necessary accessibility to biomolecules to be detected in solution usually requires some spatial separation of their active sites in ordered monolayers.

Suitable molecules with biomolecular function units to be used for thin film formation may contain short and long chain thiol coupling groups for their attachment to the substrate and in addition specific endstanding groups for their mutual coupling to form a stable two-dimensional system. For the

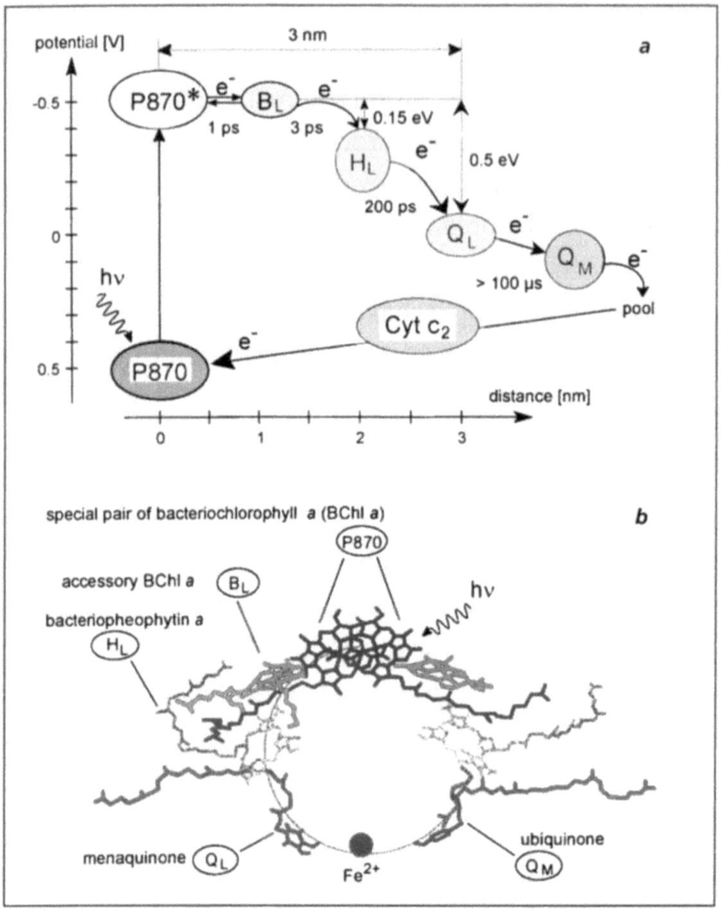

Figure 3 *Natural biomolecular function units indicate challenges for future biosensor developments: Signal transfer with picosecond (ps) time scales and electron volt (V) energies in nanometer (nm) distances. a) Energy scheme of the electron transport chain in the bacterial photosynthetic reaction center. The photo-excited electron is removed from P870* via the bacteriochlorophyll B_L and the bacteriopheophytin H_L to the menaquinone Q_L where it is stored for about 100 µs before being transferred to Q_M and carried to the pool. The spatial distance between P870* and Q_L must be large enough (3 nm) to avoid recombination by electron tunneling, and the energy level of Q_L^- must be sufficiently lower (0.5 eV) than that of P870* in order to avoid recombination by thermal activation (after Kuhn [87]). b) Scheme of the optimum spatial arrangement of the components of the reaction center. The dotted line indicates the pathway of the electron. For details see [86] and [88].*

spatial separation of active sites, a mixture of active and non-active molecules may be used. In this case, two-dimensional phase separation has to be avoided. Another possibility is the embedding of active biomolecules into suitable matrices. Examples include sol-gel materials [16] and polyethyleneglycol [17] or dextran [18] overlayers with attached biomolecular function units.

If electronic transducers are to be utilized, the direct electronic coupling of biomolecular function units to the conducting electrodes is essential. Because of the spatial limitation due to electronic tunneling currents, this requires specific geometric arrangements with a control down to the nanometer scale or the development of suitable mediators. The advantage of a variety of optical transducers is that the film geometry has to be optimized only on a larger scale with dimensions in the order of the extension of evanescent fields (provided that these thicker films exhibit reasonable diffusion rates and hence response times). Typical thicknesses in the lower micrometer range are used which are large as compared to thicknesses of those layers which exhibit controlled charge transfer and which therefore require a controlled geometric order in the nanometer range (for details, see for example, [19]).

The importance of a perfect geometric arrangement for controlled electron transfer rates may be demonstrated by considering the bacterial photosynthetic reaction center: In nature different molecules act as electron mediators with well-separated electronic energies in the order of electron volts (the order of binding energies in covalent bonds) and controlled relaxation times in the order of picoseconds (the order of molecular vibration times). They are arranged geometrically with molecular separations in the order of nanometers (the order of atomic distances), thereby allowing the generation of chemical energy by a metabolism which is triggered by photo-excitation (Fig. 3).

12.3 Transducers

Mass sensitive transducer

Over the last decade mass-sensitive signal transduction has become a feasible tool to study interface interactions. These transducers are based on acoustic waves which propagate in a piezoelectric crystal. The device usually operates as a frequency-determining element in an oscillator circuit.

Bulk acoustic wave devices like the quartz crystal microbalance (QCM) typically operate at resonance frequencies of a few 10 MHz. The measurement effect employed is a frequency decrease of the QCM with increased mass loading ($\Delta f \approx \Delta m f_0^2$). Therefore the sensitivity of QCM-devices is strongly dependent on the value of the resonance frequency. Higher operating frequencies are therefore necessary to achieve higher sensitivities. This could be realized by using very thin quartz substrates or by evaluating overtones. Typical detection limits achieved with 10 MHz QCM's are about 10 pg/mm^2 [20].

Another type of mass-sensitive transducer is known as surface acoustic wave device. In this case a surface acoustic wave (SAW) is generated on an elastic transducer surface between interdigital electrodes. In contrast to the QCM the SAW-technique only detects the surface part of the acoustic waves. These devices operate in higher frequency ranges (typically $f_0 > 100$ MHz) and lower detection limits compared to the QCM are reported (LOD = 3 pg/mm^2) [21]. However, noise problems are usually more severe and for application in aqueous solution the used SAW technique needs to be modified in order to deal with possible damping. Therefore the use of horizontal polarized shear waves is necessary combined with mass gratings on the sensor device. The resulting wave type is called surface transverse wave (STW) and causes only small energy losses in water.

A new approach in bioaffinity studies is possible by using the Love wave devices. This method combines advantages of the bulk acoustic with those of the SAW technique. A SAW-device is typically coated with a thin layer of a thickness in the range of the acoustic wavelength. A suitable layer material is SiO_2 with a thickness of about a few µm. The acoustic wave propagates in the layer with a lower wave velocity than on the piezoelectric substrate. This leads to an higher sensitivity of the device. Preliminary investigations show detection limits below 1 pg/mm^2 [22]. Disadvantages of this device concern the strict requirement of a perfect acoustic coupling of the layer which must be solved separately for each system. The different mass-sensitive transducers are shown in Figure 4.

More commonly used acoustic wave devices use surfaces of gold or silica which can be effectively modified by thiol or by silane chemistry. The active

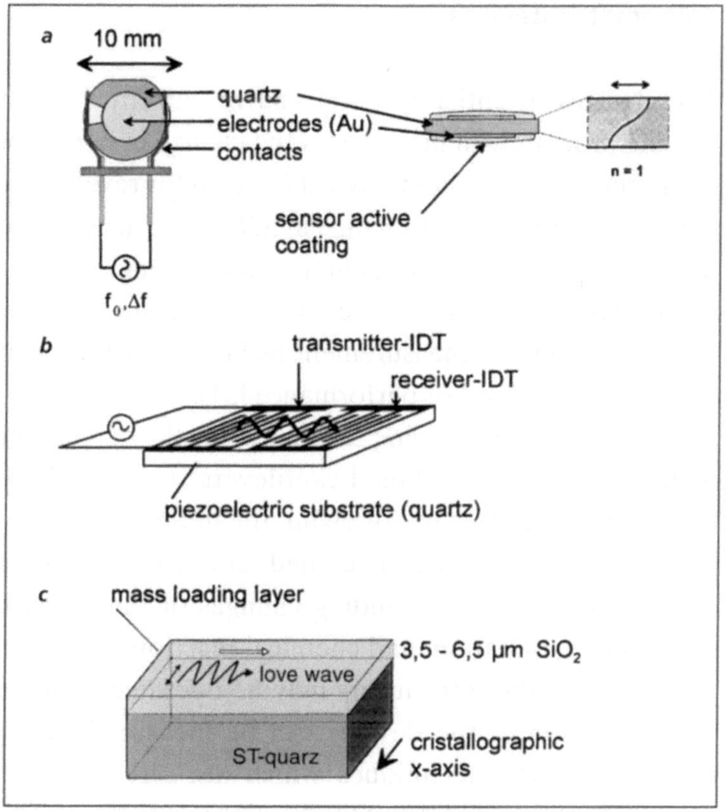

Figure 4 *Schematic drawings of mass sensitive transducers, a) Bulk acoustic wave quartz crystal microbalance; b) surface acoustic wave device; c) love-mode transducer*

surface areas required amount to a few mm^2. Further references to these techniques are given in [23–25].

The great advantage of QCM transducers is, that they are available from high volume applications in consumer electronics, that the readout circuitry is of limited complexity and imposes – other than in optics – no constraints for the geometry of the readout system. This makes the development of a transducer array from individual transducers attractive [26]. A lot of bioanalytical applications have been described already. This concerns in particular investigations of immunological interactions [27, 28] and interactions of oligonucleotides [29].

Calorimetric devices

The changes in enthalpy (ΔH) and in entropy (ΔS) may drive molecular binding processes. In many cases the change in ΔH accounts for at least half of the change in free enthalpy (ΔG) which for reversible changes amounts up to some 10 kJ/mole. The change in ΔH can be measured directly by calorimetric techniques (Fig. 5 – left part). A prominent example is the thermometric determination of enzymatic activities such as the hydrolysis of urea by urease [30]. Thermometric measurement techniques based on micro-bimorphs have achieved a high level of performance [31].

Another very promising technique for the detection of enthalpy changes is the use of micromachined cantilevers. A bimetallic cantilever measures temperature changes by following the deflection of the apex (Fig. 5 – right part) [32]. These micromachined devices translate a broad variety of different parameters, including changes of the force, heat, stress, magnetism, charge, radiation, and chemical reaction into a mechanical deflection. Furthermore, the transducers may be operated in static or dynamic mode. The deflection of the devices can be measured with picometer accuracy using those laser techniques which are already established in scanning probe microscopy. Alternatively piezo resistive effects may be utilized [33].

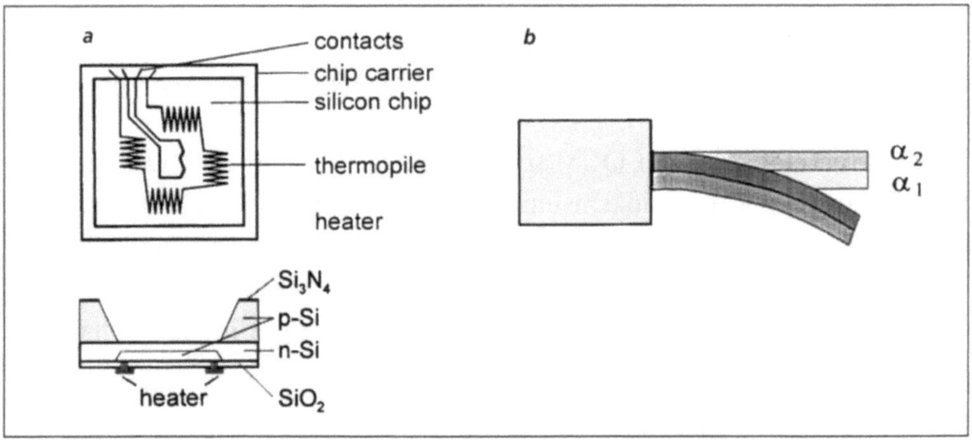

Figure 5 *Schematic drawings of calorimetric devices, a) thermopile chip; b) bimetallic cantilever with different thermal coefficients of expansion* α_1 *and* α_2

Electrochemical detection

Electrochemical techniques may be used to quantify redox-active substances. An overview of different methods applied in electrochemical analysis is given in several reviews [34, 35].

Among the electrochemical techniques, *cyclic voltammetry* is probably the most used one. It is applied for the characterization of electrode materials, surface coverages, electron transfer properties, and electron transfer mechanisms including adsorption phenomena and chemical reactions at the surface [36]. The special case of electrode reactions which occur when the redox species are confined to the surface may be studied by cyclic voltammetry. Other electrochemical techniques may also be applied for the characterization of surface redox processes, e.g. square-wave voltammetry [37] or normal pulse voltammetry.

Redox polymer films may also be investigated by impedance spectroscopy. Typical examples include poly(4′-vinylpyridine) films modified with Ru^{2+} complexes [38] or poly(vinylferrocene) films [39].

The crucial point in the development of electrochemical biosensors is the control of charge transfer rates. As an example, cytochromes coupled to an electrode may enable the controlled electron transfer between the electrode and a redox protein in future biosensor devices. In this context, many electrochemical investigations of redox proteins have been carried out. Reviews were given by [40, 41]. Cytochrome *c* adsorbed or covalently bound to a self-assembled monolayer was exposed to a saturated solution of KNO_3 [42]. The faradic response of adsorbed cytochrome *c* nearly disappeared due to desorption. In contrast, negligible changes in the cyclic voltammograms of covalently immobilized cytochrome *c* were observed.

The electrode current depends on the concentration and the diffusion coefficient of the analyte in the solution. The limiting current *i* found at a planar electrode in a not stirred solution at time *t* is given by the Cottrell equation:

$$i = nFAc\sqrt{\frac{D}{2\pi t}} \qquad (1)$$

with the number of electrons transferred per molecule *n*, the Faraday constant *F*, the electrode area *A*, and the concentration *c*. Over recent years electrochemical methods gained additional attention in modern analytical chemistry as miniaturized electrodes, microdot electrode arrays and interdigitated microelectrode arrays became available. With these devices detection limits in the lower nanomolar range are achievable without preconcentration. Furthermore, these miniaturized devices allow investigations in small sample volumes (< 0.1 µl).

An array of microdot electrodes (see Fig. 6 – left part) shows no influences by stirring the sample. Higher current densities are the result of the hemispherical diffusion of the analyte to the electrodes. Additionally the charging current is small due to the small electrode area. As a result the signal-to-noise-ratio is enhanced significantly, which leads to an increased detection limit.

Interdigitated microelectrode arrays with typical electrode gaps in the order of 5 µm and below may be operated in the redox cycling mode especially with reversible redox active substances. Figure 6 (right part) shows a typical example of such a transducer consisting of two working electrodes, counter- and reference-electrode. In the steady-state, one electrode generates electroactive species (the generator electrode) which diffuse to the neighboring electrode (the collector electrode) where the electrochemical regeneration occurs. This effect leads to an electrochemical feedback. The resulting current is typically amplified by one order of magnitude or more and therefore the detection limit is improved significantly [43].

There is a huge potential for the application of microelectrodes in biosensing because in principle only very small sample volumes are required. Furthermore the basic transducers are accessible by low-cost standard microfabrication techniques [44, 45]. Also the required readout circuitry is of limited complexity and cost. Additionally a set of electrodes can easily be arran-

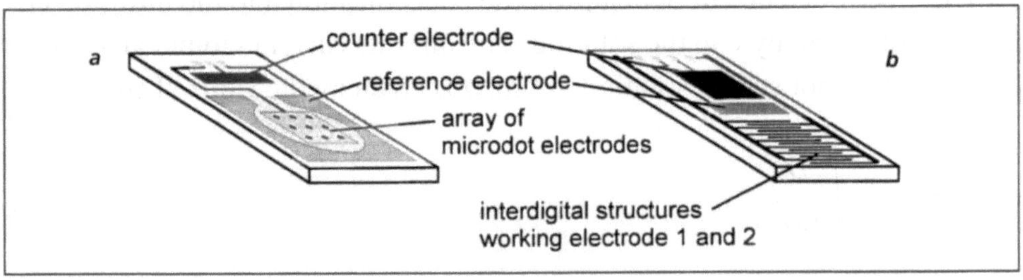

Figure 6 *Schematic drawings of typical electrochemical transducers, a) array of microdot contacts as working electrode; b) one pair of interdigital electrodes as working electrodes for cycling experiments*

ged as a comb-like structure. First investigations were now performed with parallel electrochemical devices for bioanalytical screening applications which are compatible with the established ELISA format. The compatibility of miniaturized transducers with the required microtiter format was demonstrated in recent studies [46].

Typically minimum concentrations to be detected are in the ten nanomolar range. Depending on the readout mode, the technique is considerably fast. Stepped potential techniques (e.g. chronoamperometry) require only a few seconds, while complete potential scans may be acquired within ten seconds.

Ion-sensitive field effect transistors (ISFET's)

These devices (see Fig. 7) have been investigated for many years. To overcome the well-known problems of drift, lack of stability, cross-sensitivity and lack of miniaturized reference electrodes new approaches were presented. For example, stable electrode coatings with very low baseline drifts were achieved by pulsed laser deposition techniques [47]. Another approach presents diamond-like carbon layers as a novel chemically resistant and nevertheless pH-sensitive coating [48]. Additionally several ISFETs were often arrayed and a central electrode used as a counter- or pseudo-reference electrode integrated on a chip.

New polysiloxane membranes with different functional groups were developed for ion-sensitive coatings to monitor e.g. K^+-, Na^+-, Mg^{2+}-, Ca^{2+}-, or NO_3^-- concentrations. These studies established the preparation of suitable coating materials with high stability and selectivity [49].

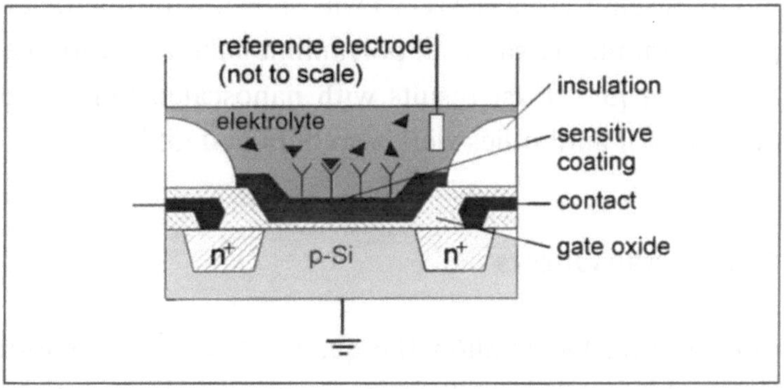

Figure 7 *Schematic drawing of an ion selective field effect transistor (ISFET)*

Impedance spectroscopy

Impedance spectroscopy at electrode surfaces may be carried out with almost any bare electrode surface. The principle is described in [50]. It is recognized as a general tool to study interfaces and molecular properties of surface bound, but also of free molecular species. In studies directed towards affinity analysis very low levels of detection have been realized [51]. The strong non-linearity in the response with increasing coverage and with increasing thickness of a (defect free) adlayer have stimulated research towards layer systems with improved and more controllable performance. Measurements at a fixed frequency are relatively simple to perform. Frequency dependent measurements to study electrodes during surface processes give very detailed information about capacitive and resistive parts to the overall impedance. This may be correlated with concentrations of free charges and layer capacities. The technique is outstanding in the simplicity of the transducer structure itself. The limits of detection vary considerably and several practical devices are under development. The integration of electrodes and drive circuitry in a microstructure will reduce requirements for shielding, referencing, etc. and hence lead to a sensitive practical impedance based device.

The use of nanostructured interdigitated electrodes (based on silicon device technologies) is a new approach in impedance spectroscopy. These electrodes can be functionalized, for example with enzymes (such as horseradish peroxidase) and the change in capacitance scales with the concentration of products (hydrogen peroxide) [52].

Immunological applications of the impedance spectroscopy are also reported. The detection of antibodies against a synthetic epitope of the foot and mouth disease virus (FMDV) was shown with microelectrodes [53]. A competitive immunoassay with polyaniline-labeled antibodies was introduced by Sergeyeva [54]. First results with nanoscaled impedimetric sensors for the detection of antibodies were also obtained [55].

Optical transducers

Optical transducers allow the quantitative determination of one or more of the fundamental characteristics of optical radiation such as amplitude, phase, frequency and polarization. For biosensing application a change of the activity, the concentration or the spatial distribution of a biochemical or biological structure has to result in a change of one of these fundamental quantities. The following discussion will briefly characterize the different transducer principles used for biomolecular studies.

Surface plasmon resonance (SPR)

Surface plasmons are collective fluctuations of the electron plasma at the surface of an electrically conducting material. Surface plasmons can be excited by evanescent light waves using the basic attenuated total reflection (ATR) configuration. Surface Plasmon Resonance (SPR) occurs when the photon momentum along the surface matches the plasmon frequency. SPR is then detected as a strong attenuation of the reflected light beam (see Fig. 8). Coupling conditions are sensitive to the surface refractive index [56]. Changes in refractive index below 10^{-6} can be resolved. A variety of optical set-ups have been proposed and in part commercialized. Detection limits below 1 pg protein/mm^2 are reported [57, 58].

Reflectometric interference spectroscopy (RIfS)

The detection of binding reactions by reflectometric interference spectroscopy (RIfS) is based on the interference occurring at thin planar transparent films. A light beam passing a weakly reflecting thin film will be reflected in part at each of the interfaces (see Fig. 8). As the two reflected beams travel different optical paths, a phase difference is introduced. Interference of the two beams leads to a modulation of reflected light intensity due to constructive and destructive interference [59]. The transducer is simply a nonstructured thin film system on a transparent substrate. Limits of detection are in the rage of 2 pg/mm^2 [60].

Integrated optical devices (IOD)

Integrated optics with monomode channel waveguides are routinely used in long range, high speed telecommunication. Monomode channel waveguides allow for a very straightforward implementation of interferometer structures. A waveguide is split into a measurement and a reference path and

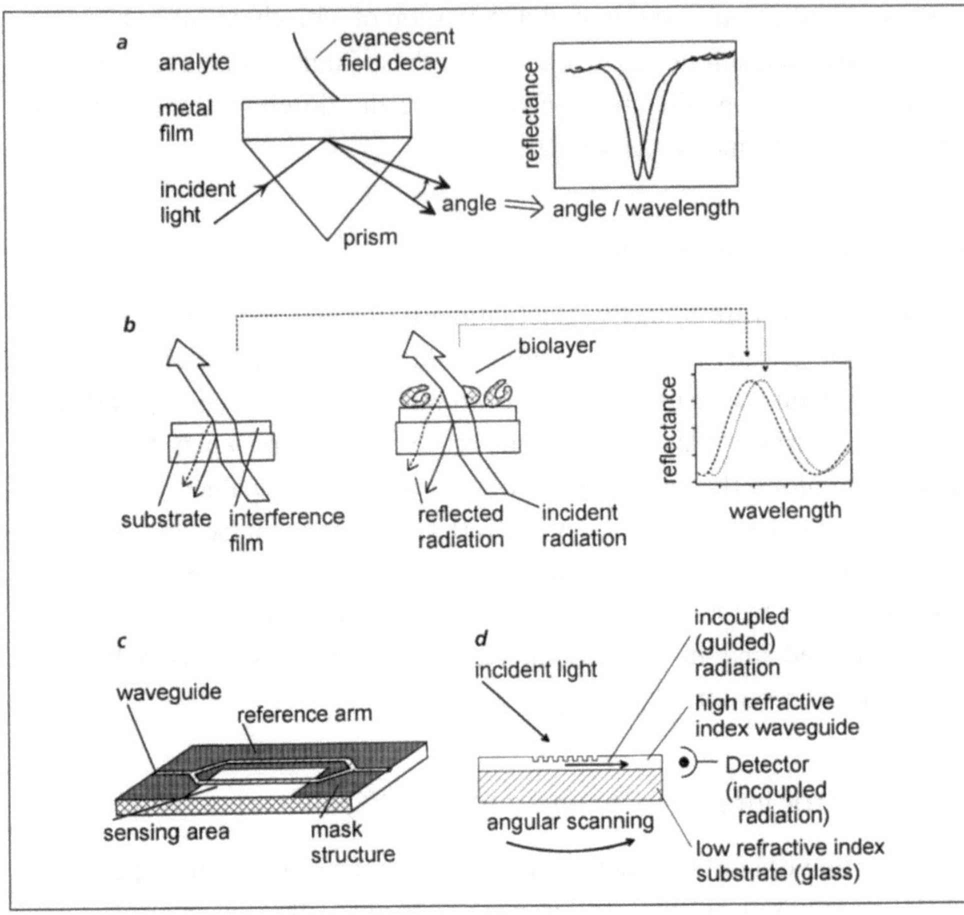

Figure 8 *Schematic drawings of optical transduction principles, a) surface plasmon resonance (SPR); b) reflectometric interference spectroscopy (RIfS); c) integrated optical device; d) grating coupler*

recombined after some distance (see Fig. 8). Any phase difference introduced while the beams were split can be detected by interference effects [61]. Binding of organic matter at the surface of the measurement arm leads to a modulation of intensity at the interferometer output. Limits of detection are in the range of 10 pg/mm^2 [62]. Today the costs of manufacturing such devices are prohibitive, but the implementation of IO devices in silicon or in polymers, driven by the needs of future information technologies, may change this situation [63]. As an example Bookham Technology (UK) develops integrated systems based on its patented ASOC silicon technology.

Grating coupler

The first integrated optical device used in bioaffinity analysis was a grating coupler structure embossed in a monomode film waveguide proposed by Lukosz and co-workers [64]. A periodic grating on the surface of a waveguide can be used for in- or outcoupling of radiation from the waveguide (see Fig. 8). The angle of deflection (coupling angle) depends on the wavelength and the grating period. Binding of organic or biological matter on top of the grating alters the coupling angle. The limit of detection is about 10 pg protein/mm^2.

Prism coupler

The second coupler device which has found application in a commercial affinity detection system is a prism coupler. In a prism coupler light is coupled into a film waveguide from a prism separated by a small gap (filled with low refractive index material) from the waveguide. Light totally internally reflected at the base of the prism can excite a mode in the waveguide. The coupling conditions of a prism coupler are sensitive to the refractive index at the film waveguide. When the coupling conditions are met, light will be coupled into the waveguide, and after some distance back into the prism. This leads to different paths between coupled light and non coupled light, which can conveniently be detected by forced interference between the p and the s state of incident light [65]. The transducer is just a prism with a thin film structure at its base. A commercial system for interaction analysis is available (FISON). The limit of detection is about 10 pg protein/mm^2.

Fluorescence correlation spectroscopy (FCS)

An elegant technique for the determination of translational diffusion coefficients of fluorescence labeled molecules was introduced to affinity applications by Eigen and Rigler [66]. An optical set-up with confocal excitation and detection beam paths is used to focus an excitation laser beam into a liquid sample. By the z-dependence of the confocal optics, the detection of emitted fluorescence light is effectively restricted to the focus of the excitation light beam. The probed volume can be as low as a few femtoliters. At concentrations recommended for the technique (nanomolar and below) only a few analyte molecules – typically less than 10 – are found within the detection volume at a time. The statistical fluctuations in the photon flux over time are correlated to the concentration and the diffusion coefficient of the analyte molecule. Autocorrelation analysis of the time resolved photon flux is used to extract information about binding state and concentrations in the monitored sample. In principle the method may operate with cycle times of 1 s, the detection volumes are very low and the concentrations required are in the nanomolar range. Commercially available instrumentation allows to detect changes in the diffusional characteristics if the molecular weight of the complex exceeds the molecular weight of the labeled compound at least by a factor of two. The high intrinsic time resolution achieved by the autocorrelation of the photon flux signal allows to resolve binding processes at their intrinsic time scale. The ability of *FCS* to probe small sample volumes makes this technique unique for *in situ* studies of binding effects, e.g. of the intracellular distribution of receptor occupation.

Fluorescence polarisation

The first binding assay with considerable success which employed changes in molecular weight was based on the increase in the rotational relaxation time of a fluorophore attached to a low molecular weight ligand on binding. Fluorophores attached to the low molecular weight compound are used as a probe. The determined parameter is the fluorescence anisotropy. The sample is illuminated with linearly polarized excitation light. Therefore only fluorophores with proper orientation will be excited. If rotational relaxation is slower than the lifetime of the excited state, the emitted fluorescence light will be polarized to some extent, too. This principle called fluorescence polarization immunoassay (FPIA) [67] has been applied to diagnostic assays for various drugs. Binding of the labeled compound to the receptor will increase the fluorescence anisotropy. The measurement of static anisotropy is based on excitation with polarized light and measurement of the orthogonally polarized compounds of the fluorescence light. This is quite straightforward and requires little technical effort. More information can be obtained by dynamic anisotropy measurements with pulsed or modulated excitation and time resolved or phase sensitive detection [68].

Ellipsometry

A sensitive technique to monitor thin film structures is ellipsometry, where the reflectance for p- and s-polarized light is used to determine the properties of a thin film system. Applications for biomolecular interaction processes have been reported [59]. An imaging (spatially resolving) ellipsometry system for binding studies was described recently by Jin et al. [69].

12.4 Selected case studies

A few examples may illustrate practical implication to use these transducers.

Enzyme biosensors

Enzyme biosensors can be classified by three categories representing three generations of development [70]:

first generation – oxygen electrode-based sensors
second generation – mediator-based sensors
third generation – directly coupled enzyme-electrodes.

The recently developed immobilization techniques enable to "wire" an enzyme directly to an electrode, facilitating rapid electron transfer and hence high current densities. These new approaches will be discussed briefly for two examples.

Ye et al. [71] used a glucose dehydrogenase containing the redox center pyrroloquinolinequinone (PQQ), which was wired to a glassy carbon electrode through a redox-conducting polymer containing an osmium complex. The observed current densities were three times higher than electrodes prepared with glucose oxidase of similar activity. The PQQ-glucose dehydrogenase was effectively electrically wired resulting in high current densities, which imply a fast electron transfer from the redox center of the enzyme to the metal complex centers in the polymer.

Using modified thiols, Willmer et al. [72] prepared an ordered monolayer of the enzyme glutathione reductase (GR) on gold electrodes. The rate of glutathione formation was enhanced when the enzyme monolayer was modified by linking a bipyridinium group covalently to the GR. This enhancement in glutathione formation is attributed to an improved electrical communication between the enzyme active site and the electrode as the bipyridinium group represents an electron relay unit. The rate of the electron transfer was depending on the chain length, which bound the bipyridinium group to the enzyme. The results indicate that the rate-limiting step in the reduction of the oxidized glutathione is the electron transfer from the intramolecular electron relay, the bipyridinium group, to the electron-accepting active site of the enzyme.

Bioaffinity sensors

Optical detection

Reflectometric interference spectroscopy (RIfS) was used for direct kinetically controlled detection of equilibrium free antibody concentration in a flow system. The binding signal was varied between the maximum response and total inhibition by titration of the concentration of free binding sites using analyte concentrations over a range of 2–3 orders of magnitude. Affinity constants are determined by fitting a model to the titration curves taking the valence of the antibody into account. Using monoclonal antibodies against atrazin and immobilized atrazin analyte-antibody systems with interaction constants within $4 \cdot 10^7\ M^{-1}$ and $2 \cdot 10^9\ M^{-1}$ could be characterized [73]. This principle was also applied to the determination of pesticides in water. Concentrations at the EU-drinking water limit of 0.1 ppb could be determined [74].

Mass sensitive detection

The binding of thrombin to a dextran hydrogel which were modified with thrombin inhibitors was registered as a function of time by monitoring the resonance frequency of a QCM. As the binding is diffusion controlled the binding curves are proportional to $t^{1/2}$. A square root fit of the response curve was used to quantify the rate of signal change. For low concentrations of thrombin (< 10 µg/ml) a linear relationship between rate of signal change and thrombin concentration was obtained. The inhibition of thrombin by different inhibitors was investigated by preincubating the thrombin solution with different concentrations of inhibitor. After the preincubation period the test solutions were transferred into the measurement cell and time-dependent frequency changes recorded as shown in Figure 9 (left part). The signals obtained are a measure for the concentration of thrombin, which is unoccupied by the added inhibitor. Binding inhibition assays (4 µg/ml/100 nM thrombin, 0–850 nM thrombin inhibitor in PBS) led to titration curves as shown in Figure 9 (right part). A binding constant of $3 \cdot 10^9$ l/mol could be estimated by fitting of an equilibrium binding model to the data [20].

Electrochemical detection

Electrochemical transducers have been used in a variety of bioaffinity applications. The electrochemical characterization of affinity interactions may be based on a homogeneous competitive assay. Since only a few test-substances are electrochemically active at a given potential, a competitive assay format was presented in which it is necessary to label a reference compound with a suitable redoxmarker [46]. The detection principle is based on a change in the diffusion characteristics of the reference substance. As the mass increases when the ligand is bound to the target, the diffusion coefficient decreases and so does the diffusion limited current.

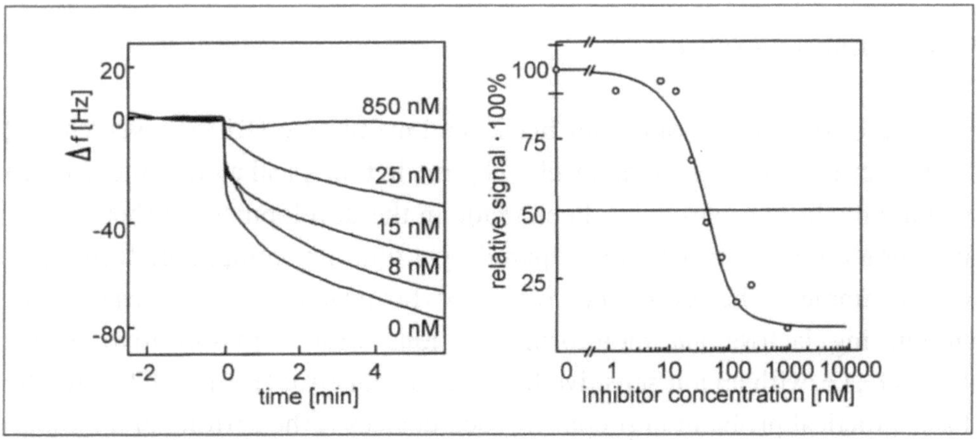

Figure 9 *Left: Characteristic frequency changes Δf in a competitive binding assay with thrombin (0.1 µM) and a thrombin inhibitor. Right: Relative signals Δf/Δf (100%) achieved from the binding assay as function of inhibitor concentration. Solid line: fit of measured data by equilibrium binding model*

Different affinity assays are described in the literature that fulfill this requirement to a certain extent [75, 76], but they lack the possibility of a simple scale-up procedure because the surface of the electrode has to be renewed after each measurement involving the analyte adsorption. So a new marker has to be established. In this context dopamine was chosen as redox label because of the well-known good detection limit under physiological conditions [77].

Chronoamperometry as detection principle with an initial potential of – 100 mV and a conversion potential of + 350 mV was used in these studies [46]. After integration of the current from 5 to 15 s after application of the conversion potential (with a duration of one measurement of 40 s) a reproducible detection limit of less than 50 nM for dopamine was achieved.

First investigations on a mobility-based affinity assay were also performed with the biotin/streptavidin system. To estimate the response of the electrochemical assay to specific binding, increasing amounts of biotin-conjugate (biotin coupled to dopamine) were added either to working buffer, or to a solution containing streptavidin. From theoretical considerations a maximum decrease of current to 20% of the initial value (unbound conjugate) is expected. The presence of streptavidin was found to reduce the current response by about this amount and hence proves the principal feasibility of this concept. The results are illustrated in Figure 10.

Transmembrane sensors

In spite of the success of earlier black lipid membrane (BLM) research [78] and in spite of the large number of receptors in biological membranes known so far, only little progress has been made in the development of those transmembrane sensors which are prepared by a thin film compatible technique. One example is the use of the co-transporter protein lactose permease for monitoring lactose *via* monitoring H^+ ions with a pH-sensitive device (Fig. 11), i.e. with an ion-sensitive field effect transistor (ISFET) [79, 80]. The basic practical problem in preparing such devices is the spriting of a biological membrane without destroying its activities and is the achievement of a sufficient electrical insulation of the planar membrane at the transducer. Therefore, the black lipid membrane (BLM) approach has often been pre-

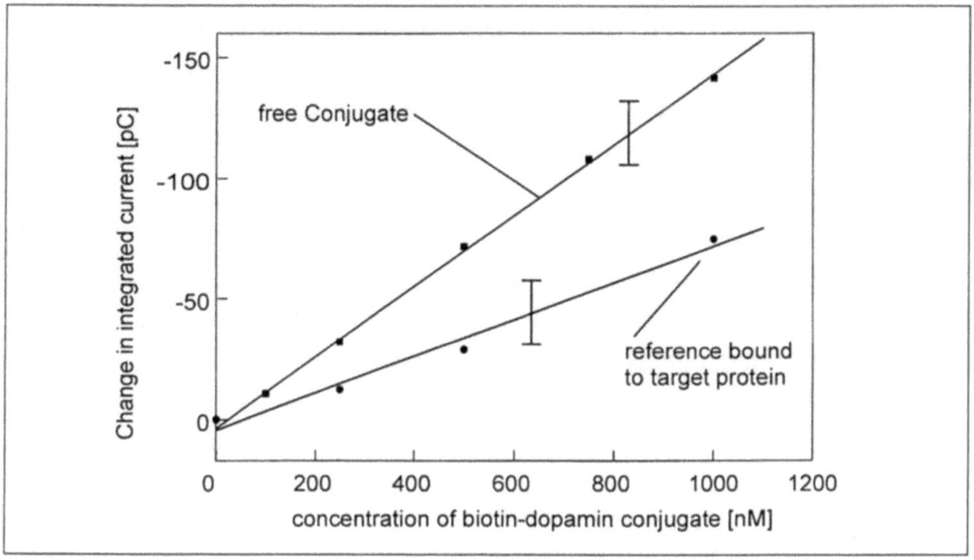

Figure 10 *The total change of charge (integrated diffusion-limited current) is plotted as a function of the concentration of the reference substance (here biotin-dopamine-conjugate) and of the target protein (here streptavidin). The upper curve shows the unbound reference with a slope of – 0.15 pC/nM. At the lower curve the reference is bound to the target protein (1000 nM). The slope decreased therefore to – 0.06 pC/nM and – 0.08 pC/nM, respectively, due to the decreased diffusion coefficient*

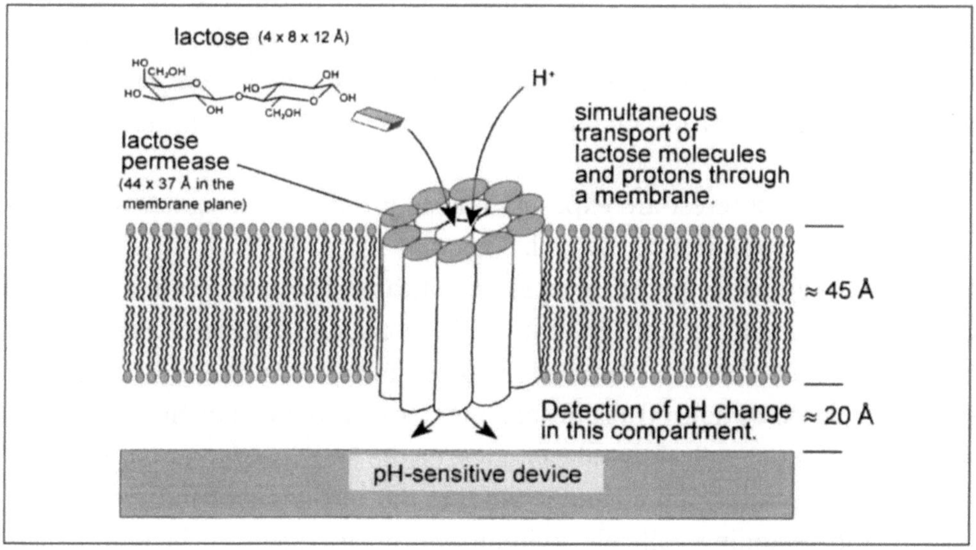

Figure 11 *Principle of a lactose-sensor based on the membrane protein lactose permease. One lactose molecule and one proton are transported simultaneously across the membrane. Thus, the presence of lactose in the test solution yields a shift of pH in the compartment between the membrane and the transducer [80]*

ferred. As an example, the Na^+/D-glucose co-transporter protein from small intestine of guinea pigs was chosen [81]. As the sensor response, the flux of Na^+ ions was measured at a constant potential. This flux was triggered by D-glucose. The BLM was formed in a small aperture in a Teflon film by the folding method.

Electrical signals generated by those transport proteins which are embedded in a solid supported lipid membrane were reported by Seifert et al. [82]. They used bacteriorhodopsin, Na,K-ATPase, H,K-ATPase, and Ca-ATPase and observed similar signals with conventionally prepared BLM as they were observed with membranes of fragments or with vesicles adsorbed at hydrophobic surfaces.

Cell sensors

Perfect spriting of biological membranes onto thin film substrates is extremely difficult to achieve. An alternative development of more sophisticated biosensors is therefore based on the use of whole cells which are specialized towards certain metabolic functions. The different possibilities to obtain intra- and extracellular electrical signals from such cells in different analytes include extracellular recording of ions with glass micropipettes, intracellular recording to monitor the transmembrane potential, patch-clamp techniques to monitor individual ion channels or microelectrodes for extracellular recording.

Of particular interest are experimental setups by which signals are deduced without destruction of the cell. This may either be realized electrically (see Fig. 12 for a few typical setups) or optically. The most commonly used optical method is to monitor the intracellular calcium transient with fluorescent labels. The Ca^{2+} ions are involved for instance in the contraction of muscles, release of neurotransmitters, metabolic processes of the cell by the activation of biochemical signal cascades, or the activity of ion pumps. More sophisticated set-ups aim at understanding the responses of individual cells and at understanding cell-cell communication, for instance in neural networks.

Biological neuronal networks are very sensitive to changes in their chemical environment. The response of the neuronal network is often substance-

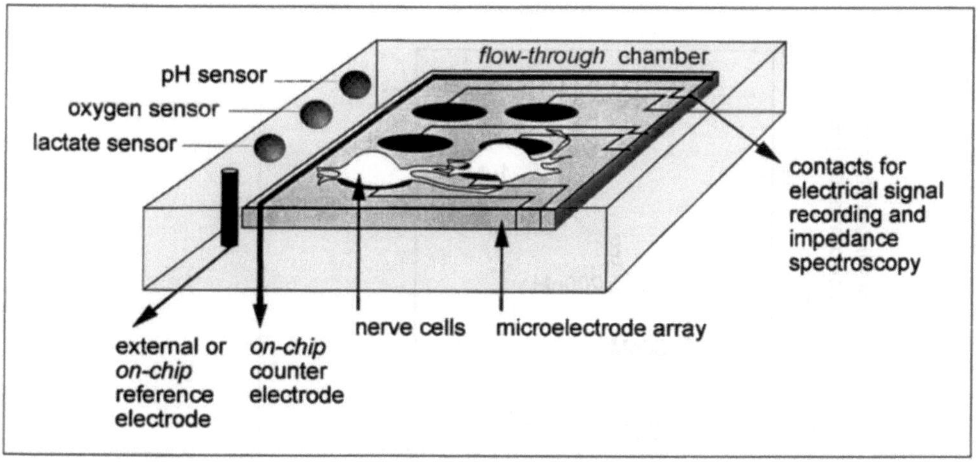

Figure 12 *Schematic presentation of sensors based on electrogenic cells. Cell activities may either be monitored indirectly through the fluid (pH sensor, oxygen sensor, lactate sensor etc.) or via extracellularly recorded electrical signals. Even for non-electrogenic cells, impedance spectroscopy may be applied by measuring the real and imaginary part of the frequency-dependent conductivity between an electrode close to the cell and a counter electrode in the liquid*

and concentration- specific. Employing appropriate data processing and analysis, these biological systems may potentially be used for certain sensory tasks as network biosensors. Electrophysiological network activity patterns of spinal cord cell cultures were recorded via an array of 64 photoetched electrodes [83]. Changes in the spontaneous activity due to the addition of different concentrations of strychnine could be detected. Evaluating the data by principal component analysis a qualitative correlation between the concentration and recorded signals could be obtained [84]. By artificial neural network analysis a quantitative correlation of network signals and strychnine concentration could be evaluated over broad range of concentrations of strychnine (see results in Fig. 13) [85]. Since a huge amount of molecules influences electrical activities of cells with certain receptors in a specific way, this approach offers a large flexibility for future work. This approach is expected to lead to significant improvements in both the different preparation and technology steps to build reliable long-term stable interfaces and in the basic understanding of bioelectronics.

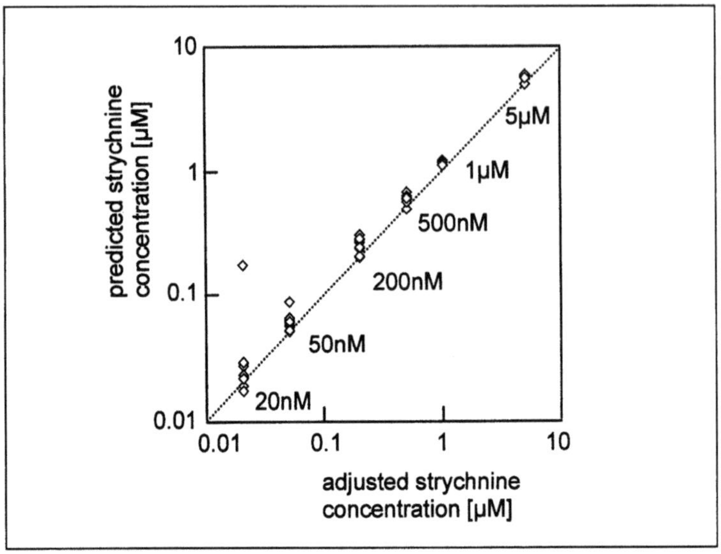

Figure 13 *Example for biosensing with a (natural) neural network by extracellularly recording signals from an array of spinal cord nerve cells of embryonic mice. Extracellularly recorded bursts have been characterized by their patterns (frequency, amplitudes etc.) and subsequently evaluated in an artificial neural network approach. A backpropagation net with 11 input neurons (burst parameters), two hidden layers with six and seven neurons, and one output neuron was used. This leads to a correlation between the predicted strychnine concentration and the adjusted concentration [84, 85]*

12.5 Conclusions

Biochemical sensors provide complimentary and additional information to the well-established bioanalytical techniques. Particular advantages of bioelectronic sensors concern the possibility

- to miniaturize the setup in principle down to the molecular scale,
- to use well-established microsystem technologies in the manufacturing (of at least certain sensor components),
- to integrate signal preprocessing steps on a chip, and
- to build arrays for more complex pattern recognition analysis.

Major progress relies on improvements in our molecular understanding of surface modifications with particular emphasis on artificial biosensing interfaces. The latter link biosensor-active coatings (containing either synthetically manufactured "biomimetic" or natural biological function units) with inorganic transducer substrates.

A variety of transducer principles has already been tested for their use in biosensing. Further progress is to be expected from merging the classical sensor technologies with new technologies in scanning probe techniques. The tip of an atomic microscope, for instance, may be used to monitor forces between complementary biomolecules. As a next step sensors based on this principle will be designed.

In view of microelectronic compatibility, electrical signal transduction principles should be favored. In many cases, however, the corresponding interface problems cannot be solved satisfactorily yet. In this situation, optical transducers become extremely important, although in the long term electrical transducers involving not only electrons, but also ions will become of great importance. The latter is evident, for example, for the often desired direct link between manmade computers and nerve systems. Bridging this interface requires to build reliable hybrid systems. This requires a fundamental understanding of bioelectronics principles including signal generating, transferring, recording, and storing [86].

References

1 Göpel W, Hesse J, Zemel JN (eds) (1991) *Sensors: a comprehensive survey*. Vol. 2 and 3: Chemical and Biochemical Sensors. VCH, Weinheim

2 Göpel W (1995) Controlled signal transduction across interfaces of "Intelligent" molecular systems. *Biosens Bioelectron* 10: 35–60

3 Göpel W, Heiduschka P (1994) Introduction to bioelectronics: "Interfacing biology with electronics". *Biosens Bioelectron* 9: iii–xiii

4 Connolly P (1994) Bioelectronic interfacing: micro- and nanofabrication techniques for generating predetermined molecular arrays. *Trends Biotechnol* 12: 123–127

5 a) Allara DL (1995) Critical issues in applications of self-assembled monolayers. *Biosens Bioelectron* 10: 771–784
b) Chaudhury MK (1995) Self-assembled monolayers on polymer surfaces. *Biosens Bioelectron* 10: 785–786
c) Löfås S, Johnsson B, Edström Å, Hansson A, Lindquist G, Müller Hillgren R-M, Stigh L (1995) Methods for site controlled coupling to carboxymethyldextran surfaces in surface plasmon resonance sensors. *Biosens Bioelectron* 10: 813–822
d) Morales P, Pavone A, Sperandei M, Leter G, Mosiello L, Nencini L (1995) A laser assisted deposition techniques suitable for the fabrication of biosensors and molecular

electronic devices. *Biosens Bioelectron* 10: 847–852

e) Ratner BD (1995) Surface modification of polymers: chemical, biological and surface analytical challenges. *Biosens Bioelectron* 10: 797–804

6 Decher G, Lehr B, Lowack K, Lvov Yu, Schmitt J (1994) New nanocomposite films for biosensors: layer-by-layer adsorbed films of polyelectrolytes, proteins or DNA. *Biosens Bioelectron* 9: 677–684

7 Geckeler KE, Müller B (1993) Polymer materials in biosensors. *Naturwissenschaften* 80: 18–24

8 Bartlett PN, Cooper JM (1993) A review of the immobilization of enzymes in electropolymerized films. *J Electroanal Chem* 362: 1–12

9 Moser I, Schalkhammer Th, Mann-Buxbaum E, Hawa G, Rakohl M, Urban G, Pittner F (1992) Advanced immobilization and protein techniques on thin film biosensors. *Sens Actuators B* 7: 356–362

10 a) Fodor SPA, Read JL, Pirrung C, Stryer L, Lu AT, Solas D (1991) Light-directed, spatially addressable parallel chemical synthesis. *Science* 251: 767–773

b) Calvert JM (1993) Lithographic patterning of self-assembled films. *J Vac Sci Technol B* 11: 2155–2163

c) Morgan H, Pritchard DJ, Cooper JM (1995) Photo-patterning of sensor surfaces with biomolecular structures: Characterization using AFM and fluorescence microscopy. *Biosens Bioelectron* 10: 841–846

d) Morales P, Pavone A, Sperandei M, Leter G, Mosiello L, Nencini L (1995) A laser assisted deposition techniques suitable for the fabrication of biosensors and molecular electronic devices. *Biosens Bioelectron* 10: 847–852

11 Mittler-Neher S, Spinke J, Liley M, Nelles G, Weisser M, Back R, Wenz G, Knoll W (1995) Spectroscopic and surface-analytical characterization of self-assembled layers on Au. *Biosens Bioelectron* 10: 903–916

12 Pantano P, Kuhr WG (1993) Dehydrogenase-modified carbon-fiber microelectrodes for the measurement of neurotransmitter dynamics. 2. Covalent modification utilizing avidin-biotin technology. *Anal Chem* 65: 623–630

13 Müller W, Ringsdorf H, Rump E, Wildburg G, Zhang X, Angermaier L, Knoll W, Liley M, Spinke J (1993) Attempts to mimic docking processes of the immune system: Recognition-induced formation of protein multilayers. *Science* 262: 1706–1708

14 de Alwis U, Wilson GS (1989) Strategies for the reversible immobilization of enzymes by use of biotin-bound anti-enzyme antibodies. *Talanta* 36: 249–253

15 Spinke J, Liley M, Guder H-J, Angermaier L, Knoll W (1993) Molecular recognition at self-assembled monolayers: The construction of multicomponent multilayers. *Langmuir* 9: 1821–1825

16 a) Dave BC, Dunn B, Valentine JS, Zink JI (1994) Sol-gel encapsulation methods for biosensors. *Anal Chem* 66: 1120–1127

b) Avnir D, Braun S, Lev O, Ottolenghi M (1994) Enzymes and other proteins entrapped in sol-gel materials. *Chem Mater* 6: 1605–1614

17 Harris JM (ed) (1992) Poly(ethylene glycol) chemistry. *Biotechnical and Biomedical Applications*. Plenum Press, New York and London

18 a) Lundström I (1994) Real-time biospecific interaction analysis. *Biosens Bioelectron* 9: 725–736

b) Löfås S, Johnsson B, Edström Å, Hansson A, Lindquist G, Müller Hillgren R-M, Stigh L (1995) Methods for site controlled coupling to carboxymethyldextran surfaces in surface plasmon resonance sensors. *Biosens Bioelectron* 10: 813–822

19 Brecht A, Gauglitz G (1995) Optical probes and transducers. *Biosens Bioelectron* 10: 923–936

20 Rickert J, Brecht A, Göpel W (1997) Quartz microbalances for quantitive biosensing and characterizing protein multilayers. *Biosens Bioelect* 12: 567–575

Rickert J, Weiß T, Kraas, Jung G, Göpel W (1996) A New Affinity Biiosensor: Self-Assembled Thiols as Selective Monolayer Coatings of Quartz Microbalances. Conf. Proc. Eurosensors IX, Stockholm (S) (6/1995); Conf. Proc. ABI Workshop, Tampere (SF) 6/1995 and *Biosens Bioelect* 11: 591–598

Rickert J, Brecht A, Göpel W (1997) QCM Operation in Liquids: Constant Sensitivity Found During Formation of Extended Protein Multilayers by Affinity. Analyt. Chem. 69: 1441–1448

Rickert J, Hayward GL, Cavic BA, Thompson W, Göpel W (1999) Biosensors Based on Acoustic Wave Devices. In: Baltes H, Göpel W, Hesse J (eds) "Sensors Update", Vol. 5, Wiley-VCH, Weinheim (D), *in press*

21 Tom-Moy M, Baer RL, Spira-Solomon D, Doherty TP (1995) Atrazine measurements using surface transverse wave. *Anal Chem* 67: 1510–1516

22 Du J, Harding, GL, Ogilvy JA, Dencher PR, Lake M (1996) An experimental study of love wave acoustic sensors operating in liquids. *Sens Actuators A* 56: 211–219

23 Nieuwenhuizen MS, Venema A, in: Göpel G, Hesse J, Zemel N (ed) (1991) Sensors Volume 2: Chemical and Biochemical Sensors Part I, Verlag Chemie Weinheim pp 647–680

24 Suleiman AA, Guilbault GG (1994) Recent developments in piezoelectric immunosensors. *Analyst* 119: 2279–2282

25 Ballantine DS, Martin SJ, Ricco AJ, Frye GC, Wohltjen H, White RM, Zellers ET (1997) Acoustic wave sensors-theory, design and physico-chemical applications. Academic Press San Diego

26 Rickert J, Weiß T, Kraas W, Jung G, Göpel W (1996) A new affinity biosensor: self-assembled thiols as selective monolayer coatings of quartz crystal microbalances. *Biosens Bioelectron* 11: 591–598

27 Rickert J, Weiß T, Göpel W (1996) Self-assembled monolayers for chemical sensors: Molecular recognition by immobilized supramolecular structures. *Sens Actuators B* 31: 45–50

28 Rickert J, Brecht A, Göpel W (1996) QCM operation in liquids: Constant sensitivity during formation of extended protein multilayers by affinity. *Anal Chem* 69: 1441–1448

29 Su H, Chong S, Thompson M (1997) Kinetics of hybridization of interfacial RNA homopolymer studied by thickness shear mode acoustic wave sensor. *Biosens & Bioelect* 12: 161–173

30 Lerchner J, Oehmgen R, Wolf G (1999) Supermicrocalorimetric devices for the investigations of small samples. Sens. and Act., *in press*

31 Jelesarov I, Leder L, Bosshard HR (1996) Methods (Orlando) 9: 533–541

32 Berger R, Gerber C, Gimzewski JK, Meyer E, Günthrodt H-J (1996) Thermal analysis using a micromechanical calorimeter. *Appl Physics Let* 69: 40–42

33 Berger R, Gerber C, Gimzewski JK (1996) Nanometers picowatts, femtojoules: Thermal analysis and optical spectroscopy using micromechanics. Analytical Methods & Instrumentation, Special Issue μTAS'96

34 Bard AJ, Faulkner, LR (1980) Electrochemical methods, fundamentals and applications. John Wiley & Sons New York

35 Kissinger PT, Heineman WR (ed) (1996) Laboratory techniques in electroanalytical chemistry. Marcel Dekker New York

36 Rusling JF, Suib SL (1994) Characterizing materials with cyclic voltammetry. *Adv Mater* 6: 922–930

37 O'Dea JJ, Osteryoung JG (1993) Characterization of quasi-reversible surface processes by square-wave voltammetry. *Anal Chem* 65: 3090–3097

38 Gabrielli C, Takenouti H, Haas O, Tsukadá A (1991) Impedance investigations of the charge transport in film-modified electrodes. *J Electro Anal Chem* 302: 59–89

39 Láng G, Bácskai J, Inzelt G (1993) Impedance analysis of polymer film electrodes. *Electrochim Acta* 38: 773–780

40 Armstrong FA (1990) Probing metalloproteins by voltammetry. *Structure and Bonding* 72: 137–221

41 Guo L-H, Hill HAO (1991) Direct electrochemistry of proteins and enzymes. *Adv Inorg Chem* 36: 341–375

42 Collinson M, Bowden EF (1992) Voltammetry of covalently immobilized cytochrome c on self-assembled monolayer. *Langmuir* 8: 1247–1250

43 Niwa O, Xu Y, Halsall HB, Heineman WR (1993) Small-volume voltammetric detection of 4-aminophenol with interdigitated array electrodes and ist application to electrochemical enzyme immunoassay. *Anal Chem* 65: 1559–1563

44 Fiaccabrino GC, Tang X-M, Skinner N, de Rooij NF, Koudelka-Hep M (1996) Electrochemical characterization of thin-film carbon interdigitated electrode arrays. *Anal Chim Acta* 326: 155–161

45 Hintsche R, Paeschke M, Wollenberger U, Schnakenberg U, Wagner B, Lisec T (1994) Microelectrode arrays and application to biosensing devices. *Biosens & Bioelectron* 9: 697–705

46 Brecht A, Burckardt R, Rickert J, Stemmler I, Schuetz A, Fischer S, Friedrich Th, Gauglitz G, Goepel W (1996) Transducer based approaches for parallel binding assays in HTS. *J Biomol Screen* 1: 191–201

47 Schöning MJ, Tsarouchas D, Beckers L, Schubert J, Zander W, Kordos P, Luth H (1996) A highly long term stable silicon based pH-sensor fabricated by pulsed laser deposition technique. *Sensors & Actuators B* 35: 228–233

48 Voigt H, Schitthelm F, Lange T, Kullick T, Ferretti R (1996) Diamond-like carbon ISFET. Conf Proc Eurosensors X, Leuven Belgium

49 Hogg G, Lutze O, Cammann K (1996) Novel membrane material for ione selective field effect transistors with extended lifetime and improved selectivity. *Anal Chim Acta* 335: 103–109

50 Bone S, Zaba B (1992) Bioelectronics, John Wiley & Sons Chichester

51 Maupas H, Saby C, Martelet C, Jaffrezic-Renault N, Soldatkin AP, Charles MH, Delair T, Mandrand B (1996) Impedance analysis of Si/SiO_2 heterostrucutres grafted with antibodies: an approach for immunosensor development. *J Electroanal Chem* 406: 53–58

52 Tegenfeldt J (1997) Nanofabrication and characterization for applications in biochemistry and molecular electronics. Thesis, Lund University Sweden

53 Rickert J, Göpel W, Beck W, Jung G, Heiduschka P (1996) A 'mixed' self-assembled monolayer for an impedimetric immunosensor. *Biosens Bioelectron* 11: 757–768

54 Sergeyeva TA, Lavrik NV, Piletsky SA, Rachkov AE, El'skaya AV (1996) Polyaniline label-base conductometric sensor for IgG detection. *Sens Actuators B* 34: 283–288

55 van Gerwen P, Varlan A, Huyberechts G, op de Beek M, Baert K, Sansen W, Hermans L, Mertens R (1997) Nanoscaled impedimetric sensors for multiparameter testing of biochemical samples. Proc. of 1st Conf. on

microreaction technology, Frankfurt/Main Germany

56 Liedberg B, Nylander C, Lundström I (1983) Surface plasmon resonance for gas detection and biosensing. *Sens Actuators* 4: 299–304

57 Jorgenson RC, Yee SS (1993) Optical biosensors. *Sens. Actuators B* 12: 213–220

58 Lavers CR, Wilkinson JS (1994) Design principles of biosensors. *Sens Actuators B* 22: 75–81

59 Striebel Ch, Brecht A, Gauglitz G (1994) Characterization of biomembranes by spectral ellipsometry, surface plasmon resonance and interferometry with regard to biosensor application. *Biosens Bioelectron* 9: 139–146

60 Piehler J, Brecht A, Gauglitz G (1996) Affinity detection of low molecular weight analytes. *Anal Chem* 68: 139–143

61 Lukosz W (1995) Integrated optical chemical and direct biochemical sensors. *Sens Actuators B* 29: 37–50

62 Gauglitz G, Ingenhoff J (1994) Design of new integrated optical substrates for immuno-analytical applications. *Fres J Anal Chem* 349: 355–359

63 a) Gale MT, Baraldi LG, Kunz RE (1994) Proc SPIE-Int Soc Opt Eng, 2213: 2–10
b) Lai Q, Bachmann M, Hunziker W, Besse PA, Melchior H (1996) Arbitrary ratio power spitters using anglied silica on silicon multimode interference couplers. *Elect Letters* 32: 1576–1577

64 Nellen Ph, Lukosz W (1993) Integrated optical input grating couplers as direct affinity sensors. *Biosens Bioelectron* 8: 129–147

65 Edwards PR, Gill A, Pollard-Knight DV, Hoare M, Buckle PE, Lowe PA, Leatherbarrow RJ (1995) Kinetics of protein-protein interactions at the surface of an optical biosensor. *Anal Biochem* 231: 210–217

66 Eigen M, Rigler R (1994) Sorting single molecules: Application to diagnostics and evolutionary biotechnology. *Proc Natl Acad Sci USA* 91: 5740–5747

67 Dandliker WB, de Saussure VA (1970) Fluorescence polarization in immunochemistry. *Immunochemistry* 7: 799–828

68 Maliwal BP, Lakowicz JR, Kupryszewski G, Rekowski P (1993) Fluorescence study of conformational flexibility of RNase S-peptide: Distance-distribution, end-to-end diffusion, and anisotropy decays. *Biochemistry* 32: 12337–12345

69 Jin G, Tengvall P, Lundstrom I, Arwin H (1995) A biosensor concept based on imaging ellipsometry for visualization of biomolecular interactions. *Anal Biochem* 232: 69–72

70 Eggins B (1996) Biosensors: An Introduction. Wiley/Teubner, Chichester/Stuttgart

71 Ye L, Hämmerle M, Olsthoorn AJJ, Schuhmann W, Schmidt H-L, Duine JA, Heller A (1993) High current density "wired" quinoprotein glucose dehydrogenase electrode. *Anal Chem* 65: 238–241

72 Willner I, Lapidot N, Riklin A, Kasher R, Zahavy E, Katz E (1994) Electron-transfer communication in glutathione reductase assemblies: Electrocatalytic, photocatalytic, and catalytic systems for the reduction of oxidized glutathione. *J Am Chem Soc* 116: 1428–1441

73 Piehler J, Brecht A, Giersch T, Hock B, Gauglitz G (1997) Assessment of affinity constants by rapid solid phase detection of equilibrium binding in a flow system. *J Immun Meth* 201: 189–206

74 Brecht A, Piehler J, Lang G, Gauglitz G (1995) A direct optical immunosensor for atrazin detection. *Anal Chim Acta* 311: 289–299

75 Le Gal La Salle A, Limoges B, Rapicault S, Degrand C, Brossier P (1995) New immunoassay techniques using NAFION-modified electrodes and cationic redox labels or enzyme labels. *Anal Chim Acta* 311: 301–308

76 Sugawara K, Tanaka S, Nakamura H (1995) Electrochemical assay of avidin and biotin

using a biotin derivative labeled with an electroactive compound. *Anal Chem* 67: 299–302

77 Gonon FG, Frombarlet CM, Buda MJ, Pujol JF (1981) Electrochemical treatment of pyrolytic carbon fiber electrodes. *Anal Chem* 53: 1386–1389

78 Thompson M, Krull UJ, Worsfold PJ (1980) The analytical potential of chemoreception at bilayer lipid membranes. *Anal Chim Acta* 117: 121–132

79 Kindervater R, Göpel W, Ottenbacher D, Knichel M (1992). Biosensors based on receptors: Electrical transducer principles. GBF-Monographs (VCH, Weinheim), 17: 178–186

80 Ottenbacher D, Jähnig F, Göpel W (1993) A prototype biosensor based on transport proteins: Electrical transducers applied to lactose permease. *Sens Actuators B* 13–14 173–175

81 Sugao N, Sugawara M, Minami H, Uto M, Umezawa Y (1993) Na^{+}/D-Glucose cotransporter based bilayer lipid membrane sensor for D-glucose. *Anal Chem* 65: 363–369

82 Seifert K, Fendler K, Bamberg E (1993) Charge transport by ion translocating membrane proteins on solid supported membranes. *Biophys J* 64: 384–391

83 Gross GW (1994) Internal dynamics of randomized mamalian neuroanl networks in culture. In: DA Sterger, TM McKenna (eds): *Enabling technologics for cultured neruonal networks*. Academic Press NY 277–317

84 Harsch A, Ziegler C, Göpel W (1997) Strychnine analysis with neuronal networks *in vito*: Extracellular array recording of network response. *Biosens Bioelectron* 12: 827–835

Ziegler Ch, Harsch A, Göpel W. Natural Neural Networks for Quantitative Sensing of Neurochemicals: an Artificial Neural Network Analysis. (7/1998) Conf. Proc. 7th IMCS, Beijing (China) 801–803

85 Harsch A (1997) Extrazelluläre Ableitungen an neuronalen Netzwerken für die Biosensorik. Thesis, University of Tübingen, Germany

86 Göpel W, Ziegler Ch, Breer H, Schild D, Apfelbach R, Joerges J, and Malaka R (1998) Bioelectronic Noses: A Status Report, Part I. *Biosens Bioelect* 13: 479–493

Ziegler Ch, Göpel W, Hämmerle H, Jung G, Laxhuber L, Schmidt HL, Schütz S, Vögtle F, Zell A (1998) Bioelectronic Noses: A Status Report, Part II. *Biosens Bioelect* 13: 539–571

Göpel W (1998) Chemical Imaging. I. Concepts and Visions for Electronic and Bioelectronic Noses. *Sensors and Actuators* B52: 125–142

Weimar U, Göpel W (1998) Chemical Imaging: II. Trends in Practical Multiparameter Sensor Systems. *Sensors and Actuators* B52: 143–161

87 Kuhn H (1994) Reflections on biosystems motivating supramolecular engineering. *Biosens Bioelectron* 9: 707–717

88 Göpel W (1996) Nanosensors and Molecular Recognition. Nanotechnology, Spec. Issue of Microelectronics Engineering 32: 75–110

Göpel W (1998) Bioelectronics and Nanotechnologies. Conf. Proc. Biosensors 96, Bangkok (Thail.) (5/1996); *Biosens Bioelect* 13: 723–728

13 Fluorescence detection of single molecules applicable to small volume assays

Jörg Enderlein, W. Patrick Ambrose, Peter M. Goodwin and Richard A. Keller

13.1 Introduction

In this chapter we give an overview of optical detection of single molecules with respect to possible application to high throughput screening of substance libraries. We restrict ourselves to single molecule detection (SMD) because the field of ultrasensitive detection is too broad to be covered by a single chapter; thus we will concentrate on the ultimate detection limit. In the future SMD will be the method of choice because of the ultimate small amount of analyte needed in connection with extremely high speed screening of vast libraries. Besides the topic of high-speed DNA sequencing (see below), the high sensitivity received attention also in the context of combinatorial chemistry (Eigen, 1984; Plückthun, 1991; Eigen, 1994; Xiang, 1995) and molecular computation (Adleman, 1994; Lipton, 1995). The basic idea of combinatorial chemistry is to synthesize large arrays of slightly different molecules, followed by screening and selection of molecules with desired properties. In the case of molecular computation, a random chemical synthesis is used to find solutions of numerically difficult problems; again, a fast and ultrasensitive screening method is needed to extract the result of the "computation".

Within the domain of SMD, the emphasis will be on optical detection, i.e. laser induced fluorescence (LIF) detection. LIF is one of the most sensitive detection methods in existence today. It is easy to implement, highly characteristic, and mostly non-invasive, which is important for the detection (and possible separation) of biomolecules. Moreover, based on the broad application of LIF in many fields of analytical chemistry and biochemistry, established physical technologies and a huge variety of fluorescence dyes are available for LIF.

With respect to SMD on surfaces and interfaces, we will discuss only SMD at room temperature. High-resolution spectroscopy of single molecules in and on solids at low temperatures is a very broad field in itself, but probably of minor interest for applications in library screening. Although of fundamental interest for studying molecule-host interactions, the technological complexity of low-temperature, high-resolution experiments will prevent their application in biolabs and related facilities. For a recent overview of SMD and high-resolution spectroscopy at low temperatures see, for example, (Orrit, 1996; Basché, 1997). Finally, we note that the emphasis of this chapter will be the physics of SMD for room temperature applications rather than its chemical aspects (e.g. design of fluorescence labels).

An overview of all current optical SMD methods at room temperature, potentially applicable to high-speed screening of biomolecular libraries, is presented in Table 1. In the following sections, each of these methods will be discussed and referenced in more detail.

There exists a large variety of other non-optical SMD methods applicable to biomolecule recognition and screening, like conventional STM, see e.g. (Cricenti, 1991; Hansma, 1991; Heckl, 1991; Li, 1991; Youngquist, 1991; Heckl, 1992; Cooper, 1994; Guckenberger, 1993; Frisbie, 1994; Thimonier, 1994; Kasaya, 1995; Venkataraman, 1995; Zuccheri, 1995; Hinterdorfer, 1996; Tanaka, 1996; Walba, 1996); STM with surface adsorbed water (Guckenberger, 1994; Heim, 1996); scanning force microscopy, see for example (Schaper, 1993; Radmacher, 1994 a–c; Gunning, 1995; Muzzalupo, 1995); electrochemical or ion-channel SMD (Fan, 1995; Bard, 1996; Fan, 1996; Kasianowicz,

Table 1 SMD methods at room-temperature with potential use in biomolecular applications

In fluids	Laser induced fluorescence	Fluid flow
		Micro-capillary electrophoresis
		Fluorescence correlation spectroscopy
		Levitated micro-droplets
	Laser induced Raman scattering	Single molecules adsorbed on colloidal metal particles in solution
On surfaces	Far-field methods	Confocal microscopy
		Wide-field fluorescence microscopy with low noise CCD cameras
	Near-field methods	Near-field scanning microscopy

1996); and mass spectroscopy (Fenn, 1989; Jacobson, 1991). However, none of these methods combines high molecular specificity, relative technical simplicity and non-invasiveness in such a unique way as optical SMD based on laser induced fluorescence (or possibly Raman scattering).

13.2 Single molecule detection in fluids

Before the age of optical detection of single molecules in solution, the first SMD in liquids was done indirectly by monitoring the enzymatic activity of single protein molecules (Rotman, 1961; Rotman, 1973). This method is exploited successfully even today, see for example (Xue, 1995; Craig, 1996). The method relies on the amplification of sample by repetitive chemical transformations.

The first successful optical detection of a single molecule (with multiple fluorescent labels) in a liquid was reported by Hirschfeld (Hirschfeld, 1976). By contrast, this method relies on repetitive emission of photons by a single molecule. In subsequent years, great progress was made in the refinement of the methodology (Dovichi, 1983, 1984; Mathies, 1986; Nguyen, 1987a–b; Peck, 1989; Mathies, 1990; Rigler, 1990; Shera, 1990), making single molecule detection (SMD) of single fluorescence labels in fluids a nearly routine procedure today (Hahn, 1991; Soper, 1991a; Rigler, 1992; Soper, 1992; Castro, 1993; Goodwin, 1993a; Soper, 1993; Wilkerson, 1993; Lee, 1994; Mets, 1994; Nie, 1994; Tellinghuisen, 1994; Li, 1995; Funatsu, 1995; Mertz, 1995; Soper 1995a; Berland, 1996; Chiu, 1996; Edman, 1996; Sauer, 1996; Wu, 1996; Zander, 1996). Comprehensive overviews of SMD in fluids can be found in (Barnes, 1995; Goodwin, 1996a, c; Keller, 1996).

The standard fluid-flow SMD system is similar to common flow cytometry systems. A sample stream containing the analyte molecules is injected into a surrounding sheath flow, providing hydrodynamic focusing (Kachel, 1990) of the sample stream. The sample stream is transported to the detection region, where a tightly focused laser beam excites the molecules (picoliter detection volume). Fluorescence is monitored by highly efficient collection optics and a single photon sensitive detector. Single fluorophores can emit ca. 10^8 photons/s; the main problem in SMD is not so much the detection of the molecule's fluorescence but the efficient rejection of the background signal. The two main sources of background are fluorescence from contaminants, and Rayleigh/Raman scattering of the exciting laser beam by the solvent. The use of ultrasmall volumes and efficient optical filters is effective in reducing the background. In addition to the small volume and optical filters, other methods of background rejection have been applied. One of the most common methods is the application of pulsed laser excitation together with a time gate in the detection channel that is used to reject prompt scatter, see for example (Harris, 1983; Shera, 1990). Guenard (1996) investigated the use of a highly efficient narrow band metal vapor filter for blocking the laser light and its applicability in SMD. For reducing impurity fluorescence of the sheath flow, in-line photobleaching before the detection region was found to be effective (Affleck, 1996). Another approach is the exploitation of two-photon excitation (Mertz, 1995; Berland, 1996; Overway, 1996), which was found to be useful for the reduction of background. Soper et al. are promoting the application of near-infrared dyes (Soper, 1995a–b), since there is a strong decrease of light scattering intensity and impurity fluorescence at longer wavelengths.

Besides the already mentioned suppression of scattered laser light by time-gating, the detection of the fluorescence decay characteristics of single molecules provides a convenient tool for distinguishing between different molecules. Recently, the application of the time-resolved single-photon counting (TCSPC), see (O'Connor, 1984), for lifetime measurements in SMD has received considerable interest. The first successful life-time measurements at the single molecule level were reported in (Soper, 1992; Wilkerson, 1993; Tellinghuisen, 1994). In (Enderlein, 1995b–c; Erdmann, 1995), new TCSPC-electronics, allowing for the continuous detection of TCSPC curves in millisecond intervals, was described and its application for SMD discussed. In (Müller, 1996), a continuous TCSPC technique was successfully applied to distinguish between molecules with different fluorescence decay times at the single molecule level. In (Seidel, 1996), nucleobase specific quenching of fluorescent dyes was studied, which could be important for the application of TCSPC-SMD to DNA sequencing (see below). Finally, (Sauer, 1996) reported the use of a diode laser as a light source in SMD, which will be of great importance for future broad biological and chemical applications of SMD, requiring simple, low cost, and compact operation.

A number of recent papers are dedicated to the theoretical study of SMD, mainly its statistics, maximum possible efficiency, and the usefulness of TCSPC in SMD (Stevenson, 1992a–b; Whitten, 1992; Köllner, 1992; Köllner, 1993; Tellinghuisen, 1993; Enderlein, 1995a), see also (Köllner, 1996) for a comparison between theory and preliminary experiments.

An already realized application of SMD in fluid flow is DNA fragment sizing (Ambrose, 1993; Castro, 1993; Goodwin, 1993b; Johnson, 1993; Petty, 1995; Huang, 1996). DNA sizing at the single molecule level became especially feasible after the introduction of a new class of intercalating dyes (Rye, 1992; Rye, 1993), which show extremely low fluorescence in their unbound state. In (Huang, 1996), fluid flow SMD was applied for the first time to the sizing of human DNA (bacterial artificial chromosome clones).

One of the most exciting potential applications of SMD in fluid flow is DNA sequencing (Jett, 1989; Davis, 1991; Soper, 1991b; Harding, 1992; Ambrose, 1993; Goodwin, 1993c; Eigen 1994; Goodwin, 1995, 1996b). Although much progress has been made in recent years to achieve this goal, no group has yet reported the successful sequencing of a single DNA molecule. Nonetheless, SMD promises to be a high speed method for reading long (> 10 kbase) DNA sequences.

In addition to SMD in fluid flow, a number of groups have reported SMD in gel electrophoresis experiments (Guo, 1992; Castro, 1995a–b; Haab, 1995; Soper 1995b; Chen, 1996). Guo, Castro and Haab applied the method to DNA sizing. Chen studied the limitations of quantitative analysis at the single molecule level.

A completely different method of SMD was applied by Ramsey and coworkers. They used levitated diluted microdroplets for SMD (Whitten, 1991; Kin, 1992; Ng, 1992; Barnes, 1993; Barnes, 1996; Hell, 1996). An advantage of the method is the low background level due to the small volume of illuminated liquid. The main applications of their technique are ultrasensitive chemical analysis, and investigations of quantum confinement effects.

Related to SMD in fluid flows is SMD in fluorescence correlation spectroscopy (FCS). The main setup of FCS is similar to SMD in fluid flows, but without hydrodynamic flow. The molecules move in and out of the detection region by diffusion. For a comprehensive review of FCS see e.g. (Thompson, 1991). One advantage of FCS is the use of a much smaller laser focus (of the order of 0.5 μm) and thus detection volume of femtoliters, reducing significantly the background signal. This is in contrast to fluid flow SMD where the laser focus is set large enough to detect all molecules in the sample stream. The disadvantage of FCS is its intrinsically "non-sequential" character – one has to wait until a specific molecule diffuses into the detection region. Nonetheless, FCS was applied successfully to kinetic studies at a single molecule level, such as probe-target binding and triplet state kinetics (Rigler, 1992, 1993; Mets, 1994; Widengren, 1994; Edman, 1995; Kinjo, 1995; Rigler, 1995; Widengren, 1995; Edman, 1996), see also (Nie, 1994, 1995).

At the end of this section, we mention a new technique of optical SMD in fluids: the excitation and detection of surface-enhanced Raman signals from single molecules adsorbed on colloidal metal particles diffusing through a focused laser beam. First experimental studies approaching this technique were reported in (Kneipp, 1994; Kneipp, 1995a-d). Because of the adsorption of the molecules on metal particles, this method can be considered as a hybrid between SMD on surfaces (which is discussed below) and SMD in fluids.

13.3 Single molecule detection on surfaces

As mentioned in the Introduction, we will consider SMD at room temperature only. The two main methods for optically detecting single molecules on surfaces are far-field and near-field microscopy.

Far-field microscopy

In far-field microscopy two different approaches have been used for SMD: confocal microscopy and conventional wide-field microscopy. In confocal microscopy, the sample is illuminated by a tightly focused laser beam that is scanned over the surface for recording a complete image. For recent reviews of confocal optical microscopy see (Inoue, 1995; Webb, 1996). The first successful detection of single Rhodamine-6G molecules by a confocal scanning system was reported in (Dapprich, 1995). Ambrose (1996) and Macklin (1996) used this technique to measure time-resolved fluorescence of single molecules. The advantage of the technique is the relatively low background due to the small illuminated area combined with spatial filtering, and the possibility to obtain time-resolved fluorescence data. This is not possible in conventional wide field microscopy due to the current absence of commercially available single-photon sensitive cameras with sub-nanosecond time-resolution, but see (Ho, 1993), and (Köllner, 1994; Kalusche, 1995) for plans of a construction of such a camera for SMD applications.

Many groups are using conventional wide field microscopy together with high-sensitivity low-noise optical cameras. The first report of disodium fluorescein detection on a silicon single-crystal wafer with such a system was (Ishikawa, 1994). In (Schmidt, 1995, 1996a–b), this technique was applied to the detection and tracking (on a millisecond time scale) of single molecules at an air-liquid interface. Fluorescence collection by a conventional wide field microscope objective was also the basis of a three-dimensional monitoring of single molecules in a gel layer (Dickson, 1996), where fluorescence excitation was achieved by the evanescent field of total internal reflection. Ueda reported the monitoring of single DNA molecule phase transitions (Ueda, 1996).

Near-field microscopy

Another form of optical microscopy that has been used widely for SMD is near-field microscopy. The idea is to illuminate the sample with a light source of sub-wavelength spatial extent, thus circumventing the Abbe limit of spatial resolution in conventional microscopy. This approach was first described in papers by Synge (1928, 1932). Today, there are several different techniques for near-field microscopy. The most frequently used in SMD is the transmission near-field scanning optical microscope (NSOM or SNOM). In NSOM, the sample is illuminated using a tapered metal coated optical fiber. A small aperture in the metal coating at the apex provides a light source of approximately 100 nanometers across. In transmission NSOM, the excited fluorescence of the sample is monitored by a conventional microscope through the optically transparent support of the sample. By scanning the exciting fiber over the sample surface, a spatially resolved image of the sample is generated. For recent reviews of the NSOM technique see (Harris, 1994; Kopelman, 1994; Paesler, 1996; Trautman, 1997). With this method, the detection of single molecules has been widely investigated (Betzig, 1993; Ambrose, 1994a–b;

Betzig, 1994; Dunn, 1994; Trautman, 1994; Xie, 1994; Ambrose, 1995; Bian, 1995; Dunn, 1995; Meixner, 1995; Bopp, 1996; Lu, 1997). A recent review can be found in (Xie, 1996). Again, as in far-field confocal microscopy, the point probe character of the NSOM allows for time-resolved detection of the molecules fluorescence (Ambrose, 1994b; Dunn, 1994; Trautman, 1996). Ha et al. used this technique for monitoring the fluorescence polarization of single molecules, and energy transfer between two different single molecules (Ha, 1996a–b).

At present, SMD with NSOM has been reported only for apertures ≥ 100 nm in diameter. Quenching of fluorescence by the metal coating at smaller diameters and a larger relative background (Trautman, 1997) may prevent further size reduction for SMD. A possible improvement in future generation NSOMs could be the introduction of new optical probes, like the tetrahedral tip of (Koglin, 1996a–b), using surface plasmons for generating a sub-wavelength light source, or the exploitation of micro-photodiodes (Davis, 1995; Akamine, 1996).

Besides the NSOM technique, the so called apertured photon scanning tunneling microscope (apertured PSTM) was used for detecting Rhodamine-6G molecule aggregates (Tsai, 1995). In this technique, the sample surface is illuminated by the evanescent field of a totally reflected light wave, which is incident from beneath the transparent sample support. The fluorescence is then collected by a metal coated tapered fiber, which is equivalent to the excitation probe in an NSOM. The low collection efficiency of this setup may prevent its application to real single fluorophore detection.

There are two promising alternatives to the NSOM and PSTM technique, which are worthy of mention in the context of SMD. The first class of new techniques can be called near-field disturbance methods. The idea is to use a small (nanometer-range) metallic probe disturbing a near-field configuration (and thus generating extremely confined electromagnetic fields), and to measure the interaction of disturbance with the sample (Pedarnig, 1992; Specht, 1992; Pedarnig, 1993; Bachelot, 1994; Inoyue, 1994; Zenhausern, 1994; Bachelot, 1995a–b; Zenhausern, 1995; Wickramasinghe, 1996). The advantage of this method is the potential very high spatial resolution, which can be better than a nanometer. It remains to be seen whether it has the sensitivity to detect single molecules.

The second class of techniques uses the emission of photons in an scanning tunneling microscope (STM) (Gimzewski, 1989; Berndt, 1993, 1994, 1995).

One expects that this emission will depend critically on the close environment of the STM metal probe, including single molecules. Again, the achievable spatial resolution could be in the sub-nanometer range comparable to standard STM.

13.4 Conclusion

In the present chapter, we presented an overview of techniques for SMD at room temperature, potentially applicable to high-speed and high-throughput screening of large molecular libraries. The detection speed and throughput of fluid flow SMD, FCS, and wide-field microscopy have the greatest potential for such applications. Already, fluid flow SMD is successfully applied to DNA fragment sizing. A promising application of SMD is DNA sequencing, which could lead to a method of rapid sequencing of long DNA fragments. In addition, imaging techniques with single molecule sensitivity have the possibility for interrogating large libraries of molecules.

References

Adleman LM (1994) Molecular computation of solutions to combinatorial problems. *Science* 266: 1021–4

Affleck RL, Ambrose WP, Demas JN, Goodwin PM, Schecker JA, Wu JM, Keller RA (1996) Reduction of luminescent background in ultrasensitive fluorescence detection by photobleaching. *Anal Chem* 68(13): 2270–6

Akamine S, Kuwano H, Yamada, H (1996) Scanning near-field optical microscope using an atomic force microscope cantilever with integrated photodiode. *Appl Phys Lett* 68(5): 579–81

Ambrose WP, Goodwin PM, Jett JH, Johnson ME, Martin JC, Marrone BL, Schecker JA, Wilkerson CW, Keller RA, Haces A, Shih P-J Harding JD (1993) Application of single molecule detection to DNA sequencing and sizing. *Ber Bunsenges Phys Chem* 97(12): 1535–42

Ambrose WP, Goodwin PM, Martin JC, Keller RA (1994a) Single molecule detection and photochemistry of a surface using near-field optical excitation. *Phys Rev Lett* 72(1): 160–3

Ambrose WP, Goodwin PM, Martin JC, Keller RA (1994b) Alterations of single molecule fluorescence lifetimes in near-field optical microscopy. *Science* 265: 364–7

Ambrose WP, Affleck RL, Goodwin PM, Keller RA, Martin JC, Petty JT, Schecker JA, Wu M (1995) Imaging biological molecules with single molecule sensitivity using near-field scanning optical microscopy. *Exp Tech Phys* 41 (2): 237–48

Ambrose WP, Goodwin PM, Enderlein J, Semin DJ, Martin JC, Keller, RA (1997) Fluorescence photon antibunching from single molecules on a surface. *Chem Phys Lett*, 269: 365–70

Bachelot R, Gleyzes P, Boccara AC (1994) Near-field optical microscopy by local perturbation of a diffraction spot. *Microsc Microanal Microstruct* 5 (4–6): 389–97

Bachelot R, Gleyzes P, Boccara AC (1995a) Apertureless near-field optical microscopy by local perturbation of a diffraction spot. *Ultramicroscopy* 61 (1–4): 111–6

Bachelot R, Gleyzes P, Boccara AC (1995b) Near-field optical microscope based on local perturbation of a diffraction spot. *Opt Lett* 20 (18): 1924–6

Bard AJ, Fan FRF (1996) Electrochemical detection of single molecules. *Acc Chem Res* 29 (12): 572–8

Barnes MD, Ng KC, Whitten WB, Ramsey JM (1993) Detection of single rhodamine 6G molecules in levitated microdroplets. *Anal Chem* 65 (17): 2360–5

Barnes MD, Whitten WB, Ramsey JM (1995) Detecting Single Molecules in Liquids. *Anal Chem* 67 (13): 418A–23A

Barnes MD, Kung C-Y, Whitten WB, Ramsey JM (1996) Fluorescence of oriented molecules in a microcavity. *Phys Rev Lett* 76 (21): 3931–4

Basché T, Moerner WE, Orrit M, Wild UP (eds) (1997) Single molecule optical detection, imaging and spectroscopy. VCH Weinheim, Chap. 1

Berland K, So PTC, Ragant T, Yu WM, Gratton E (1996) 2-photon excitation for low-background fluorescence microscopy : detection of single molecules and single chromophores in solution. *Biophys J* 70 (2/pt.2): WP295–WP295

Berndt R, Gaisch R, Gimzewski JK, Reihl B, Schlittler RR, Schneider WD, Tschudy M (1993) Photon emission at molecular resolution induced by a scanning tunneling microscope. *Science* 262: 1425–7

Berndt R, Gaisch R, Schneider WD, Gimzewski JK, Reihl B, Schlittler RR, Tschudy M (1994) Sub-nanometer lateral resolution in photon emission from C60 molecules on Au (110). *Surface Sci* 309 (Pt.B): 1033–7

Berndt R, Gaisch R, Schneider WD, Gimzewski JK, Reihl B, Schlittler RR, Tschudy M (1995) Atomic resolution in photon emission induced by a scanning tunneling microscope. *Phys Rev Lett* 74 (1): 102–5

Betzig E, Chichester RJ (1993) Single molecules observed by near-field scanning optical microscopy. *Science* 262: 1422–5

Betzig E, Chichester RJ (1994) Single molecules observed by near-field scanning optical microscopy (NSOM). *Biophys J* 66 (2/pt.2): A277

Bian RX, Dunn RC, Xie XS (1995) Single molecule emission characteristics in near-field microscopy. *Phys Rev Lett* 75 (26): 4772–5

Bopp MA, Meixner AJ, Tarrach G, Zschokke-Gränacher I, Novotny L (1996) Direct imaging of single molecule diffusion in a solid polymer host. *Chem Phys Lett* 263 (6): 721–6

Castro A, Fairfield FR, Shera EB (1993) Fluorescence detection and size measurement of single DNA molecules. *Anal Chem* 65 (7): 849–52

Castro A, Shera EB (1995a) Single-molecule detection: Applications to ultrasensitive biochemical analysis. *Appl Opt* 34 (18): 3218–22

Castro A, Shera EB (1995b) Single-molecule electrophoresis. *Anal Chem* 67 (18): 3181–6

Chen DY, Dovichi NJ (1996) Single-molecule detection in capillary electrophoresis: Molecular shot noise as a fundamental limit to chemical analysis. *Anal Chem* 68 (4): 690–6

Chiu DT, Zare RN (1996) Biased diffusion, optical trapping, and manipulation of single molecules in solution. *JACS* 118 (27): 6512–3

Cooper JM, Shen J, Young FM, Connolly P, Barker JR, Moores G (1994) The imaging of streptavidin and avidin using scanning tunnelling microscopy. *J Mater Sci-Mater Electron* 5(2): 106–10

Craig DB, Arriaga EA, Wong JCY, Lu H, Dovichi NJ (1996) Studies on single alkaline phosphatase molecules: Reaction rate and activation energy of a reaction catalyzed by a single molecule and the effect of thermal denaturation – The death of an enzyme. *JACS* 118(22): 5245–53

Cricenti A, Selci S, Chiarotti G, Amaldi F (1991) Imaging of single-stranded DNA with the scanning tunneling microscope. *J Vac Sci Technol B* 9(2): 1285–7

Dapprich J, Mets Ü, Simm W, Eigen M, Rigler R (1995) Confocal scanning of single molecules. *Exp Tech Phys* 41 (2): 259–64

Davis LM, Fairfield FR, Harger CA, Jett JH, Keller RA, Hahn JH, Krakowski LA, Marrone BL, Martin JC, Nutter HL, Ratliff RL, Shera EB, Simpson DJ, Soper SA (1991) Rapid DNA sequencing based upon single molecule detection. *GATA* 8(1): 1–7

Davis RC, Williams CC, Neuzil P (1995) Micromachined submicrometer photodiode for scanning probe microscopy. *Appl Phys Lett* 66(18): 2309–11

Dickson RM, Norris DJ, Tzeng Y-L, Moerner WE (1996) Three-dimensional imaging of single molecules solvated in pores of poly(acrylamide) gels. *Science* 274: 966–8

Dovichi NJ, Martin JC, Jett JH, Trukula M, Keller RA (1983) An approach to single molecule detection by laser-induced fluorescence. *Proc SPIE* vol 426: 71–3

Dovichi NJ, Martin JC, Jett JH, Trukula M, Keller RA (1984) Laser-induced fluorescence of flowing samples as an approach to single-molecule detection in liquids. *Anal Chem* 56(3): 348–54

Dunn RC, Holtom GR, Mets L, Xie XS (1994) Near-field fluorescence imaging and fluorescence lifetime measurement of light harvesting complexes in intact photosynthetic membranes. *J Phys Chem* 98(12): 3094–8

Dunn RC, Allen EV, Joyce SA, Anderson GA, Xie XS (1995) Near-field fluorescent imaging of single proteins. *Ultramicroscopy* 57(2–3): 113–7

Edman L, Mets Ü, Rigler R (1995) Revelation of intramolecular transitions in single molecules in solution. *Exp Tech Phys* 41(2): 157–63

Edman L, Mets Ü, Rigler R (1996) Conformal transitions monitored for single molecules in solution. *Proc Natl Acad Sci USA* 93: 6710–5

Eigen M, Gardiner W (1984) Evolutionary molecular engineering based on RNA replication. *Pure & Appl Chem* 56(8): 967–78

Eigen M, Rigler R (1994) Sorting single molecules: Application to diagnostics and evolutionary biotechnology. *Proc Natl Acad Sci USA* 91: 5740–7

Enderlein J (1995a) Maximum likelihood criterion and single molecule detection. *Appl Opt* 34(3): 514–26

Enderlein J, Erdmann R, Krahl R, Klose E, Ortmann U (1995b) Time correlated single photon counting and single molecule detection. *Exp Tech Phys* 41 (2): 183–7

Enderlein J, Krahl R, Ortmann U, Wahl M, Erdmann R, Klose E (1995c) Lifetime-based identification of single molecules. *Proc SPIE* vol. 2388: 71–8

Erdmann R, Enderlein J, Becker W, Wahl M (1995) High-speed electronics for fast detection of time-resolved fluorescence in single molecule experiments. *Exp Tech Phys* 41 (2): 189–93

Fan FRF, Bard AJ (1995) Electrochemical detection of single molecules. *Science* 267: 871–4

Fan FRF, Kwak J, Bard AJ (1996) Single-molecule electrochemistry. *JACS* 118(40): 9669–75

Fenn JB, Mann M, Meng CK, Wong SF, Whitehouse CM (1989) Electrospray ionization for mass spectrometry of large biomolecules. *Science* 246: 64–70

Frisbie CD, Rozsnyai LF, Noy A, Wrighton MS, Lieber CM (1994) Functional group imaging by chemical force microscopy. *Science* 265: 2071–4

Funatsu T, Harada Y, Tokunaga M, Salto K, Yanagida T (1995) Imaging of single fluorescent molecules and individual ATP turnovers by single myosin molecules in aqueous solution. *Nature* 374: 555–9

Gimzewski JK, Sass JK, Schlitter RR, Schott J (1989) Enhanced photon emission in scanning tunneling microscopy. *Europhys Lett* 8(5): 435–40

Goodwin PM, Wilkerson CW, Ambrose WP, Keller RA (1993a) Ultrasensitive detection of single molecules in flowing sample streams by laser-induced fluorescence. *Proc SPIE* vol 1895: 79–89

Goodwin PM, Johnson ME, Martin JC, Ambrose WP, Marrone BL, Jett JH, Keller RA (1993b) Rapid sizing of individual fluorescently stained DNA fragments by flow cytometry. *Nucl Acids Res* 21(4): 803–6

Goodwin PM, Schecker JA, Wilkerson CW, Hammond ML, Ambrose WP, Jett JH, Martin JC, Marrone BL, Keller RA (1993c) DNA Sequencing by single molecule detection of labeled nucleotides sequentially cleaved from a single strand of DNA. *Proc SPIE* vol. 1891: 127–31

Goodwin PM, Affleck RL, Ambrose WP, Demas JN, Jett JH, Martin JC, Reha-Krantz LJ, Semin DJ, Schecker JA, Wu M, Keller RA (1995) Progress towards DNA sequencing at the single molecule level. *Exp Tech Phys* 41(2): 279–94

Goodwin PM, Ambrose WP, Keller RA (1996a) Single-molecule detection in liquids by laser-induced fluorescence. *Acc Chem Res* 29(12): 607–13

Goodwin PM, Cai H, Jett JH, Ishaug-Riley SL, Machara NP, Semin DJ, Van Orden A, Keller RA (1996b) Application of single molecule detection to DNA sequencing. *Nucleosides and Nucleotides; in press*

Goodwin PM, Affleck RL, Ambrose WP, Jett JH, Johnson ME, Martin JC, Petty JT, Schecker JA, Wu M, Keller RA (1996c) Detection of single fluorescent molecules in flowing sample streams. In: SD Brown (ed): *Computer assisted analytical spectroscopy*. John Wiley, New York; 61–80

Guckenberger R, Hartmann T, Wiegräbe W, Baumeister W (1993) in: Scanning tunneling microscopy II – The scanning tunneling microscope in biology. R Wiesendanger, H Güntherodt (eds): Springer, New York, 51–97

Guckenberger R, Heim M, Cevec G, Knapp HF, Wiegräbe W, Hillebrand A (1994) Scanning tunneling microscopy of insulators and biological specimens based on lateral conductivity of ultrathin water films. *Science* 266: 1538–40

Guenard RD, Lee YH, Bolshov M, Hueber D, Smith BW, Winefordner JD (1996) Characteristics of a rubidium metal vapor filter for laser scatter rejection in single molecule detection. *Appl Spectr* 50(2): 188–98

Gunning AP, Kirby AR, Morris VJ, Wells B, Brooker BE (1995) Imaging bacterial polysaccharides by AFM. *Polym Bull* 34(5–6): 615–9

Guo X-H, Huff EJ, Schwartz DC (1992) Sizing single DNA molecules. *Nature* 359: 783–4

Ha T, Enderle Th, Chemla DS, Selvin PR, Weiss S (1996a) Single molecule dynamics studied by polarization modulation. *Phys Rev Lett* 77(19): 3979–82

Ha T, Enderle Th, Ogletree DF, Chemla DS, Selvin PR, Weiss S (1996b) Probing the interaction between two single molecules: Fluorescence resonance energy transfer between a single donor and a single acceptor. *Proc Natl Acad Sci USA* 93: 6264–8

Haab BB, Mathies RA (1995) Single molecule fluorescence burst detection of DNA fragments separated capillary electrophoresis. *Anal Chem* 67 (18): 3253–60

Hahn JH, Soper SA, Nutter HL, Martin JC, Jett JH, Keller RA (1991) Laser-induced fluorescence detection of Rhodamine-6G at 6 × 10E-15M. *Appl Spectr* 45 (5): 743–6

Hansma HG, Weisenhorn AL, Gould SA, Sinsheimer RL, Gaub HE, Stucky GD, Zaremba CM, Hansma PK (1991) Progress in sequencing deoxyribonucleic acid with an atomic force microscope. *J Vac Sci Technol* B9 (2): 1282–4

Harding JD, Keller RA (1992) Single-molecular detection as an approach to rapid DNA sequencing. *Trends in Biotech* 10 (1/2): 55–61

Harris TD, Lytle FE (1983) Analytical applications of laser absorption and emission spectroscopy. Ultrasensitive laser spectroscopy. In: DS Kliger (ed): Academic Press, New York, Chap. 7

Harris TD, Grober RD, Trautman JK, Betzig E (1994) Super-resolution imaging spectroscopy. *Appl Spectr* 48 (1): 14–21A

Heckl WM, Smith DPE, Binnig G, Klagges H, Hänsch TW (1991) 2-dimensional ordering of the DNA base guanine observed by scanning tunneling microscopy. *Proc Natl Acad Sci USA* 88: 8003–5

Heckl WM (1992) Scanning tunneling microscopy and atomic force microscopy on organic and biomolecules. *Thin Solid Films* 210/211: 640–7

Heim M, Eschrich R, Hillebrand A, Knapp HF, Guckenberger R, Cevc G (1996) Scanning tunneling microscopy based on the conductivity of surface adsorbed water. Charge transfer between tip and sample via electrochemistry in a water meniscus or via tunneling? *J Vac Sci Technol B* 14 (2): 1498–1502

Hell SW, Saleheen HI, Barnes MD, Whitten WB, Ramsey JM (1996) Modeling fluorescence collection from single molecules in microspheres: Effects of position, orientation, and frequency. *Appl Opt* 35 (31): 6278–88

Hinterdorfer P, Baumgartner W, Gruber HJ, Schilcher K, Schindler H (1996) Detection and localization of individual antibody-antigen recognition events by atomic force microscopy. *Proc Nat Acad Sci USA* 93 (8): 3477–81

Hirschfeld T (1976) Optical microscopic observation of single small molecules. *Appl Opt* 15 (12): 2965–6

Ho C, Priedhorsky WC, Baron MH (1993) Detecting small debris using a ground-based photon counting detector. *Proc SPIE* vol. 1951: 67–75

Huang Z, Petty JT, O'Quinn B, Longmire JL, Brown NC, Jett JH, Keller RA (1996) Large DNA fragment sizing by flow cytometry: application to the characterization of P1 artificial chromosome (PAC) clones. *Nucl Acids Res* 24 (21): 4202–9

Inoue S (1995) Foundations of confocal scanned imaging in light microscopy. In: JB Pawley (ed): *Handbook of biological confocal microscopy* (2nd. edition). Plenum Press, New York, 1–17

Inoyue Y, Kawata S (1994) Near-field scanning optical microscope with a metallic probe tip. *Opt Lett* 19 (3): 159–61

Ishikawa M, Hirano K Hayakawa T, Hosoi S, Brenner S (1994) Single-Molecule detection by laser-induced fluorescence technique with a position-sensitive photon-counting apparatus. *Jpn J Appl Phys* Pt. 1, 33 (3A): 1571–6

Jacobson KB, Arlinghaus HF, Buchanan MV, Chen C-H, Glish GL, Hettich RL, McLuckey SA (1991) Application of mass spectroscopy to DNA sequencing. *GATA* 8 (8): 223–9

Jett JH, Keller RA, Martin JC, Marrone BL, Moyzis RK, Ratliff RL, Seitzinger NK, Shera EB, Stewart CC (1989) High-speed DNA sequencing: An approach based upon fluorescence detection of single molecules. *J Biomolec Struct, Dyn* 7 (2): 301–9

Johnson ME, Goodwin PM, Ambrose WP, Martin JC, Marrone BL, Jett JH, Keller RA (1993) Sizing of DNA fragments by flow cytometry. *Proc SPIE* vol. 1895: 69–78

Kachel V, Fellner-Feldegg H, Menke E (1990) Hydrodynamic properties of flow cytometry instruments. In: MR Melamed, T Lindmo, ML Mendelsohn (eds): *Flow cytometry and sorting*. Wiley-Liss, New York, 27–44

Kalusche G, Mathis H, Köllner M, McCaskill J (1995) Steps towards two-dimensional single molecule detection in solution. *Exp Tech Phys* 41 (2): 265–76

Kasaya M, Tabata H, Kawai T (1995) Scanning tunneling microscopy observation and theoretical calculation of the adsorption of adenine on Si(100)2 × 1 surfaces. *Surface Sci* 342 (1–3): 215–23

Kasianowicz JJ, Brandin E, Branton D, Deamer DW (1996) Characterization of individual polynucleotide molecules using a membrane channel. *Proc Natl Acad Sci USA* 93 (24): 13770–3

Keller RA, Ambrose WP, Goodwin PM, Jett JH, Martin JC, Wu M (1996) Single-molecule fluorescence analysis in solution. *Appl Spectr* 50 (7): 12–32 A

Kin CN, Whitten WB, Arnold S, Ramsey JM (1992) Digital chemical analysis of dilute microdroplets. *Anal Chem* 64: 2914–9

Kinjo M, Rigler R (1995) Ultrasensitive hybridization analysis using fluorescence correlation spectroscopy. *Nucl Acids Res* 23: 1795–9

Kneipp K, Dasari RR, Wang Y (1994) Near-infrared surface-enhanced raman scattering (NIR SERS) on colloidal silver and gold. *Appl Spectr* 48 (8): 951–5

Kneipp K, Kneipp H, Seifert F (1995 a) Near-Infrared excitation profile study of surface-enhanced hyper-Raman scattering and surface-enhanced Raman scattering by means of tunable mode-locked Ti-sapphire laser excitation. *Chem Phys Lett* 233: 519–24

Kneipp K, Wang Y, Dasari RR, Feld MS (1995 b) Approach to single molecule detection using surface-enhanced resonance Raman scattering (SERRS): A study using Rhodamine 6G on colloidal silver. *Appl Spectr* 49 (6): 780–4

Kneipp K, Wang Y, Dasari RR, Feld MS (1995 c) Near-infrared surface-enhanced Raman scattering (NIR-SERS) of neurotransmitters in colloidal silver solutions. *Spectrochimica Acta* 51 A (3): 481–7

Kneipp K, Wang Y, Kneipp H, Dasari RR, Feld MS (1995d) An approach to single molecule detection using surface-enhanced Raman scattering. *Exp Tech Phys* 41 (2): 225–34

Kneipp K, Wang Y, Kneipp H, Itzkan I, Dasari RR, Feld MS (1996) Population pumping of excited vibrational-states by spontaneous surface-enhanced Raman-scattering. *Phys Rev Lett* 76 (14): 2444–7

Koglin J, Fischer UC, Fuchs H (1996 a) 6 nm lateral resolution in scanning near-field optical microscopy with the tetrahedral tip. In: M Nieto-Vesperinas, N Garcia (eds): *Optics at the Nanometer Scale*. Kluwer, Dordrecht, 247–56

Koglin J, Fischer UC, Fuchs H (1996b) Scanning near-field optical microscopy with a tetrahedral tip at a resolution of 6 nm. *J Biomed Opt* 1 (1): 75–8

Köllner M, Wolfrum J (1992) How many photons are necessary for fluorescence-lifetime measurements? *J Chem Phys Lett* 200: 199–204

Köllner M (1993) How to find the sensitivity limit for DNA sequencing based on laser-induced fluorescence. *Appl Opt* 32(6): 806–20

Köllner M, Fischer P, McCaskill J (1994) Detector for two-dimensional time-resolved photon counting. *Proc SPIE* vol. 2328: 112–8

Köllner M, Fischer P, Arden-Jacob J, Drexhage K-H, Müller R, Seeger S, Wolfrum J (1996) Fluorescence pattern recognition for ultrasensitive molecule identification: comparison of experimental data and theoretical predictions. *Chem Phys Lett* 250: 355–60

Kopelman R, Tan WH (1994) Near-field optical microscopy, spectroscopy, and chemical sensors. *Appl Spectr Rev* 29 (1): 39–66

Lee YH, Maus RG, Smith BW, Winefordner JD (1994) Laser-induced fluorescence detection of a single molecule in a capillary. *Anal Chem* 66 (23): 4142–9

Li L-Q, Davis LM (1995) Rapid and efficient detection of single chromophore molecules in aqueous solution. *Appl Opt* 34 (18): 3208–17

Li M-Q, Zhu,J-D, Zhu J-Q, Hu J, Gu M-N, Xu Y-L, Zhang L-P, Huang Z-Q, Xu L-Z, Yao X-W (1991) Direct observation of B-form and Z-form DNA by scanning tunneling microscopy. *J Vac Sci Technol B* 9 (2): 1298–1303

Lipton RJ (1995) DNA Solution of Hard Computational Problems. *Science* 268: 543–5

Lu HP, Xie XS (1997) Single-molecule spectral fluctuations at room-temperature. *Nature* 385: 143–6

Macklin JJ, Trautman JK, Harris TD, Brus LE (1996) Imaging and time-resolved spectroscopy of single molecules at an interface. *Science* 272: 255–8

Mathies RA, Stryer L (1986) Single-molecule fluorescence detection: A feasibility study using phycoerythrin. In: DL Taylor, AS Waggoner, RF Murphy, F Lanni, RR Birge (eds): *Applications of fluorescence in biomedical sciences*. Alan R. Liss, New York, 129–40

Mathies RA, Peck K, Stryer L (1990) High-sensitivity single-molecule fluorescence detection. *Proc SPIE* vol. 1205: 52–9

Meixner AJ, Zeisel D, Bopp MA, Tarrach G (1995) Super-resolution imaging and detection of fluorescence from single molecules by scanning near-field optical microscopy. *Opt Eng* 34 (8): 2324–32

Mertz J, Xu C, Webb WW (1995) Single-molecule detection by two-photon-excited fluorescence. *Opt Lett* 20 (24): 2532–4

Mets Ü, Rigler R (1994) Submillisecond detection of single rhodamine molecules in water. *J Fluores* 4 (3): 259–64

Müller R, Zander C, Sauer M, Deimel M, Ko DS, Siebert S, Arden-Jacob J, Deltau G, Marx NJ, Drexhage KH, Wolfrum J (1996) Time-resolved identification of single molecules in solution with a pulsed semiconductor diode-laser. *Chem Phys Lett* 262 (6): 716–22

Muzzalupo I, Nigro C, Zuccheri G, Samori B, Quagliariello C, Buttinelli M (1995) Deposition on mica and scanning force microscopy imaging of DNA molecules whose original B structure is retained. *J Vac Sci Technol A* 13 (3/pt.2): 1752–4

Ng KC, Whitten WB, Arnold S, Ramsey JM (1992) Digital chemical-analysis of dilute microdroplets. *Anal Chem* 64 (23): 2914–9

Nguyen DC, Keller RA, Jett JH, Martin JC (1987a) Detection of single molecules of phycoerythrin in hydrodynamically focused flows by laser-induced fluorescence. *Anal Chem* 59: 2158–61

Nguyen DC, Keller RA, Trkula M (1987b) Ultrasensitive laser-induced fluorescence detection in hydrodynamically focused flows. *JOSA B*, 4 (2): 138–43

Nie S, Chiu DT, Zare RN (1994) Probing individual molecules with confocal fluorescence microscopy. *Science* 266: 1018–21

Nie S, Chiu DT, Zare RN (1995) Real-time detection of single-molecules in solution by confocal fluorescence microscopy. *Anal Chem* 67 (17): 2849–57

O'Connor DV, Phillips D (1984) Time-correlated single photon counting, Academic Press, London

Orrit M, Bernard J, Brown R, Lounis B (1996) Optical spectroscopy of single molecule in solids. In: E Wolf (ed): Progress in optics, vol. XXXV, Elsevier, Amsterdam, 63–144

Overway KS, Duhachek SD, Loeffel-Mann K, Zugel SA, Lytle FE (1996) Blank-free two-photon excited fluorescence detection. *Appl Spectr* 50(10): 1335–7

Paesler MA, Moyer PJ (1996) Near-field optics – theory, instrumentation, and applications. John Wiley, New York

Peck K, Stryer L, Glazer AN, Mathies RA (1989) Single-molecule fluorescence detection: Autocorrelation criterion and experimental realization with phycoerythrin. *Proc Natl Acad Sci USA* 86: 4087–91

Pedarnig JD, Specht M, Heckl WM, Hänsch TW (1992) Scanning plasmon near-field microscopy on dye clusters. *Appl Phys A*, 55: 476–7

Pedarnig JD, Specht M, Heckl WM, Hänsch TW (1993) Scanning Plasmon Near-Field Microscope. In: DW Pohl, D Courjon (eds): *Near-field Optics*. NATO ASI 242, Kluwer, Dordrecht 273–80

Petty JT, Johnson ME, Goodwin PM, Martin JC, Jett JH, Keller RA (1995) Characterization of DNA size determination of small fragments by flow cytometry. *Anal Chem* 67(10): 1755–61

Plückthun A, Ge LM (1991) The rationality of random screening: efficient methods of selection of peptides and oligonucleotide ligands. *Ang Chem* 30(3): 296–8

Radmacher M, Cleveland JP, Fritz M, Hansma HG, Hansma PK (1994a) Mapping interaction forces with the atomic-force microscope. *Biophys J* 66(6): 2159–65

Radmacher M, Fritz M, Cleveland JP, Walters DA, Hansma PK (1994b) Imaging adhesion forces and elasticity of lysozyme adsorbed on mica with the atomic-force microscope. *Langmuir* 10(10): 3809–14

Radmacher M, Fritz M, Hansma HG, Hansma PK (1994c) Direct observation of enzyme activity with the atomic force microscope. *Science* 265: 1577–80

Rigler R, Widengren J (1990) Ultrasensitive detection of single molecules by fluorescence correlation spectroscopy. *Bioscience* 3: 180–3

Rigler R, Mets Ü (1992) Diffusion of single molecules through a Gaussian laser beam. *Proc SPIE* vol. 1921: 239–48

Rigler R, Mets Ü, Widengren J, Kask P (1993) Fluorescence correlation spectroscopy with high count rate and low background: analysis of translational diffusion. *Europ Biophys J* 22: 169–75

Rigler R (1995) Fluorescence correlation, single molecule detection and large number screening: Applications in biotechnology. *J Biotechnol* 41: 177–86

Rotman B (1961) Measurement of activity of single molecules of (β-D-galactosidase. *Proc Nat Acad Sci USA* 47(12): 1981–91

Rotman B (1973) Measurement of single molecules of antibody by their ability to activate a deflective enzyme. In: AA Thaer, M Sernetz (eds): *Fluorescence Techniques in Cell Biology*. Springer, New York, 333–7

Rye HS, Yue S, Wemmer DE, Quesada MA, Haugland RP, Mathies RA, Glazer AN (1992) Stable fluorescent complexes of double-stranded DNA with bis-intercalating asymmetric cyanine dyes: properties and applications. *Nucl Acids Res* 20(11): 2803–12

Rye HS, Dabore JM, Quesada MA, Mathies RA, Glazer AN (1993) Fluorometric assay using dimeric dyes for double-stranded and single-stranded DNA and RNA with picogram sensitivity. *Anal Biochem* 208: 144–50

Sauer M, Drexhage KH, Zander C, Wolfrum J (1996) Diode laser based detection of single molecules in solutions. *Chem Phys Lett* 254(3–4): 223–8

Schaper A, Pietrasanta LI, Jovin TM (1993) Scanning force microscopy of circular and linear plasmid DNA spread on mica with quarternary ammonium salt. *Nucl Acids Res* 21(25): 6004–9

Schmidt T, Schütz GJ, Baumgartner W, Gruber HJ, Schindler H (1995) Characterization of photophysics and mobility of single molecules in a fluid lipid-membrane. *J Phys Chem* 99(49): 16662–8

Schmidt T, Schütz GJ, Baumgartner W, Gruber HJ, Schindler H (1996a) Imaging of single molecule diffusion. *Proc Natl Acad Sci USA* 93(7): 2926–9

Schmidt T, Schütz GJ, Gruber HJ, Schindler H (1996b) Local stochiometrics determined by counting individual molecules. *Anal Chem* 68(24): 4397–401

Seidel CAM, Schulz A, Sauer MHM (1996) Nucleobase-specific quenching of fluorescent dyes. 1. Nucleobase one-electron redox potentials and their correlation with static and dynamic quenching efficiencies. *J Phys Chem* 100(13): 5541–3

Shera EB, Seitzinger NK, Davis LM, Keller RA, Soper SA (1990) Detection of single fluorescent molecules. *Chem Phys Lett* 174(6): 553–7

Soper SA, Shera EB, Martin JC, Jett JH, Hahn JH, Nutter HL, Keller RA (1991a) Single-molecule detection of Rhodamine 6G in ethanolic solutions using continuous wave laser excitation. *Anal Chem* 63: 432–7

Soper SA, Davis LM, Fairfield FR, Hammond ML, Harger CA, Jett JH, Keller RA, Marrone BL, Martin JC, Nutter HL, Shera EB, Simpson DJ (1991b) Rapid sequencing of DNA based on single molecule detection. *Proc SPIE* vol. 1435: 168–78

Soper SA, Davis LM, Shera EB (1992) Detection and identification of single molecules in solution. *JOSA B* 9(10): 1761–9

Soper SA, Mattingly QL, Vegunta P (1993) Photon burst detection of single near-infrared fluorescent molecules. *Anal Chem* 65(6): 740–7

Soper SA, Legendre BL, Huang JP (1995a) Evaluation of thermodynamic and photophysical properties of tricarbocyanine near-IR dyes in organized media using single-molecule monitoring. *Chem Phys Lett* 237(3–4): 339–45

Soper SA, Legendre BL, Williams DC (1995b) Single molecule detection in the near-IR: Applications in chemistry and biochemistry. *Exp Tech Phys* 41(2): 167–82

Specht M, Pedarnig JD, Heckl WM, Hänsch TW (1992) Scanning plasmon near-field microscope. *Phys Rev Lett* 68(4): 476–9

Stevenson CL, Winefordner JD (1992a) Estimating detection limits in ultratrace analysis – 2. Detecting and counting atoms and molecules. *Appl Spectr* 46(3): 407–19

Stevenson CL, Winefordner JD (1992b) Estimating detection limits in ultratrace analysis – 3. Monitoring atoms and molecules with laser-induced fluorescence. *Appl Spectr* 46(5): 715–24

Synge EH (1928) A suggested method for extending microscopic resolution into the ultra-microscopic region. *Phil Mag* 6: 356–62

Synge EH (1932) An application of Piezo-electricity to Microscopy. *Phil Mag* 13: 297–300

Tanaka H, Yoshinobu J, Kawai M, Kawai T (1996) Imaging of nucleic acid base molecules on Pd(110) surfaces by scanning tunneling microscopy. *Jpn J Appl Phys* Pt. 2. 35(2B): L244–6

Tellinghuisen J (1993) Effect of background on the least-square estimation of exponential decay parameters. *Anal Chem* 65(9): 1277–80

Tellinghuisen J, Goodwin PM, Ambrose WP, Martin JC, Keller RA (1994) Analysis of fluorescence lifetime data for single rhodamine molecules in flowing sample streams. *Anal Chem* 66: 64–72

Thimonier J, Chauvin JP, Barbet J, Roccaserra J (1994) Scanning tunneling microscopy of monoclonal immunoglobulin G. *Microsc Microanal Microstruct* 5(4–6): 341–9

Thompson NL (1991) Fluorescence correlation spectroscopy. In: JR Lakowicz (ed): *Topics in fluorescence spectroscopy I*. Plenum Press, New York, 337–78

Trautman JK, Macklin JJ, Brus LE, Betzig E (1994) Near-field spectroscopy of single molecules at room temperature. *Nature* 369(6475): 40–2

Trautman JK, Macklin JJ (1996) Time-resolved spectroscopy of single molecules using near-field and far-field optics. *Chem Phys* 205(1–2): 221–9

Trautman JK, Ambrose WP (1997) Near-field optical imaging and spectroscopy of single molecules. In: T Basché, WE Moerner, M Orrit, UP Wild (eds): *Single molecule optical detection, imaging and spectroscopy*. VCH Weinheim, 191–222

Tsai DP, Kovacs J, Moskovits M (1995) Applications of apertured photon scanning tunneling microscopy (APSTM). *Ultramicroscopy* 57(2–3): 130–40

Ueda M, Yoshikawa K (1996) Phase transition and phase segregation in a single double-stranded DNA molecule. *Phys Rev Lett* 77(10): 2133–6

Venkataraman B, Flynn GW, Wilbur JL, Folkers JP, Whitesides GM (1995) Differentiating functional groups with the scanning tunneling microscope. *J Phys Chem* 99(21): 8684–9

Walba DM, Stevens F, Clark NA, Parks DC (1996) Detecting molecular chirality by scanning tunneling microscopy. *Acc Chem Res* 29(12): 591–7

Webb RH (1996) Confocal optical microscopy. *Rep Progr Phys* 59(3): 427–71

Whitten WB, Ramsay JM, Arnold S, Bronk BV (1991) Single-molecule detection limits in levitated microdroplets. *Anal Chem* 63(10): 1027–31

Whitten WB, Ramsey JM (1992) Photocount probability distributions for single fluorescent molecules. *Appl Spectr* 46(10): 1587–9

Wickramasinghe HK, Martin Y, Zenhausern F (1996) Scanning interferometric apertureless microscopy at 10 Ångstrom resolution. In: M Nieto-Vesperinas, N Garcia (eds): *Optics at the Nanometer Scale*. Kluwer, Dordrecht, 131–41

Widengren J, Rigler R, Mets Ü (1994) Triplet-state monitoring by fluorescence correlation spectroscopy. *J Fluoresc* 4(3): 255–8

Widengren J, Mets Ü, Rigler R (1995) Fluorescence correlation spectroscopy of triplet states in solution: A theoretical and experimental study. *J Phys Chem* 99(36): 13368–79

Wilkerson CW, Goodwin PM, Ambrose WP, Martin JC, Keller RA (1993) Detection and lifetime measurement of single Molecules in flowing sample streams by laser-induced fluorescence. *Appl Phys Lett* 62: 2030–2

Wu M, Goodwin PM, Ambrose WP, Keller RA (1996) Photochemistry and fluorescence emission dynamics of single molecules in solution: B-phycoerythrin. *J Phys Chem* 100(43): 17406–9

Xiang X-D, Sun X, Briceño G, Lou Y, Wang K-A, Chang HY, Wallace-Freedman WG, Chen S-W, Schultz PG (1995) A combinatorial approach to materials discovery. *Science* 268: 1738–40

Xie XS, Dunn RC (1994) Probing Single Molecule Dynamics. *Science* 265: 361–4

Xie XS (1996) Single molecule spectroscopy and dynamics at room-temperature. *Acc Chem Res* 29(12): 598–606

Xue Q, Yeung ES (1995) Differences in the chemical reactivity of individual molecules of an enzyme. *Nature* 373: 681–3

Youngquist MG, Driscoll RJ, Coley TR, Goddard WA, Baldeschwieler JD (1991) Scanning tunneling microscopy of DNA: Atom-resolved imaging, general observations and possible contrast mechanism. *J Vac Sci Technol B*, 9(2): 1304–8

Zander C, Sauer M, Drexhage KH, Ko D-S, Schulz A, Wolfrum J, Brand L, Eggeling C, Seidel CAM (1996) Detection and characterization of single molecules in aqueous solution. *Appl Phys B*, 63(5): 517–24

Zenhausern F, Oboyle MP, Wickramasinghe HK (1994) Apertureless near-field optical microscope. *Appl Phys Lett* 65(13): 1623–5

Zenhausern F, Martin Y, Wickramasinghe HK (1995) Scanning interferometric apertureless microscopy: Optical imaging at 10 Ångstrom resolution. *Science* 269: 1083–5

Zuccheri G, Ranieri GA, Nigro C, Samori B (1995) Writhing number of supercoiled DNA from its scanning force microscopy imaging. *J Vac Sci Technol B* 13(1): 158–60

14 Fluorescence correlation spectrometry (FCS): Measuring biological interactions in microstructures

Gabriele Gradl, Rolf Guenther and Silvia Sterrer

14.1 Introduction

The measurement of biochemical interaction in very small volumes has recently become a key issue in biotechnological applications in such fields as the life sciences, agriculture, ecology, to name but a few. The cost of biochemical substances in macroscopic amounts, not to mention their scarcity, is one of the driving forces behind the need to miniaturize sample preparation, handling and analysis. Moreover, recently developed, highly parallel experimental approaches, such as combinatorial chemistry or evolutionary strategies, employed in the search for active compounds contribute to the need for miniaturization. The sheer number of similar samples, often surpassing 100 000 samples per batch, prohibits a macroscopic approach. Another field which is presently subject to a combinatorial explosion of numbers is genomic science. Here, hybridization of oligomers for sequencing or comparison purposes is carried out in highly parallel approaches.

Thus, miniaturization reducing sample volumes to the submicroliter scale and beyond is highly desirable. To fully capitalize on the advantages offered by miniaturization, it is necessary not only to address sample preparation and liquid handling, but also the measurement of biochemical interaction.

One approach to addressing the measurement of biochemical interactions in miniaturized formats is to integrate the sensors into the overall micromechanical design of the sample cavity or the reaction vessel. In this case, the sensor is produced along with the sample carrier and can thus be fabricated in large quantities. No complicated and expensive assembly needs to be performed. However, the variety of sensor design is quite limited and only allows for the most simplistic carrier architecture. Examples of this sensor type are evanescent wave sensors based on waveguides, ISFETs, and conductivity sensors. They all can be easily integrated into the surface of the cavity wall.

Integrated scanning probes, such as the STM or AFM, are not yet state of the art, though STM cantilevers are often etched using microfabrication techniques.

Another way to measure samples in microstructures is to design a macroscopic sensor which measures in miniaturized volumes. FCS, *fluorescence correlation spectroscopy*, is one of the most well-known representatives of this class of sensor. In FCS, the idea of an integrated sensor is sacrificed in favour of high-precision sensor architecture which consists of a complete confocal microscope and additional signal processing equipment. FCS not only monitors fluorescence intensity, but even concentrations and binding constants.

Thus, FCS offers many advantages compared to other sensors designed to operate in a miniaturized setting:

- FCS measures *far from the surface*. Surface effects, such as non-specific adsorption of molecules to walls, concentration effects, ionic exchange, etc., do not affect the measurement process itself. The measured parameters, therefore, represent the real values for the bulk solution with no need for correction.
- FCS measures in a *homogeneous solution*. The system's biochemical parameters thus remain unaltered, an occurrence which cannot be completely discounted when assay components are immobilized.
- FCS is *intrinsically miniaturized*. As demonstrated later in more detail, the actual measurement volume in FCS is only about 1 femtoliter; the size of an *E. coli* cell! Thus, FCS data obtained from macroscopic samples does not differ from that obtained from equivalent samples measured in microstructures.
- FCS offers *single-molecule sensitivity*. In verw low volumes, the absolute number of molecules can be quite low at physiological concentrations. Single-molecule sensitivity is thus a prerequisite for any measurement technique to be used with microstructures.
- FCS measures interaction *specifically*. Many other techniques, such as conductivity measurements ("microphysiometers") just monitor an overall change of a physical parameter. The pH of a cell environment, for example, is affected by any change in the metabolism of a cell, whatever the origin may be. Often this change is not caused by the interaction under investigation, but by uncontrolled environmental parameters. A fluorescently-labelled tracer substance and its measurement by FCS, on the other hand, is a means to specifically measure an interaction at the molecular level.

The versatility of FCS as a tool for monitoring biochemical activity is documented in the following chapters. It will become clear that, using FCS, complete biochemical assays can be performed in miniaturized formats in very low volumes. These kinds of assays have formerly been restricted to the macroscopic world of ELISA and radioactive binding assays. FCS opens the door to the measurement of a wide variety of labelled probes – from small organic molecules to peptides, proteins, oligosaccharides, RNA and DNA – and analytes from macromolecules up to micelles, liposomes and even whole cells.

The detection and quantification of molecular interactions require the use of specific probes in combination with an appropriate detection system. This system needs to be fast, sensitive with respect to analyte concentration, but unaffected by other assay components or disturbing factors such as impurities of the biological material to be assayed or background noise. For almost 40 years techniques based on radioactivity ruled the assay world as a result of the exquisite sensitivity of radionuclide detection which made them superior to, for example, colorimetric systems. However, handling radionuclides is problematic with respect to safety, licensing requirements, short half-lives and waste disposal. Therefore, a great deal of effort has been put into the development of alternative techniques with comparable or even greater sensitivity [1, 2].

One feature which is desirable in future assay techniques is the ability to obtain the information from a homogeneous assay system ("mix and read"). For example, the need to remove non-bound probe is a major disadvantage of ELISAs (Enzyme Linked Immuno Sorbent Assays) and radioligand-binding assays. In homogeneous assays, not only can time-consuming separation steps be avoided, but low affinity binding reactions can also be observed without perturbing equilibrium.

The majority of such homogeneous, non-radioactive assays are based on fluorescence. Fluorescence anisotropy, for example, and especially fluorescence polarization (FP) are widely used for the analysis of the interaction between small molecules and proteins, between peptides and proteins, and proteins and DNA [3, 4]. While fluorescence anisotropy is applied to molecules in solution, the application of fluorescence resonance energy transfer (FRET) overcomes the size limitation associated with FP, and can be extended to the study of plasma membrane proteins and large structural biomolecules [5]. It is a technique for the measurement of distances between molecules with near Ångstrom resolution. A disadvantage of FRET is the requirement of two fluorophores with well defined energy transfer characteristics. The method of fluorescence photobleaching recovery (FPR/FRAP) is especially useful for investigations of membrane protein mobility in artificial or natural biomembranes [6]. The need to measure analyte concentrations without interfering with the equilibrium of a given system is a major constraint on the control and evaluation of bioreactors. Here, fluorescence detection, such as lifetime-based phase modulation fluorimetry, has emerged as an alternative to electrochemical sensing of ion concentrations [7].

Fluorescence correlation spectroscopy (FCS) [8, 9] combines features of all the above fluorescence techniques. Like FP it is a homogeneous assay technique; like FRET it can be used to measure differences in specific fluorescence of molecules or complexes. Measurements of molecular diffusion in membrane systems are also feasible, a process formerly only accessible to FPR/FRAP, but compared to FPR/FRAP much lower concentrations of the reporter molecules are required [10]. In addition, analyte concentrations can be deduced directly from measurement parameters using FCS without the need for a calibration standard. The unique feature of FCS compared to all other assay techniques described above is that the detection volume represents only a fraction of the sample volume so that signals are recorded from single molecules within the detection volume and not from bulk solution or from entire compartment surfaces.

14.2 Principle of the FCS technology

Fluorescence Correlation Spectroscopy (FCS), introduced 25 years ago [11, 12, 61], has been a powerful, though difficult to perform method for monitoring molecular aggregation and mobility ever since. Wider use of FCS has been hindered by the complexity of the experimental set-up and by the length of data collection times needed to achieve an acceptable signal-to-noise ratio. Technological developments, such as new, more stable and reliable laser sources, the use of photon counting avalanche photodiodes as high quantum yield photon detectors, and high speed data processing electronics have had a revolutionary effect on FCS. A commercial FCS instrument, the ConfoCor (ZEISS/EVOTEC) is now available.

In an FCS spectrometer, a sharply focused laser beam illuminates a volume element of 10^{-15} l (Fig. 1). Single fluorescent molecules diffusing through this volume are excited and emit light quanta. The result is a fluctuating fluorescence signal representing single molecules entering and leaving the detection volume. The photons emitted are recorded in a time-resolved manner by a single-photon detection device. All signals resulting from the diffusion of a series of particles through the confocal volume over the course of the measurement are recorded and the fluorescence fluctuations are directly converted into a measurement curve by an autocorrelation process (Fig. 2). For evaluation, the curve is fit using a correlation function which is based on a model of translational diffusion (for detailed information regarding the mathematical background see [8]. Several parameters can be computed from this function, such as the number and brightness of molecules in the illuminated volume

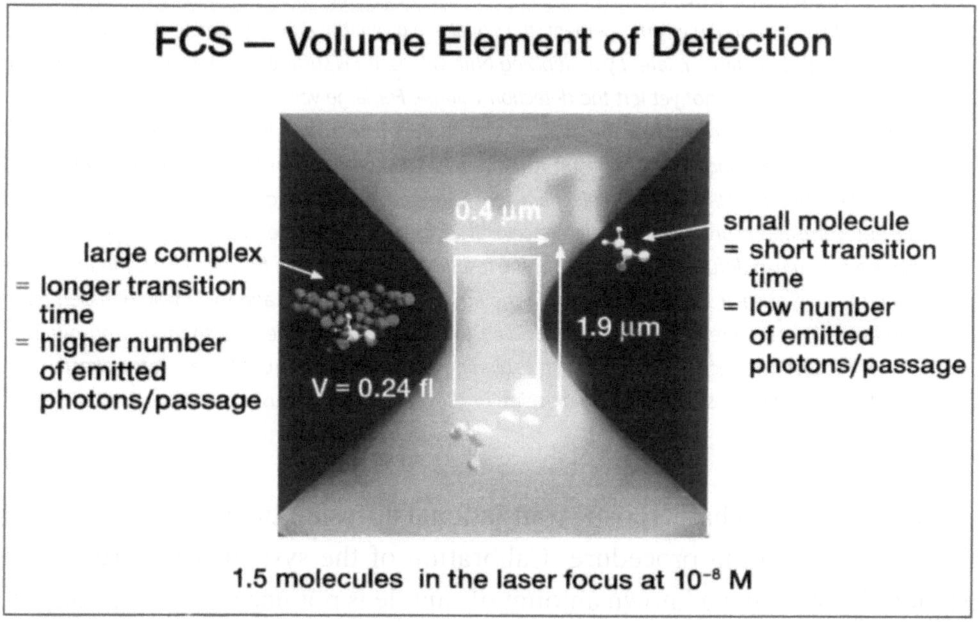

Figure 1 *The principle of FCS. In FCS, a sharply focused laser beam illuminates a volume element of one femtoliter. Molecules diffusing through the illuminated volume element by Brownian motion give rise to bursts of fluorescent light quanta. The photons are recorded in a time-resolved manner by a highly sensitive single-photon detection device. The measured transition time is related to the size and shape of the particle (see colour plate 6)*

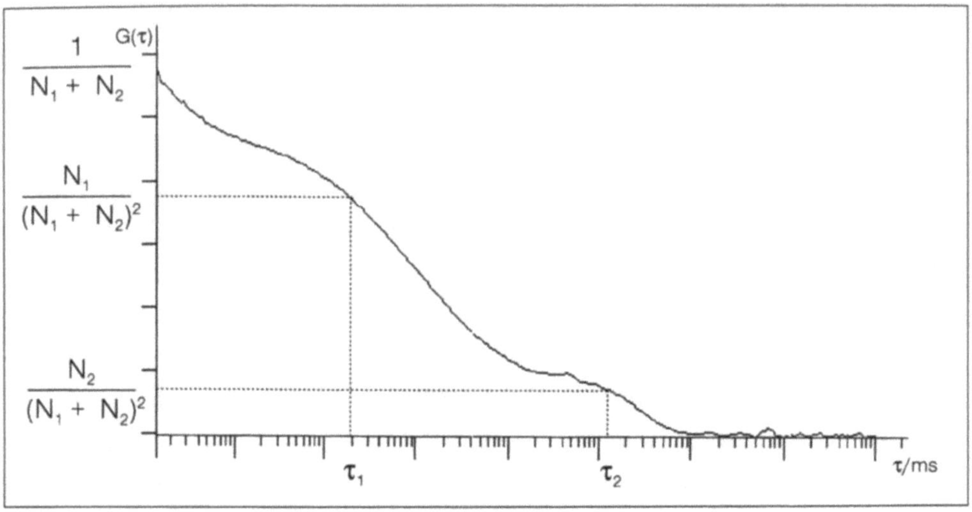

Figure 2 *FCS analysis. Consider a single molecule diffusing through the sample. The time-dependent fluorescence I(t) is high as long as this molecule stays inside the sample volume element. When the molecule leaves this element the intensity drops. The length of time of the molecule is present in the volume element depends on its translational diffusion coefficient. In order to extract information about the time required for a molecule to diffuse through the volume element, the signal at a given (but arbitrary) time t, I(t), is compared to the intensity $I(t + \tau)$ some time, τ, later by multiplying both values. If τ is sufficiently small, the product will be high since the molecule has not yet left the detection volume. For large values of τ, when the molecule has left the volume, $I(t + \tau)$ is zero and so is the product. The product is calculated for various time intervals τ and averaged over the measuring time. The result is the autocorrelation function $G(\tau) = \langle I(t)I(t+\tau)\rangle_t / \langle I(t)I(t)\rangle_t$. The number of molecules, N, as well as their characteristic translational diffusion times, τ_D, can be computed from the autocorrelation curve. When there are several species present in the sample, each represented by their characteristic mean diffusion times τ_1, τ_2,..., the resulting autocorrelation function $G(\tau)$ is the sum of the correlation function of each species. As shown in the figure, two amplitudes and two mean diffusion times are represented in the correlation function providing information about the concentration and diffusion constants of two distinct species. Properties such as binding or catalytic activity can be calculated directly from the ratio of faster/smaller to slower/larger molecules, or the difference in diffusion time*

element and their characteristic translational diffusion times. Therefore, FCS is an autocalibrating procedure. Calibration of the system with a reference standard containing a known amount of sample is not necessary. The concentration of fluorescent molecules can be calculated directly from the number of particles and the well-defined size of the volume element.

The diffusion time required for the passage of a fluorescent particle through the volume element is determined by its diffusion coefficient and is related to the size and shape of a particle. Thus, FCS allows smaller/faster and larger/slower diffusing particles to be differentiated. It makes a whole spectrum of molecular information accessible, such as the degree of binding from the ratio of faster to slower molecules, aggregation or fragmentation by examining the particle number, and conformational change if accompanied by a change in fluorescence intensity. Measurement time ranges from a matter of seconds to a few minutes, depending on the concentration of the molecules. Theoretically, there is no lower limit of analyte concentration for FCS measurements – as long as measurement time can be accordingly prolonged so that the increasingly rare passage of a molecule through the detection volume can be detected. Intrinsic fluorescence of sample material other than the analyte can contribute to the signal, especially in the case of biological probes or cellular material. In the majority of applications this biological background will predominate over mere physical limitations, such as scattered laser light, Raman scattering from the solvent, or autofluorescence of optical equipment [8]. Thus the dynamic concentration range for biological FCS assays is typically 10^{-11} to 10^{-7} M.

The sensitivity of fluorescence intensity detection is restricted by the photon flux per analyte molecule in relation to background noise. In FCS, the high specificity of the signal with respect to noise is achieved by two characteristics of the technology: a reduction of the detection volume to a space small enough to host single analyte molecules, and the development of a principle of data evaluation based on the analysis of fluorescence fluctuations arising from individual molecules instead of averaging intensity over the entire sample.

14.3 Examples of FCS applications

Nucleic acids

Over the past 20 years, a few publications have demonstrated the use of FCS for the determination of hydrodynamic and biochemical characteristics of DNA. The molecular weight of DNA molecules has been determined by rotating the sample [13]. The binding characteristics of ethidium bromide molecules was also studied using FCS [14, 15a, b].

The general applicability of FCS for the observation of the stoichiometry, kinetics, and thermodynamics of hybridization under physiological conditions was demonstrated by investigating the interaction of an 18-base oligonucleotide used in sequencing and M13 template DNA [16]. In the case of *in vivo* anti-sense strategies target sequences in cells might be inaccessible as a result of secondary structure hindrances or protein complexing. Reliable qualitative and quantitative information regarding *in vivo* hybridization will thus depend on appropriate model systems. FCS enables the design of homogeneous hybridization assays under physiological conditions. An example with implications for anti-sense research is given with the hybridization of DNA probes to HIV-1 replication primer binding site RNA [17].

The detection of specific nucleotide sequences of interest, for example in the diagnostics of pathogens, can be accomplished by hybridization-based approaches. Very often it is necessary to quantitatively amplify DNA or RNA under investigation from single copies in order to achieve detectable concentrations. FCS can easily be combined with amplification techniques such as PCR (Polymerase Chain Reaction) [18] or NASBA (Nucleic Acid-Based Sequence Amplification) [19] to detect minute amounts of target DNA or RNA in a one-step procedure [20, 21]. The method which is called APEX (Amplified Primer EXtension) FCS [20] is based on the principle of endpoint titration. The endpoint is reached when a fluorescentlylabelled detection primer present in a finite concentration, below that of the amplification primers, is fully incorporated into the correctly synthesized strands. Upon initiation of the PCR reaction, all detection primer is present in a state of high mobility, whereas when incorporation reaches its endpoint it is present only in a state of low mobility. This mobility shift can be followed by FCS measurement while the sample remains in a sealed compartment and thus the risk of cross-contamination by sample carry-over can be eliminated.

Transcription factors

Many advances have been made in the last 5 years in understanding the mechanisms responsible for the assembly and function of multicomponent transcription complexes for activation or repression of gene expression [22]. Examples of such conventional protein/DNA binding assays are filter binding or Electrophoretic Mobility Shift Assays (EMSA) [23]. They are time-consuming techniques and usually employ radioactively-labelled probes. Here, two examples of ways to study the interaction of transcription factors with their respective target sequences using FCS are given.

The first is the binding of recombinant heterodimers of the thyroid hormone receptor (TRα) and the retinoid-X-receptor (RXRα) to double-stranded oligonucleotides containing a consensus sequence [24]. The oligonucleotide was labelled with both a radioactive phosphate and a fluorescent tag so that FCS and EMSA experiments carried out in parallel could be directly compared. Although the difference in diffusion time through the focus of free vs. bound DNA was only two-fold, complicating the separation of the signal into bound and free components, the percentage of complex formed could be calculated reproducibly and was in good agreement with gel retardation data (Fig. 3).

In many cases recombinant proteins are not available and, hence, crude nuclear protein extracts are used. These extracts are sometimes problematic with respect to non-specific binding and solid phase artefacts. The binding of Sp1 protein, present in nuclear protein extracts prepared from rat liver, to a GC rich sequence [25] was studied proving that protein/DNA binding assays can be carried out with FCS using non-purified material (Fig. 4). Comparable amounts of complex were formed in samples measured by FCS and in control samples not subjected to FCS measurement.

In both experiments a slightly lower percentage of DNA complexes were measured by gel retardation when compared to FCS. This may be explained by the emergence of non-equilibrium conditions caused by gel electrophoresis.

Membrane receptors

In order to understand signal transmission between cells one has to unravel the interplay of ligands with their receptors. The material commonly used in ligand binding assays is solubilized receptor preparations, or tissue homogenates. Since FCS monitors Brownian motion populations of molecules or particles of interest should possess well-defined diffusion coefficients in order to yield reliable results. Therefore, the method of receptor preparation must guarantee a high degree of homogeneity. Sufficient homogeneity can general-

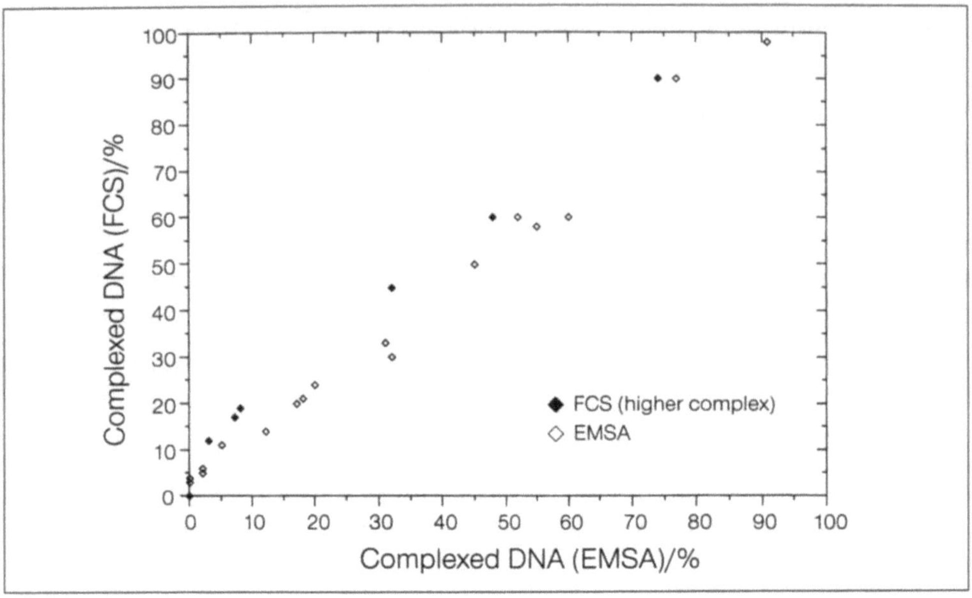

Figure 3 *Comparison of FCS (ordiante) and EMSA (abscissa) results from the association of heterodimers of the retinoid-X-receptor (RXR) and the thyroid hormone receptor (TXR) with bound cofactor T3 and a double-stranded oligonucleotide containing the TRE consensus sequence. Material: Recombinant RXR/TXR heterodimers were expressed in HeLa cells. Nuclear extracts from vaccinia virus infected cells carrying both receptors were prepared according to [56] except that a buffer of 30 mM Tris pH 8.0 containing 400 mM sodium chloride, 0.5 mM Magnesium chloride, 10% glycerol, 10 mM dithiothreitol, 2 mM phenylmethylsulfonylfluoride and 200 U/ml Trasylol was used for extraction of the nuclear pellet. The receptors were purified according to [57]. The oligonucleotides (AGCTTCAGGTCACTCCAGGTCAAGCT and AGCTTGACCTGAAGTGACCTGAAGCT) containing the TRE (DR4) consensus sequence were obtained from NAPS (Goettingen, Germany). One oligonucleotide strand was labelled with tetramethylrhodamine at the 5′ end. After annealing with the complementary strand in 180 mM NaCl, the oligonucleotide was labelled with ^{32}P using T4 polynucleotide kinase according to standard protocols. Radioactively-labelled oligonucleotides were separated from free nuclide by gel electrophoresis. Binding reaction: 1 nM oligonucleotide and 0.5 to 10 µg in 20 µl buffer containing 60 mM KCl, 5 mM $MgCl_2$, 10 mM HEPES pH 7.5. The mixture was incubated for 40 min on ice. Measurement: Eight-chamber cover glasses (Nunc); room temperature; data acquisition time for all experiments was 60 s for each measurement. Samples were analyzed on a 4.5% polyacrylamide gel together with control samples. Equipment: ConfoCor with an helium neon laser (Uniphase, 1.5 mW, 543.5 nm), 40 x /1.2 W objective (Zeiss), 580DF30 interference filter (Omega). Phospho Imager (Molecular Devices).*

ly be achieved with vesicular systems without great effort, but can be difficult for tissue preparations. Current advances being made in the FCS technology will enable less homogeneous samples to be tolerated. With these limitations in mind, the opportunity to directly observe the interaction of

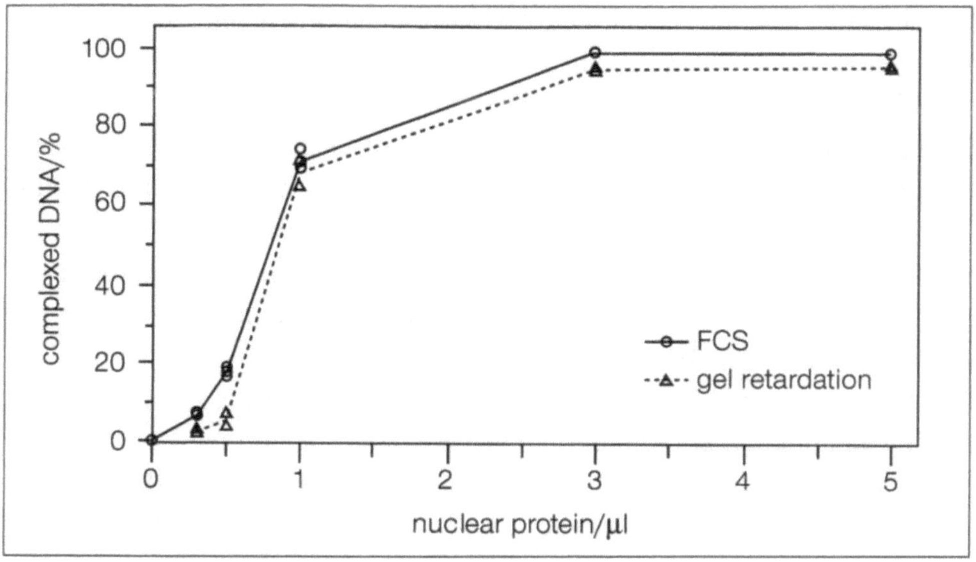

Figure 4 *Association curve of various amounts of nuclear protein extract containing Sp1 protein incubation with 6.8 nM target oligonucleotide. FCS measurements were performed in triplicate (circles) and compared with duplicate EMSA assay samples (triangles). Material: Nuclear protein extracts from rat liver were prepared [58]; the protein concentration of the crude extracts was 12 µg/ml. The oligonucleotides (ATTCGAAGGTCGGGGCGGGGCGAGC and GCTCGCCCCGCCCCGACCTTCGAAT) containing the GC rich consensus sequence were obtained from NAPS (Goettingen, Germany). One oligonucleotide strand was labelled with 5-carboxy-X-rhodamine at the 5'end. After the annealing of the complementary strand in 180 mM NaCl, the oligonucleotide was labelled with ^{32}P by using T4 polynucleotide kinase according to standard protocols. Radioactively-labelled oligonucleotides were separated from free nuclide by gel electrophoresis. Binding reaction: 6.8 nM oligonucleotide in 20 µl bffer containing 60 mM KCl, 5 mM $MgCl_2$, 10 mM HEPES pH 7.9, 250 µM $ZnSO_4$, 0.1 µg/µl poly dI dC, and 0.3–3 µl nuclear extract. The mixture was incubated for 40 min on ice. Measurement: Eight-chamber cover glasses (Nunc); room temperature; data acquisition time for all experiments was 60 s. Samples were analyzed on a 4.5% polyacrylamide gel together with control samples. Equipment: ConfoCor with an helium neon laser (Uniphase, 1.5 mW, 543.5 nm), 40x/1.2 W objective (Zeiss), 610W100 interference filter (Omega). Phospho Imager (Fujix BAS 2000)*

receptors with ligands at the molecular level in their native environment is invaluable. Two examples of the application of FCS on ligand-receptor interactions are given.

The dynamics of the interaction of tetramethylrhodamine-labelled α-bungarotoxin with the detergent solubilized nicotinic acetylcholine receptor (AChR) of *Torpedo californica* electric organ was characterized [26]. Both the association of fluorescently-labeled α-bungarotoxin (TMR-α-bungarotoxin), as well as the dissociation of bound TMR-α-bungarotoxin initiated by adding an excess of non-labelled α-bungarotoxin, was investigated. Since mono- and multi-liganded receptor macromolecules are counted as only one type of diffusing fluorescent species from the viewpoint of FCS, detailed data analysis was unambiguous only for the AChR monomer with its two α-binding sites. Complex formation of the receptor with α-bungarotoxin was shown to be monophasic. The association rate coefficient for the mono-liganded species as well as the rate constants of the biphasic dissociation reaction of bound TMR-α-bungarotoxin were determined.

In the second example, a membrane vesicle preparation from a cell line bearing a high number of epidermal growth factor receptors (> 10^5 per cell) was used in binding studies with tetramethylrhodamine-labelled EGF. The dissociation constant K_D could be deduced using unlabelled EGF (Fig. 5). The homogeneous set-up of the FCS assay used in binding experiments enables the determination of affinity constants in the presence of free ligand. This option is extremely valuable for low-affinity interactions because their equilibrium is excessively disturbed in assays which include washing steps.

Second messengers

Many intracellular second messengers can associate with cellular membranes and this association is thought to be important for their recruitment to target molecules with which they interact [27]. In some cases specific lipid molecules act as cofactors for the activity of second messengers. The activation of PKC by phosphatidylserine in the presence of calcium and by diacylglycerol is one example of such an interaction. Using FCS, the dependency of the association of PKC with phosphatidylserine micelles on calcium was investigated by using labelled mixed micelles of Triton X-100 and lipids [28]. The affinity of PCK for the lipid was found to be higher in the presence of calcium. FCS simultaneously also offers a very elegant method for determining the size of the micelles with great precision. The experiments were supplemented with resonance energy transfer measurements to determine the dependency of the interaction on lipid cofactors. This provides a nice example of how FCS and other methods can complement each other.

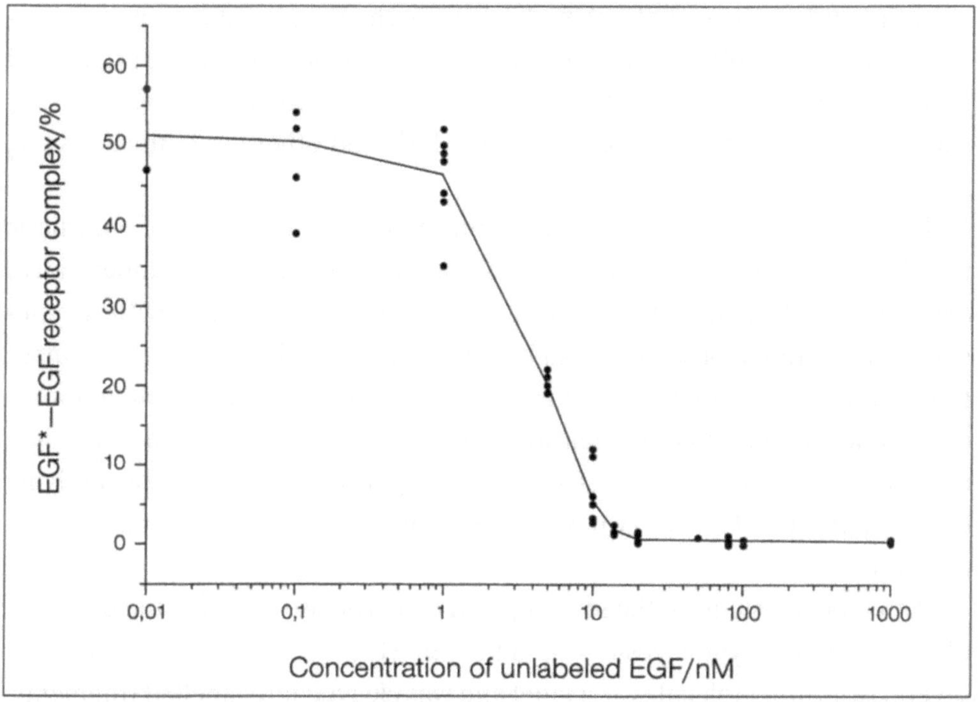

Figure 5 *FCS analysis of competition between fluorescently-labelled EGF (10 nM, Molecular Probes) and non-labelled EGF (10 pM to 100 nM, SIGMA) for receptor binding sites on membrane vesicles. The fraction of the bound fluorescent EGF conjugate was plotted versus the concentration of the non-labelled competitor. Material: Membrane vesicles from A431 cells were prepared according to standard protocols [59]. Tetramethyl-rhodamine-labelled EGF was purchased from Molecular Probes, unlabelled EGF was obtained from SIGMA. Incubation was performed in 20 mM HEPES pH 7.4 containing 140 mM NaCl, 5 mM $MgCl_2$, 1.8 mM CaCl, 4.2 mM $NaHCO_3$ and 5.5 mM glucose for 40 min at room temperature. Measurement: Eight-chamber cover glasses (Nunc); room temperature; data acquisition time was 60 s. Equipment: ConfoCor with an helium neon laser (Uniphase, 1.5 mW, 543.5 nm), 40x/1.2 W objective (Zeiss), 610W100 interference filter (Omega). A K_D of 5×10^{-9} mol l^{-1} was calculated which is in accordance with published data [60]*

Enzymes

Enzymes represent key components in biotechnological processes. They are not only used in food production and biomaterial processing, but also for chemical synthesis and diagnostics [29]. The ever-growing number of enzyme applications in biotechnology together with the high standard of performance required of them results in the need for good characterization of such enzymes. In addition, the characterization and modulation of enzymatic activity is a continuously growing field of medical research. The identification of a number of disease conditions in which enzymes, expecially proteases, are critical for disease pathology continues to grow and includes conditions such as metastasis, inflammation, cytokine processing, HIV and other viral infections [30, 31, 32].

FCS is a rapid and reliable method for the real-time analysis of enzymatic reactions. It enables a scientist to investigate enzymatic systems under native conditions with nanomolar substrate concentrations. Kinetics with a time-course in the range of seconds can be resolved due to the short measurement time required. Another large advantage when working with low amounts of sample is that kinetic measurement can be performed in real time using a single sample. All this makes FCS particularly useful for studying the catalytic activity of enzymes such as proteases, kinases, phosphatases or nucleases.

As a model system for the investigation of a protease reaction, the digestion of casein with chymotrypsin was analyzed (Fig. 6). The substrate carried several resorufin molecules as a label and was cleaved into labelled fragments. The resulting increase in fluorescent particle number was directly followed using FCS during the course of the enzymatic reaction.

FCS analysis can also provide insights into the mechanism of enzymatic reactions. This is demonstrated by FCS analysis of β-galactosidase assembly [62]. A bacterial β-galactosidase analogue consisting of two subunits – a large dimeric polypeptide, denoted enzyme acceptor (EA), and a small polypeptide, denoted enzyme donor (ED) – were used for a diagnostic system. EA and ED are enzymatically inactive but they spontaneously associate to form enzymatically active tetramers. In CEDIA assays, the hapten or analyte is covalently linked to the ED, and an analyte-specific antibody is used to inhibit the assembly of the enzymatically active tetramers [33]. Analyte in a patient's serum competes with the analyte in the analyte-ED conjugate for antibody, modulating the amount of active β-galactosidase formed. Characterization of the tetramer association process is required in order to investigate contamination which may cause perturbations of the assay. Further optimization of CEDIA assays requires insight into the reaction mechanism. Therefore, a detailed study of association was performed using FCS in order to elucidate the reaction mechanism prior to enzymatic conversion of the substrate. The experimental data led to the assumption

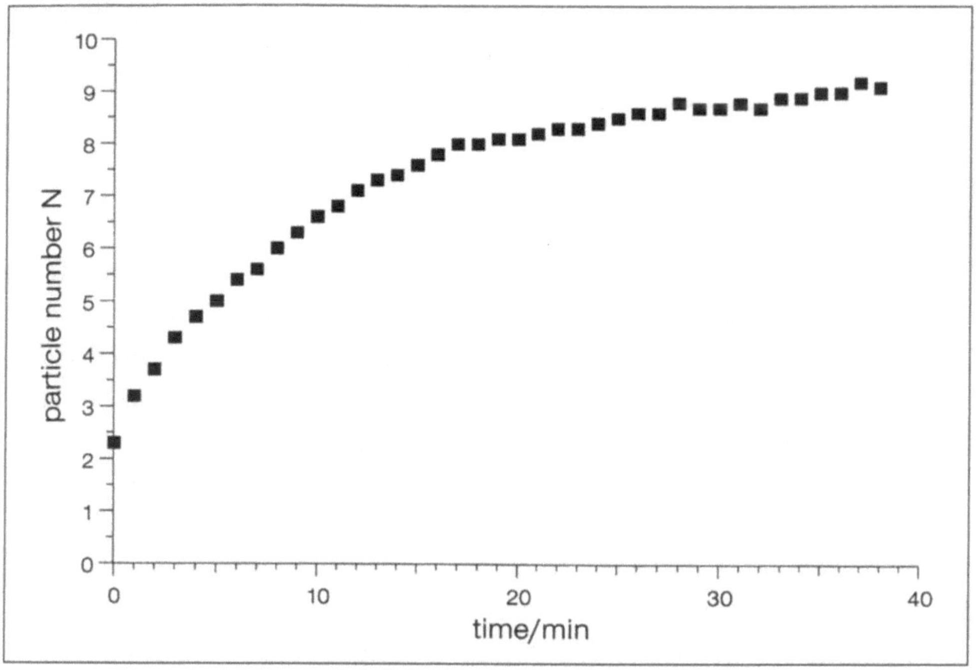

Figure 6 *Digestion of casein with chymotrypsin. Digestion with chymotrypsin resulted in smaller fragments of the substrate (casein), some of which carried a fluorophore. As the digestion reaction proceeded, the number of fluorescent particles in the detection volume, N, thus increased. The time-course of the digestion reaction was followed by time-resolved FCS. The graph shows the increase in particle number N as a function of time. Note that the conditions prevailing in this FCS experiment were different from those usually chosen when monitoring enzymatic reactions. The enzyme concentration was in excess of the substrate concentration so that the turnover rate was not constant for the period of time observed. Material: Resorufin-labelled casein was obtained from Merck. Chymotrypsin was purchased from SIGMA. Digestion of 2 nM resorufin-casein with 333 nM chymotrypsin was carried out in 0.1 M Tris buffer pH 8.0 containing 0.02 M calcium chloride and 0.1% (w/v) polyethyleneglycol 6000. Measurement: Eight-chamber cover glasses (Nunc); room temperature; FCS measurements were performed every 20 s, data acquisition time was 20 s. Equipment: ConfoCor with an helium neon laser (Uniphase, 1.5 mW, 543.5 nm), 40 x/1.2 W objective (Zeiss), 610W100 interference filter (Omega)*

that the kinetics of the reaction were best described by a fast pre-equilibrium state and a subsequent slower formation of the enzymatically active complex. Comparison of native and inactivated $(EA)_2$ revealed similar association kinetics for both forms. Hence, it could be concluded that inactivation of the protein did not affect the binding domain, but rather the catalytic site. The outcome of the FCS experiments was that, in order to optimize the effectiveness of $(EA)_2$, the conformation of the catalytic site of the enzyme must be stabilized rather than the binding domain.

Cells

FCS already has a long history of application in membrane research. Membrane dynamics, such as the lateral diffusion of lipids and the aggregation of ligand-receptor or antigen-antibody complexes, have been followed using FCS [34–46]. FCS measurements are also feasible in the cytoplasm of living mammalian cells [10]. When translational diffusion is measured within the cytoplasm of cells one has to take into account that the behaviour of molecules might differ as compared to that observed in "test tube" experiments. The mobility of molecules and particles is often reduced within the cytoplasm due to high protein concentration, collisions with intracellular obstacles, such as the cytoskeletal network, and potential binding to cellular macromolecules or cytoplasmatic components diffusing slower than the molecule of interest [47, 48].

With FCS, the apparent dynamic viscosity, η^c, of the cytoplasm of human embryonic kidney cells compared to medium, was observed to be four times greater for fluorescein and 14 nm lates beads (Tab. 1). The measurements were performed in different cells at various cytoplasmatic locations. The high standard deviation observed demonstrates the heterogeneity of these locations and supports the assumption that depending on the cellular location, different local effects diminish the mobility of the particles under observation to various degrees. Using FCS, translational diffusion studies using fluorescent beads in living mouse embryonic fibroblasts yielded a diffusion coefficient with a value just over half of that observed in aqueous solution [10]. The rate of diffusion slowed down further over time indicating that the beads might have bound to cellular components.

The confocal volume element can be positioned at any position desired, thus preventing signal interference from the extra-cellular space, neighbouring cells, or locations other than the cellular compartment of interest. However, fluorescence of cellular components has to be taken into account. The autofluorescence of a variety of cells was found to be due to endogenous flavoproteins [49], a factor which perturbs cellular FCS mea-

Table 1

	Aparent dynamic viscosity/10^{-3} $kgm^{-1}s^{-1}$	Relative dynamic viscosity
Fluorescein[1]	3.31 ± 1.95	3.7 ± 2.19
14 nm FluoSpheres[2]	3.82 ± 1.66	4.29 ± 1.87

[1] 15 single measurements.

[2] 31 single measurements.

Cytoplasmatic viscosity observed for fluorescein and 14 nm beads within HEK 293 cells. The apparent viscosity of the cytoplasm was calculated using the equation:

$$\eta^c = \eta^m \frac{\tau^c_{diff}}{\tau^m_{diff}}$$

with $\tau_{diff}{}^m$ = mean diffusion time of the fluorescent probe through the detection volume in medium.

τ^c_{diff} = mean diffusion time of the fluorescent probe through the detection volume in the cytoplasm.

η^m = dynamic viscosity in medium (8.9×10^{-4} $kgm^{-1}s^{-1}$ [Pas] for water at 25°C).

η^c = apparent dynamic viscosity in the cytoplasm.

Material: HEK 293 cells were cultured in RPMI 1640 medium containing 10% fetal bovine serum. They were seeded into eight-chamber cover glasses (NUNC), washed three times with PBS and incubated with 12 nM fluorescein diacetate in PBS just prior to measurement. For the beads experiment, 10^7 HEK 293 cells were harvested by trypsinization and resuspended in 800 µl PBS containingn 1.3×10^{14} FluoSpheres. Electroporation was performed with 0.2 kV, 960 µF with a 26 ms pulse. Ten minutes after electroporation cells were washed with 10 ml PBS and transferred to the wells at 5×10^4 cells per well in 20 mM HEPES pH 7.4 containing 140 mM sodium chloride, 5 mM magnesium chloride, 1.8 mM calcium chloride, 4.2 mM sodium bicarbonate and 5.5 mM glucose. Equipment: ConfoCor (Zeiss/EVOTEC) with 40x/1.2 W objective (Zeiss), an Ar^+ laser, 488 nm, with an excitation energy of 132 µW in aqueous solution and 7.7 µW in cells and a 550W50 interference filter from Zeiss (ZL488 505 DRLP/23) for measurement of fluorescein; a helium neon laser (Uniphase, 543.5 nm, measurement in solution with an excitation energy of 340 µW in solution and 320 nW in cells) combined with a 610W100 interference filter (Omega) for the beads experiment; A Gene Pulser combined with capacity extender from Bio-Rad was used for electroporation. Measurement: A ConfoCor (Zeiss/EVOTEC) was fitted for phase contrast illumination of the sample area to localize individual cells. Positioning of the confocal volume element in the cytoplasm was done by manually adjusting the x-y-scanning table and the z-position of the objective. Measurements were done at room temperature; data acquisition time for all experiments was 30 s.

surements, just as it interferes with other cellular fluorescence labelling procedures [50].

FCS is a non-invasive method allowing second messenger pathways or gene regulation to be followed in real-time by employing appropriate fluorescent analytes. Recent advances in the development of fluorescent probes for the investigation of signal transduction events will have a large impact on analytical cell biology. The number of probes sensitive for second messengers like calcium [51] or cyclic adenosine monophosphate [52] is still increasing. Green fluorescent protein, originally derived from the jellyfish *Aequoria victoria*, is a suitable analyte for FCS measurements [53]. It is a very attractive tool for monitoring gene expression [54] and protein trafficking [55] in living cells. It is evident that these developments will open up new possibilities for the use of FCS inside living cells.

Conclusion

In summary, FCS enables the observation of molecular interactions in their native environment, on the cell surface or within individual, living cells. It will be a very useful tool for the future study of biomolecular processes *in vitro* and in cellular systems.

Acknowledgement

We would like to thank G. Bauer and R. Turner for critically reading the manuscript and all our colleagues for providing information and engaging in valuable discussions.

References

1 Jeffcoate S (1996) The role of bioassays in the development, licensing and batch control of biotherapeutics. *TIBTech* 14: 121–124

2 Kessler C (1992) General aspects of non-radioactive labeling and detection. In: *Non-radioactive labeling and detection of biomolecules*. Springer-Verlag, Berlin Heidelberg, 21–24

3 Jameson DM, Sawyer WH (1995) Fluorescence anistoropy applied to biomolecular interactions. *Methods in Enzymology* 246: 283–300

4 Chekovich WJ, Bolger RE, Burke T (1995) Fluorescence polarization – a new tool for cell and molecular biology. *Nature* 375: 254–256

5 Selvin PR (1995) Fluorescence resonance energy transfer. *Methods in Enzymology* 246: 300–334

6 Cherry RJ (1979) Rotational and lateral diffusion of membrane proteins. *Biochim Biophys Acta* 559: 289–327

7 Bambot SB, Lakowicz JR, Rao G (1995) Potential applications of lifetime-based, phase-modulation fluorimetry in bioprocess and clinical monitoring. *TIBTech* 13: 106–115

8 Eigen M, Rigler R (1994) Sorting single molecules: Applications to diagnostics and evolutionary biotechnology. *Proc Natl Acad Sci USA* 91: 5740–5747

9 Rigler R (1995) Fluorescence correlations, single molecule detection and large number screening. Applications in biotechnology. *J of Biotechnology* 41: 177–186

10 Berland KM, So PT, Gratton E (1995) Two-photon fluorescence correlation spectroscopy: method and application to the intracellular environment. *Biophys J* 68: 694–701

11 Elson EL, Magde D (1974a) Fluorescence correlation spectroscopy. I. Conceptual basis and theory. *Biopolymers* 13: 1–27

12 Elson EL, Magde D (1974b) Fluorescence correlation spectroscopy. II. An experimental realization. *Biopolymers* 13: 29–61

13 Weissmann M, Schindler H, Feher G (1976) Determination of molecular weights by fluctuation spectroscopy: Application to DNA. *Proc Natl Acad Sci USA* 73: 2776–2770

14 Icenogle RD, Elson EL (1983a) Fluorescence correlation spectroscopy and photobleaching recovery of multiple binding reactions. I. Theory and FCS measurements. *Biopolymers* 22: 1919–1948

15 Icenogle RD, Elson EL (1983b) Fluorescence correlation spectroscopy and photobleaching recovery of multiple binding reactions. II. *Biopolymers* 22: 1949–1966

16 Kinjo M, Rigler R (1995) Ultrasensitive hybridization analysis using fluorescence correlation spectroscopy. *Nuc Acids Res* 10: 1795–1799

17 Schwille P, Oehlenschläger F, Walter NG (1996) Quantitative hybridization kinetics of DNA probes to RNA in solution followed by diffusional fluorescence correlation analysis. *Biochemistry* 35: 10182–10193

18 Saiki RK, Scharf S, Faloona F, Mullis KB, Horn GT, Erlich HA, Arnheim N (1985) Enzymatic amplification of β-globin genomic sequences and restriction site analysis for diagnosis of sickle cell anemia. *Science* 230: 1350–1354

19 Compton S (1991) Nucleic acid sequence-based amplification. *Nature* 350: 91–92

20 Walter NG, Schwille P, Eigen M (1996) Fluorescence correlation analysis of probe diffusion simplifies quantitative pathogen detection by PCR. *Proc Natl Acad Sci* 93: 12806–12810

21 Oehlenschläger F, Schwille P, Eigen M (1996) Detection of HIV-1 RNA by nucleic acid sequence-based amplification combined

with fluorescence correlation spectroscopy (FCS). *Proc Natl Acad Sci* 93: 12811–12816

22 Hill CS, Treisman R (1995) Transcriptional regulation by extracellular signals: Mechanisms and specificity. *Cell* 80: 199–211

23 Novak U, Paradiso L (1995) Identification of proteins in DNA-protein complexes after blotting of EMSA gels. *Biotechniques* 19: 54–55

24 Wahlström GM, Sjöberg M, Andersson M, Nordström K, Vennström B (1992) Binding characteristics of the thyroid hormone receptor homo- and heterodimers to consensus AGGTCA repeat motifs. *Mol Endo* 6: 1013–1022

25 Briggs MR, Kadonaga JT, Bell SP, Tjian R (1986) Purification and biochemical characterization of the promoter-specific transcription factor, Sp1. *Science* 234: 47–52

26 Rauer B, Neumann E, Widengren J, Rigler R (1996) Fluorescence correlation spectrometry of the interaction kinetics of tetramethylrhodamine α-bungarotoxin with *Torpedo californica* acetylcholine receptor. *Biophysical Chemistry* 58: 3–12

27 Cross M, Dexter TM (1991) Growth factors in development, transformation and tumorigenesis. *Cell* 64: 271–280

28 Bastiaens PIH, Pap EHW, Widengren J, Rigler R, Visser AJWG (1994) Fluorescence method to study lipid-protein association. *J Fluorescence* 4: 377–383

29 Rutloff H (ed) (1994) Industrielle Enzyme. Behr's Verlag, Hamburg

30 Dinarello CA, Margolis NH (1995) Cytokine-processing enzymes. Stopping the cuts. *Curr Biol* 5: 587–590

31 Himmelstein BP, Canete-Soler R, Bernhard EJ, Dilks DW, Muschel RJ (1994/95) Metalloproteinases in tumor progression: the contribution of MMP-9. *Invasion – Metastasis* 14: 246–258

32 Neuzil KM (1994) Pharmacologic therapy for human immunodeficiency virus infection: a review: *Am J Med Sci* 307: 368–373

33 Henderson DR, Friedman SB, Harris JD, Manning WB, Zoccoli MA (1986) CEDIA, a new homogeneous immunoassay system. *Clin Chem* 32/9: 1637–1641

34 Webb WW (1976) Applications of fluorescence correlation spectroscopy. *Quarterly Reviews of Biophysics* 9: 49–68

35 Koppel DE, Axelrod D, Schlessinger J, Elson EL, Webb WW (1976) Dynamics of fluorescence marker concentration as a probe of mobility. *Biophys J* 43: 345–354

36 Fahey PF, Koppel DE, Barak LS, Wolf DE, Elson EIL, Webb WW (1977) Lateral diffusion in planar lipid bilayers. *Science* 195: 305–306

37 Thompson NL, Axelrod D (1983) Immunoglobulin surface-binding kinetics studied by total internal reflection with fluorescence correlation spectroscopy. *Biophys J* 43: 103– 114

38 Petersen NO (1986) Scanning fluorescence correlation spectroscopy. I. Theory and simulation of aggregation measurements. *Biophys J* 49: 809–815

39 Petersen NO (1986) Scanning fluorescence correlation spectroscopy. II. Application to virus glycoprotein aggregation. *Biophys J* 49: 817–820

40 Palmer AG, Thompson NL (1989) High-order fluorescence fluctuation analysis of model protein clusters. *Proc Natl Acad Aci USA* 86: 6148–6152

41 Palmer III AG, Thompson NL (1989) Fluorescence correlation spectroscopy for detecting submicroscopic clusters of fluorescent molecules in membranes. *Chemistry and Physics of Lipids* 50: 253–270

42 Quian H, Elson EL (1990) Distribution of molecular aggregation by analysis of fluctuation moments. *Proc Natl Acad Sci USA* 87: 5479–5483

43 St-Pierre PR, Petersen NO (1990) Relative ligand binding to small or large aggregates

measured by scanning correlation spectroscopy. *Biophys J* 58: 503–511

44 St-Pierre PR, Petersen NO (1992) Average density and size of microclusters of epidermal growth factor receptors on A431 cells. *Biochemistry* 31: 2459–2463

45 Petersen NO, Hoddelius PL, Wiesmann PW, Seger O, Magnusson K-E (1993) Quantitation of membrane receptor distributions by image correlation spectroscopy: concept and application. *Biophys J* 65: 1135–1146

46 Koppel DE, Morgan F, Cowan AE, Carson JH (1994) Scanning concentration correlation spectroscopy using the confocal laser microscope. *Biophys J* 66: 502–507

47 Pin Kao H, Abney JR, Verkman AS (1993) Determinants of the translation mobility of a small solute in cell cytoplasm. *J Cell Biol* 120: 175–184

48 Luby-Phelps K (1994) Physical properties of cytoplasm. *Curr Opin Cell Biol* 6: 3–9

49 Benson RC, Meyer RA, Zaruba ME, McKhann GM (1979) Cellular autofluorescence – is it due to flavins? *J Histochem Cytochem* 27: 44–48

50 Aubin JE (1979) Autofluorescence of viable cultured mammalian cells. *J Histochem Cytochem* 27: 36–43

51 Mason WT (ed) (1993) Fluorescent and luminescent probes for biological activity. A practical guide to technology for quantitative real-time analysis. Academic Press, Harcourt Brace & Company, Pub., London

52 Adams SR, Harootunian AT, Buechler YJ, Taylor SS, Tsien RY (1991) Fluorescence ratio imaging of cyclic AMP in single cells. *Nature* 349: 694–697

53 Terry BR, Matthews EK, Haseloff J (1995) Molecular characterization of recombinant green fluorescent protein by fluorescence correlation microscopy. *Biochem Biophys Res Commun* 217: 21–27

54 Cheng L, Fu J, Tsukamoto A, Hawley RG (1996) Use of green fluorescent protein variants to minor gene transfer and expression in mammalian cells. *Nature Biotechnology* 14: 606–609

55 Ogawa H, Inouye S, Tsuj FI, Yasuda K, Umesono K (1995) Localization, trafficking, and temperature-dependence of the *Aequoria* green fluorescent protein in cultured vertebrate cells. *Proc Natl Acad Sci* 92: 11899–11903

56 Dignam JD, Lebovitz RM, Roeder RG (1983) Accurate transcription initiation by RNA polymerase II in a soluble extract from isolated mammalian nuclei. *Nucleic Acids Res* 11: 1475–1489

57 Hochuli E (1990) in "Genetic Engineering, Principle and Methods", JK Setlow (ed): Plenum Press, New York, Vol. 12: 87–98

58 Lichtsteiner S, Wuarin J, Schibler U (1987) the interplay of DNA-binding proteins on the promoter of the mouse albumin gene. *Cell* 47: 963–973

59 Carraway III KL, Koland JG, Cerione RA (1989) Visualization of epidermal growth factor (EGF) receptor aggregation in plasma membranes by fluorescence resonance energy transfer. *J Biol Chem* 264: 8699–8707

60 Kawamoto T, Sato JD, Le A, Polikoff J, Sato GH, Mendelsohn J (1983) Growth stimulation of A431 cells by epidermal growth factor: Identification of high-affinity receptors for epidermal growth factor by an anti-receptor monoclonal antibody. *Proc Natl Acad Sci* 80: 1337–1341

61 Magde D, Elson E, Webb WW (1972) Thermodynamic fluctuations in a reacting system. Measurement by fluorescence correlation spectroscopy. *Physical Review Letters* 29: 705–708

62 Meyer-Almes F-J, Wyzgol K, Powel MJ (1998) Mechanism of the α-complementation reaction of *E. coli* β-galactosidase deduced from fluorescence correlation spectroscopic measurements. *Biophysical Chemistry* 75: 151–160

15 Scanning force microscopy: A microstructured device for imaging, probing, and manipulation of biomolecules at the nanometer scale

Wolfgang Fritzsche

15.1 Scanning force microscope

Principle

Scanning force microscopy (SFM, also known as atomic force microscopy, AFM) probes the surface topography by raster-scanning a sharp tip (mounted at the end of a flexible cantilever) over the sample (Fig. 1). A piezoelectric scanner moves the sample relative to the tip and allows movements with Angstrom precision in x, y, and z direction. The deflection of the cantilever

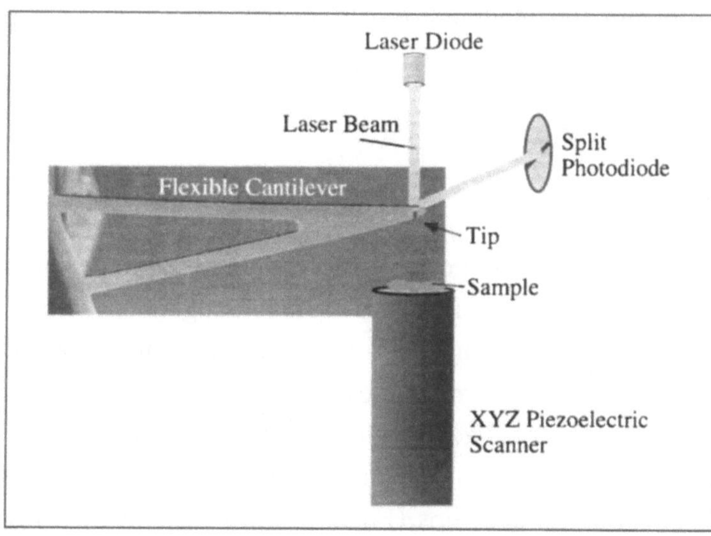

Figure 1 *Scheme of a scanning force microscope. The sample is usually mounted on top of a piezoelectric tube scanner, and is scanned below a sharp tip. The tip-sample contact results in a deflection of the flexible cantilever, which is detected by a split photodiode. This signal can be used as height information, or is the input for a height feedback of the sample*

due to changing surface topography is usually monitored using a light-pointer: a laser beam is reflected from the back side of the cantilever top toward a divided photo diode. The signal of the photo diode can be used by a feedback mechanism for keeping the deflection of the cantilever constant by adjusting the z-height of the scanner. This mode is called "constant force" (also "isoforce" or "constant deflection"), thereby the height movement of the cantilever represents the sample topography and is used for visualization. Turning the feedback off and monitoring the changing deflection using the diode signal results in an image in the "constant height" mode, which can improve the detection of edge features.

Two modes differing in the duration of tip-sample contact are in general use in biological applications. The first bases on a steady contact (therefore denoted as "contact mode") which results in significant shear forces applied to the specimen. Soft molecules or samples with low substrate adsorption (as in the case of molecules adsorbed in buffer) are often unstable using this mode of operation. The "tapping mode" minimizes the shear forces by decreasing the tip-sample interaction, which is achieved by oscillation of the tip normal to the surface. In this case the amplitude of the oscillation is monitored and the feedback mechanism bases on the damping due to surface contact (as monitored by the photo diode).

Tips

Due to the topographic contrast the detection limit is mainly influenced by the ratio of specimen height and surface roughness. The SFM has in principle atomic resolution, which was demonstrated on crystalline surfaces (Giessibl, 1995). For single biomolecules with their complex structures the resolution is limited by the tip size, which is a crucial parameter in SFM. The first commercially available cantilever with integrated tips is based on silicon technology. The pyramidal tips (Fig. 2a) were formed by using an etch pit on the (100) surface of Si as a mold (Albrechts et al., 1990). The image of the tip reveals the bluntness at the very top, which elucidates a fundamental problem of scanning force microscopy: The visualization process involves a complex interaction between tip and sample, and the resulting image is influenced by the tip geometry. The extent of this influence is minor for a

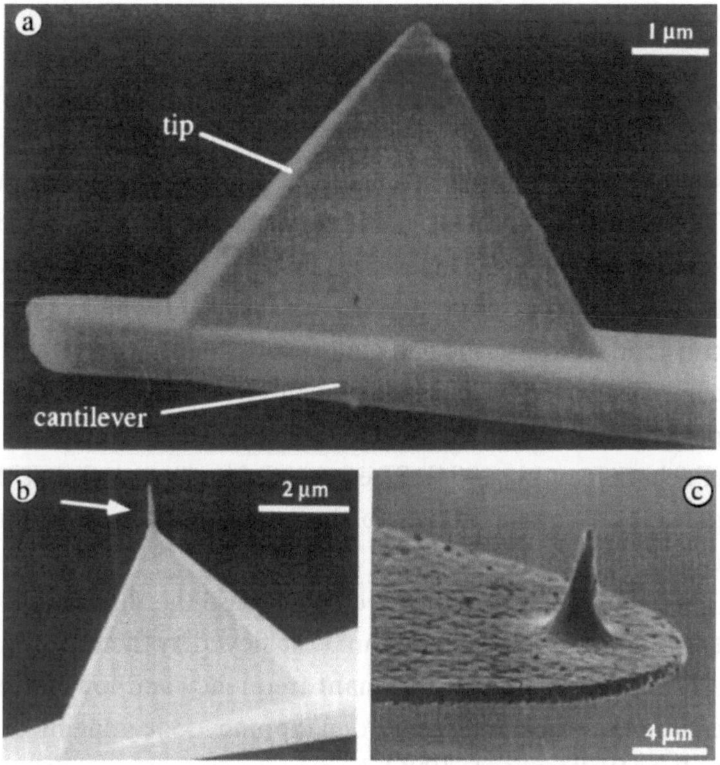

Figure 2 *Scanning electron micrographs of tips used in scanning force microscopes (after Fritzsche et al., 1996a). a) Silicon nitride cantilever with integrated pyramidal tip. b) Enhancement of a pyramidal tip by electron beam deposition (arrow). c) Etched silicon tip*

high ratio of sample size and tip diameter. However, it is significant if the tip is in the size range of the specimen. In the case of biomolecules in the lower nanometer range there is a definite need for sharp tips, which resulted in the introduction of electron beam deposited tips for SFM (Keller and Chou 1992). Based on the phenomenon of electron beam assisted deposition in the scanning electron microscope (SEM), conventional SFM tips were enhanced by the growth of carbonaceous needles under the electron beam of a SEM (Fig. 2b). Today, a variety of tips based on the described procedures or using other sharpening methods (for a review see Tortonese (1997); Fig. 2c) are available.

Cantilevers

A basic feature of the SFM is the flexible cantilever. Typical dimensions are 3 × 20 × 100 µm (thickness, width, height), which yields for silicon a force constant of 20 N/m and a resonance frequency of 400 kHz (Wolter et al., 1991). The spring constant of cantilevers should be low in order to achieve high sensitivity, whereas a high resonance frequency minimizes excitation of vibrations by ambient noise and allows sufficiently high scan rates. A low spring constant, which is especially interesting for imaging of soft biological samples, can be achieved by using softer cantilever materials. A cantilever made from polymeric material has been developed and tested on biological samples (Pechmann et al., 1994). The results demonstrated that thin film technology provides a new approach to the batch fabrication of highly sensitive force transducers for SFM.

A high scan rate is needed to increase the throughput or for the observations of fast processes, and can be achieved by the use of short cantilevers. The advantages of the high resonant frequency and low noise of a short (30 µm) cantilever were demonstrated in tapping mode imaging of a protein sample in buffer (Walters et al., 1996). Low noise images were collected at a scan rate of about 65 lines/s (0.5 frames/s).

15.2 Imaging of biomolecules

Detection of single molecules

The high lateral resolution of the SFM provides the base for detection of single biomolecules on surfaces. DNA studies started early in the history of SFM, due to the great importance of DNA detection and structural characterization in biology and medicine. Adaptation and refinement of sample preparation methods known from electron microscopy resulted in high resolution images of the three-dimensional structure of DNA adsorbed onto substrates in air and in physiological conditions (Bustamante et al., 1993; Hansma et al., 1996; Mou et al., 1995; Schaper and Jovin 1996; Vesenka et al., 1992). The unique potential of SFM to provide Angstrom resolution in physiological conditions opens the way for the high resolution observation of DNA and other biomolecules in a native environment. First demonstrations were the visualization of the dynamics of DNA adsorption (Hansma et al., 1996) and transcription (by monitoring single RNA polymerase molecules in real time) (Kasas et al., 1997).

Imaging of defined DNA sequences

A variety of approaches has been developed to identify the sequence of imaged DNA. One possibility is the use of sequence-dependent enzymes, e.g., restriction enzymes. The EcoRI restriction sites of linearized DNA plasmids were mapped using the SFM (Allison et al., 1996). Therefore, EcoRI enzyme mutants (which only bind but do not cut) were incubated with the DNA, then the DNA-protein complexes were adsorbed onto mica substrates, and visualized by SFM. Other approaches with sequence-specific binding involved antibodies against z-DNA (Pietrasanta et al., 1994) and triple helix formation (Pfannschmidt et al., 1996). The same detection technique, namely the topographic contrast of the biomolecular complexes, was also applied for the visualization of *in situ* hybridization of metaphase chromosomes (Putman et al., 1993; Rasch et al., 1993) or whole cells (Kalle et al., 1996). Therefore, complementary DNA probes were labeled with an enzyme (peroxidase). After hybridization an enzymatic reaction results in a build-up of amorphous material (3,3′-diaminobenzidine) on the binding site, which induces the topographic signal. A parallel fluorescent and SFM detection scheme applied fluorescently labeled DNA probes, which were hybridized to spread chromatin samples (Fritzsche et al., 1996b).

The visualization of defined DNA sections is furthermore possible by imaging genes in their active mode (transcribing or active genes), which yields the well-defined triangular Christmas-tree pattern of a DNA axis complexed by polymerases and their "tails" of transcribed RNA exhibiting a length gradient (Miller and Beatty, 1969). Especially for repeated genes a typical pattern occurs. This is the case for ribosomal genes of newt oocyte chromatin, which were visualized by SFM (Fritzsche et al., 1998). Ribosomal genes could be identified based on their specific ultrastructure (i.e., gene length of ~2.4 µm, polarity, repeating unit of ~3.25 µm).

Image processing

Studies of biomolecular assemblies and their dynamics are usually based on a quantification of the biomolecular ultrastructure (e.g., height, conformation, volume). Image processing is a helpful tool in this field, providing the means of extracting quantitative data from SFM images. Image processing of single molecules involves the isolation of the molecules of interest from a larger image. Due to the topographic character of SFM images, a separation of the molecules from a usually flat surface is possible based on the height (gray) value (for a discussion of different approaches see Fritzsche, 1998). A suitable procedure for a shape determination from isolated three-dimensional structures was developed and applied to ~100 nucleosomes (Fritzsche and Henderson, 1996a). Therefore, virtual cross-sections parallel to the substrate level were conducted using defined threshold values. A sectioning nearby the substrate level includes nearly the complete molecules but is strongly influenced by the background noise (due to salt residues, etc.), resulting in a less defined geometry compared to sections at higher heights (Fritzsche and Henderson, 1996b). For comparable and reproducible measurements a cross-section at half maximum height was chosen. The geometry of the sectioning planes was probed using a best fit ellipse and their axis ratio. The axis ratio distribution peaked at ~1.30. This ratio reproduces the literature value and demonstrates therefore the reliability of SFM-based image processing methods for conformational analysis of biomolecules. Other objects of similar approaches were the orientation of nucleosomes along the nucleosomal chain (Fritzsche and Henderson, 1997a) and the ribosome ultrastructure (Fritzsche and Henderson, 1998).

A special field is the use of the unique topographical information of SFM images for volume determinations. The structures of biomolecules isolated from the background (as explained before) contain already a volume information due to the topographic character of the SFM data set. Subtracting the substrate (background) level from the volume yields the apparent volume of the visualized molecule, which is exaggerated by the influence of the tip-sample convolution. This problem has to be addressed in SFM-based volume determinations. A possible solution is the use of relative measurements for specimens of similar kind (e.g., with the same height/width ratio), where the influence of the tip shape is comparable. One example for such samples are metaphase chromosomes. SFM-based volume determination of air-dried chromosomes yielded less than 1/4 of the values known from electron microscopy (EM) of embedded and serial sectioned samples (Fritzsche and Henderson, 1996c). After rehydration a swelling of the chromosomes was observed, and the volumes determined from SFM-images were comparable to the results from EM.

15.3 Determination of elastic properties

Principle

A basic part of a SFM is the flexible cantilever, which is used for detection of tip-sample contact by a changing deflection. For ideally rigid samples the deflection is solely based on topographic changes of the surface. This is not the case if the sample exhibits elastic behavior. Then both the sample and the cantilever can be deflected, or only the sample is deformed. The former case is valid if the elastic behavior (e.g., expressed as a force constant) of sample and cantilever are of the same magnitude, the latter is achieved when the cantilever is much more rigid compared to the sample. The elastic behaviour of the tip and/or the sample can be deduced from the force-distance curve. Such a curve gives the cantilever deflection in dependence from the tip-sample distance for approach/retraction cycles staying at one point of the sample surface (Fig. 3). The ideal behavior for a rigid sample is given in Fig. 3a: The approach of the tip toward the surface yields no deflection until the point of contact (non-contact line). From this point on further movement results in a proportional deflection (contact line). Under ideal conditions, the retraction curve runs along the same course back. Real measurements of rigid samples differ usually at the region around the point of contact (distance = 0), where negative deflections can be observed (empty arrowhead in Fig. 3b). In the case of the approach curve, this “snap into contact” results mostly from long-range attractive interactions. The negative regions of the retraction curve are induced by tip-sample adhesion, and can be used for the characterization of adhesive interactions (filled arrowhead). This effect is significantly reduced when scanning in liquid. Elastic behavior of samples causes a significant

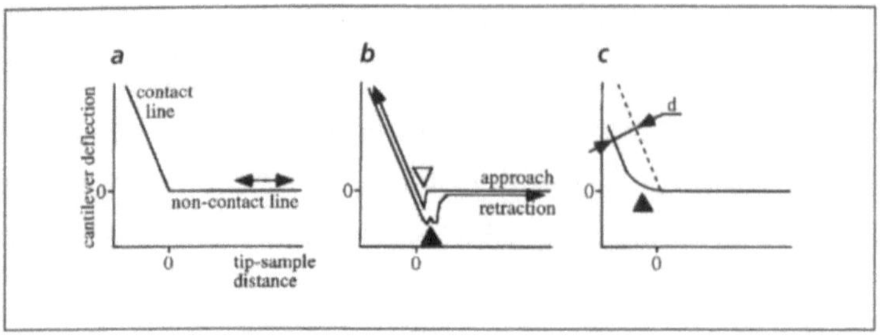

Figure 3 *Force-distance curves for characterization of tip-sample interactions. a) A curve for ideal behavior. b) Real curves measured for rigid samples in air exhibit typical features: The approach curve (top curve) snaps into contact (empty arrowhead), and the retracting tip sticks on the surface due to adhesive interactions (filled arrowhead). c) Samples with elastic behavior exhibit a typical curvature after contact (arrowhead), which approaches a straight line parallel to the ideal contact line (dotted). The deviation from the ideal line gives the indentation of the sample (arrows)*

change of the curve after contact: the straight line (from rigid samples, e.g., Fig. 3 a,b) changes to a more complex pattern (c), before it reaches the straight contact line with further approach. This phenomenon results from the indentation of the elastic sample surface. The resistance increases with the indentation, which is mainly due to growing contact area. At one point the resistance allows no further sample intention and the further curve approaches the straight non-contact line typical for rigid surfaces.

Mapping of the viscoelastic behavior

Force-distance curves were used for the characterization of the elastic behavior of biological samples, as reported for cells (Hoh and Schoenenberger, 1994; Radmacher et al., 1996), chromosomes (Ikai et al., 1997), or bones (Tao et al., 1992). Mapping techniques were introduced to extend the investigated area from the point probed by the force curve, for example, to cover whole molecules in a measurement. In the following, an approach is explained to map viscoelastic properties of biomolecules. The sample is repeatedly scanned with different deflections of the cantilever each time, which results in changed imaging forces. A description of the viscoelastic properties of bio-

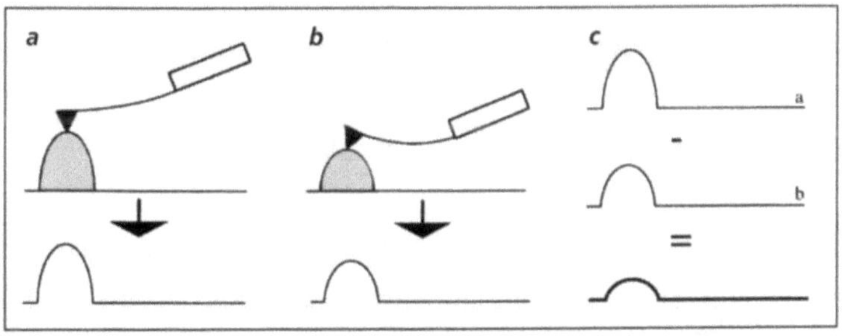

Figure 4 *Mapping viscoelastic properties by a differential approach (Fritzsche and Henderson 1997b). a) The sample is scanned with a defined loading force (top), the resulting image is given below. b) A repeated scan with a changed (higher) loading force induces a different deformation of the sample, therefore the height in the image changes. c) A map of the viscoelastic response of the surface can be obtained by subtracting the high force image (b) from the image obtained with lower loading force (a)*

molecules is based on the response of the molecule structure (i.e., height) to variations in the imaging force, yielded by comparing images of the same structure under increasing imaging forces. The resulting differential image maps the response of the local surface structure to changes in the imaging force. In principle, this is a simplified version of force mapping, which includes the recording of a force curve on every point of the scanned surface (Radmacher et al., 1994), and which was recently implemented in commercial SFMs (force volume). Instead of recording a force curve for each sample point, only one point of the force curve is recorded for each pixel in every original image, and the difference of both values is visualized in the resulting differential image (Fig. 4). In terms of force curves, the recorded points lay along the contact part of the approaching curve, and their location is determined by the chosen imaging force (which corresponds to the force in the force-distance plot). Regions in the difference image with higher values exhibit higher viscoelasticity, and regions with values of ca. 0 show no response. The described mapping technique was used to investigate metaphase chromosome spreads that were air-dried and then rehydrated in aqueous buffer (Fritzsche and Henderson, 1997b). A strong viscoelastic response for the rehydrated structures was revealed by mapping the viscoelasticity. These results were supported by results from conventional force-distance scans.

15.4 Probing of intermolecular forces

Characterization of intermolecular bonds

The force-distance curves monitor interactions between tip and sample by measuring the resulting cantilever deflection. This principle can be used for the investigation of intermolecular interactions. Studied systems include biotin-(strept)avidin (Florin et al., 1994), complementary DNA-strands (Lee et al., 1994), and cell adhesion proteoglycan (Dammer et al., 1995). Therefore, the complementary molecules were bound to tip and substrate surface, res-

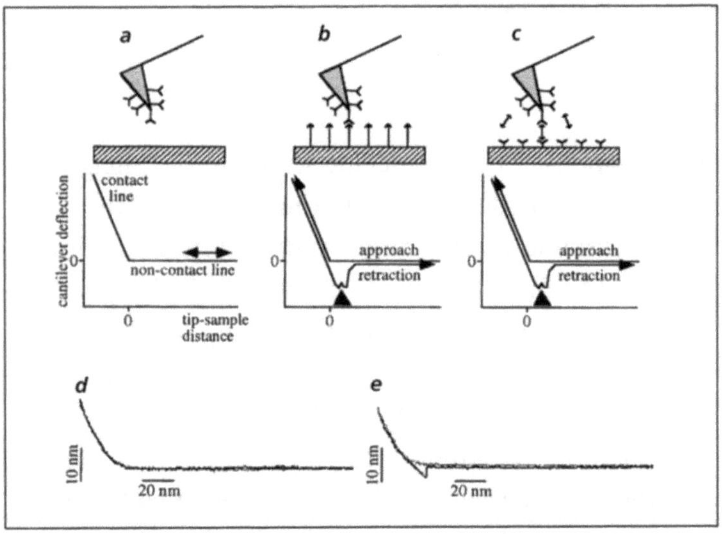

Figure 5 *Detection of intermolecular interactions by force distance curves. a) In the absence of intermolecular binding events (e.g., no antigen on the surface) the corresponding curve exhibits no significant difference from the ideal behavior (bottom, cf. Fig. 3a). b) In the case of binding events due to antigens bound to the surface the curve changes significantly, an adhesion peak becomes apparent (arrowhead). c) A similar curve is the result for a sandwich assay, where the antigen bridges antibodies adsorbed both on tip and surface. d,e) Experimental curves for an antibody-antigen pair: The tip is coated with rabbit IgG, and approached toward a surface of goat anti-rabbit IgG (C. Mosher & E. Henderson, unpublished results). d) Control experiment: The antibodies on the tip were blocked with free rabbit IgG, the curve shows no interactions. e) A clear adhesion peak can be observed in case of the antibody-antigen interaction*

pectively; and the tip is brought into contact with the substrate. Now the tip is retracted and the peak of negative deflection in the retraction curve is studied (Fig. 5). This peak is explained by bonds, which break with increasing retraction (resulting in increasing forces applied to the bonds). Single events are usually obscured due to the great number of simultaneously breaking bonds, only the very last process could result in a single resolvable step before reaching the straight non-contact line. By repeating the experiment many times a distribution of the value for this last step is obtained, which should reveal maximums at the multiples of the value for a single binding event. Such a quantitative study requires a careful analysis of possible artifacts induced by the SFM instrument, and is usually done using additional equipment.

Screening of biomolecules

The qualitative use of the information from the adhesion peak opens a broad application in respect to screening applications. One general idea is to prepare an array of spots containing different types of molecules on the substrate, and to adhere a probing molecule to the tip. Now it should be possible to scan the substrate and look for the tip-sample adhesion in every scanned spot of the array. This signal would reveal, for example, antigens in the case of an antibody, or complementary DNA-sequences in the case of a DNA probe. By using spot sizes in the lower micrometer range an array of more than 1000 spots could be probed in one SFM scan in a couple of minutes.

A similar mechanism could be applied for flow detection of biomolecules using a "sandwich" approach (Fig. 5). In general, biomolecules can only be imaged by SFM if they are somehow adsorbed to the substrate. But for flow detection, molecules complementary to the molecule of interest are bound to substrate and tip (which requires at least two binding sites). If the molecule of interest exists in the solution, it will be sandwiched between the tip and the substrate. This event is detected as adhesion peak. Control solutions free of the molecules of interest show no adhesion peak (C. Mosher and E. Henderson, unpublished results).

Another detection scheme bases on the increase of mass of an oscillator. An oscillating tip is coated with molecules complementary to the molecule of interest, and any binding event is detected by the difference in oscillation amplitude or phase (Baselt et al., 1996).

15.5 Manipulation at the molecular scale

Nanoindentation

If the forces applied by the tip exceed a defined value, an irreversible change of the sample takes place. This effect is used for indentation measurement by SFM, where a hard tip (usually diamond) is pressed into the substrate (for an introduction see (VanLandingham et al., 1997)). After retraction the induced hole is imaged, and the size of the hole is the base for hardness characterization of the investigated material. Such measurements are of special interest in material sciences, but less important in biology due to the complex (and therefore inhomogenous) composition of biological specimens.

Nanodissection of biomolecules

For biological SFM applications, irreversible changes of the sample are often observed but seldom intended. These changes are mainly due to shear forces, which sweep away specimens with low substrate adsorption. The introduction of scanning modes with reduced tip-sample interaction time minimized the shear forces and opened thereby the way for the observation of samples with weak adsorption. Another type of sample damage is the dissection of strongly bound specimen. Thereby the tip is usually scanned multiple times along a scan line using high loading force. This effect was used to cut single DNA molecules (Henderson, 1992), another application was the dissection of gap junctions (Hoh et al., 1991). The defined manipulation of metaphase chromosomes for extracting genetic material using SFM is discussed in the context of higher throughput DNA sequencing (Thalhammer et al., 1997; E. Henderson, unpublished results). Therefore, specific regions of chromosomes are chosen, for example, by *in situ* hybridization. This region is then scratched by the SFM to remove the chromosomal material from the substrate. Both imaging and manipulating of the chromosome can be accomplished by SFM using different loading forces: Imaging is done with minimized forces, whereas higher forces result in removal of material. In Figure 6a a metaphase chromosome is imaged, and a cross-section of the lower left chromatid along the dotted line is given in the inset. Subsequently, the scanning was repeated, and the loading force was maximized at two defined regions (given by the arrows in Fig. 6b). Then, the chromosome was imaged again with low forces. Comparing the cross-section before (a) and after (b) manipulation a removal of material became apparent. An accumulation of the removed genetic material at the tip can be accomplished by variation of parameters as scan speed, tip shape, tip material, environment (air, buffer, pH), and loading force. In a final step, the accumulated material can be amplified by a PCR prior to sequencing.

Revealing differences in material strength

The above mentioned examples describe the complete removal of the sample material at the scanned location. However, the SFM can be also used for a selective removal, which is material-strength dependent. An example is given

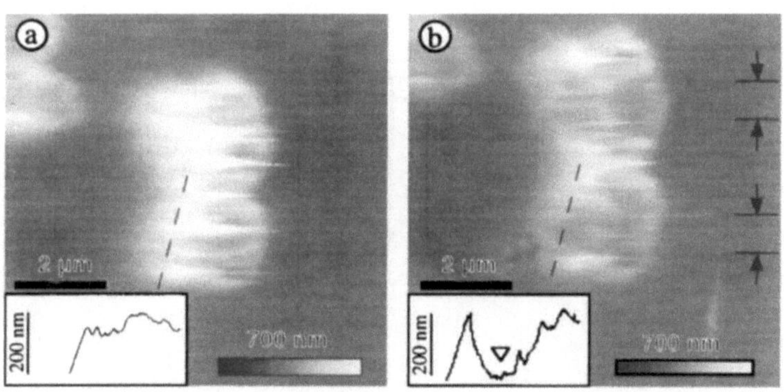

Figure 6 *SFM-extraction of genetic material from metaphase chromosomes (Fritzsche, 1994). The imaging/manipulation was conducted in 5 ×phosphate buffered saline. a) SFM image of a metaphase chromosome using minimal loading forces. The inset shows a cross-section along the lower left chromatid along the dotted line. b) The chromosome was scanned again, and the loading force was maximized at two defined regions (arrows). Another scan with minimized forces resulted in the shown image. The cross-section (inset) demonstrates the successful removal of chromosomal material at the defined region*

in Figure 7, which demonstrates the manipulation of a cell preparation. Chicken erythrocytes were hypotonic lysed to extract the chromatin (Fritzsche et al., 1995). Therefore, the cells were incubated in low salt/high pH buffer, prior to centrifugation through a sucrose/formaldehyde solution onto glass substrates. Besides the chromatin empty cell membranes ("ghost cells") and membrane patches were observed. In Figure 7a such a membrane is shown. The soft appearance of the sample points to a coverage by a layer which smoothes the edges. To test for this hypothesis, a squared region was repeatedly scanned using increased forces applied by the tip. This resulted in an image full of disturbed scan lines, suggesting movement of material. After machining, an overview comparable to the picture given in Figure 7a was imaged using lower forces (Fig. 7b). The manipulated region is clearly visible as a square hole in the center of the image, which points to a removal of material in this region. The elevated edges at the right and the left of the hole suggest the new location of the moved material (Fig. 7c). The scanning tip scans line for line (fast scan direction), thereby pushing the material to the side of the scanned region. The lower and upper border of the region is different (Fig. 7d): only the top part is limited by a wall. This is explained by the slow scan: The imaging process started at the bottom of the imaged region,

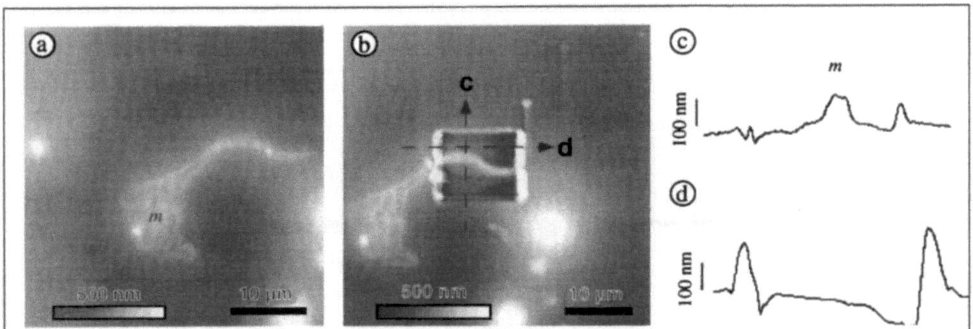

Figure 7 *Contrast-enhancement by SFM-based milling (Fritzsche, 1994). a) SFM image of a cell membrane preparation. A membrane patch is labeled with m. b) Applying high loading force by the scanning tip in a squared region (ca. 12 µm in the center region) resulted in an irreversible surface modification, which is apparent as a squared hole. c) The cross-section in slow scan direction reveals that the membrane patch inside the whole was hardly influenced by the milling, in contrast to the surrounding material which was removed. d) A cross-section along the fast scanning direction (horizontal) shows that the removed material is deposited at the turning points*

and material was moved towards this slow scan direction to the top. An interesting result from this experiment is the enhanced contrast of the membrane patch in the machined hole (*m* in Fig. 7c). In the case of the topographic contrast of the SFM, an enhancement is due to an increased height difference between sample and background, which is illustrated by the cross-section. This can be explained by different resistance of the membrane patch compared to the surrounding material, which results in different rates of removal by the scanning tip. The height of the patch in the hole (relative to the unaffected substrate level) is comparable to the height before milling, pointing to a high stability of the membranous material.

15.6 Conclusions

While much progress has been made in the application of SFM to the study of single biomolecules, many exciting discoveries and applications remain ahead. The nanomanipulation capabilities of the SFM will be exploited for single molecule gene mapping with resolution in the 10 to 100 bp range, and with extremely high throughput. Advances in SFM instrumentation and integration of methods related to SFM will expand the possibilities significantly. A promising area is opened by the development of optical proximal probe methods usually referred to as Near Field Optical Microscopy (NSOM, also SNOM) for visualization of optical signals (transmission, fluorescence, reflectance) at sub-diffraction limited distances (Van Hulst et al., 1993). This method connects the efficient fluorescence labeling techniques from molecular biology with the high resolution of scanning probe techniques for identification of biomolecules. Finally, direct detection of forces between biologically or chemically functionalized SFM tips and specimen surfaces is just beginning to be explored and may open the door to high throughput screening applications.

References

Albrechts TR, Akamine S, Carver TE, Quate CF (1990) Microfabrication of cantilever styli for the atomic force microscope. *J Vac Sci Tech A* 8: 3386–3396

Allison DP, Kerper PS, Doktycz MJ, Spain JA, Modrich P, Larimer FW, Thundat T, Warmack RJ (1996) Direct atomic force microscope imaging of EcoRI endonuclease site specifically bound to plasmid DNA molecules. *Proc Natl Acad Sci USA* 93: 8826–8829

Baselt DR, Lee GU, Colton RJ (1996) Biosensor based on force microscope technology. *J Vac Sci Tech B* 14: 789

Bustamante C, Keller D, Yang GL (1993) Scanning force microscopy of nucleic acids and nucleoprotein assemblies. *Current Opinion in Structural Biol* 3: 363–372

Dammer U, Popescu O, Wagner P, Anselmetti D, Güntherodt H-J, Misevic GN (1995) Binding strength between cell adhesion proteoglycans measured by atomic force microscopy. *Science* 267: 1173–1175

Florin E-L, Moy VT, Gaub HE (1994) Adhesion forces between individual ligand-receptor pairs. *Science* 264: 415–417

Fritzsche W (1994) Scanning force microscopy of chromatin (in German) (PhD Thesis). University Göttingen

Fritzsche W (1998) Biomolecular ultrastructure revealed by image processing of scanning force microscopy data. *Scan Microsc; in press*

Fritzsche W, Henderson E (1996a) Scanning force microscopy revealed ellipsoid shape of chicken erythrocyte nucleosomes. *Biophys J* 71: 2222–2226

Fritzsche W, Henderson E (1996b) Ultrastructural characterization of chicken erythrocyte nucleosomes by scanning force microscopy. *Scanning* 18: 138–139

Fritzsche W, Henderson E (1996c) Volume determination of human metaphase chromosomes by scanning force microscopy. *Scanning Microscopy* 10: 103–110

Fritzsche W, Henderson E (1997a) Chicken erythrocyte nucleosomes are oriented along the linker DNA – a scanning force microscopy study. *Scanning* 19

Fritzsche W, Henderson E (1997b) Mapping elasticity of rehydrated metaphase chromosomes by scanning force microscopy. *Ultramicroscopy* 69: 191–200

Fritzsche W, Henderson E (1998) Ribosome substructure investigated by scanning force microscopy and image processing. *J Microsc* 189: 50–56

Fritzsche W, Schaper A, Jovin TM (1995) Scanning force microscopy of chromatin fibers in air and in liquid. *Scanning* 17: 148–155

Fritzsche W, Schaper A, Jovin TM (1996a) Scanning force microscopy in structural biology: Probing DNA and chromatin structure. *Microscopy and Analysis* May: 5–7

Fritzsche W, Spring H, Jovin TM (1998) Ultrastructural characterization of transcriptionally active chromatin by scanning force microscopy. *Probe Microscopy; in press*

Fritzsche W, Takacs L, Vereb G, Schlammadinger J, Jovin TM (1996b) Combination of fluorescence *in situ* hybridization and scanning force microscopy for the ultrastructural characterization of defined chromatin regions. *Journal of Vacuum Science & Technology B* 14: 1399–1404

Giessibl F (1995) Atomic resolution of the silicon (111)-(7×7) surface by atomic force microscopy. *Science* 267: 68–71

Hansma H, Revenko I, Kim K, Laney DE (1996) Atomic force microscopy of long and short double-stranded, single-stranded and triple stranded nucleic acids. *Nucleic Acids Research* 24: 713–720

Henderson E (1992) Imaging and nanodissection of individual supercoiled plasmids by atomic force microscopy. *Nucleic Acids Research* 20: 445–447

Hoh JH, Lal R, John SA, Revel JP, Arnsdorf MF (1991) Atomic force microscopy and dissection of gap-junctions. 253: 1405–1408

Hoh JH, Schoenenberger C-A (1994) Surface morphology and mechanical properties of MDCK monolayers by atomic force microscopy. *J Cell Sci* 107: 1105–1114

Ikai A, Mitsui K, Xu XM (1997) Mechanical measurements of a single protein molecule and human chromosomes by atomic force microscopy. Materials science & engineering. *C Biomimetic* 4: 233–237

Kalle WHJ, Macville MVE, van de Corput MPC, de Grooth BG, Tanke HJ, Raap AK (1996) Imaging of RNA *in situ* hybridization by atomic force microscopy. *J Microsc* 182:192–199

Kasas S, Thomson NH, Smith BL, Hansma HG, Zhu X, Guthold M, Bustamante C, Kool ET, Kashlev M, Hansma PK (1997) *Escherichia coli* RNA polymerase activity observed using atomic force microscopy. *Biochemistry* 36: 461–468

Keller DJ, Chou CC (1992) Imaging steep, high structures by scanning force microscopy with electron-beam deposited tips. *Surface Sciences* 268: 333–339

Lee GU, Chrisey LA, Colton RJ (1994) Direct measurement of the forces between complementary strands of DNA. *Science* 266: 771–773

Miller OLJ, Beatty BR (1969) Visualization of Nucleolar Genes. *Science* 164: 955–957

Mou J, Czajkowsky DM, Zhang Y, Shao Z (19959 High-resolution atomic-force microscopy of DNA: the pitch of the double helix. *FEBS Letters* 371: 279–282

Pechmann R, Köhler JM, Fritzsche W, Schaper A, Jovin TM (1994) The novolever: A new cantilever for scanning force microscopy mcrofabricated from polymeric materials. *Review of scientific instruments* 65: 3702–3706

Pfannschmidt C, Schaper A, Heim G, Jovin TM, Langowski J (1996) Sequenc-specific labeling of superhelical DNA by triple helix formation and psoralen crosslinking. *Nucleid Acids Research* 24: 1702–1709

Pietrasanta LI, Schaper A, Jovin TM (1994) Probing specific molecular conformations with the scanning force microscope. Complexes of plasmid DNA and anti-Z antibodies. *Nucleid Acids Research* 22: 3288–3292

Putman CAJ, De Grooth BG, Wiegant J, Raap AK, Van der Werf KO, Van Hulst NF, Greve J (1993) Detection of *in situ* hybridization to human chromosomes with the atomic force microscope. *Cytometry* 14: 356–361

Radmacher M, Cleveland JP, Fritz M, Hansma HG, Hansma PK (1994) Mapping interaction forces with the atomic force microscope. *Biophys J* 66: 2159–2165

Radmacher M, Fritz M, Kacher CM, Cleveland JP, Hansma PK (1996) Measuring the viscoelastic properties of human platelets with the atomic force microscope. *Biophys J* 70: 556–567

Rasch P, Wiedemann U, Wienberg J, Heckl WM (1993) Analysis of banded human-chromosomes and *in situ* hybridization patterns by scanning force microscopy. *Proc Natl Acad Sci USA* 90: 2509–2511

Schaper A, Jovin TM (1996) Striving for atomic resolution in biomolecular topography: the scanning force microscope (SFM). *BioEssays* 18: 925–935

Tao NJ, Lindsay SM, Lees S (1992) Measuring the microelastic properties of biological material. *Biophys J* 63: 1165–1169

Thalhammer S, Stark RW, Heckl WM (1997) The atomic force microscope as a new microdissecting tool for the generation of genetic probes. *J Struct Biol* 119: 232

Tortonese M (1997) Cantilevers and tips for atomic force microscopy. *IEEE Engineering in Medicine and Biology* March/April: 28–33

Van Hulst NF, Moers MHP, Bölger B (1993) Near-field optical microscopy in transmission and reflection modes in combination with force microscopy. *J Microsc* 171: 95–105

VanLandingham MR, McKnight SH, Palmese GR, Elings JR, Huang X, Bogetti TA, Eduljee RF, JW Gillespie J (1997) Nanoscale indentation of polymer systems using the atomic force microscope. *Journal of Adhesion* 64: 31–59

Vesenka J, Guthold M, Tang CL, Keller D, Delaine E, Bustamante C (1992) Substrate preparation for reliable imaging of DNA molecules with the scanning force microscope. *Ultramicroscopy* 42: 1243–1249

Walters DA, Cleveland JP, Thomson NH, Hansma PK, Wendman MA, Gurley G, Elings V (1996) Short cantilevers for atomic force microscopy. *Revue of Scientific Instruments* 67: 3583–3590

Wolter O, Bayer T, Greschner J (1991) Micromachined silicon sensors for scanning force microscopy. *J Vac Sci Tech B* 8: 1353–1357

16 DNA resequencing, mutation detection and gene expression analysis by oligonucleotide microchips

Andres Metspalu and John M. Shumaker

16.1 Introduction

The genetic information of an organism is encoded as DNA within the chromosomes. The Human Genome Project (HGP) is expected to produce, among others, the DNA sequence of a human genome. The long-term effect of the HGP is envisioned to be an understanding of gene interaction that occurs during both developmental and diseases processes. Clearly, an efficient and accurate characterization of both DNA and RNA at any phase during the life cycle of a cell, tissue or organism is required if these tasks of the HGP are to be successful.

DNA sequencing technology is today still based on dideoxy terminator method (Sanger et al., 1977). Volume of the resequencing is increasing every day for the diagnostic, therapeutic, drug development and research purposes and it would be too expensive and time consuming to continue with gel based DNA sequencing technology. Radically new ideas towards DNA sequencing were put forward already some 8 – 9 years ago by several groups (Southern, 1988, 1989; Drmanac et al., 1989; Lysov et al., 1988; Khrapko et al., 1989). The next breakthrough for the DNA chip was a few years later when photolithographic combinatorial *in situ* synthesis of oligonucleotides was published (Fodor et al., 1991). However, original goals for rapid *de novo* DNA sequencing have not been realized and in recent years new applications have been introduced for the oligonucleotide arrays (Schena et al., 1995; Shumaker et al., 1996) and modified goals for the original DNA chips (Lipshutz et al., 1995, Yershov et al., 1996, Fodor 1997). The driving force behind DNA analysis is genetic testing. Although in its early stage, it is likely to revolutionize our basic concepts of what medicine is and how we should develop better diagnostics and drugs for the treatment of human disease.

This chapter will describe DNA chip-based methods for nucleic acid sequence analysis and discuss potential advantages and problems of these methods as it stands in August 1997.

16.2 Oligonucleotide chips

A DNA chip is a substrate (i.e., glass or plastic material) with nucleic acid (NA) probes attached to the surface. The type of attached NA (i.e., oligonucleotides or cDNA clones) defines the chip. Moreover, DNA chips with large clones, such as cosmids, BACs or YACs, for chromosomal analysis are also in preparation, but are not the subject of this review. Finally, arrays based on peptide nucleic acid (PNA) probes have been proposed (Weiler et al., 1997). The major advantage of the DNA chip approach is the enabling of a platform for massive parallel analysis, thereby providing for (1) minimal amounts of required reagents, (2) error checking due to redundancy, (3) potential for complete automation and (4) a procedure that is robust, rapid and economic. Once fully developed, DNA chip technology has the potential to meet the requirements of genomic resequencing and mutation detection in a cost- and time-efficient manner.

An alternative approach to the universally used Sanger approach for DNA sequencing was proposed in several laboratories (Lysov et al.,1988; Bains and Smith 1988; Drmanac et al., 1989, 1993; Southern, 1989; Southern et al., 1992; Pease et al., 1994, for review see Mirzabekov, 1994 and Southern, 1996). The method was called sequencing by hybridization (SBH). The DNA sequence was to be reconstructed from multiple hybridization signals between the target DNA and oligonucleotides. Two concepts for SBH were proposed originally. Drmanac and coworkers proposed that multiple target DNA clones are to be attached to a solid support (e.g., nylon filter) and sequentially hybridized with labeled probes of known sequence (oligonucleotides). Unfortunately, this method involves large number of manipulations – the use of octanucleotide probes (8-mers) requires thousands of hybridization experiments to be performed before the sequence can be reconstructed. A similar approach has been used for fingerprinting arrayed genomic clones on high density filters for mapping purposes (Lennon and Lehrach, 1991; Gress et al.,

1992). A second and more commonly used strategy is based upon the fabrication of an oligonucleotide array with all possible sequences of a given length. The sequence of each probe in the array is identified by its location within the array. The target DNA is added to the array of probes (DNA chip), and hybridization pattern is recorded. Overlapping oligonucleotides can theoretically reveal the sequence of the target DNA as it is outlined in Figure 1. Unfortunately, large numbers of oligonucleotides are required to produce an informative array – there are 65536 8-mers, 262144 9-mers, 1048576 10-mers, etc.

The feasibility of producing dense oligonucleotide arrays *in situ* was solved by two laboratories (Fodor et al., 1991; Pease et al., 1994; Southern et al.,

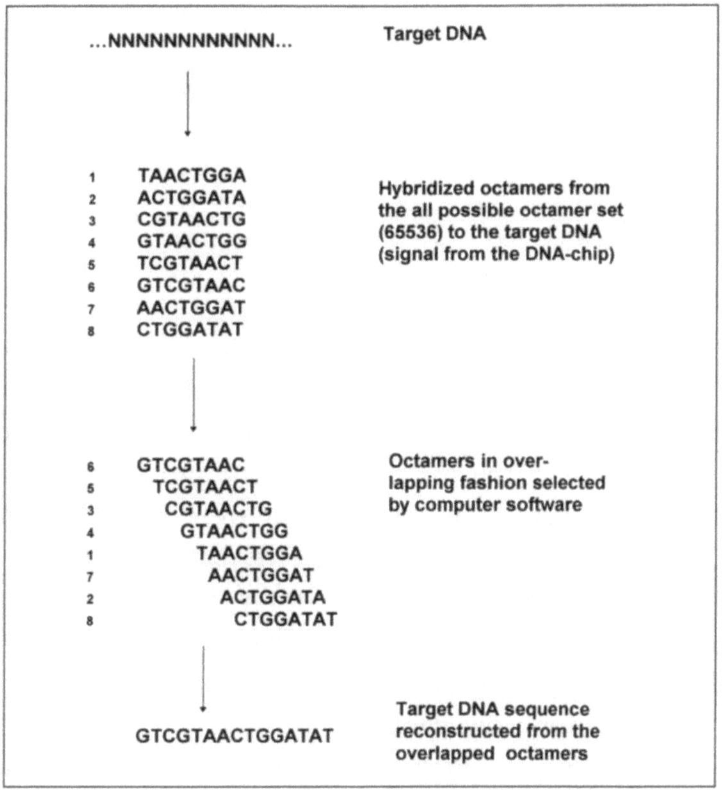

Figure 1 *SBH method. Sequence reconstruction by the maximal overlap of constituent DNA sequences. The unknown sequence (...NNN...) is determined by ordering the 8-base segments by unique 7-base overlaps (n – 1)*

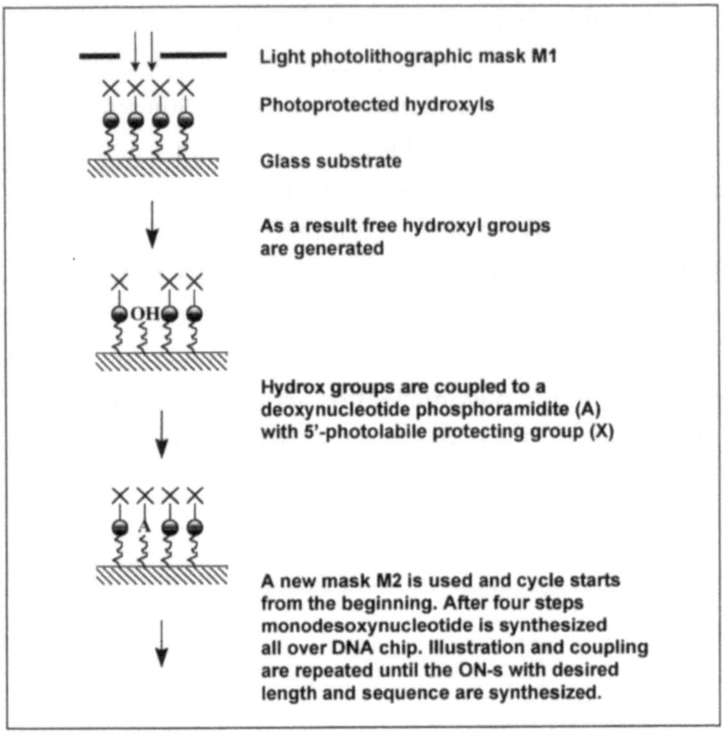

Figure 2 *Light-directed synthesis of oligonucleotides*

1992). Affymetrix (USA) developed the most elegant solution for the production of oligonucleotide arrays. Photolithographic masks were adapted from semiconductor fabrication technology to oligonucleotide synthesis technology through the use of photolabile protecting groups. Affymetrix has synthesized an array consisting of over 400000 20-mers in an area of 1.28 cm × 1.28 cm (Fodor, 1997). The basic strategy for light-directed oligonucleotide synthesis is schematically given in Figure 2, and the synthesis of all possible N-mers is outlined in Figure 3. Note, the synthesis of all possible decamers (4^{10}) requires only 40 steps – which is four times the number of steps required for only one decamer.

Using physical masking a different approach for *in situ* synthesis of oligonucleotide arrays was designed (Maskos and Southern, 1992b, 1993b; Southern et al., 1992, 1994). This device has currently a probe area around mm^2 which limits the number of probes that one can synthesize on a substrate of a manageable size. New strategies are emerging where solid-phase oli-

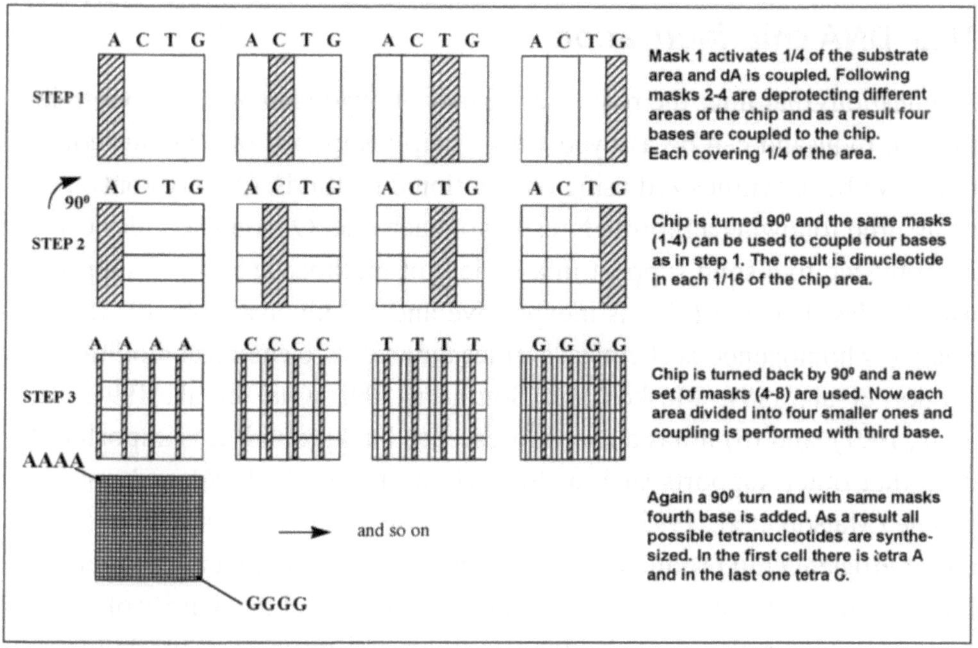

Figure 3 *Combinatorial synthesis of all N-mer oligonucleotides using masking technology. For synthesis 4^{10} oligonucleotides 4 × 10 steps are needed*

gonucleotide synthesis is combined with polymeric photoresist films serving as the photoimageable component. This opens the way to very high density and better quality oligonucleotide array synthesis with unit cell size ≤1 µm (McGall et al., 1996). Conventional oligonucleotide phosphoramidite chemistry for oligonucleotide synthesis in both directions can be fused with micro stamping technology providing high density oligonucleotide grids *in situ* (H.-P. Saluz, Jena, FRG, personal communication).

The complete n-mer chip (universal chip) is theoretically required, but not sufficient, for *de novo* sequencing. The prevalence of repeat units in the human genome and unequal representation of oligonucleotides (Burge et al., 1992) makes the application SBH using the universal chip for *de novo* sequencing very difficult. However, disease gene specific or mutation specific DNA-chips are possible using arrays with limited numbers of oligonucleotides. The oligonucleotides can be attached onto either polyacrylamide (Khrapko et al., 1989; Yershov et al., 1996; Timofeev et al., 1996) or glass surfaces (Lamture et al., 1994; Shumaker et al., 1996, Guo et al., 1994).

16.3 DNA chip fabrication

Applications that limit the required number of oligonucleotide probes to less than 20 thousand can be assayed with a chip comprised of oligonucleotides that have been synthesized using conventional methods. Quality control can be assured following a thorough testing of each one. Oligonucleotides can be immobilized to a solid support in a variety of methods. The solid support is usually glass because (1) it is inexpensive and readily available, (2) it has a relatively homogeneous chemical surface amenable to chemical modification (Fodor et al., 1991; Maskos and Southern 1992; Guo et al., 1994, Joos et al., 1997), and (3) it has a low fluorescence background. It is important to note that other supports such as SiO_2 (Lamture et al., 1994) and polypropylene (Matson et al., 1995) have also been used. Parameters to be considered for efficient chip production include the stability, length and charge of the linker between the oligonucleotide and the surface, and the density of the oligonucleotides on the chip. A high density could promote steric hindrance (Guo et al. 1994). Shchepinov et al., 1997 concluded that the optimal spacer is 30–60 atoms in length, hydrophilic and slightly negatively charged. Amino-modified oligonucleotide can be immobilized to the glass surface by secondary amine formation between an epoxysilane monolayer and the amino linkage at the 3′ end of the oligonucleotide, yielding a surface density of 10^{10} molecules/mm^2 at added probe concentration of 50 µM (Lamture et al., 1994). Phenylisothiocyanate groups can be used on the glass for coupling of 5′ amino-modified oligonucleotide providing a 23 atom linker between the glass surface and the oligonucleotide, which can be further extended by addition of a non-specific poly dT spacer sequence at the 5′ end of the oligonucleotide (Guo et al., 1994; Pastinen et al., 1997). Assuming that (i) the diameter of dsDNA is 20 Å, (ii) the hybridized 15-mer duplex length is approximately 50 Å, and (iii) the linker length is 50 Å; then the minimum area covered by one oligonucleotide would be approximately 300 $Å^2$ and free rotation of the 15-mer oligonucleotide would need an area of approx. 7500 $Å^2$. In fact, optimal area occupied by one oligonucleotide in hybridization experiment was ~500 $Å^2$ (Guo et al., 1994). This means that free rotation of the oligonucleotide is not required for optimal hybridization, however with increasing the target size hybridization efficiency is decreased, or in other words, longer targets need more space on the grid for optimal hybridization.

The optimal conditions for oligonucleotide immobilization into polyacrylamide gel have been established (Timofeev et al., 1996). It appears that immobilization of amino-modified oligonucleotide on aldehyde gel supports gave the highest yield, is stable and can be considered as the most effective method of oligonucleotide chip preparation for this particular application.

16.4 DNA hybridization to the DNA chip

Physical chemistry rules appear to govern nucleic acid hybridization both in solution and on a solid support. Because of the potential complexity of the large number of different oligonucleotides on a chip, it could be practically impossible to find a set of hybridization conditions in which each oligonucleotide can hybridize and dissociate under optimal conditions. Perfectly complementary AT-rich oligonucleotides can be as stable as mismatched GC-rich oligonucleotide. Mismatch discrimination at each end of the oligonucleotide is a particularly difficult problem, especially the G-T mismatch (Yershov et al., 1996). Several methods to decrease the effects on stability from the nucleotide composition of the oligonucleotide have been published. The base composition dependence of duplex formation can be reduced by adding tetramethylammonium chloride, which presumably increases the AT base-pair stability, to the hybridization solution (Maskos and Southern, 1992 b, 1993 c). Oligonucleotide arrays provide for additional possibilities: The oligonucleotide length and/or concentration can be varied to compensate partially for the oligonucleotide base composition. For example, the apparent oligonucleotide dissociation temperature rises approx. 5°C upon either increasing the concentration of immobilized oligonucleotide threefold (Khrapko et al., 1991), or using 9-mers for AT-rich and 8-mers for GC-rich oligonucleotide in hybridization with filter bound target DNA (Drmanac et al., 1993). However, these additional modifications do not necessarily equalize the duplex forming ability of each oligonucleotide. Moreover, nucleotide sequence can also be important; therefore, the search of the optimal oligonucleotides for RNA binding is based upon empirical studies and not theoretical ΔG calculations (Milner et al., 1997).

An octanucleotide chip with 65 536 oligonucleotide is technically possible to synthesize, but is limited in utility due to difficult discrimination of mismatches at the oligonucleotide ends and the repetitive content of DNA. Theoretical calculations show that the length of random DNA sequence that can be analyzed by a complete set of oligonucleotides of a given length is approximately the square root of the number of oligonucleotides in the set (Southern et al., 1992). For 8-mer chip with 65 536 oligonucleotides the useful range would be 200–300 bp. All possible 13-mers ($\sim 67 \times 10^6$ oligonucleotide) or even 15-mers (10^9 oligonucleotide) can sequence more but are too complex to synthesize at present time. Moreover, calculations show that the optimal length of the oligonucleotide where one base-pair mismatch can still be discriminated is 10-mer (Bains, 1994). One way to increase the efficiency of SBH is to use two adjacent oligonucleotides capable to hybridize to the same target just next to each other – contiguous stacking hybridization (Lysov et al., 1988; Parinov et al., 1996). A second oligonucleotide, e.g. 5-mer can be added to the hybridization solution together with the 8-mer. Stability of the 5-mer is guaranteed by the stacking energy between the adjacent 8-mer without the phosphodiester bond. Ligation of these stacked oligonucleotides, of course, can be of additional value, permitting more confidence and extra labels for detection (Broude et al., 1994). However, this approach will complicate the original stacking-based method without clear advantage over it.

16.5 Template preparation

Several methods have been demonstrated using either DNA or RNA as a template for the chip. The DNA (e.g., patient sample) can be amplified by PCR, transcribed into labeled RNA using labeled rNTPs. Thereafter RNA transcript can be fragmented by a treatment of heat and $MgCl_2$ and used in the hybridization assay (Kozal et al., 1996). Another method can be to include the dUTPs during first PCR, thereafter incorporate fluorescein-12-dGTP during the next, asymmetric PCR step and finally use uracil-N-glycosylase for fragmentation of the target ssDNA (Cronin et al., 1996). A third method is utilizing fragmentation of the PCR product with Maxam-Gilbert sequencing chemistry (Yershov et al., 1996). Although double stranded DNA produces poor signal compared to ssDNA (Guo et al., 1994), it would be advantageous to use dsDNA as a template. Efforts are currently directed towards this goal.

16.6 DNA chip image analysis

No universal chip reader exists. Each chip group is developing a chip system with its own incorporated chip reader. For example, confocal-based laser scanning systems have been developed specifically to read the Affymetrix chips. Several laboratories have been using commercially available PhosphoImagers or FluorImagers (Guo et al., 1994; Pastinen et al., 1997, Metspalu et al., 1998). Mirzabekov's group has built an image system based upon an epifluorescence microscope and a CCD camera (Yershov et al., 1996). Finally, the Shumaker laboratory has taken a different approach by using the glass DNA chip as two-dimensional optical wave guide. The evanescent wave created by total internal reflection of an excitation source extends only a few hundred nanometers ($\lambda/2$) from the wave guide surface. The emitted fluorescence of the labelled oligos or DNA can be imaged onto a CCD chip (Stimpson et al., 1995; Shumaker et al., 1997).

16.7 Mutation analysis and resequencing based on oligonucleotide hybridization on DNA chips

SBH approach was initially proposed to analyze an unknown polynucleotide sequence (Southern 1989 patent; Drmanac et al., 1988, 1993; Bains and Smith 1988, Lysov et al., 1988). However, the recent application of SBH is the resequencing of the known DNA and mutation analysis. An Affymetrix DNA chip was designed to analyse HIV-1 clade B protease gene nucleotide sequence, and the obtained data were compared to results of conventional Sanger dideoxy DNA sequencing (Kozal et al., 1996). HIV-1 protease is a 99 aa protein and authors selected a 382 bp contiguous region of the HIV genome for testing, which includes a 297 nt of the protease gene and small portions of the flanking gag (18 nt) and rt genes (67 nt). A chip comprised of 12224 different oligonucleotides were synthesized onto the 1.28×1.28 cm glass surface – DNA chip. Each base in the protease gene is analyzed with a set of 16 oligonucleotides and to get sequence information from both strands 32 oligonucleotide are needed. Each oligonucleotide covered the area of 95×95 µm and contained 10^8 copies of a specific oligonucleotide immobilized onto the glass surface. According to the proposed strategy the minimal number of the oligonucleotide necessary to interrogate one nt in a given nucleic acid sequence length L is 4L as given in Figure 4 (tiled array). As seen from the scheme at the central position (in 15-mer case 7th nt from the 3′-end) all four possible nt are available for binding, so that depending on the identity the nt at this particular site in the template DNA, one out of four oligonucleotide will be correctly matched as demonstrated on Figure 5. However, in reality more oligonucleotides are used and probably required for redundant error checking – 16 in the study by Kozal et al., 1996 and 14 in BRCA1 case for the one strand only (Hacia et al., 1996). The net result of a mutation is that the surrounding sequence cannot be determined because of duplex mismatches. Unfortunately, analyzing the complementary strand of the same sequence does not improve the result, because the pair of probes for each strand are themselves complementary – both probes hybridized to the same region in the DNA. The situation becomes even more complicated in cases where two mutations can occur within a 15 bp region, such as mtDNA D-loop region (Stoneking et al.,

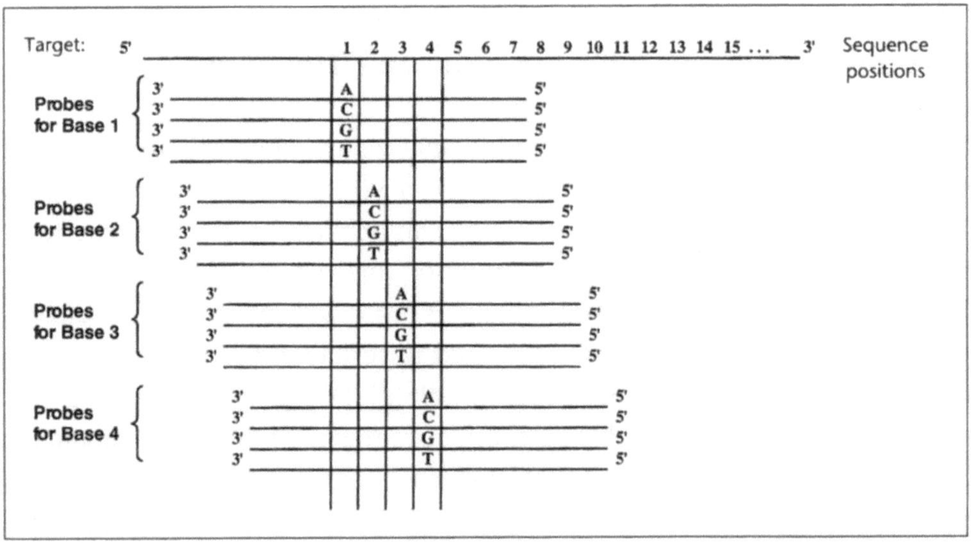

Figure 4 *Tiling strategy for sequence determination using 15-mer probes. A set of four 15-mer oligonucleotide probes differing only at the 7 nucleotide position is used to determine the identity of each base in the gene. Oligonucleotides are synthesized according to the "wild type" sequence, however, in case of point mutation one oligonucleotide out of four should match perfectly. For detection of deletions and insertions additional probes are necessary. (Figure is reprinted from Kozal et al., 1996, with permission)*

1991). Figure 6 depicts the scheme of the template preparation for the DNA chip analysis. Compared to the DNA chip analysis which takes up to 20 min., template preparation with its 2.5 hours is the main area for improvement.

A major barrier to the acceptance of hybridization-based arrays in the diagnostic setting is the low signal to noise ratio (S/N). In the study cited above (Kozal et al., 1996) and in (Lipshutz et al., 1995) the ratio of highest to next highest intensities for each set of four oligonucleotides interrogating one nucleotide of the protease gene is determined, and if this ratio is greater than 1.2 then a nucleotide determination is made. A ratio of at least 10:1 would be required for complete and reliable results, including the heterozygous cases

Figure 5 *False-color image of HIV-1 protease gene DNA-chip. a) DNA-chip image 1.28 × 1.28 cm. Area outlined in white contains 12 224 oligonucleotides interrogating 383-bp region; b) Each base is analyzed with 16 probes 11, 14, 17 and 20 nucleotides in length. The brightest squares are used for base calling; c) Each probe cell is 93 μm² and contains ≈ 10⁸ copies of a particular oligonucleotide. (Figure is reprinted from Kozal et al., 1996, with permission) (see colour plate 7)*

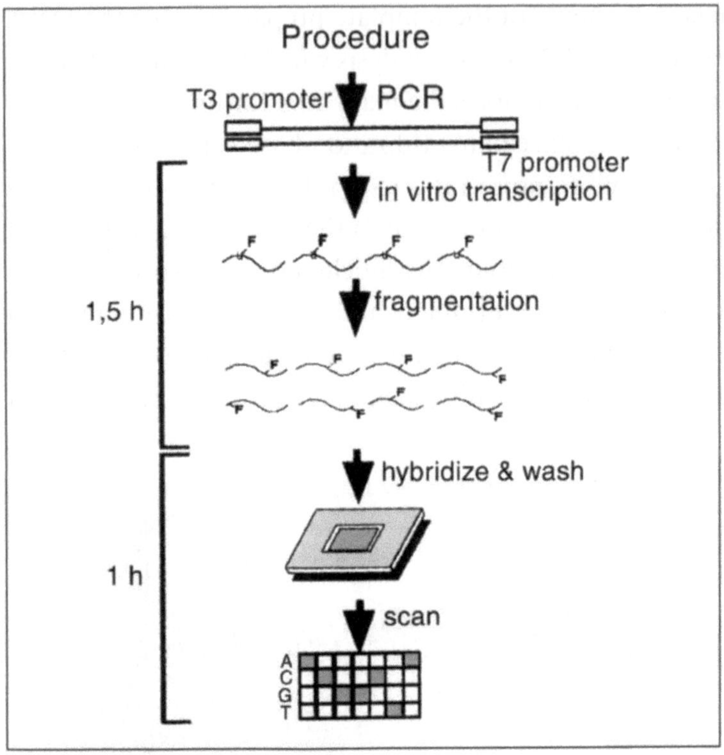

Figure 6 *Template preparation for DNA-chip. The PCR product is transcribed with T7 or T3 RNA polymerase in the presence of fluoresceine-labelled rUTP, fragmented by heat treatment and hybridized to the chip*

and tumor samples with normal cell background. Nevertheless, overall accuracy using DNA chip for HIV-1 protease gene genotyping was reported to be greater than 98% which should be increased in future by further development of the DNA chip technology.

The field of mutation analysis could benefit greatly through a proliferation of DNA chip usage. Cronin and coworkers (1996) have tested the DNA chip on a well studied model – human cystic fibrosis transmembrane conductance regulator (CFTR) gene. Two types of 15-mer arrays were tested. One array was designed to scan through the length of CFTR exon 11 to identify differences from the wild type sequence. The second oligonucleotide array was developed to test for the 37 most common known mutations in the CFTR gene. The results show that single base changes are difficult to distinguish unequivocally by using single sets of oligonucleotide probes to check every nucleotide position in the template DNA. This was particularly true for heterozygous cases. As stated previously, the DNA chip cannot identify insertions and deletions in the target sequence. If the mutation has not been characterized, only the region and not the sequence of the mutation is determined. A sequencing-based method is required for mutation characterization. Although potentially more effective than a gel-based method, a chip approach would benefit greatly if it could characterize mutations in a given gene without the necessity of requiring prior knowledge about the mutations.

This problem was approached using BRCA1 gene as a model (Hacia et al., 1996), and using a two-color analysis system that is in principle similar to comparative genomic hybridization (CGH). A DNA chip with over 96600 oligonucleotide was manufactured to detect all possible single base substitutions, single base insertions and 1–5 bp deletions from both strands in a 3.45 kb region from exon 11. Even with a redundancy of 28 oligonucleotide probes per base analyzed, 15 of 16 mutations were identified, with an error rate of 6.7%. There are several possible mechanisms causing false positive and false negative results. Oligonucleotide in the array or template DNA can form secondary structure in certain parts of the DNA sequence under study, and any hybridization system has some intrinsic noise level due to the stability of certain types of mismatches. Template preparation is another source of errors due to fragmentation which is difficult to control.

An impressive example of DNA chip usage is the mtDNA analysis of Chee et al., 1996. Due to the limited capacity of the gel-based methods, mtDNA sequencing is usually limited to the mitochondrial D-loop (Stoneking et al., 1991). MtDNA analysis is, however, one of the basic methods used to study human evolution (Krings et al., 1997). Mark Chee and coworkers have used DNA chips with 135000 oligonucleotides to resequence the entire 16.6 kb of human mtDNA. Compared to the speed of gel-based system (e.g. one ABI 373 type instrument, 48 lanes, run twice per day with 400 nucleotide reads) the DNA chip method was approx. 25 times more rapid.

In many cases the disease process may not be caused by changes in the DNA sequence, but rather by altered expression of the genes. DNA chips have the potential to be very useful in gene expression studies, diagnostics and therapy. Lockhart et al., in 1996, have demonstrated the sensitivity of the gene expression chips in a model system – one mRNA out of 300000 can be detected, and detection is quantitative over three orders of magnitude. Oligonucleotide arrays and cDNA arrays (see the next section) each have advantages. The oligonucleotide array approach has the potential of providing both RNA expression level analysis, as well as an analysis of differential splicing, promotor structures, closely related gene families, exon presence, etc. The major advantage of the cDNA approach is that the sequences of the cDNA probes are not required for cDNA arrays. Other applications for DNA chips include the phenotype analysis of yeast deletion mutants (Shoemaker et al., 1996), the ordering of clones from genomic libraries (Sapolsky and Lipshutz, 1996), the detection of β-thalassemia mutations (Yershov et al., 1996; Drobyshev et al., 1997), and the selection of antisense oligonucleotides (Milner et al., 1997). Southern's group used combinatorial method (Southern et al., 1994 b) for producing an oligonucleotide array complementary to the first 122 bases of rabbit β-globin mRNA. Surprisingly, only one 15-mer gave high duplex yield and this correlated with antisense activity measured in a RNase H assay and by *in vitro* translation. Moreover, there was no apparent correlation between the estimated ΔG of heteroduplex formation and the extent of hybrid formed on the array (Milner et al., 1997). This work may explain the random success when selecting antisense oligonucleotide for mRNA inhibitory experiments.

16.8 cDNA microarrays for gene expression analysis

Although cDNA arrays on the filters are common in genomics laboratories (Lennon and Lehrach 1991; Gress et al., 1992), the large EST databases have pushed gene function analysis to a point such that cDNA arrays are printed onto the glass at densities more than 1000 cDNAs per cm^2 (Schena, 1996). A custom-built robot to spot few nl spots spaced 100 – 500 μm apart and a custom-built laser scanner to detect the two-color fluorescence hybridization signals were fabricated. The cDNAs were immobilized onto poly-L-lysine-coated microscope. The technology was first applied to the differential expression measurements of 45 *Arabidopsis* genes (Schena et al., 1995). Polyadenylated RNA was isolated from the total *Arabidopsis* RNA (e.g. root tissue and leaf tissue), and fluorescent target DNA was prepared by reverse transcription. Fluorescein-12-dCTP and lissamine-5-dCTP were used to label the pools of cDNAs. The results showed that the specific expression levels covered three orders of magnitude (Fig. 7). This technology was developed

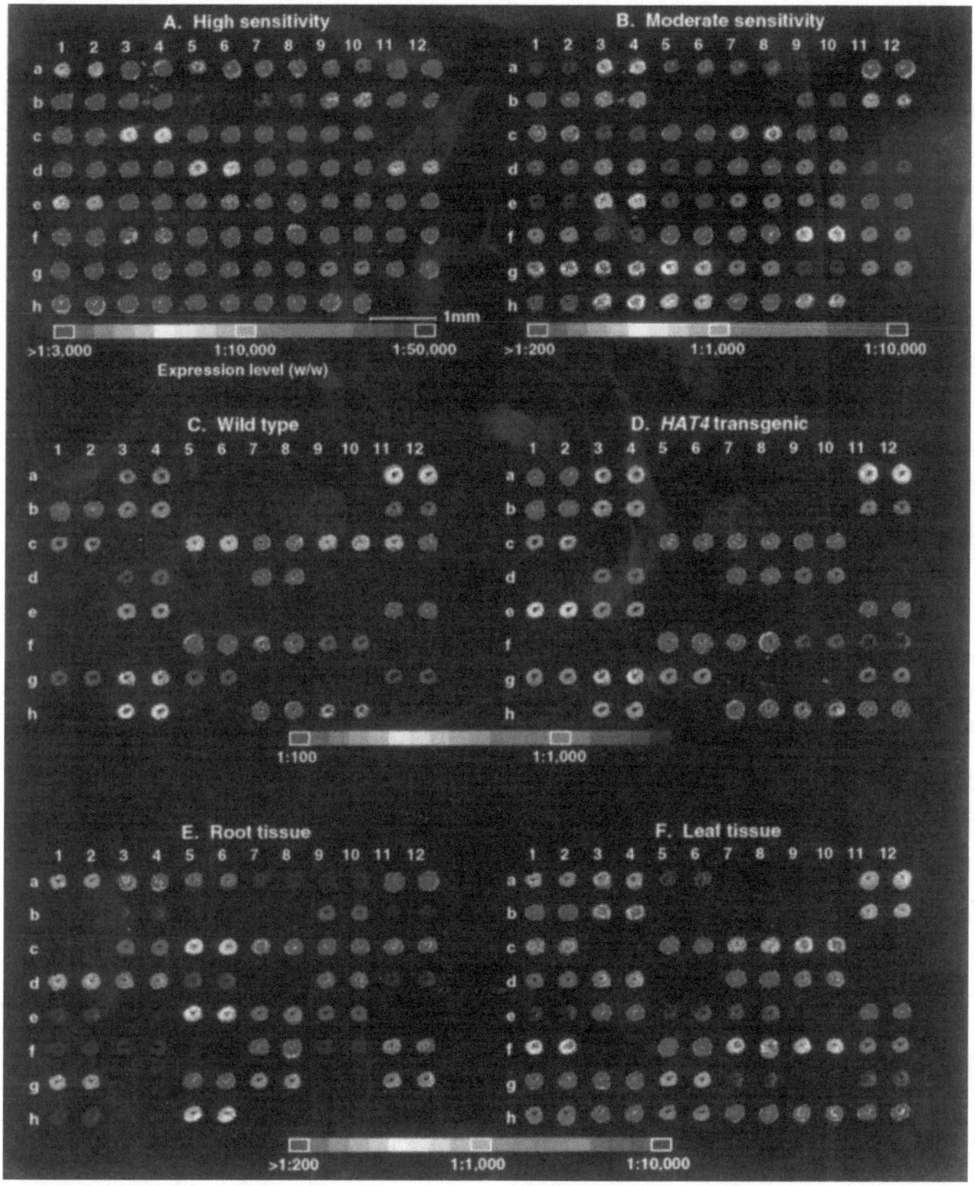

Figure 7 *Gene expression monitored with the use of cDNA microarrays. Fluorescent scans represented in pseudocolour correspond to hybridization intensities. E and F: A single array was probed with a 1:1 mixture of fluorescein-labelled cDNA from root tissue and histamine-labelled cDNA from leaf tissue. Array was then scanned successively to detect the mRNAs expressed in roots (E) and in leaves (F). (Figure is reprinted from Kozal et al., 1996, with permission) (see colour plate 8)*

further to analyze the expression of more than 1000 human genes, and the detection limit was increased 10 fold to monitor transcripts that represent approx. 1:500000 of the total mRNA (Schena et al., 1996). The cDNAs were attached to silanylated microscope slides via 5′ end amino group to produce hybridization signals with greater affinity, specificity, and lower background. Moreover, cDNA arrays can be more specific in mRNA expression studies than oligonucleotide arrays (see above) and cDNAs or EST clones can be used without prior sequencing (Schena et al., 1996; Shalon et al., 1996). The potential of the technology was demonstrated in that a parallel analysis of 1046 human cDNAs of unknown sequence revealed known and novel heat shock and phorbol ester regulated genes in human T-cell cells. Moreover, a second DNA chip with 1161 elements was used to analyze the difference in gene expression between tumorigenic UACC-903 and non-tumorigenic UACC-903 (+ 6) cell lines. The results showed some levels of expression were increased (n = 15), and some were decreased (n = 63) by introducing a tumor suppressor with a fragment of chromosome 6 (DeRisi et al., 1996). Combined with Zhang et al., 1997, these results begin to illustrate the potential magnitude of alteration in gene expression during tumor formation. cDNA microchips can also reveal molecular mechanisms and pathways underlying malignant process. By analyzing gene expression of all genes of the pathway, albeit it a cell or tissue, the potential for diagnostic and therapeutic application should expand.

16.9 DNA chips and enzymes

SBH is, in principle, an indirect method for DNA sequencing; only the presence or absence of a particular DNA sequence can be inferred. Thus, many oligonucleotide probes are required for sequence determination. Dideoxysequencing is based upon primer extension complementary to the template DNA (Sanger et al., 1977). This method is based upon the elongation of a sequencing primer with a template-dependent DNA polymerase and a mixture of ddNTPs/dNTPs. Suppose, for example that the Sanger approach is modified so that multiple sequencing primers are extended by only one base using ddNTPs. In addition, suppose that the sequencing primers are attached to a surface via their 5′ end, enabling a free 3′ end for DNA polymerase extension. The result is arrayed primer extension method -APEX (Caskey et al., 1994; Metspalu et al., 1995). Outline of the method is in Figure 8. Primer extension method or minisequencing has been developed for gel-based separation or for microplates several years ago and proved to be very specific, sensitive and robust (Sokolov, 1989; Syvänen et al., 1990, and 1993; Nikiforov et al., 1994; Livak and Hainer, 1994). Further development of this assay was presented (Shumaker et al., 1996; Pastinen et al., 1996, 1997; Metspalu et al., 1998). In these studies DNA chip format was used together with DNA polymerase extension of labeled ddNTPs to demonstrate the feasibility of resequencing and mutation detection methods. Oligonucleotides are immobilized through 5′ end via spacer to the silanized glass, and they are designed to scan the whole DNA fragment to be resequenced in single base

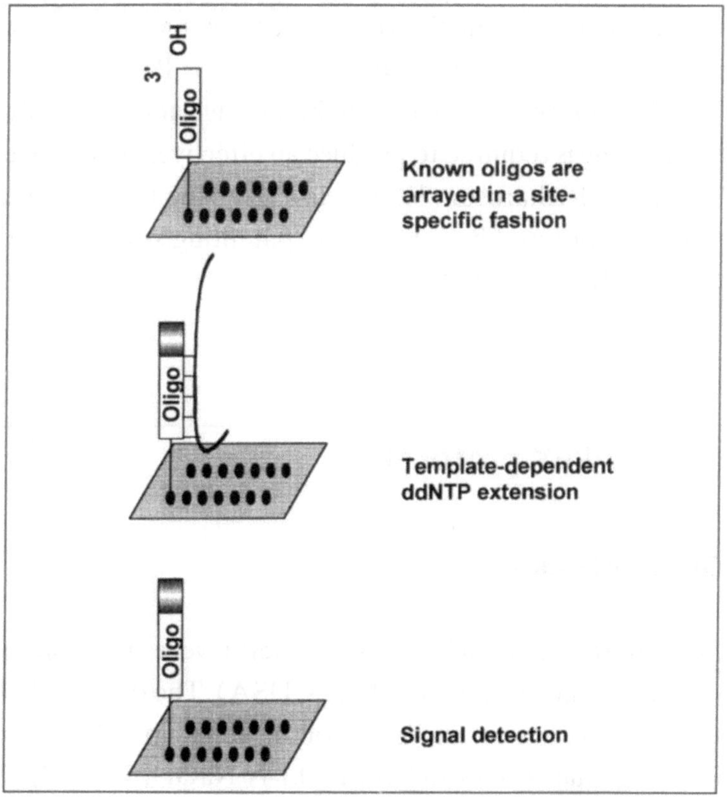

Figure 8 *Three basic steps in arrayed primer extension (APEX)*

steps. Each oligonucleotide on the DNA chip directs a template-dependent polymerase extension of fluorescently labeled dideoxynucleotide terminators and identifies one position within the template, the DNA. The color of the labeled oligonucleotide identifies the added terminator, and thus the DNA sequence can be obtained. APEX application for mutation detection and fidelity of different DNA polymerases was studied (Pastinen et al., 1997).

Single nucleotide primer extension method can also be used for mRNA expression analysis (Lombardo and Brown, 1996). Good signal yields and S/N ratios make APEX potentially very attractive in mRNA quantitation and mutation analysis (Pastinen et al., 1997, Schwartz et al., 1997).

Research in the field of chips is only in its infancy period. Many problems have yet to be solved: (i) the level of expertise needed is too great for a typical lab to utilize it at present, (ii) the inability to detect repeat sequences, (iii) parallel multiplex template preparation, and (iv) the optimal combinations of DNA polymerase and fluorescent ddNTPs have yet to resolved. The last point is especially crucial since fluorescent ddNTPs are not the natural substrates for DNA polymerases and more knowledge about the polymerases and protein engineering is required to produce an effective, error-free enzyme (Joyce, 1997; Tabor and Richardson, 1995). Moreover, the fluorescent dye problem is even more complicated because of the patenting situation and limited number of suitable dye molecules.

16.10 APEX protocol

Oligonucleotides

Six hundred oligonucleotides (25-mers) were synthesized with an amino link on their 5′ end (Life Technologies, USA). The oligonucleotides were designed to scan every base of exons 5–6 of the human p 53 gene as depicted in Figure 9. The oligonucleotides (50 µM in 0.1 N NaOH, vol 5 nl) were spotted 200 µm apart by a custom-built robot (Synteni, CA, USA). The arrays were washed extensively in boiling water before use.

```
                         p53 template (exon 5)

3' ATGAGGCCACGGGAGTTGTTCTACAAAACGGTTGAC   5'

5' TACTCCCCTGCCCTCAACAAGATGT
     ACTCCCCTGCCCTCAACAAGATGTT
      CTCCCCTGCCCTCAACAAGATGTTT           + ddNTP^F
       TCCCCTGCCCTCAACAAGATGTTTT
        CCCCTGCCCTCAACAAGATGTTTTG
         CCCTGCCCTCAACAAGATGTTTTGC

   oligonucleotides on the chip
```

Figure 9 *Partial sequence of the exon 5 of the p53 gene and associated scanning 25-mer oligonucleotides. Each oligonucleotide scans one base after template-dependent primer extension using DNA polymerase and fluorescent ddNTP-s*

Preparation of the DNA arrays

The surface of a number 2 micro cover glass (CMS, Inc., Houston, TX) was cleaned with acetone and then with 0.1 M NaOH prior to chemical modification. The glass surface was functionalized by treating with 2% glycidoxypropyltrimethoxysilane in 95% ethanol (pH 5.0) for 2 min. The slides were washed with 100% ethanol and dried in an oven.

Preparation of the template and APEX assay

Exons 5–6 of p53 were amplified by AmpliTaq Gold polymerase (Perkin Elmer, CA) using the following primers:
exon V 5′-GCCGTGTTCCAGTTGCTTTATC and
5′-TCAGTGAGGAATCAG AGGCC;
exon VI 5′-CTGGAGAGACGACAGGGCTG and
5′-GCCACTGACAACCACCCTTA.
The PCR consisted of using dATP, dCTP, dGTP, and a mixture of dUTP/dTTP (1:4). The amplified product was purified using a HyPure Spin column (Boehringer-Mannheim, USA) and an ultrafiltration membrane from Microcon. Following purification, the dsDNA template was digested with uracil-N-

glycosylase according to Cronin et al. (1996). Fluorescent-labeled ddNTPs (DuPont/NEN, Boston, MA) were used as substrates for polymerase extension. The labels included fluorescein, Cy3, Cy5, and Cy7. A typical reaction consisted of 50 nM probe, 0.625 µM ddNTPs, 0.01 units Taquenase (ScienTech, St. Louis, MO). The reaction mixture was heated to 95°C, then placed on the primer array and incubated at 42°C for 10 min. The array was washed in 95°C water, dried, covered with antifade (Molecular Probes, Beaverton, OR) and imaged.

Image capture and analysis

Fluorescent detection was accomplished with a custom built CCD system. Four lasers were used for excitation: 488 nm argon ion (NEC), 532 nm (Pheonix), 635 laser diode, and a 755 laser diode. The emission filters were matched to the dyes for maximal signal separation. Image analysis was performed by a software package from Media Cybernetics.

16.11 APEX results

Four-color DNA chip image after APEX assay is shown on Figure 10.

The p53 sequence can be reconstructed from a translation of the fluorescent tag of the primer to the respective dye terminator. Following the spotting pattern the p53 nucleotide sequence can be read from the image.

16.12 Discussion

The APEX technology provides several distinct advantages over hybridization based methods. First, the use of allele-specific oligonucleotide hybridization for either mutation detection or resequencing of DNA requires many oligonucleotide probes – as many as 28 per base pair analyzed to increase the discrimination power between the perfect and mismatched duplex (Hacia et al., 1996; Guo et al., 1997). In contrast, APEX requires only two oligonucleotides per base pair to be analyzed. Second, the oligonucleotides in an

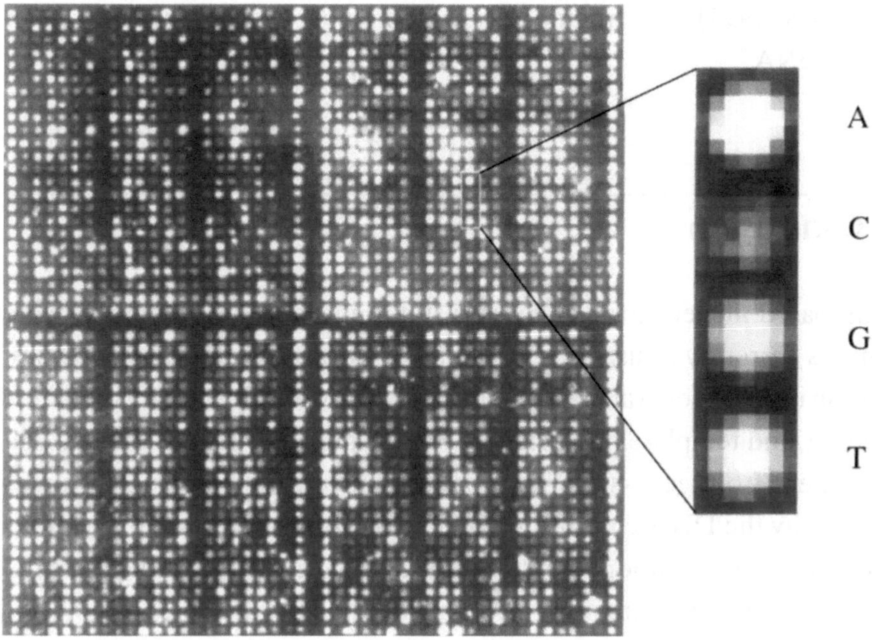

Figure 10 *Combined image of 300 bp region of p53 exons 5 and 6 probing both sense and anti-sense strands simultaneously. All primers were spotted in a three-fold redundant pattern which can be seen clearly in the image. The primers are spotted on 200 μm centres (see colour plate 9)*

SBH approach overlap by n−1 bases, whereas in APEX they overlap by n bases plus one base added by DNA polymerase; thus, direct sequence is provided without the requirement of a software reconstruction algorithm. Third, the utilization of a template-dependent DNA polymerase increases the discrimination against the 3′ end mismatches between the template and the oligonucleotide, thereby increasing the specificity and decreasing the associated noise. This improved signal to noise ratio can be expressed also as a power of discrimination between the genotypes (Pastinen et al., 1997). The power of discrimination in hybridization based assay is typically only in the range of 1.1 to 5 (Pastinen et al., 1977; Cronin et al., 1996; Lipshutz et al., 1995), but in APEX is in the range of 10–150. Fourth, APEX can be used for analysis of the known mutations as well as to scan for the unknown mutations and for analyzing heterozygous mutations and tumor samples with normal cell background (Pastinen et al., 1997; Shumaker et al., 1996, 1997). Finally, the target preparation is more robust since it does not need target labeling. With "long

PCR" (Barnes, 1994) and fragmentation, templates can be prepared from cosmid size DNA.

16.13 Conclusions

DNA chip based nucleic acids analysis is still in its early development stage, although it is currently in use in research laboratories for different purposes. Proliferation into the general market may not occur without advances in detection, spotting, and template preparation technologies. The long-term goal will be a movement from gene-specific chips to a universal chip using an n-mer chip. In this way the DNA chips fulfill their great promise and will make enormous amounts of information about DNA sequence available for health care, drug development, environment protection and food production.

Acknowledgements

The authors are grateful to Dr. Rolf Ehrnström from Amersham Pharmacia Biotech (Uppsala) for support, Dr. Per Johan Ulfendal for patent information, Dr. Jeff Tollett for critical reading of the manuscript and Mrs. Krista Liiv for the artwork and typing. A.M. is supported by ESF grant no. 2492 and EC grant no. CIPA-CT 94-0148 J.S. is supported by NIH grant no. 1 K01 HG001401 and NIST ATP program.

References

Bains W, Smith GC (1988) A novel method for nucleic acid sequencing determination. *J theor Biol* 135: 303–307

Bains W (1994) Selection of oligonucleotide probes and experimental conditions for multiplex hybridization experiments. *GATA* 11: 49–62

Barnes WM (1994) PCR amplification of up to 35-kb DNA with high fidelity and high yield from λ bacteriophage templates. *Proc Natl Acad Sci USA* 91: 2216–2220

Broude NE, Sano T, Smith CL, Cantor CR (1994) Enhanced DNA sequencing by hybridization. *Proc Natl Acad Sci USA* 91: 3072–3076

Burge C, Campbell AM, Karlin S (1992) Over- and under-representation of short oligonucleotides in DNA sequences. *Proc Natl Acad Sci USA* 89: 1358–1362

Caskey TC, Shumaker JM, Metspalu A (1994) "Parallel primer extension approach to nucleic acid sequence analysis". US Patent No. PCT/US94/07086. WO 95/00669

Chee M, Yang R, Hubbell E, Berno A, Huang XC, Stern D, Winkler J, Lockhart DJ, Morris MS, Fodor SPA (1996) Accessing genetic information with high-density DNA arrays. *Science* 274: 610–614

Cronin MT, Fucini RV, Kim SM, Masino RS, Wespi RM, Miyada CG (1996) Cystic fibrosis mutation detection by hybridization to light-generated DNA probe arrays. *Hum Mutat* 7: 244–255

DeRisi J, Penland L, Brown PO, Bittner ML, Meltzer PS, Ray M, Chen Y, Su YA, Trent JM (1996) Use of a cDNA microarray to analyse gene expression patterns in human cancer. *Nature Genetics* 14: 457–460

Drmanac R, Drmanac S, Strezoska Z, Paunesku T, Labat I, Zeremski MT, Snoddy MJ, Funkhouser WK, Koop B, Hood L, Crkvenjakov R (1993) DNA sequence determination by hybridization: A strategy for efficient large-scale sequencing. *Science* 260: 1649–1652

Drmanac R, Labat I, Brukner I, Crkvenjakov R (1989) Sequencing of megabase plus DNA by hybridization: Theory of the method. *Genomics* 4: 114–128

Drobyshev A, Mologina N, Shic V, Pobedinskaja D, Yersov G, Mirzabekov A (1997) Sequence analysis by hybridisation with oligonucleotide microchip – identification of β-thalassemia mutation. *Gene* 188: 45–52

Fodor SPA, Read JL, Pirrung MC, Stryev L, Lee AT, Solas D (1991) Light-directed, spatially addressed parallel chemical synthesis. *Science* 251: 767–773

Fodor SPA (1997) Massively parallel genomics. *Science* 277: 393–395

Gress TM, Hoeisel JD, Lennon GG, Zehltner G, Lehrach H (1992) Hybridization fingerprinting of high-density cDNA library arrays with cDNA pools derived from whole tissues. *Mammalian Genome* 3: 609–619

Guo Z, Guilfoyle RA, Thiel AJ, Wang R, Smith LM (1994) Direct fluorescence analysis of genetic polymorphisms by hybridization with oligonucleotide arrays on glass supports. *Nucleic Acids Res* 22: 5456–5465

Guo Z, Liu Q, Smith LM (1997) Enhanced discrimination of single nucleotide polymorphisms by artificial mismatch hybridization. *Nature Biotech* 15: 331–335

Hacia JG, Brody LG, Chee MS, Fodor SPA, Collins FS (1996) Detection of heterozygous mutations in BRCA1 using high density oligonucleotide arrays and two-colour fluorescence analysis. *Nature Genetics* 14: 441–447

Joos B, Kuster H, Cone R (1997) Covalent attachment of hybridizable oligonucleotides to glass supports. *Analytical Biochemistry* 247: 96–101

Joyce CM (1997) Choosing the right sugar: How polymerases select a nucleotide substrate. *Proc Natl Acad Sci USA* 94: 1619–1622

Khrapko KR, Lysov YP, Khorlin AA, Ivanov IB, Yershov GM, Vasilenko SK, Florentiev VL, Mirzabekov AD (1991) A method for DNA sequencing by hybridization with oligonucleotide matrix. *DNA Sequence J. DNA Sequencing Mapping* 1: 375–388

Khrapko KR, Lysov YP, Khorlin AA, Schick VV, Florentiev VL, Mirzabekov AD (1989) An oligonucleotide hybridization approach to nucleic acid sequence determination. *FEBS Lett* 256: 118–122

Kozal MJ, Shah N, Shen N, Yang R, Fucini R, Merigan TC, Richman DD, Morris D, Hubbell E, Chee M, Gingeras TR (1996) Extensive polymorphisms observed in HIV-1 clade B protease gene using high-density oligonucleotide arrays. *Nature Medicine* 2: 753–759

Krings M, Stone A, Schmitz RW, Krainitzki H, Stoneking M, Pääbo S (1997) Neandertal DNA sequences and the origin of modern humans. *Cell* 90: 19–30

Lamture JB, Beattie KL, Burke BE, Eggers MD, Ehrlich DJ, Fowler R, Hollis MA, Kosicki BB, Reich RK, Smith SR, Varma RS, Hogan ME (1994) Direct detection of nucleic acid hybridization on the surface of a charge coupled device. *Nucleic Acids Res* 22: 2121–2125

Lennon GG, Lehrach H (1991) Hybridization analyses of arrayed cDNA libraries. *TIG* 7: 314–317

Lipshutz RJ, Morris D, Chee M, Hubbell E, Kozal MJ, Shah N, Shen N, Yang R, Fodor SPA (1995) Using oligonucleotide probe arrays to access gentic diversity. *Biotechniques* 19: 442–447

Livak KJ, Hainer JW (1994) A microtiter plate assay for determining apolipoprotein E genotype and discovery of a rare allele. *Hum Mutat* 3: 379–385

Lockhart DJ, Dong H, Byrne MC, Follettie MT, Gallo MV, Chee MS, Mittmann M, Wang C, Kobayashi M, Horton H, Brown EL (1996) Expression monitoring by hybridization to high-density oligonucleotide arrays. *Nature Biotechnology* 14: 1675–1680

Lombardo AJ, Brown GB (1996) A quantitative and specific method for measuring transcript levels of highly homologous genes. *Nucleic Acids Res* 24: 4812–4816

Lysov YP, Florentiev AA, Khvapko KR, Shick VV, Florentiev VL, Mirzabekov AD (1988) *Proc USSR Acad Sci* 303: 1508–1511

Maskos U, Southern EM (1993) A novel method for the analysis of multiple sequence variants by hybridisation to oligonucleotides. *Nucleic Acids Res* 21: 2267–2268a

Maskos U, Southern EM (1993) A novel method for the parallel analysis of multiple mutation in multiple samples. *Nucleic Acids Res* 21: 2269–2270b

Maskos U, Southern EM (1993) A study of oligonucleotide reassociation using large arrays of oligonucleotides synthesised on a glass support. *Nucleic Acids Res* 21: 4663–4669 c

Maskos U, Southern EM (1992) Oligonucleotide hybridisation on glass supports: a novel linker for oligonucleotide synthesis and hybridisation properties of oligonucleotides synthesised *in situ*. *Nucleic Acids Res* 20: 1679–1684a

Maskos U, Southern EM (1992) Parallel analysis of oligodeoxyribonucleotide (oligonucleotide) interactions. I. Analysis of factors influencing oligonucleotide duplex formation. *Nucleic Acids Res* 20: 1675–1678b

Matson RS, Rampal J, Pentoney SL Jr, Anderson PD, Coassin P (1995) Biopolymer synthesis on polypropene supports: Oligonucleotide arrays. *Analytical Biochemistry* 224: 110–116

McGall G, Labadie J, Brock P, Wallraff G, Nguyen T, Hinsberg W (1996) Light-directed synthesis of high-density oligonucleotides arrays using semiconductor photoresists. *Proc Natl Acad Sci USA* 93: 13555–13560

Metspalu A, Saulep H, Kurg A, Tõnisson N, Shumaker JM (1998) Primer extension from two-dimensional oligonucleotides grid for DNA sequencing analysis. In: Genomics: Commercial opportunities from a scientific revolution. *BIOS Scientific Publishers*, pp. 217–219

Metspalu A, Shumaker JM, Caskey CT (1995) Direct mutation detection by fluorescent solid phase primer detection. *Med Genetik* 2: 108

Milner W, Mir KU, Southern EM (1997) Selecting effective antisense reagents on combinatorial oligonucleotide arrays. *Nature Biotech* 15: 537–541

Mirzabekov AD (1994) DNA sequencing by hybridization – a megasequencing method and a diagnostic tool? *TIBTECH* 12: 27–32

Nikiforov TT, Rendle RB, Grolet P, Rogers YH, Kotewicz ML, Andreson S, Trainor GL, Knapp MR (1994) Genetic bit analysis: A solid phase method for typing single nucleotide polymorphisms. *Nucleic Acids Res* 22: 4167–4175

Parinov S, Barsky V, Yershov G, Kirillov E, Timofeev E, Belgovskiy A, Mirzabekov A (1996) DNA sequencing by hybridization to microchip octa- and decanucleotides extended by stacked pentanucleotides. *Nucleic Acids Res* 24: 2998–3004

Pastinen T, Kurg A, Metspalu A, Peltonen L, Syvänen A-C (1997) Minisequencing: A specific tool for DNA analysis and diagnostics on oligonucleotide arrays. *Genome Research* 7: 606–614

Pastinen T, Partanen J, Syvänen A-C (1996) Multiplex, fluorescent, solid-phase minisequencing for efficient screening of DNA sequence variation. *Clin Chem* 42: 1391–1397

Pease AC, Solas D, Sullivan EJ, Cronin MT, Holmes CP, Fodor SPA (1994) Light-generated oligonucleotide arrays for rapid DNA sequence analysis. *Proc Natl Acad Sci USA* 91: 5022–5026

Sanger F, Nicklen S, Coulson R (1977) DNA sequencing with chain-terminating inhibitors. *Proc Natl Acad Sci USA* 74: 5463–5467

Sapolsky RJ, Lipshutz RJ (1996) Mapping genomic library clones using oligonucleotide arrays. *Genomics* 33: 445–456

Schena M, Shalon D, Davis RW, Brown PO (1995) Quantitative monitoring of gene expression patterns with a complementary DNA microarray. *Science* 270: 467–470

Schena M, Shalon D, Heller R, Chai A, Brown PO, Davis RW (1996) Parallel human genome analysis: microarray-based expression monitoring of 1000 genes. *Proc Natl Acad Sci USA* 93: 10614–10619

Schena M (1996) Genome analysis with gene expression microarrays. *BioEssays* 18: 427–431

Schwartz M, Sorensen N, Hansen FJ, Hertz JM, Norby S, Tranebjerg L, Skovby F (1997) Quantification, by solid-phase minisequencing, of the telomeric and centromeric copies of the survival motor neuron gene in families with spinal muscular atrophy. *Hum Mol Genet* 6: 99–104

Shalon D, Smith SJ, Brown PO (1996) A DNA microarray system for analyzing complex DNA samples using two-color fluorescent probe hybridization. *Genome Research* 6: 639–645

Shchepinov MS, Case-Green SC, Southern EM (1997) Steric factors influencing hybridisation of nucleic acids to oligonucleotide arrays. *Nucleic Acids Res* 25: 1155–1161

Shoemaker DD, Lashkari DA, Morris D, Mittmann M, Davis RW (1996) Quantitative phenotypic analysis of yeast deletion mutants using a highly parallel molecular bar-coding strategy. *Nature Genetics* 14: 450–456

Shumaker JM, Metspalu A, Caskey TC (1996) Mutation Detection by Solid Phase Primer Extension. *Hum Mutat* 7: 346–354

Shumaker JM, Tollett JJ, Shah A, Roa BB, Richards CS, Nye SN, Staub R, Pirrung M (1997) Rapid and accurate genetic screening by arrayed primer extension. On: HUGO Mutation Detection 4th International Workshop, Brno, 34

Sokolov BP (1989) Primer extension technique for the detection of single nucleotide in genomic DNA. *Nucleic Acids Res* 18: 3871

Southern EM, Case-Green SC, Elder JK, Jonson M, Mir KU, Wang L, Williams JC (1994) Arrays of complementary oligonucleotide for analysing the hybridisation behaviour of nucleic acids. *Nucleic Acids Res* 22: 1368–1373

Southern EM, Maskos U, Elder JK (1992) Analyzing and comparing nucleic acid sequences by hybridization to arrays of oligonucleotides: Evaluation using experimental models. *Genomics* 13: 1008–1017

Southern EM (1989) Analysing polynucleotide sequences. International Patent Application, PCT WO89/10977

Southern EM (1996) DNA chips: analyzing sequence by hybridization to oligonucleotides on a large scale. *TIG* 12: 110–115

Stimpson DI, Hoijer JV, Hsieh WT, Jou C, Gordon J, Theriault T, Gamble R, Baldeschwieler JD (1995) Real-time detection of DNA hybridization and melting on oligonucleotide arrays by using optical wave guides. *Proc Natl Acad Sci USA* 92: 6379–6383

Stoneking M, Hedgecock D, Higuchi RG, Vigilant L, Erlich HA (1991) Population variation of human mtDNA control region sequences detected by enzymatic amplification and sequence-specific oligonucleotide probes. *Am J Hum Genet* 48: 370–382

Syvänen A-C, Aalto-Setälä K, Harju L, Kontula K, Söderlund H (1990) A primer-guided nucleotide incorporation assay in the genotyping of apolipoprotein E. *Genomics* 8: 684–692

Syvänen A-C, Sajantilex A, Luhka M (1993) Identification of individuals by analysis of biallelic DNA markers, using PCR and solid-phase minisequencing. *Am J Hum Genet* 52: 46–59

Tabor S, Richardson CC (1995) A single residue in DNA polymerases of the *Escherichia coli* DNA polymerase I family is critical for distinguishing between deoxy- and dideoxyribonucleotides. *Natl Acad Sci USA* 92: 6339–6343

Timofeev EN, Kochetkova SV, Mirzabekov AD, Florentiev VL (1996) Regioselective immobilization of short oligonucleotides to acrylic copolymer gels. *Nucleic Acids Res* 24: 3142–3148

Weiler J, Gausepohl H, Hauser N, Jensen ON, Hoheisel JD (1997) Hybridisation based DNA screening on peptide nucleic acid (PNA) oligomer arrays. *Nucleic Acids Res* 25: 2792–2799

Yershov G, Barsky V, Belogovskiy A, Kirillov E, Kreindlin E, Ivanov I, Parinov S, Guschin D, Drobishev A, Dubiley S, Mirzabekov A (1996) DNA analysis and diagnostics on oligonucleotide microchips. *Proc Natl Acad Sci USA* 93: 4913–4918

Zhang L, Zhou W, Velculescu E, Kern SE, Hruban RH, Hamilton SR, Vogelstein B, Kinzler KW (1997) Gene expression profiles in normal and cancer cells. *Science* 276: 1268–1272

Key patents

Caskey TC, Shumaker JM, Metspalu A (1994) Parallel primer extension approach to nucleic acid sequence analysis. US94/07086. PCT WO 95/00669

Dower W, Fodor SPA (1992) Sequencing of surface immobilized polymers utilizing microfluorescence detection. PCT WO 92/10587

Drmanac RT, Crkvenjakov RB (1993) Method of sequencing of genomes by hybridization of oligonucleotide probes. US 5,202,231

Fodor SPA, Solas D, Dower WJ (1992) Sequencing by hybridization of a target nucleic acid to a matrix of defined oligonucleotides. PCT WO 92/10588

Gouelet P, Knapp MR, Andreson ST (1992) Nucleic acid typing by polymerase extension of oligonucleotides using terminator mixtures. PCT WO 92/15712

Gouelet P, Knapp MR (1995) Single nucleotide polymorphisms and their use in genetic analysis. PCT WO 95/12607

Nikiforov T, Knapp MR (1995) Method for the immobilization of nucleic acid molecules. PCT WO 95/15970

Pirrung MC, Read J, Fodor SPA, Stryer L (1990) Very large scale immobilized peptide synthesis. PCT WO 90/15070

Pirrung MC, Durham NC, Read JL, Fodor SPA (1992) Large scale photolithographic solid phase synthesis of polypeptides and receptor binding screening thereof. US 5,143,854

Southern EM (1989) Analysing polynucleotide sequences. PCT WO 89/10977

Söderlund H, Syvänen A-C (1991) Method and reagent for determining specific nucleotide variations. PCT WO 91/13075

17 Miniaturized arrays for DNA analysis

Maryanne J. O'Donnell and Hubert Köster

17.1 Introduction

Currently, standard protocols for DNA sequencing and diagnostics rely on gel electrophoretic based assays which require many hours for preparation and analysis, but the need exists for much more rapid sample preparation and data acquisition; miniaturized arrays are being explored as a viable solution. Since the arrays do not employ electrophoresis, the analysis time is already greatly reduced. As well, the miniaturized arrays offer the prospect of massive parallelization in sample manipulations and/or sample analysis, could potentially be employed in a fully automated detection system, and result in less waste of costly reagents.

In the more common array hybridization system, a sample of unknown target DNA is applied to an ordered array of immobilized oligodeoxynucleotide probes whose sequence is known, and the hybridization pattern is analyzed to produce many pieces of sequence information simultaneously (Fig. 1a) [1, 2]. Alternatively, the unknown target DNA's may be immobilized in an array format, then the array can be interrogated with a set of labeled oligonucleotide probes one at a time to reveal the sequence of the target [3, 4]. The hybridization pattern on an array may be monitored using radiolabelled DNA with autoradiographic detection [3] or storage phosphor screen technology (phosphorimager) [5]. More recently, however, radiographic detection is being replaced by fluorescent detection. It is possible to use these hybridization arrays with a large number of probe oligonucleotides for *de novo* sequencing a long stretch of DNA [3], or a smaller number of probes can be employed to interrogate local DNA sequence with diagnostic applications [6].

Although results are rapidly being produced using straight-forward hybridization detection, these systems are limited by the fact that base pair mismatches in the hybrid may not be discriminated very well. Since all posi-

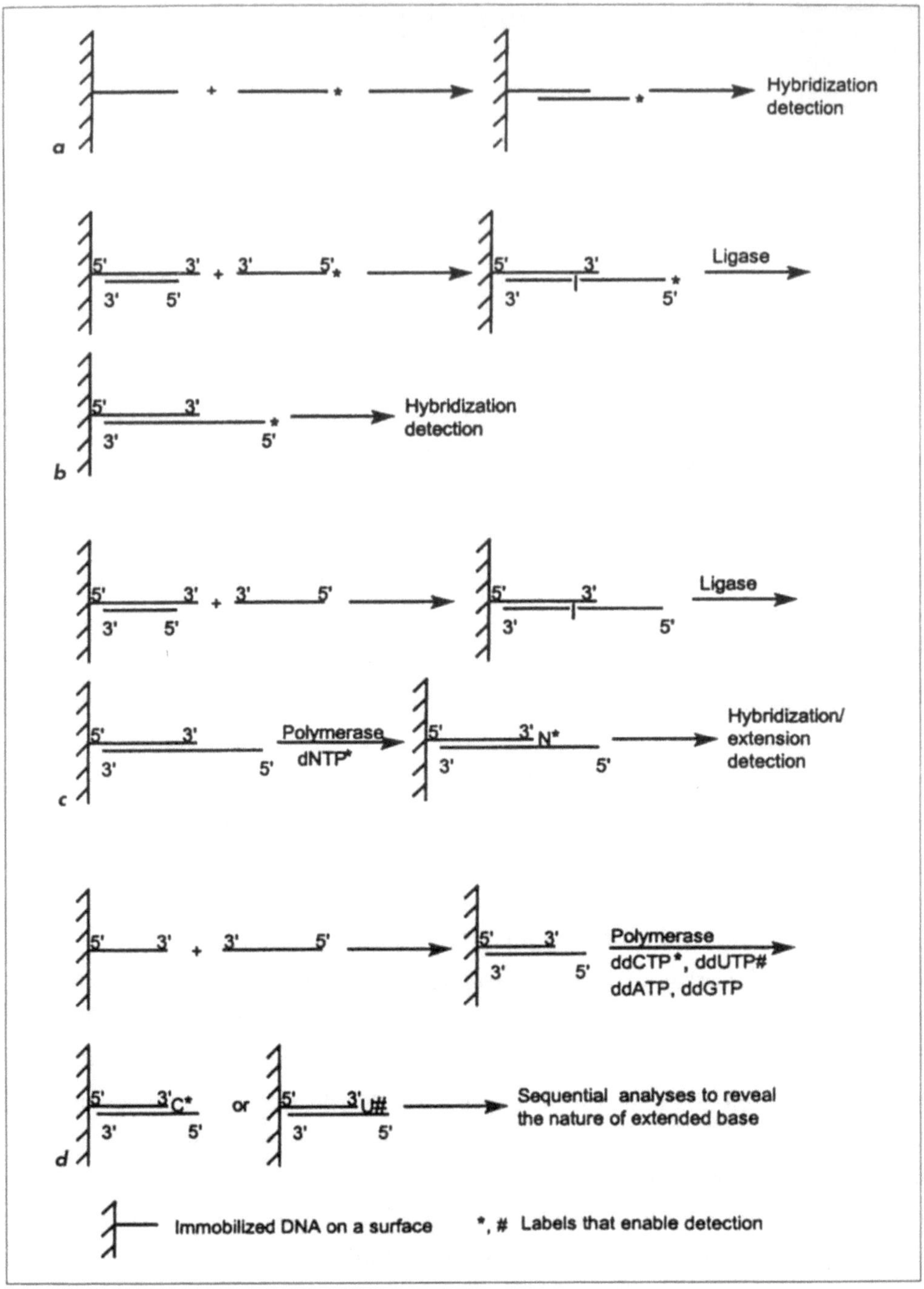

Figure 1 *Immobilized DNA probes for (a) direct hybridization detection, (b) hybridization detection using ligation, (c) hybridization detection using ligation and extension, (d) genetic bit analysis*

tions of the array are processed simultaneously, stringent conditions cannot be applied to individual elements; therefore, the occurrence of mismatches especially at the ends of the hybrids ("end mismatches") is frequent and a major disadvantage of this approach. The process of hybridization may be made more stringent by using electronic control on the array, as will be described later in this chapter. In addition to this more advanced hybridization detection, enzyme-mediated hybridization has also been explored. For example, positional sequencing by hybridization uses duplex probes containing 3′ single-stranded overhangs to capture the target, followed by enzymatic ligation of the target to the duplex probe (Fig. 1b) [7]. The stacking interactions that result from the hybridization of the single-stranded target to the duplex probe and the enzyme-catalyzed step help to minimize mismatches.

Although array hybridization detection may be elegant in its simplicity and its high through-put capabilities, it suffers from a number of drawbacks such as mismatch hybridization especially at the termini or branched hybridizations. In the latter case, different fragments bind by way of the same or similar complementary sequence to the one element of the array ("branched hybridizations"). Both types of these imprecise hybridizations can be reduced using enzyme mediation, however, they cannot be fully eliminated. Due to the nature of the detection system employing labeled templates or probes which only reveal a positive or negative hybridization result, but do not reveal the identity of the hybridized molecule, mismatch or branched hybrids will not be identified as such; mass spectrometric (MS) detection, on the other hand, produces a result of the highest analytical sensitivity, the molecular weight of the molecule, without using any labels. In addition, matrix assisted laser desorption/ionization (MALDI) mass spectrometry enables the separation and detection of a mixture of biomolecules in a fraction of a millisecond without gel electrophoresis and labeling [8–11]. For DNA molecules, the detection range of MALDI time of flight (TOF) MS has been extended to fragments 500 bases long [12]. Because of these advantages, mass spectrometry of DNA is now emerging as an extremely powerful analytical tool. Since it is such a new approach to diagnostics, many of the MS procedures have not yet been fully integrated with array technology. However, many of the protocols described herein are performed on solid surfaces such as beads and are presently being translated into a two-dimensional array format; a manuscript such as this would not be complete without reporting on the very latest technology, therefore such data have been included.

17.2 Protocols

Array development

Implementation of array hybridization detection systems necessitates the development of arrays of immobilized oligonucleotides. For successful performance, the immobilized DNA must be stable and not desorb during hybridization, washing, or analysis. The density of the immobilized oligodeoxynucleotide must be sufficient for the ensuing analyses; however, there must be minimal non-specific binding of non-target DNA to the surface. The immobilization process should not interfere with the ability of immobilized probes to hybridize; therefore, it is often best for only one point, ideally a terminus, of the DNA to be immobilized. In considering development of arrays, there are essentially two methods of immobilization: post-synthetic attachment and *in situ* synthesis.

Post synthetic attachment involves automated synthesis of oligodeoxynucleotides [13], removal of the strands from the synthesis support, deprotection, purification, and immobilization at specific positions on the surface of a functionalized chip. Numerous immobilization methods have been investigated including: streptavidin coated silicon chips that are complexed with biotinylated oligonucleotides [14] a glass slide functionalized with 1,4-phenylene diisothiocyanate that is conjugated to amino-functionalized oligonucleotides [15], an iodoacetamido-functionalized silicon surface that is linked to thiolated DNA [16], and an epoxide coated chip that is coupled to amino-functionalized oligonucleotides [17, 18].

DNA has been synthesized *in situ* on chips by two different means. In the first method, synthesis is performed on glass slides in a manner analogous to *in situ* synthesis on CPG beads using automated procedures [13] except the linkage to the support is stable to all deprotecting solutions and remains intact [19, 20]. The second method to synthesize oligodeoxynucleotides *in situ* combines solid phase chemistry, photolabile protecting groups, and photolithography, and results in a set of spatially distinct and highly diverse chemical products. This method was originally devised for peptides, but has proven to be applicable to the synthesis of oligodeoxynucleotides as well [21]. It was possible to assemble on one glass surface all 65536 different octanucleotides (4^8) in only 32 chemical steps [22]. In this method, light is employed to cleave photolabile protecting groups from the surface, then the entire surface is exposed to a phosphoramidite, but coupling only takes place where the protecting groups have been removed. Exposure to illumination determines which regions of the support are activated for cemical coupling; this process is controlled by a series of masks. The cycles continue until a diverse set of 65536 spatial defined oligodeoxynucleotides has been synthesized. One drawback of this method is that purification and quality control of the oligonucleotides synthesized on the array are impossible; for example, incomplete or failure sequences resulting from non-quantitative chemical reactions remain undetected. Additionally, deprotection of the exocyclic amine functions can be either incomplete or can produce side reactions such as deamination of cytosine.

Sequence analysis by direct array hybridization detection

Ordinarily, the pattern of hybridization on an array of immobilized probes is examined using fluorescent detection [23]. For example, PCR of the target amplicon may be perfomed using one biotinylated primer and one fluorescently labeled primer. The product can then be denatured on streptavidin-coated magnetic beads, and the fluorescently labeled strand hybridized on the array; the hybridization pattern is then detected by fluorescence scanning [15]. Depending on the number of elements in the array, the data can then be compiled and analyzed either manually or using computer programs.

An alternative to fluorescent detection uses a charge coupled device (CCD) to obtain data rapidly [17, 18]. In this case, a ^{32}P radiolabeled target is hybridized to an immobilized probe on a silicon wafer. The wafer is then placed upon the CCD surface and a signal is generated. Using the combination of ^{32}P labeled target molecules with CCD detection described above, hybridization signals can be obtained ~10 fold faster than on a gas phase detector and ~100 fold faster than using autoradiography.

Sequence analysis by advanced array hybridization detection

Although a robust system, direct hybridization detection on an array still suffers from obstacles arising from mismatch hybrids. Hybridization can be made more stringent using a technology termed APEX (automated programmable electronic matrix) in which an electric field is applied to the chip during hybridization to concentrate target DNA at array elements; after allowing hybridization to proceed, unhybridized DNA is expelled by reversing the polarity of the field [24]. This technique has proven to greatly reduce mismatch hybrids on an array.

Enzyme-catalyzed steps can also be used to help minimize mismatches on an array. Ligation detection has been modified to incorporate an extension of the immobilized probe using DNA polymerase which additionally minimizes mismatches in the capture (Fig. 1c) [25]. A novel method for detecting single nucleotide polymorphisms employing an array of immobilized DNA probes and enzyme assisted hybridization detection is genetic bit analysis (Fig. 1d) [26]. In this technique, specific fragments of genomic DNA that contain polymorphic sites are amplified by PCR, the amplicon rendered single stranded, and that strand is then captured by hybridization to an array of immobilized probes. The probe is designed to hybridize to the target adjacent to the polymorphic site of interest. The 3′ end of the immobilized probe is then extended by one base using a polymerase in the presence of two differently labeled dideoxynucleoside triphosphates, and two unmodified dideoxynucleoside triphosphates. In a parallel experiment, the labeled and unlabeled ddNTP's were reversed. Because of the different haptenated-ddNTPs incorporated, two enzyme-assisted colorimetric assays may then be performed sequentially to reveal the nature of the extended base.

Sequence analysis by mass spectrometry

Because mass spectrometry produces molecular weight data from the samples analyzed, the result is far more accurate and informative than simple hybridization detection via labels which only reveals a positive or negative signal. Prior to combining chip-based samples with MS detection, the MS technique has been used as a means of detection for existing sample preparation protocols, such as direct sizing of PCR fragments [27–29] and restriction digests [30, 31], but the results are generated faster and the analysis is more precise than gel-based methods. For example, point mutations in the cystic fibrosis gene can be detected by PCR amplifying the DNA template with two sets of different sized primers that overlap at the mutation site [32]. Depending on the presence or absence of a mutation, one set of primers will hybridize and produce an amplicon. Direct analysis of the PCR product on MS reveals the size of the amplicon, and thus, the nature of the mutation site. MS has also been used as a detection method for standard restriction fragment length polymorphisms where the products were either single [33] or double stranded fragments [31].

MS has also recently been introduced into the realm of sequencing DNA templates; however, in this case again, it is not yet a fully array-based technology. MS can replace the common separation/detection method of gel electrophoresis for Sanger dideoxy sequencing reactions, but the analysis is easier without labels or gel preparation and proceeds more quickly. Short synthetic templates (40 to 50 bases) have been sequenced using conventional (solution phase) Sanger sequencing protocols [34] or Sanger cycle sequencing [35] with MS detection of the ladders. An improved scheme employs solid phase Sanger sequencing reactions on streptavidin beads which enable removal of dNTP's, ddNTP's, enzyme, and buffer salts [36]. Such a scheme is successful since it had been shown that duplex DNA, in which one strand is immobilized either through a biotin-streptavidin linkage [37] or through the iodoacetamido linkage [16], can be analyzed by MALDI-TOF MS; only the hybridized strand is directly desorbed off the solid surface. In this case, the Sanger ladders can be directly desorbed from the surface [38]. Alternatively, a similar format is being employed to generate the ladders, however, prior to analysis, the sequencing products are melted off the beads and approximately 10 nl are transferred to the surface of a chip for MS analysis.

A novel technique called primer oligo base extension (PROBE) can be used in combination with MS analysis to definitively reveal the identity of a point mutation [27, 39] or a polymorphic site [40] (scheme shown in Fig. 2). The initial non-array based one-tube protocol involves PCR amplification of the target region with one biotinylated primer, immobilization on a solid surface via streptavidin, denaturation of the hybridized strand, annealing of the

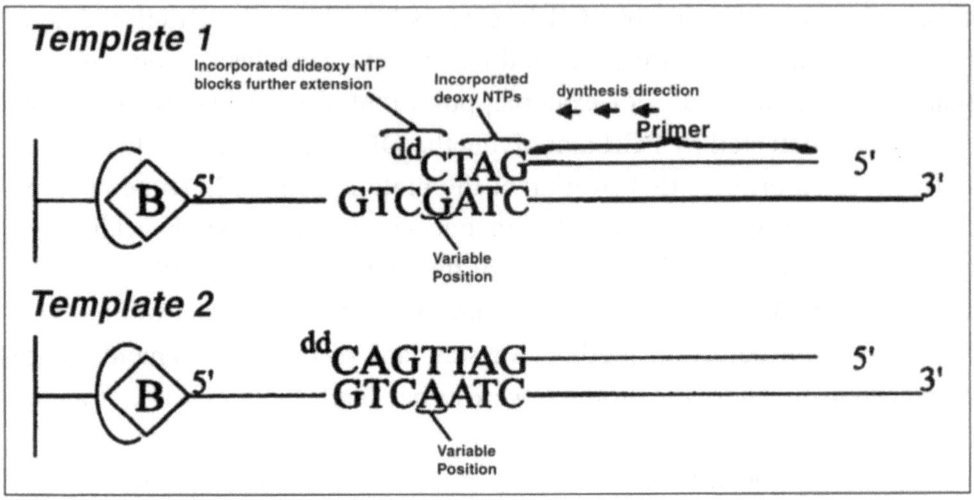

Figure 2 *Scheme of DNA mutation detection by the primer oligo base extension (PROBE) reaction*

PROBE primer upstream of the mutation site of interest, temperature cycled extension by a base composition specific to the identity of the variable base, conditioning of the PROBE products, and MS analysis to reveal the nature of the mutation. Similarly, short tandem repeats (microsatellites) can be analyzed via PROBE and MS by extending through the repeat region [41].

Although analysis time is already greatly reduced compared to electrophoresis, it is desirable to further expedite the process of PROBE reactions analyzed by MS by moving to a miniaturized array format. For the high through-put applications of DNA diagnostics, it is important that MS fragment separation and detection be combined with an array format ("DNA Mass Array Technology"), and that it is amenable to full automation. For example, PCR amplification followed by isothermal PROBE can be performed from 384 different samples and/or patients in 384 wells of a microtiter plate at once. After denaturation of the diagnostic product from the solid support, each sample is transferred to a nanowell in a silicon "SpectroChip" using various nanodispensing techniques. Automated MS can then scan the wells [42], and software programs are employed to compile and analyze the data to reveal the nature of the mutation, polymorphism, or repeat length from each well of the SpectroChip.

17.3 Results and Discussion

Array development

The results of the synthesis of the arrays described previously can be tabulated by the amount of DNA that is conjugated to the flat surface as shown in Table 1. The table is not intended for a comparative analysis of immobilization density since a high density of immobilized probe may not be the ultimate goal of the experiment; each array described has produced successful results for its particular ensuing application.

Sequence analysis by direct array hybridization detection

Arrays of immobilized probes have recently emerged as a commanding technology in diagnostics of the future. Although many varieties of arrays are being utilized for basic research and diagnostics, the most prevalent chips are those synthesized *in situ* by combining solid phase chemistry, photolabile protecting groups, and photolithography [21]. Arrays are currently available to detect literally hundreds of mutations in a plethora of different genes. By analyzing the hybridization pattern of fluorescently labeled target DNA to these

Table 1 Immobilization of DNA on flat surfaces

Solid surface	Oligodeoxy-nucleotide	Coupling agent	Amount of DNA immobilized (fmole mm^{-2} of surface)
Streptavidin-coated silicon wafer	Biotinylated oligomer	–	20
Isothiocyanated-coated glass	NH_2-oligomer	–	130
Iodoacetamide-coated silicon wafer	SH-oligomer	–	250
Epoxide-coated silicon wafer	NH_2-oligomer	–	16
Epoxide-coated glass	*In situ* synthesis	Standard "PC"	110
Amine-coated	*In situ* synthesis	Photolithography	nr

Abbreviations; nr, not reported; "PC", phosphoramidite chemistry.

arrays of densely packed elements, many high throughput arrays have been developed to screen for mutations in the following genes: CFTR gene (cystic fibrosis) [43] HIV-1 reverse transcriptase and protease genes [6, 44], the β-globin gene [45] the mitochondrial genome [46]; and BRCA1 gene (breast cancer) [23]. Due to the lack of quality control of the chip-generated oligonucleotides as described above, the usefulness of this technology will be limited to research applications; for diagnostic applications, a validated array that assures the quality and identity of the immobilized oligonucleotides is required.

Sequence analysis by advanced array hybridization detection

Despite the fact that straight-forward hybridization detection on arrays is being utilized in numerous laboratories for basic research and *de novo* sequencing, the need still exists for more stringent hybridization conditions to reduce mismatches. The technique termed APEX which employs electronic stringency has proven to greatly reduce mismatch hybrids on an array [24]. Additionally, enzyme mediated detection such as genetic bit analysis which involves the capture of target strands, extension of the probe with labeled ddNTP's, and fluorescent analysis to reveal the nature of the extended base has been successfully used to detect single nucleotide polymorphisms [26].

Sequence analysis by mass spectrometry

Whereas the use of array hybridization detection by fluorescence or radiolabels is most frequently used, one must recall that the result is not of the highest possible analytical sensitivity and accuracy. It is possible to envision the use of MS detection from any of the arrays described above to catapult the technology to the next level. Already however, MS detection coupled with either direct PCR amplicon sizing or restriction fragment length polymorphism have successfully been employed for mutation detection of single stranded templates in the CFTR gene [30, 32] the RET proto-oncogene [27], and Apolipoprotein E [33, 40], while double stranded templates have been analyzed in Apolipoprotein E [31].

In addition to MS detection being applied to pre-existing diagnostic assays, it is also a viable detection method for Sanger sequencing reactions. Easily interpreted data can be obtained from conventional Sanger reactions [34] and cycle Sanger reactions [35] performed on synthethic templates in the range of 50 bases in length. By combining solid phase Sanger reactions with MS detection, the sequence of a 39 base synthetic template could be easily determined (Fig. 3) with better resolution than the solution reactions, and it has been

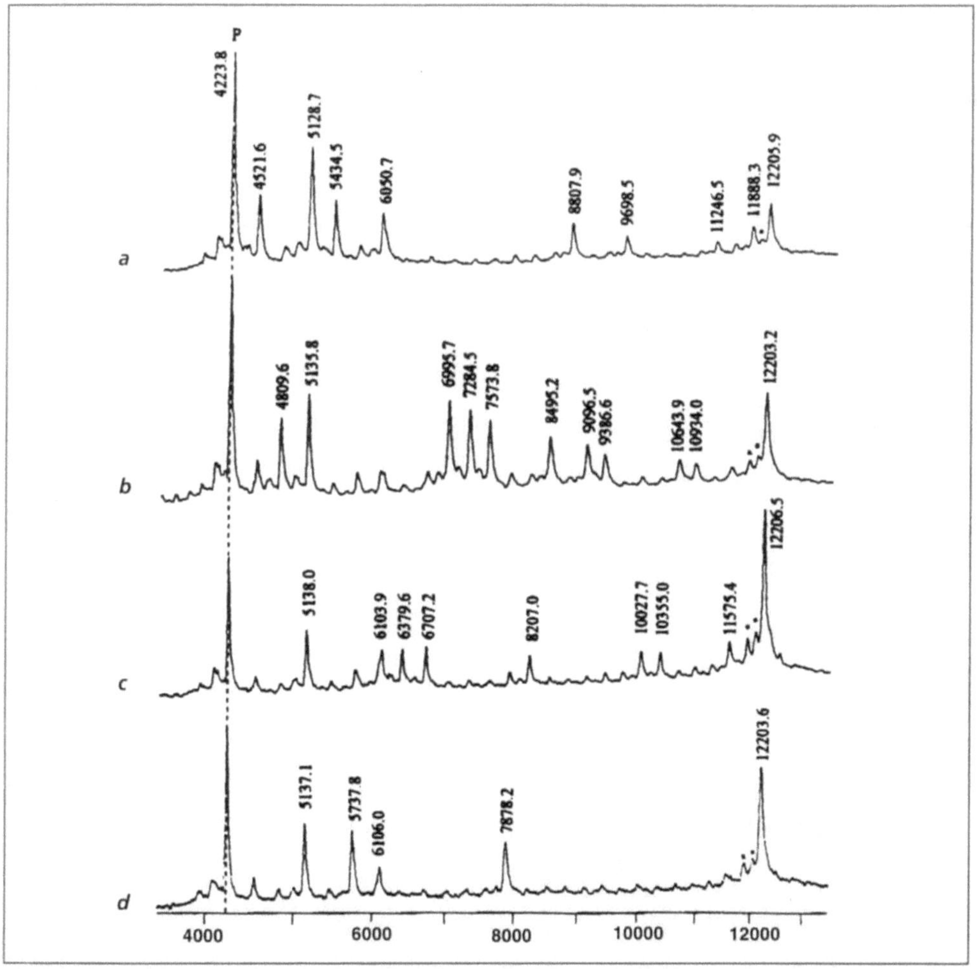

Figure 3 *MALDI-TOF mass spectra of the sequencing ladders generated from immobilized 39-base template strand d(TCT GGC CTG GTG CAG GGC CTA TTG TAG TTG TGA CGT ACA-[A]$_n$). P refers to the primer d(TGT ACG TCA CAA CT). The peaks resulting from depurination are labeled by an asterisk. (a) A-reaction, (b) C-reaction, (c) G-reaction, and (d) T-reaction*

possible to extend the analysis to approximately 63 bases in length [38]. The facile nature of this approach has also enabled to sequence exons 5, 6, 7, and 8 of the p53 gene using Sanger sequencing with primer walking [47]. In addition, this approach has been miniaturized by nanodispensing the sequencing ladders taken from the beads to the surface of a SpectroChip prior to MS analysis. As shown in Figure 4, data measured from such miniaturized arrays is comparable to larger scale preparations, while still enabling smaller reactions and automated MS scanning. Since the reactions can be robustly performed on the solid surface of a bead, this method is now at the point where it can be fully translated onto the surface of a miniaturized chip using various nanodispensing techniques and parallel processing, and the MS detection of ladders on the SpectroChip can be fully automated to rapidly reveal the sequence. Compared to gel electrophoresis, MS not only significantly increases the speed of separating and detecting sequencing ladders, but also

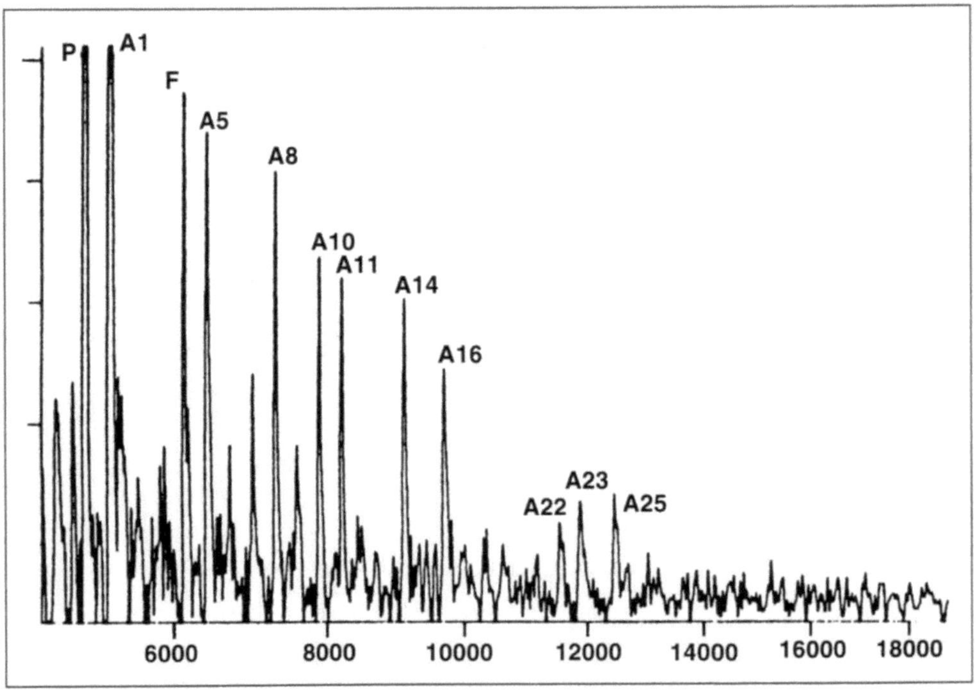

Figure 4 *Mass spectra of the Sanger sequencing ladder generated from the A-reaction for a region of exon 7 of the p53 gene; 5 nl of the sequencing product were dispensed to the surface of a chip for MS analysis. P refers to the sequencing primer; F refers to a false stop*

eliminates gel-based artifacts such as band compression and provides a more informative signal than a labeled band in a gel.

The PROBE reaction described above in conjunction with non-array based MS detection has been streamlined into a "one-tube" process and used to perform a pentaplex reaction; five different loci on three different genes related to lipid metabolism (Apo E, Apo B, Apo A4) were successfully analyzed simultaneously [48]. In addition to the well characterized, published

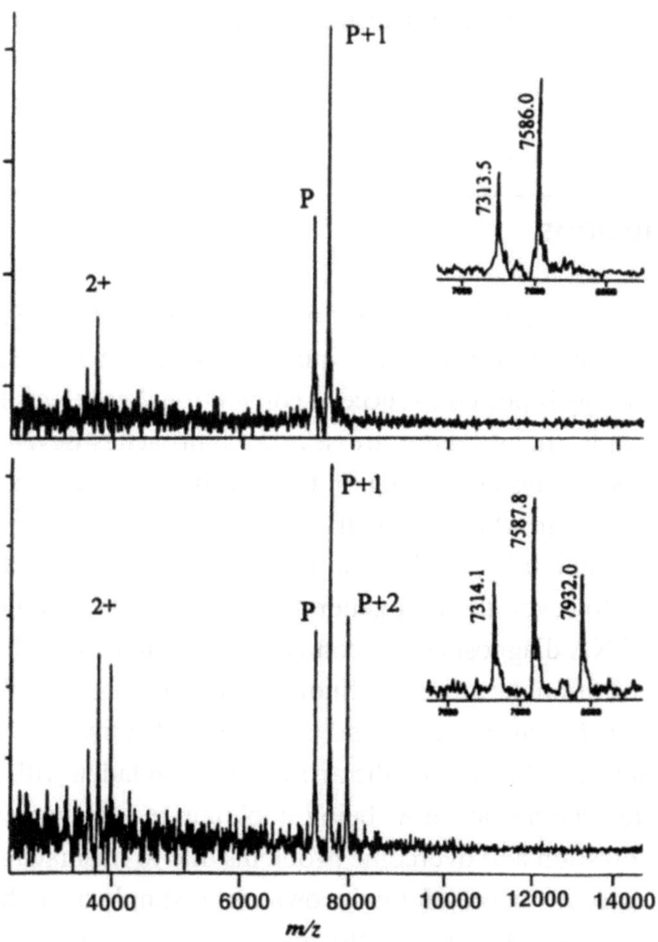

Figure 5 *Differentiation of (top) homozygous and (bottom) heterozygous alleles of the apolipoprotein B gene position 3500 using the PROBE reaction; 5 nl of the diagnostic product were dispensed to the surface of a chip for MS analysis. P refers to the PROBE primer*

results described above, protocols for investigating greater than 100 sites in genes/loci containing insertion, deletion, and base substitution mutations/polymorphisms or short tandem repeats have been developed in our laboratory. Because of the countless genetic sites which could be examined by the PROBE assay, it is desirable to miniaturize the scale of these reactions similar to the miniaturization which was carried out on the sequencing samples. Following diagnostic product generation, nanoliters of sample can be transferred to the surface of a SpectroChip and hundreds of samples analyzed rapidly by MS; one such example is shown in Figure 5. Additionally, the PROBE technology provides for results which are easy to interpret and unambiguous.

17.4 Conclusions

Array hybridization is a powerful tool for rapidly analyzing large numbers of DNA fragments and producing large quantities of DNA sequence information; this technology is poised to succeed conventional gel electrophoresis for a number of applications. There are several approaches developed for the synthesis of these arrays depending on the density and complexity required. To date, the manner in which dense arrays can be synthesized in a cost effective way while maintaining quality control over the oligonucleotides has not been resolved. However, the first generation of the arrays are now being investigated for DNA diagnostics and comparative sequencing. Most commonly, direct hybridization detection via fluorescent scanning is employed for the diagnostic assays, but mismatched and branched hybridizations can produce erroneous results; for this reason, the arrays are often laden with redundancy. More elaborate schemes are now being implemented to increase the stringency of hybridization and overcome problems due to mismatch hybrids.

Although currently successful in its own right, simple array hybridization detection does not result in data of the highest accuracy; a positive result for hybridization can be quite ambiguous since the researcher is not certain of the identity of the molecule that is hybridized. Mass spectrometric detection, however, adds an element of accuracy that is not available with other means, i.e. the molecular weight of the molecule. Such a detection method in conjunction with miniaturized arrays (i.e. DNA Mass Array Technology) is emerging as a tool of the future for diagnostics and sequencing. The combination of parallel processing, rapid detection, and the ability to automate the entire process will enable the technology to successfully compete in laboratories of the 21st century.

References

1 Bains W, Smith GC (1988) A novel method for nucleic acid sequence determination. *J Theor Biol* 135: 303–307

2 Khrapko KR, Lysov YP, Khorlyn AA, Shick VV, Florentiev VL, Mirzabekov AD (1989) An oligonucleotide hybridization approach to DNA sequencing. *FEBS Letts* 256: 118–122

3 Strezoska Z, Paunesku T, Radosavljevic D, Labat I, Drmanac R, Crkvenjakov R (1991) DNA sequencing by hybridization: 100 bases read by a non-gel-based method. *Proc Natl Acad Sci USA* 88: 10089–10093

4 Drmanac R, Labat I, Brukner I, Crkvenjakov R (1989) Sequencing of megabase plus DNA by hybridization: Theory of the method. *Genomics* 4: 114–128

5 Maskos U, Southern EM (1993) A study of oligonucleotide reassociation using large arrays of oligonucleotides synthesized on a glass support. *Nucl Acids Res* 21: 4663–4669

6 Kozal MJ et al (1996) Extensive polymorphisms observed in HIV-1 clade B protease gene using high density oligonucleotide arrays. *Nature Med* 2: 753–759

7 Broude NE, Sano T, Smith CL, Cantor CR (1994) Enhanced DNA sequencing by hybridization. *Proc Natl Acad Sci USA* 91: 3072–3076

8 Karas M, Hillenkamp F (1988) Laser desorption ionization of proteins with molecular masses exceeding 10000 daltons. *Anal Chem* 60: 2299–2301

9 Cotter RJ (1992) Time-of-flight mass spectrometry for the structural analysis of biological molecules. *Anal Chem* 64: 1027A–1039A

10 Vestling MM, Fenselau C (1994) *Anal Chem* 66: 472–477

11 Nordhoff E, Cramer R, Karas M, Hillenkamp F, Kirpekar F, Kristiansen K, Reopstorff P (1993) Ion stability of nucleic acids in infrared matrix-assisted laser desorption/ionization mass spectrometry. *Nucleic Acids Res* 21: 3347–3357

12 Tang K, Tarenenko NI, Allman SL, Chang LY, Chen CH (1994) Detection of 500-Nucleotide DNA by laser desorption mass spectrometry. *Rapid Commun Mass Spectrom* 8: 727–730

13 Sinha ND, Biernat J, Köster H (1983) Beta-Cyanoethyl N,N-Dailkylamino/N-Morpholinomonochloro Phophoamidites, New Phophitylating Agents Facilitating Ease of Deprotection and Work-Up of Synthesized Oligonucleotides. *Tetrahedron Lett* 24: 5843–5846

14 O'Donnell MJ, Köster HK, Cantor CR, unpublished results

15 Guo Z, Guilfoyle RA, Thiel AJ, Wang R, Smith LM (1994) Direct fluorescence analysis of genetic polymorphisms by hybridization with oligonucleotide arrays on glass supports. *Nucl Acids Res* 22: 5456–5465

16 O'Donnell MJ, Tang K, Köster H, Smith CL, Cantor CR (1997) High-density, covalent

attachment of DNA to silicon wafers for analysis by MALDI-TOF mass spectrometry. *Anal Chem* 69: 2438–2443

17 Lamture JB, Beattie KL, Burke BE, Eggers MD, Ehrlich DJ, Fowler R, Hollis MA, Kosicki BB, Reich RK, Smith SR, Varma RS, Hogan ME (1994) Direct detection of nucleic acid hybridization on the surface of a charge coupled device. *Nucl Acids Res* 22: 2121–2125

18 Eggers MD, Hogan ME, Reich RK, Lamture JB, Ehrlich DJ, Hollis MA, Kosicki BB, Powdrill T, Beattie KL, Smith SR, Varma RS, Gangadharan R, Mallik A, Burke BE, Wallace D (1994) A microchip for quantitative detection of molecules utilizing luminescent and radioisotope reporter groups. *Bio Techniques* 17: 516–524

19 Maskos U, Southern EM (1992) Oligonucleotide hybridizations on glass supports: A novel linker for oligonucleotide synthesis and hybridization properties of oligonucleotides synthesized *in situ*. *Nucl Acids Res* 20: 1679–1684

20 Southern EM, Case-Green SC, Elder JK, Johnson M, Mir KU, Wang L, Williams JC (1994) Arrays of complementary oligonucleotides for analysing the hybridization behavior of nucleic acids. *Nucl Acids Res* 22: 1368–1373

21 Fodor SPA, Read JL, Pirrung MC, Stryer L, Lu AT, Solas D (1991) Light-directed, spatially addressed parallel chemical synthesis. *Science* 251: 767–773

22 Jacobs JW, Fodor SPA (1994) Combinatorial chemistry – applications of light-directed chemical synthesis. *Trends in Biotech* 12: 19–26

23 Hacia JG, Brody LC, Chee MS, Fodor SP, Collins FS (1996) Detection of heterozygous mutations in BRCA1 using high density oligonucleotide arrays and two-color fluorescence. *Nature Genetics* 14: 441–447

24 Heller MJ (1997) Proceedings of the third annual biochip arrays conference, San Diego, CA. "Integrated electronic device for multiplex multiple gene DNA hybridization diagnostics

25 Kuppuswamy MN, Hoffman JW, Kasper CK, Spitzer SG, Groce SL, Bajaj SP (1991) Single nucleotide primer extension to detect genetic diseases; experimental application to hemophilia B (factor IX) and cystic fibrosis genes. *Proc Natl Acad Sci USA* 88: 1143–1147

26 Nikiforov TT, Rendle RB, Goelet P, Rogers YH, Kotewicz ML, Anderson S, Trainor GL, Knapp MR (1994) Genetic bit analysis: A solid phase method for typing single nucleotide polymorphisms. *Nucl Acids Res* 22: 4167–4175

27 Little DP, Braun A, Darnhofer-Demar B, Frilling A, Li Y, McIver RT, Köster H (1997) DNA point mutation detection schemes for the RET proto-oncogene using low versus high resolution MALDI mass spectrometry. *J Mol Med* 75: 745–750

28 Jurinke C, Zollner B, Feucht HH, Jacob A, Kirchübel J, Luchow A, van den Boom D, Laufs R, Köster H (1996) Detection of hepatitis B virus DNA in serum samples via nested PCR and MALDI-TOF mass spectrometry. *Genet Anal* 13: 67–71

29 Siegert CW, Jacob A, Köster H (1996) Matrix assisted laser desorption/ionixation time-of-flight (MALDI-TOF) mass spectrometry for the detection of polymerase chain reaction (PCR) products containing 7-deazapurine moities. *Anal Biochem* 243: 55–65

30 Liu YH, Bai J, Zhu Y, Liang X, Siemieniak D, Venta PJ, Lubman DM (1995) Rapid screening of genetic polymorphisms using buccal cell DNA with detection by MALDI mass spectrometry. *Rapid Commun Mass Spectrom* 9: 735–743

31 Little DP, Jacob A, Becker T, Braun A, Darnhofer-Demar B, Jurinke C, van den Boom D, Köster H (1997) Direct detection of syn-

thetic and biologically generated double-stranded DNA by MALDI-TOF MS. *Int J Mass Spectrom Ion Processes* 169/170: 133–140

32 Taranenko NI, Matteson KJ, Chung CN, Zhu YF, Chang LY, Allman SL, Haff L, Martin SA, Chen CH (1996) Laser desorption mass spectrometry for point mutation detection. *Genetic Anal: Biomol Eng* 13: 87–94

33 Little DP, Braun A, Higgins GS, Darnhofer B, Leppin L, Craven F, Lough DM, Köster H (1996) Proceedings of the 44th ASMS conference on mass spectrometry and allied topics, Portland, OR, may 12–16, 1996. "Mass spectrometric methods for genotyping"

34 Shaler TA, Tan Y, Wickham JN, Wu KJ, Becker CH (1995) Analysis of enzymatic DNA sequencing reactions by MALDI-TOF mass spectrometry. *Rapid Commun Mass Spectrom* 9: 942–947

35 Roskey MT, Juhasz P, Smirnov I, Takach EJ, Martin SA, Haff LA (1996) DNA sequencing by delayed extracton MALDI TOF mass spectrometry. *Proc Natl Acad Sci USA* 93: 4724–4729

36 Stahl S, Hultman T, Olsson A, Moks T, Uhlen M. Solid phase DNA sequencing using the biotin-streptavidin system. *Nucleic Acids Res* 16: 3025–3038

37 Tang K, Fu DJ, Cotter RJ, Cantor CR, Köster H (1995) MALDI mass spectrometry of immobilized duplex DNA probes. *Nucleic Acids Res* 23: 3126–3131

38 Köster H, Tang K, Fu DJ, Braun A, van den Boom D, Smith CL, Cotter RJ, Cantor CR (1996) A strategy for rapid and efficient DNA sequencing by mass spectrometry. *Nature Biotech* 14: 1123–1128

39 Braun A, Little DP, Köster H (1997) Detecting CFTR gene mutations by using primer oligo base extension and mass spectrometry. *Clin Chem* 43: 1151–1158

40 Little DP, Braun A, Darnhofer-Demar B, Köster H (1997) Improved apolipoprotein E genotyping specificity using mass spectrometry. *J Mol Med* 75: 745–750

41 Braun A, Little DP, Reuter D, Mueller B, Köster H (1997) Improved analysis of microsatellites using mass spectrometry. *Genomics* 46: 18–23

42 Little DP, Cornish T, O'Donnell MJ, Braun A, Cotter R, Köster H (1997) MALDI on a chip: Analysis of low- to subfemtomole quantities of oligonucleotides and DNA diagnostic products dispensed by a "piezoelectric pipette". *Anal Chem* 69: 4540–4546

43 Cronin MT et al (1996) Cystic fibrosis mutation detection by hybridization to light generated DNA probe arrays. *Hum Mut* 7: 244–255

44 Lipshutz RJ et al (1995) Using oligonucleotide arrays to probe genetic diversity. *Biotechniques* 19: 442–447

45 Yerchov G et al (1996) DNA analysis and diagnostics on oligonucleotide microchips. *Proc Natl Acad Sci USA* 93: 4913–4918

46 Chee MS et al (1996) Accessing genetic information with high density DNA arrays. *Science* 274: 610–614

47 Fu DJ, Tang K, Braun A, Reuter D, Darnhofer-Demar B, Little DP, O'Donnell MJ, Cantor CR, Köster H (1998) Sequencing Exons 5 to 8 of the p53 gene by MALDI-TOF mass spectrometry. *Nature Biotechnol* 16: 381–384

48 Braun A et al, in preparation

18 Manipulation of particles, cells and liquid droplets by high frequency electric fields

Thomas Schnelle, Torsten Müller and Günter Fuhr

18.1 Introduction

Physics has a long tradition for manipulation of single particles or particle ensembles. Penning (1936) and Paul and Raether (1955) introduced electromagnetic cages for the trapping of atoms, ions and elementary particles *in vacuo*. This revolutionised physics and allowed spectroscopic investigation of single ions or atoms over days, weeks or months (Paul, 1990). Later, a similar technique based on laser beams was developed (Ashkin, 1978; Ashkin et al., 1986). These methods can be combined and are still the basis of exciting physical experiments (Cornell, 1995; Monroe et al., 1996). Particles can be both trapped and accurately manipulated *in vacuo*. Using a tunnelling microscope as tweezers, Meyer et al. (1996), were able to manipulate a single Cu-atom after removing it from a crystal.

The manipulation of single cells and micro particles in liquids has to take account of Brownian motion and avoid damagingly strong forces. This sets limits, especially for the handling of single macro-molecules. On the other hand, biological cells show an astonishing reliability in manipulating sub-units and even single macro-molecules such as DNA.

Dielectric forces can be used for manipulation of particles in liquids. Alternatives are mechanical forces (conventional tweezers), hydrodynamic flow, ultrasound, magnetic forces and laser tweezers. Due to the low difference in magnetic susceptibility of cells and water, magnetic fields are only suitable if the cells are bound to ferromagnetic particles. Hydrodynamic flow and ultrasound are difficult to localise and can hardly be used for single cell handling although they may provide useful "background forces". Currently, laser tweezers and high frequency electric fields are being studied for single cell manipulation. Laser traps are based on differences in the refractive indices between particle and medium and on the radiation pressure. Dielectrophoresis

(DEP), travelling wave dielectrophoresis (TWD) and electrorotation (ER) utilise differences in the electric polarizability of particles and the surrounding liquid.

If a dielectric particle is exposed to an external electric field it polarises. The size and direction of the induced dipole depend on field frequency and dielectric properties (conductivity σ and permittivity ε). This is schematically shown in Figure 1.

Since a dipole experiences a force in an inhomogeneous external field, particles are either attracted to regions of high field strength (positive DEP, particle 4) or repelled from them (negative DEP, particles 1–3). Similarly, dielectric particles can be moved in electric fields due to a gradient in the field phase. This effect is used in TWD and ER and is due to a phase lag between the external field and the induced dipole.

Local field minima can be created within a liquid, far from any surfaces, by using appropriate electrode configurations. This allows negative DEP to create closed field cages for particles and cells without feedback. Unlike elementary particle traps, dielectrophoretic cages do not require particle spinning or the application of additional forces. Particle spinning induced in rotating electric fields (Arnold and Zimmermann, 1982) can, however, be used for particle characterisation (dielectric spectroscopy, Zimmermann and Neil, 1996).

The situation becomes more complex in dense particle suspension due to the interaction of the induced dipoles. Particles can attract each other (e.g. particle 1 and 2, or 2 and 4 in Fig. 1) or repel each other (e.g. particle 3 and 4,

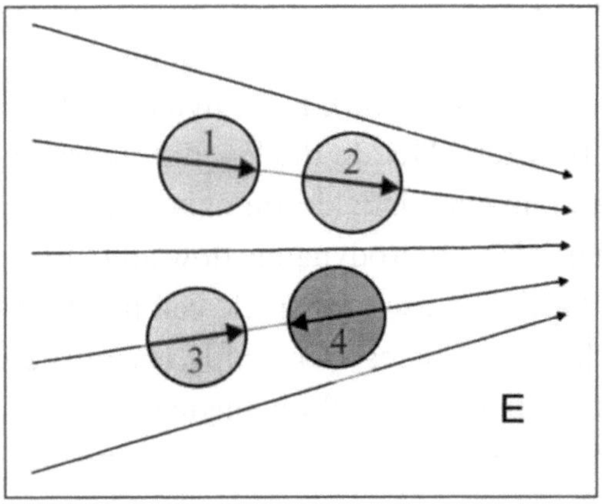

Figure 1 *Principle of dielectrophoresis. The polarisation of four particles in an external electric field is shown. Depending on field frequency and passive dielectric properties of particles and the external medium the particles polarise in direction (symbolised by arrows in the particles 1–3) of the electric field, E, or opposite to it (particle 4). In an inhomogeneous field this causes a force due to the interaction of induced dipole and external field. The particles 1–3 show negative and particle 4 shows positive DEP*

or 1 and 3). As a consequence, particles can form the so called pearl-chains or compact aggregates. Additionally, particle interactions can produce spinning (in Fig. 1 this is obvious for the particles 1 and 4, or 2 and 3).

Travelling electric fields can be used both for particle manipulation and for electrohydrodynamic (EHD) pumping of slightly conducting liquids. Melcher (1966) demonstrated that waves of electric fields, travelling perpendicular to the liquid surface or to a conductivity/permittivity-gradient, induce charges at the interface or in the liquid bulk, respectively. The interaction of induced charges and the travelling electric field can result in liquid pumping. The induction of pumping is a general phenomenon in microstructures – especially if highly conductive liquids such as physiological media are used. It can be

Table 1 Development of particle manipulation using DEP and TWD

Field-mapping by chaining of quinine sulphate particles	(Weiler, 1893)
Pearl chaining of fat particles in ac fields	(Muth, 1927)
Positive DEP of large macromolecules	(Debye et al., 1954a,b)
Separation of dead and live cells by DEP	(Pohl and Hawk, 1966)
EHD-pumps: requirement for an inhomogeneous fluid	(Melcher, 1966)
Single cell DEP	(Chen and Pohl, 1974)
Stable levitation of bubbles by negative DEP	(Jones and Bliss, 1977)
Positive and negative DEP of micro-organisms in flow	(Pohl et al., 1981)
Particle separation in synchronous travelling electric fields	(Masuda et al., 1987)
Electrostatic handling of biological molecules	(Washizu, 1991)
Microfabricated arrays for asynchronous TWD of microparticles and cells	(Fuhr et al., 1991)
DEP of viable and non-viable yeast cells	(Huang et al., 1992)
Positive and negative DEP in castellated electrodes	(Pethig et al., 1992)
High frequency dielectric field cages	(Fuhr et al., 1992, Schnelle et al., 1993)
Feedback controlled levitation	(Kaler et al., 1992)
Electrostatic stretch and positioning of DNA	(Washizu et al., 1993)
Micropump for aqueous liquids	(Müller et al., 1993a)
Particle microtools (micromanipulator)	(Fuhr et al., 1994a/b)
Chromatography based on positive DEP	(Washizu et al., 1994)
DEP in physiological media	(Fuhr et al., 1994c)
Virus enrichment by DEP	(Fuhr et al., 1995, Green et al., 1995)
Separation of cancer cells from blood	(Becker et al., 1995)
Trapping of sub-micron particles by negative DEP	(Müller et al., 1996)

utilised for the enrichment of microparticles (filtering) and for the supply of closed microsystems.

DEP and TWD are also known as ponderomotive forces and the development of their use for particle manipulation is summarised in the Table 1.

Currently, whole microsystems for cell handling are being developed and extended for use with sub-micron particles. These trends and new possibilities for particle manipulation will be discussed in this paper. Special attention is given to the influence of streaming and higher multipoles in field cages and traps.

18.2 Protocols and notes

Experimental

The general experimental set-up was as depicted in Figure 2. A PC (486 at 33 MHz) controls a generator (Hewlett Packard HP 8131 A, USA) at frequencies from 1 kHz to 250 MHz and a typical amplitude of 4 V. The generator is connected to a liquid filled micro chamber. Typically, the liquid volume is about 20 µl. The electrodes are fed with 90° or 180° phase shifted rectangular signals to produce rotating or non rotating electric field for traps, cages or travelling electric fields. In the presented experiments we used aqueous suspensions of mammalian cells, Sephadex spheres (G 15), viruses and latex beads and a two phase system (dextran/n-propanol).The resulting particle behaviour is monitored via a microscope (Zeiss Axioskop, Germany) and camera (Kappa 15/2, Germany) and recorded on a videotape.

Theoretical

Principles of dielectrophoresis and travelling wave dielectrophoresis

In an (inhomogeneous) ac electric field, the frequency-dependent differences in polarisability of particles and liquid causes an induced translatational force,

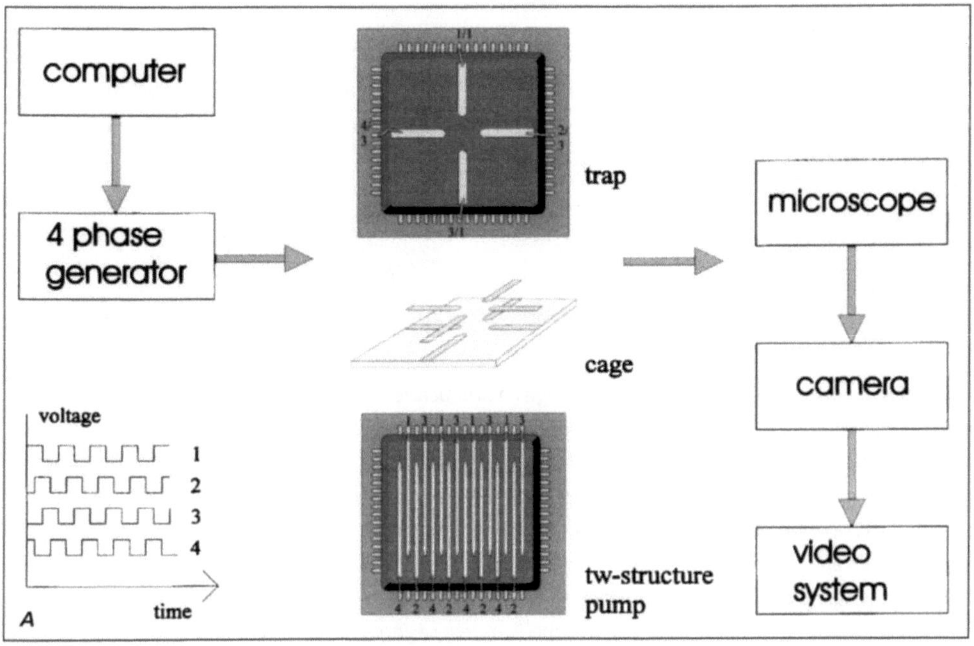

Figure 2A *Experimental set-up. A: Sketch of used structures and devices. Flow of information is symbolised by grey arrows. A computer triggers a four phase generator (definition of the used phases are given below the generator box). The generator is connected to one of the micro electrode structures shown in the middle column. Depending on the driving of the trap (upper structure) either a rotating electric field (electrode feeding with phases according to the number in front of the slash) or an ac driving of the electrodes can be realised (electrode feeding with phases according to the number after the slash). Two planar traps mounted face-to-face form a three-dimensional field cage. Again, either ac or rotating drive can be used. 180° phase difference in the signal to electrodes at adjacent vertices of the cube gives ac drive. Planar interdigitated electrodes (lower design of the middle column) driven with 90° phase shifted pulses give rise to travelling electric fields which can be used either for travelling wave dielectrophoresis of particles or for the pumping of liquids*

F, and a torque, N, in DEP and ER respectively. The dielectrophoretic forces, $\vec{F}$, acting on a dielectric particle of radius, R, in a time periodic electric field (with radian frequency ω)

$$\vec{E}(\vec{r},t) = \mathrm{Re}\,[(\vec{E}^{\mathrm{re}}(\vec{r}) + i\vec{E}^{\mathrm{im}}(\vec{r}))\exp[i\omega t]] \tag{1}$$

can be expressed in dipole approximation as

$$\vec{F}_{\mathrm{d}}(t) = (\vec{m}(t) \circ \vec{\nabla})\vec{E}(t) \tag{2}$$

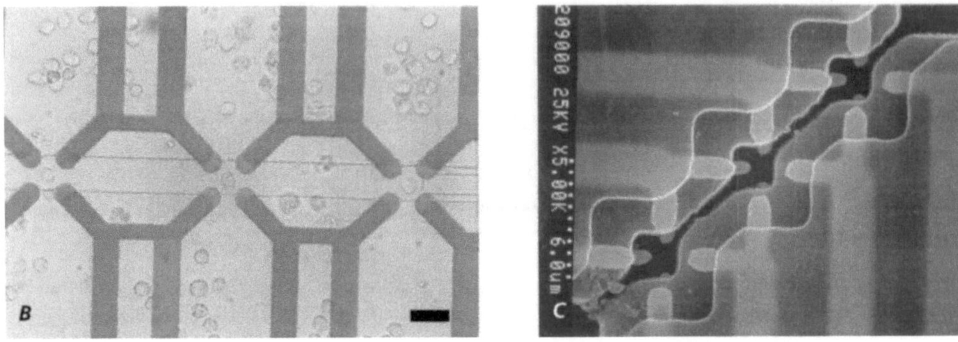

Figure 2B, C *Examples for eight electrode cages manufactured with semiconductor technology. B) On a silicon wafer and processed using photolithography, two metal electrodes planes are insulated by a silicon oxide layer. The silicon oxide is structured to form a channel through three cages (bar 20 µm). The structure was filled with fibroblasts suspended in a physiological solution (phosphate buffered saline solution, PBS). In each of the cages one fibroblast is trapped by high frequency electric fields. C) E-beam processed field cages on silicon with an electrode distance of 1.25 µm and 2.5 µm in the lower and upper electrode plane respectively. The insulating silicon oxide layer between the electrode planes has a thickness of 2 µm. Minimising the contact area between electrodes and external medium reduces non-useful current flow and heating. Therefore, the upper electrodes are covered with an additional silicon oxide layer except for the tips*

with the induced dipole moment $\vec{m}(t) = 4\pi R^3 \varepsilon_l f_{CM} \vec{E}(t)$. In rotating fields the particle experiences a torque $\vec{N}(t) = \vec{m}(t) \times \vec{E}(t)$. The subscripts l and p refer to the liquid and particle, respectively. f_{CM} represents the Clausius-Mossotti-factor for the particle (which has been calculated for multi-shelled particles; see for example Huang et al., 1992; Müller et al., 1993b). For a homogeneous sphere it has been found to be

$$f_{CM} = \frac{\tilde{\varepsilon}_p - \tilde{\varepsilon}_l}{\tilde{\varepsilon}_p + 2\tilde{\varepsilon}_l} \quad \text{with} \quad \tilde{\varepsilon}_l = \sigma_l + i\omega\varepsilon_l \tag{3}$$

where σ represents the conductivity and ε the absolute permittivity. For more complicated particles, ε_p and σ_p in Eq. (2) has to be replaced by effective frequency dependent values. Then the time averaged force on a dielectric particle can be calculated to be

$$\langle \vec{F}_d \rangle = 2\pi\,\varepsilon_l R^3 \left[\mathrm{Re}(f_{CM}) * \vec{\nabla} E^2_{rms} + \mathrm{Im}(f_{CM}) * \left(\sum_{\mu,\nu} \left[E^{re}_{\mu} \frac{\partial E^{im}_{\mu}}{\partial \nu} - E^{im}_{\mu} \frac{\partial E^{re}_{\mu}}{\partial \nu} \right] \vec{e}_{\nu} \right) \right] \tag{4}$$

with the unity vector e, and the torque

$$\langle N \rangle = 4\pi \varepsilon_l R^3 \mathrm{Im}\,(f_{CM}) \vec{E}^{\,im} \times \vec{E}^{\,re} \tag{5}$$

Usually, dielectrophoretic measurements are made by applying electric fields with no gradients in phase. Then, the second part in Eq. (4) vanishes and the time averaged dielectric force is only proportional to the real part of the Clausius-Mossotti-factor, whereas the electrorotation depends on the imaginary one (Pastushenko et al., 1985). Depending on the sign of $\mathrm{Re}[f_{CM}]$ the particle is either attracted to the electrodes (positive DEP) or repelled from them (negative DEP, $\mathrm{Re}[f_{CM}]<0$). Many cells show only negative DEP if suspended in physiological media, because both their effective conductivity and permittivity are lower than that of the liquid (Fuhr et al., 1994c). This allows low stress manipulation of biological objects. In travelling wave structures as shown in Figure 2 where the electric field travels in a direction perpendicular to the electrode strips and parallel to the electrode plane, the particles are again repelled or attracted to the electrodes according to $\mathrm{Re}[f_{CM}]$; they are also forced in the direction of field propagation or against it according to $\mathrm{Im}[f_{CM}]$.

In most cases the dipole approximation yields quantitatively correct results. However, there are situations where the induced dipole moment of the particle vanishes. Then, higher moments have to be considered. For the quadrupole part of the dielectrophoretic force, Jones and Washizu (1996) found:

$$\langle \vec{F}_q \rangle = \frac{2}{3} \pi \varepsilon_l R^5 \,\mathrm{Re}\left[f_{CM}^{II} \sum_{\mu,\nu,\gamma} \frac{\delta E_\mu}{\delta \nu} \frac{\delta^2 E_\gamma^*}{\delta \nu\, \delta \mu} \vec{e}_\gamma \right] \tag{6}$$

with the same definitions as in Eq. (2) and $f_{CM}^{II} = \dfrac{\tilde{\varepsilon}_p - \tilde{\varepsilon}_l}{2\tilde{\varepsilon}_p + 3\tilde{\varepsilon}_l}$, the asterix stands for the complex conjugate. As for the dipole approximation the force can be split into two parts – one has a potential and is proportional to the real part of f_{CM}^{II}. Higher moments can be calculated similarly (see Jones and Washizu, 1996; Jones, 1995).

Principles of liquid pumping using travelling electric fields

In an incompressible liquid of density ϱ and viscosity μ the velocity profile, v, is determined by

$$\varrho \frac{D\vec{v}}{Dt} = \varrho \tilde{g} + \nabla \circ (T^m + T^e)$$

with

$$T_{ij}^{m} = \mu\left(\frac{\partial v_j}{\partial x_i} + \frac{\partial v_i}{\partial x_j}\right) - \partial_{ij} p,$$

$$T_{ij}^{e} = \varepsilon E_i E_j - 0.5 \varepsilon \partial_{ij} E_i E_j \tag{7}$$

where p stands for the pressure and the Kronecker symbol $\partial_{ij} = \begin{cases} 1 & if \quad i = j \\ 0 & if \quad i \neq j \end{cases}$.

The Maxwell stress tensor T^e accounts both for forces attributable to free charges (of density ϱ) and polarisation charges. This can be seen from the identity:

$$\nabla \circ T^e = q\vec{E} - 0.5\,\vec{E}^2 \nabla\varepsilon$$

Therefore, EHD-pumping with harmonic electric fields (Eq. 1) requires a dielectrically inhomogeneous liquid – otherwise no charges can be induced by ac electric fields. Usually, fluids do not exhibit such inhomogeneities but stable anisotropies can be created by layering non-mixing liquids, inserting foreign microparticles or imposing a temperature gradient. At certain frequencies (propagation velocities) of the electric field the induced charges are slightly displaced and interact with the travelling field due to charge relaxation processes. As a result, forces act at the fluid interfaces, on the foreign particle surface or in the volume driving the fluid forward or backward (with or against the direction of wave propagation), respectively. Maximum pumping is to be expected at a field radian frequency that matches the inverse charge relaxation time.

Numerical approaches

Field distribution

For arbitrary electrode configurations, the calculation of electric fields cannot be simplified to an analytical solution. We use a finite difference method to solve the complex field equation:

$$\nabla[(\sigma + i\omega\varepsilon)\nabla\varphi] = 0 \tag{8}$$

with $\vec{E} = -\nabla\varphi$ and appropriate boundary conditions. Equation (8) neglects the charge transport due to streaming which would give rise to an additional term

$\nabla[\vec{v}\nabla[\varepsilon\nabla\varphi]]$ on the left-hand side and hence a coupling to Eq. (7). In general Eq. (8) is additionally coupled via the Ohmic heating $q = \sigma E_{\mathrm{ms}}$ (where ms stands for the time averaged mean square value) to the temperature, T, which is determined by:

$$\nabla[\lambda\nabla T] + q = 0 \tag{9}$$

This equation can also be solved numerically using a finite difference method. Note that due to the heating the thermal conductivity, λ, as well as electric conductivity, σ, and permittivity, ε, are spatial functions. So far, only steady-state solutions have been achieved, assuming that the system has reached a stationary temperature field and neglecting the influence of the streaming to the electric field distribution.

Particle motion

Since we are operating in aqueous media the acceleration term can be neglected. For small particles, not only deterministic forces

$$\vec{F}^{\mathrm{det}} = \vec{F}_{\mathrm{DEP}} + \vec{F}_{\mathrm{hydrodynamic}} + \vec{F}_{\mathrm{bouyancy}} + \vec{F}_{\mathrm{int\,eraction}} + \ldots$$

but also thermal forces have to be considered. The trajectory of a particle i, $\vec{r}_{\mathrm{i}}(t)$, can be determined by numerical integration of the following stochastic (Langevin) equation:

$$\dot{\vec{r}}_{\mathrm{i}}(t) = \frac{\vec{F}^{\mathrm{det}}}{\gamma} + \sqrt{\frac{2kT}{\gamma}}\,\vec{\xi}(t) \tag{10}$$

where the damping constant is given by Stokes law $\gamma = 6\pi\,\eta_1 R$ (ξ stands for independent Gaussian (white) noise).

Forces between two particles are taken into account in dipole approximation. The additional force a particle i experiences in the field that is induced by the particle j reads

$$\vec{F}_{\mathrm{ij}}^{\mathrm{int}}(t) = (\vec{m}_{\mathrm{i}}(t) \circ \vec{\nabla})\,\vec{E}_{\mathrm{ind}}(\vec{m}_{\mathrm{j}}(t)) \tag{11}$$

(for details see for example Fiedler et al. (1995 a)). In addition, at small inter-particle distances, a repulsive potential has to be introduced to avoid the overlapping of particles and to simulate the effect of surface layers for charged particles. Furthermore, in dense suspensions, the electrically induced spinning of particles yields a hydrodynamic coupling between the particles (this has so far not been incorporated into our numerical procedure; for details see Hu et al., 1994).

18.3 Results

General results

Polarisation and hence behaviour of particles in electric fields depends strongly on the field frequency. In Figure 3 examples for the dielectrophoretic and electrorotational spectra of mammalian cells (human erythrocytes in Figs. 3 A/B) and artificial particles (C) are shown in dipole approximation. Since the erythrocyte-liquid system is characterised by two interfaces, up to two ER-peaks are to be expected (Fig. 3 A) at frequencies that are inversely related to the charge relaxation times at the interfaces. Far away from these frequencies no rotation occurs.

In culture media, both the effective conductivity and the permittivity of red blood cells are lower than that of the media. Consequently, the cells are repelled from the electrodes (regions of high electric field strength, negative dielectrophoresis, Fig. 3 B). As far as biological objects are concerned, negative DEP is preferred as it minimises direct field effects (e.g., loading of the membrane which can lead to a breakdown at lower field frequencies) as well as indirect processes such as heating of the object. In addition, positive DEP has the disadvantage that the particles are attracted to surfaces (usually metal electrodes) were they are likely to remain attached even after switching off the electric field. Therefore, in the following we will exclusively consider negative DEP. It should however be noted that the ponderomotive forces are larger in the electrode vicinity (due to the higher field gradients) and therefore, dielectrophoretic trapping can be much stronger here.

The spectra depend only slightly on the field type (sinusoidal or rectangular) if compared at same square mean values of the external field but they strongly reflect field amplifications of the micro chamber (curves d in Fig. 3 A/B). Chamber resonances are more pronounced in low conductivity liquids and can be modified by external capacitance, resistance and inductance. For high amplification in conducting liquids, unused electrode areas should be covered with an insulating layer to increase the chamber resistance. For particle characterisation, rectangular electric fields are useful since they allow easy detection of chamber resonances (by identifying the amplification of higher harmonics in the spectra).

If the charge relaxation time of cell interior and external liquid are comparable, dispersions of the (homogeneous) interior became visible (Gimsa et al., 1996) otherwise they are overshadowed by the dominating Maxwell-Wagner relaxations at the interfaces. It is a well known phenomenon that biological tissue or suspensions of proteins show a dielectric dispersion in impedance measurements (for an overview see for example Pethig and Kell, 1987). The interior of a mammalian red cells consists (from a dielectrically point of view) of a dense suspension of proteins, especially hemoglobin. Typically, three dielectric relaxation processes can be seen in aqueous solutions of a globular protein (Miura et al., 1994). There is a "low" frequency relaxation in the megahertz region reflecting the overall rotation of the molecule and counterion processes. Around 100 MHz orientation of bound water molecules supplemented by fluctuations of polar side groups gives rise to a second process and at about 20 GHz dispersion of bulk water occurs. The curves c) and d) of Figure 3 A were obtained from dielectric parameters that were fitted to both electrorotational and dielectrophoretic spectroscopy on single human red cells (Gimsa et al., 1996). The agreement between the fitted dispersion frequency of 15 MHz with the theoretical value of 14.7 MHz (with K^+ as counterion) for the "low" frequency peak of hemoglobin (Miura et al., 1994) is striking.

In general, not only dipoles but also higher moments are induced and contribute to the spectra.

However, due to the similar frequency behaviour of all moments (Washizu and Jones, 1996) and the difficulty of measuring absolute values of force or torque acting on particles in micro chambers, they can hardly be evaluated from the spectra (see Fig. 3C). But they become visible under special field conditions where for example the lower moments vanishes. This will be discussed in the next chapter.

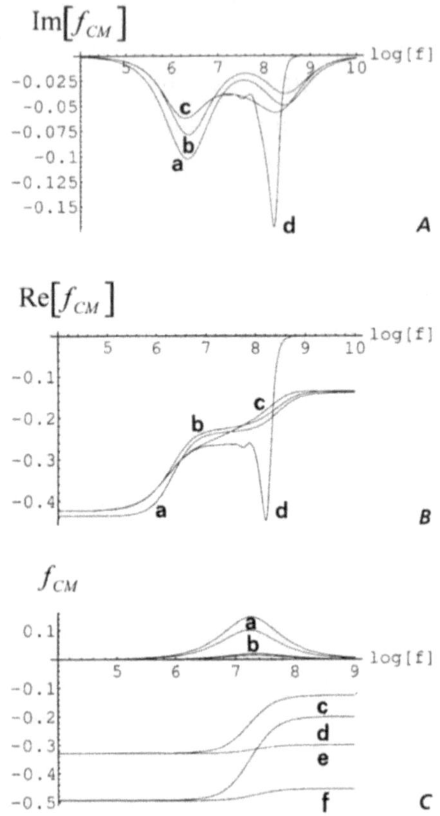

Figure 3 *Theoretical dielectrophoretic spectra of particles in micro electrode chambers. A: Electrorotational spectra of erythrocytes in physiological medium (PBS at 1.38 S/m) assuming a shelled ellipsoid with an axes ratio of 1:2 cell parameters: volume 80.9 μm³, membrane thickness 8 nm*

curve a) Sinusoidal excited electric field

dielectric parameters $\varepsilon_{membrane} = 9,\ \sigma_{membrane} = 3.84 * 10^{-6}\ S/m$

$\varepsilon_{int\ erior} = 50,\ \sigma_{int\ erior} = 0.535\ S/m$

(permittivity is given as relative permittivity)

curve b) Rectangular excited electric field same particle parameters than in curve a)

curve c) Rectangular excited electric field same particle parameters than in curve a), but with dispersion of the cell interior

$$\varepsilon_{int\ erior} = 50 + \frac{\Delta\sigma}{\varepsilon_{vac}} \frac{\tau}{1 + \omega\tau}$$

$$\sigma_{int\ erior} = 0.4\ S/m + \Delta\sigma \frac{\omega\tau}{1 + \omega\tau}$$

with $\Delta\sigma = 0.135\ S/m$ *and* $\tau = 1/(2\pi * 15 * 10^6)s$

curve d) same as in curve c, but with chamber resonance at 200 MHz

B: Dielectrophoretic spectra of erythrocytes using the same parameter as in Figure 4A

Particle behaviour in field cages and traps

A planar, ac-driven, four-electrode trap, as shown in Figure 4D, can create a sharp field funnel. Negative DEP causes those particles initially between the electrodes to be focused to the centre (Fig. 4A/B) while outlying particles are repelled from the trap. However, inward and upward liquid streaming can carry outlying particles into the trap, overcoming the potential barrier (Fig. 4A). The streaming is caused by the strong local heating between the electrodes and is of electro-hydrodynamic nature. For high enough field frequencies (>200 kHz in distilled water), cold water (having a higher permittivity) from the outside tends to replace the warmer in the trap region (Müller et al., 1996).

On the central axis the dipole moment vanishes (zero field). Consequently, small particles (compared with the electrode spacing) will sedimentate to the bottom. Experimentally, it is found that larger particles are lifted (see Fig. 4D) to a height which increases with the particle size. This can only be explained with the influence of higher induced moments. In Figure 4 C a surface of constant potential for the quadrupole part of the force is shown. It can be seen that the quadrupole potential has only a weak minimum in the central region. This results in slight focusing and lifting. Besides the focusing of particles, Figure 4C additionally shows the formation of a pearl-chain on the central axis that resists the strong liquid streaming.

In a rotating, four-electrode trap the focusing forces (Fig. 5A) are weaker than those of the ac-trap. Depending on the field frequency, particles showing negative dielectrophoresis will move on paths that are perpendicular to the contour lines of E^2, if the imaginary part of f_{CM} is neglegible. Otherwise, the trajectories are slightly twisted and the cell rotates with a velocity that

Figure 3 *(continued)*
C: Comparison of dipole and quadrupole spectrum

curves a), c) and d) for a Sephadex sphere ($\varepsilon = 40$, $\sigma = 0.9$ mS/m)
curves b), e) and f) for a latex sphere ($\varepsilon = 5$, $\sigma = 0.7$ mS/m)
suspended in water ($\sigma = 100$ mS/m)

The curves in a) and b) shows the negative of the imaginary part of the Clausius-Mossotti-factor. In both cases the upper curve correspond to the dipole approximation and the lower one to the quadrupole part. The real part of the Clausius-Mossotti-factor is plotted in c)–f) with d) and f) representing the dipole approximation and c) and e) the quadrupole part

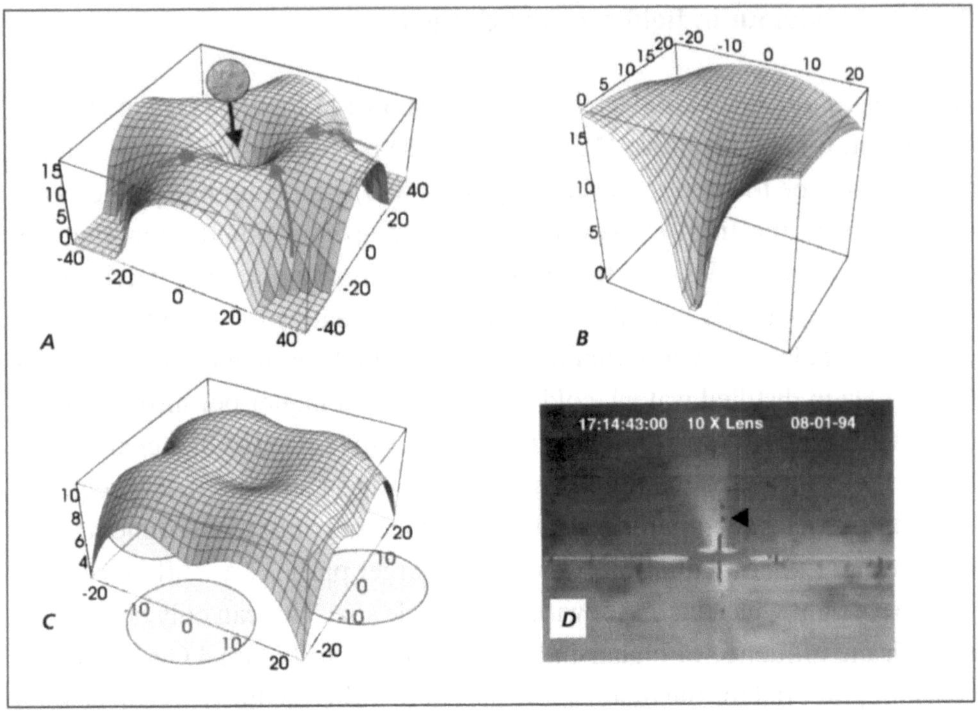

Figure 4 *Forces in ac field traps. A) A typical surface of constant time averaged square of the electric field strength E_{ms} above a planar electrode structure as shown in Figure 3 D. The phase shift between neighbouring electrodes was 180° and an ac trap is built. Co-ordinates are given in relative units defined in Figure 3 C. In dipole approximation, E_{ms} represents the potential of the DEP force. Consequently, a particle (grey circle) will experience a force perpendicular to the E_{ms}-surface. A particle showing negative DEP that is placed near the centre will be focused into the field minimum (black arrow). Green arrows symbolise typical trajectories of particles overcoming the potential wall due to centrally and upwards directed streaming (induced by the heating of the liquid). B) Cross-section through the central part of Figure 4A showing the zero field on the central axis. C) A typical surface of constant potential for the quadrupole part of the DEP-force. D) A four electrode trap with an electrode distance of 200 µm is shown. The electrode tips were processed as 1 µm high discs with a diameter of 135 µm. The chamber was filled with an aqueous solution (conductivity 1 mS/m) containing latex particles (diameter 15 µm, Serva). Electrodes were driven with an ac field of 500 kHz at a voltage between 30 V_{ptp} and 40 V_{ptp}. Shown is an intermediate stage after a few seconds. Five beads are already trapped building a 20 µm levitated particle chain. The mirror image of the particle chain on the front gold electrode is also visible. Despite a liquid streaming at 130 µm/s additional beads are attracted (see dark triangle). A pearl chain of up to 10 beads (150 µm high) can be stably held against the streaming. Outside particles are lifted (to about 150 µm) if brought near to the trap by streaming and follow the paths shown in Figure 4 A. (see colour plate 10)*

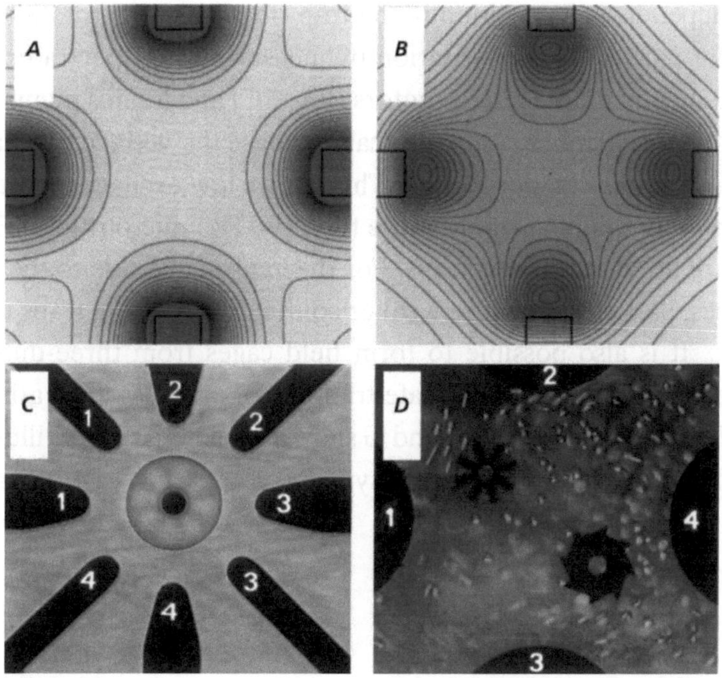

Figure 5 *Forces in rotating field traps. A) Contour plot of the mean square value of the electric field for the four phase rotating electric field in a plane 24 µm above the electrode plane. The electrode distance is 300 µm. The boundary of the underlying electrodes are indicated by thick lines. Darker (red) regions correspond to higher values. B) Contour plot of the torque (particle rotation) component perpendicular to the electrode plane (in the same plane as in Fig. 4 A). C) Planar eight electrode structure with central bearing and rotor in motion. The tips of the 20 µm high gold electrodes lie on a circle 300 µm diameter. Thickness of the bearing is 40 µm. The diameter of the 1.7 µm high aluminium rotor is 200 µm. To reduce the peak frequency for electrorotation the rotor was covered with a 300 nm silicon oxide layer. The chamber was filled with distilled water (2 mS/m) and the electrodes were driven with 4.5 Vptp at 158 kHz (phases are symbolised by the white numbers on the electrodes). The rotor operates at 100 Hz (resonance). This high rotation speed is due to a resonance in the microchamber that was enhanced by additional capacitors (660 pF) and coils (1 mH) between bond wires and connection cables amplifying the voltage by a factor 15. D) Two rotors in a four electrode structure. The two rotors are held in a trap filled by distilled water by glass capillaries. Because of the phasing of the electrodes (phases are indicated by white numbers) the rotors are driven in opposite directions. Streamings patterns are visualised by small suspended latex beads (3.4 µm)*

depends on its position in the trap (Fig. 5B). The rotation can be used either for particle characterisation by measuring ER-spectra or for the creation of long time stable micro-motors (Fig. 5C) and liquids pumps (Fig. 5D). Adjusting the chamber resonance can increase the voltage by a factor of 15 yielding a 225 times higher torque. This allows low excitation voltages (5 V) to drive rapidly spinning (up to more then 100 Hz) micro rotors. Even small particles are lifted in rotating traps. For this reason and due to the lower focusing forces, particles are more weaker confined in rotating traps than in ac ones.

It is also possible to form field cages from three-dimensional electrode arrangements allowing safe trapping of particles in rotating electric fields. In Figure 6, frames A,B,F and G show ac-cages, the others illustrate rotating ones. The ac-cage is open in its symmetry axis (Fig. 6F) according to the dipole

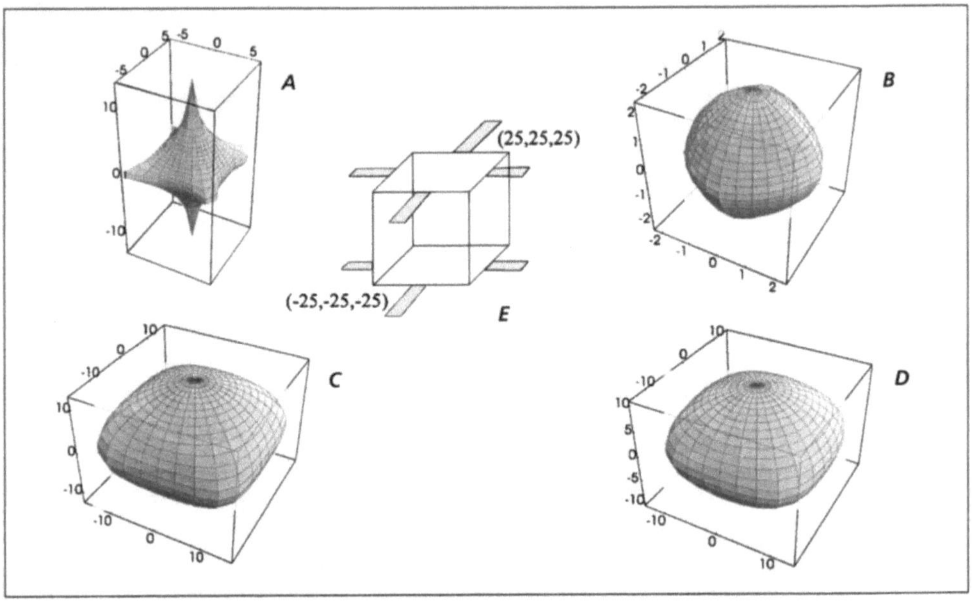

Figure 6 Forces in field cages. This figure shows typical surfaces of constant potential for the dielectrophoretic force in a cage with a geometry defined in part E (units are given in μm). A) and B) correspond to ac driving of the electrodes; C) and D) show the case with rotating fields. In A) and C), calculations where carried out for small Latex particles (radius 1 μm) and in B) and D) for larger ones (radius 10 μm) suspended in water (1 S/m) with an applied field frequency of 300 kHz. Under these conditions, the imaginary part of the Clausius-Mosotti-factor is neglectible and the DEP-force has a potential. F)-I) show the mean square value of electric field in the vertical mid-plane of E) (dark regions correspond to high values). This was determined numerically by a finite difference method. F) and G) correspond to ac driving; H) and I) were obtained for rotating fields

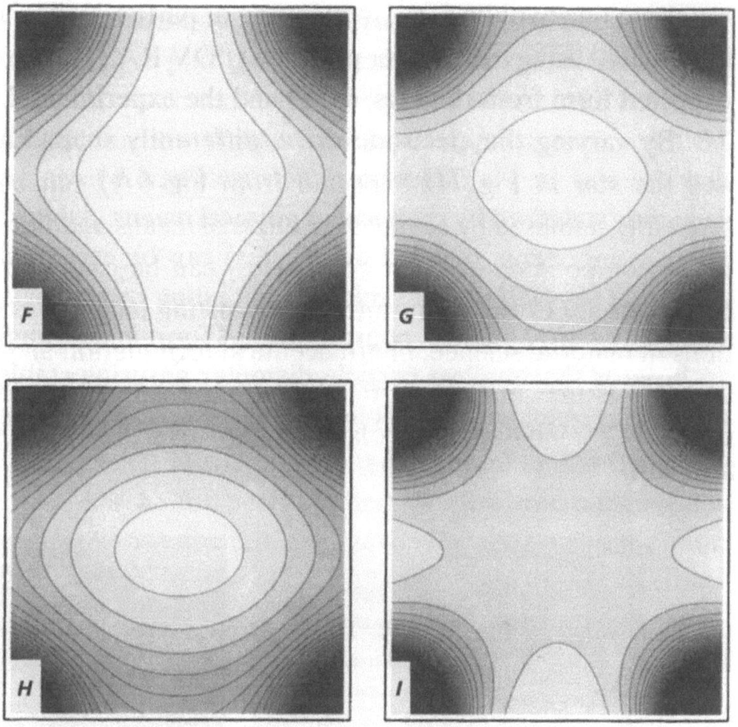

Figure 6 *(continued)*

approximation but closed by the quadrupole term (Fig. 6 G). Therefore, smaller particles will form differently shaped aggregates than larger ones. Figures 6 A–D show typical surfaces of constant force (for the rotating trap in Fig. 6 C/D it gives only the focusing part of the potential). Neglecting particle interactions, these surfaces are likely to represent the shape of aggregates, because in aggregates of this form all forces (perpendicular to the drawn surface) are balanced. This is also the case for rotating cages at appropriate field frequencies (see the curves a and b in Fig. 3 C) when the imaginary part of the Clausius-Mossotti is small. Otherwise the aggregates slightly rotate and become even more symmetric. Interestingly, for rotating cages the quadrupole part of the force is not closed in all directions. However, the gradient is fortunately comparatively weak and the cage remains closed for most particles.

In Figure 7 the formation of particle aggregates in cages is shown. In Figure 7A a rotating cage was numerically simulated also taking into account Brownian forces (Eq. 10) and the particle interaction forces (Eq. (11) and a repulsing one avoiding the overlapping of particles). The final aggregate was illuminated using a ray tracer program (POV-RAY) and is quite similar to the expected form from the Figs. 6C/D and the experimental observation in Fig. 7B. By varying the electrode drive, differently shaped aggregates (Fig. 7C and the star in Fig. 7D expected from Fig. 6A) can be formed and subsequently stabilised by chemical or physical means. Additionally, two- or multiphase components (shelled aggregates) can be cast (for further details see Fiedler et al., 1995a). This microbody casting technology could find applications in material science, pharmaceutical formulation and biotechnology.

What is the smallest particle diameter ensuring stable trapping in micro cages? For small particles, polarisation forces scale with the third power of the

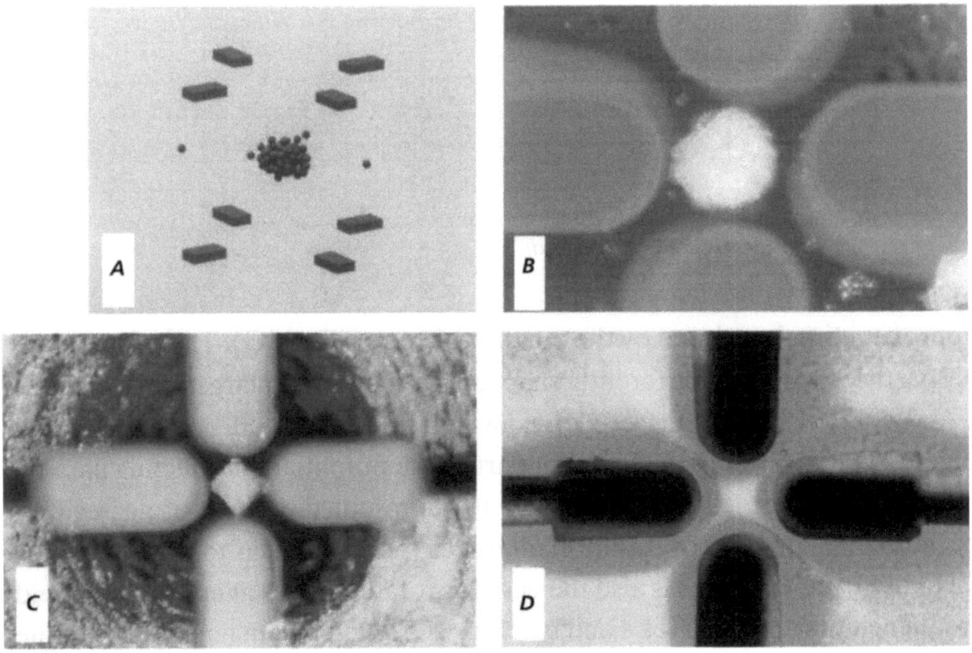

Figure 7 *Trapping of Latex beads in octopole field cages with electrode gaps of 100 µm. A) Numerical simulation for collection of particles using rotating electric fields taking into account the dipole-dipole interactions. B) Using rotating electric fields, 3.4 µm Latex beads suspended in distilled water were collected in the central part forming a dense aggregate. C) Same situation as in Figure 6B, but the electrodes in each plane were driven with 180° phase shifted signals with no phase shift between upper and lower electrode plane. D) Same situation as in Figure 6C, but with a phase shift of 180° between both electrode planes (see colour plate 11)*

particle radius, whereas thermal motions vary inversely with it. A sufficient criterion for trapping of single particles is that the potential barrier of the deterministic force exceeds the thermal energy (kT) by at least ten-fold (see e.g., Ashkin et al., 1986). For low permittivity particles, this gives a critical radius above which long term trapping reliably occurs as:

$$r_{crit} \cong \sqrt[3]{10kT/\pi\varepsilon_{H_2O}\,\partial E^2_{rms}} \qquad (12)$$

The value of the barrier (∂E^2_{rms}) depends not only on the applied field strength but also on the electrode geometry and arrangement. To estimate the smallest particle that can be stably trapped by negative DEP, the barrier ∂E^2_{rms} is assumed to be 10 times smaller than the applied value (using positive DEP it can be the applied value itself). Then, at room temperature and at the maximum field strength that has so far been applied to ultra micro-electrodes (28 MV_{rms}/m, Müller et al., 1996), the critical diameter would be about 27 nm. Keeping in mind the uncertainty of the effective voltage due to the electronic properties of the micro-structures, this gives only a rough estimation. Note that the particle radius should include any surface layers; these become important for sub-micron particles.

In many experimental conditions, confinement of particles is required only for a limited experimental time. Then, the critical radius can be even smaller. In Figure 8A numerical results for a 1 µm field cage are given. Clearly, rotating drive is superior to ac and, as expected from the discussion of the field distributions (Fig. 6), the quadrupole parts of the DEP-force enhance trapping in ac-cages and reduce it for rotating cages. But due to the small difference in the case of rotating cages, these can be well modelled within the dipole-approximation (curves e/d). At a reasonable driving voltage of 2.5 V, particles down to a radius of 25 nm can be trapped (curve f in Figure 8).

As an example of trapping sub micrometer particles, in Figure 8B the enrichment of influenza viruses in a four-electrode, ac-trap is shown. About 1 s after switching on the electric field a growing cloud of viruses appears in the central region of the trap. Under the experimental driving conditions ($E_{rms} \approx 0.28$ MV/m), Eq. (12) yields a critical radius for single particle confinement of about 125 nm, being a bit larger than the average radius of the viruses (50 nm). Interestingly, even smaller particles can be trapped in this structure and under the same field conditions (14 nm Fluospheres, Müller et al., 1996).

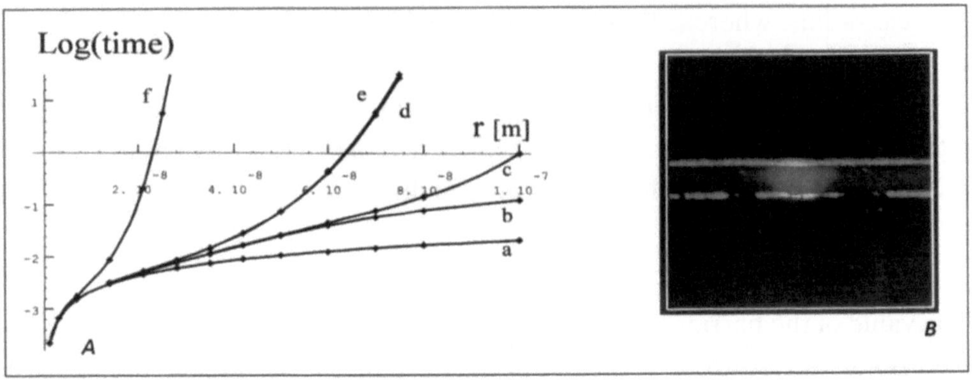

Figure 8 *Trapping of submicron particles. A) Mean first passage time for a latex beads in 1 μm field cages with the geometry of Figure 6E. Each measuring point was obtained by simulating 10^4 runs for 10 different numerical time steps δt and extrapolating $\delta t \rightarrow 0$ (see e.g. Strittmatter, 1988). a) Points correspond to a Brownian particle with no additional forces, solid curve represent the exact solution for a cube of 1 μm: t(r) = 2.1827×10^5r. In b)–f) simulations were carried out for particles with $\varepsilon = 3$ and $\delta = 0.8$ mS/m suspended in water at 3 mS/m. The applied field frequency was 1 MHz. b) Dipole approximation for ac driven electrodes at 1 V. c) Dipole and quadrupole approximation for ac drive of the electrodes at 1 V. d) Dipole and quadrupole approximation for travelling wave drive at 1 V. e) Dipole approximation for travelling wave drive at 1 V. f) Dipole and quadrupole approximation when the excitation voltage of the travelling wave is 2.5 V. B) Side view of a four electrode trap concentrating viruses from suspension. A planar structure with an electrode design as in Figure 2B (electrode distance of 25 μm) was filled with a virus suspension (see below) and covered with a glass slide (at a height of 10 μm). The electrodes were driven with ac signals at 7 V_{rms} and 7 MHz. Shown is a superposition of a fluorescence and a reflection image obtained with an xz-scan of an CLSM (40×, Aristoplan, Leica) through the central part of the trap crossing to neighbouring electrode tips. The green lines correspond to the liquid-glass interfaces. The lower line has two breaks resulting from the electrodes tips. The red cloud represent the fluorescence signal of the labelled viruses.*

Virus suspension: Viruses (Influenca Japan A, 1 mg protein/ml) were fluorescently labelled with 10 μM Octadecylrhodamine B chloride (R 18, Molecular Probes) at room temperature for 30 min in the dark, centrifuged, washed and resuspended in ice-cold phosphate-buffered 150 mM NaCl, then transferred to phosphate-buffered 300 mOsm sorbitol using a Sephadex G75 column. The final conductivity was about 74 mS/m and the final concentration was 1 mg protein/ml (2.5×10^{12} viruses/ml)

The smallest single particle that can stably trapped and manipulated in our structures is a latex bead of 0.65 μm in diameter. Due to the high energy barriers, closed cages allow long-term, stable trapping of single particles. Unfortunately, we were not able to investigate the behaviour of a single virus or Fluosphere because of the experimental difficulties in detecting a single particle of that size with our set-up and because of particle concentration due to field-induced liquid streaming (see also Fig. 4A) and particle-particle interaction. Once a few particles have been accumulated, mutual dielectrophoresis should enhance attraction of further particles from the surrounding medium. According to Saito and Schwan (1960), in homogeneous, high-frequency electric fields, the critical radius for aggregation (pearl-chaining) of low permittivity particles is found in dipole approximation to be:

$$r_{aggr} \cong 2.3 \sqrt[3]{kT/\varepsilon_{H_2O} E^2_{rms}} \qquad (13)$$

The criterion for particle aggregation is more easily fulfilled than that for stable particle trapping (Eq. 12) (at least near the electrodes). This is due to the strong gradients in our structures. In addition, higher moments contribute strongly to the interaction force. Therefore, Eq. 13 (which was obtained in dipole-approximation) is likely to underestimate the aggregation effect (Washizu and Jones, 1996).

Particle manipulation and liquid phase handling in travelling wave structures

Beside stable trapping as discussed above, micro particle manipulation requires structures allowing precise transport in multi channel systems. Travelling-wave-based designs can be used. In Figure 9A ac-drive yields alignment of growing dextran drops in the field funnels. This is an example of particle sorting with an equal distance between each other. Switching the electrode drive to a travelling wave regime results in linear motion of the droplets along the connection line. This allows a separation of particles according to their dielectric or geometric properties. In addition, Figure 9A demonstrates the possibility of collecting small macro molecules such as proteins from a two-phase system (liquid-liquid partition). Preferably, mixtures of organic solvent with large difference in dielectric properties should be used (see Eq. 3).

The simplest structure allowing linear transport in four directions (crossing) and trapping is presented in Figure 9B. The meander structure combines the travelling wave design with a trap. Using a travelling electric field a particle can be brought into the centre. By weakening the drive to one of the four electrodes accompanied by a change of the direction of field propagation, the particle can be released in any chosen one of the four channels. To force the particles along predefined narrow trajectories, tracks of an insulating material such as siliconoxinitride can be deposited (by plasma enhanced chemical vapour deposition) on the electrode array (Fuhr et al, 1994b).

Not only particles, but also inhomogeneous liquids can be moved within travelling wave structures. In Figure 10, electric field induced fluid streaming is illustrated. Figure 10A shows numerical results for the temperature profile in a glass covered 16-electrode travelling wave structure (for it Eqs (8) and (9) were solved in parallel). At an applied voltage of 10 V the maximum temperature increase is about 2 °C. The corresponding mean square value of the electric field strength in the liquid is drawn in Figure 10B. A particle showing negative DEP will strongly be repelled from the electrode array. This can be utilised for creation of adhesion inhibited surfaces (see also Schnelle et al., 1996).

For qualitative analysis of pumping behaviour we assume a simplified geometry, and consider steady-state laminar streaming only, neglect boundary effects and replace the electrodes by an excitation in the $z = d$-plane (electrodes are elongated in y-direction in this plane) by travelling sinusoidal harmonic field. The travelling-wave moves in x-direction, that is, along the axis of the channel of height d ($d \geq z \geq 0$). The liquid is assumed to behave as an "Ohmic conductor" (this is valid if the field strength is small and the

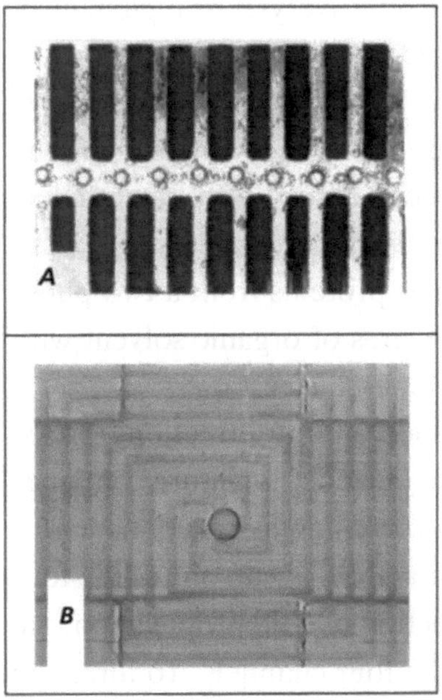

Figure 9 Manipulation of particles and liquid droplets with travelling wave electrode structures. A) Twelve electrodes (50 μm in width) on each sides of a 100 μm width channel are driven with an ac electric field of 10 V_{ptp} at 2 MHz (180° phase difference between neighbouring and opposite electrodes). The chamber is filled with 5% dextran (T250, Pharmacia) in sodium phosphate buffer (pH 5.5, 40 mS/m) mixed with 50% n-propanol. Small dextran-water droplets are focused to the field minima within the interelectrode gaps where they fuse. B) Travelling wave structures and a trap are combined in a meander structure. The structure is covered with a glass plate with two crossing channels (height 100 μm, width 600 μm). The channel was filled with weak electrolyte solution (2 mS/m). The four meanderlike electrodes are driven with rotating electric fields (1 MHz, 5 V_{ptp}) transporting a Sephadex sphere (diameter 60 μm) with a velocity of about 100 μm/s to the central trap (see colour plate 12)

frequency is high enough so that nonlinearities and diffusion processes are negligible). Due to the comparatively high conductivity of the liquids (low Debye length) liquid-solid interfaces need not to be considered. According to the heating, the temperature will rise in the channel with z. Following Melcher and Firebaugh (1966), we replace the temperature profile by a linear gradient. Than also conductivity and permittivity are approximately linearly dependent on the co-ordinate z. Now, the electric field determining Eq. (8) can be solved with the travelling-wave expressed:

$$\varphi(x,z) = \psi(z)\exp(-ikx)$$

in terms of the modified Bessel functions of the first (I_n) and second (K_n) kind to be:

$$\psi(\xi) = A\,K_0[\kappa(C+\xi)] + B\,I_0[\kappa(C+\xi)] \tag{14}$$

where A and B are to be determined from the boundary conditions, κ represent the normalised wave number $\kappa = k \times d$ and $\xi = z/d$ the normalised channel co-ordinate. The value C stands for the ratio of the complex conductivity at the channel bottom and the gradient of this value across the whole channel:

$$C = \frac{(\sigma + i\omega\varepsilon)|_{\xi=0}}{(\sigma + i\omega\varepsilon)|_{\xi=1} - (\sigma + i\omega\varepsilon)|_{\xi=0}} \tag{15}$$

Since there is now pressure gradient along the channel, the plane steady flow can be determined from the requirement that the sum of viscous and electric shear stress has to be constant (Eq. 7):

$$\frac{\eta}{d} + \frac{dv_x}{d\xi}\langle T^{e}_{xz}\rangle = \text{const}$$

with

$$\langle T^{e}_{xz}\rangle = 0.5\varepsilon\,\kappa(V/d)^2\,\mathrm{Im}\,[\psi(\xi)\,\overline{\psi}(\xi)] \tag{16}$$

The constant in Eq. 16 and the integration constant are determined by the boundary condition at top and bottom of the channel. The streaming behaviour for any given channel height and electrode spacing is therefore only determined by the complex value C. Since, $\langle T^{e}_{xz}\rangle$ is always smaller than zero across the channel (Fig. 10D) liquid will be pumped in the direction opposite to that of the propagating electric field. Maximum fluid speed is not found in the mid-plane, but a bit higher due to the increasing electric shear stress with ξ ($\xi = 1$ corresponds to the electrode plane). Formal integration of Eq. (16) shows that fluid velocity is proportional to the square of the applied field strength. This neglects, however, the change of C due to the increased heating in rising electric fields. The rise of average temperature and temperature gradient across the channel with the applied voltage can only be estimated for very simple geometries such as an homogeneously ($q = \alpha V^2$) or with an exponentially (as expected for travelling wave structures, see Schnelle et. al, 1996) heated *homogeneous* layer in a homogenous non-heated medium. There, the temperature gradient is proportional to q or to the square of the applied voltage. With this rough estimation and the knowledge that, over a wide parameter range (e.g. at fixed bottom temperature and $|C|>1$), $\mathrm{Im}[\psi(\xi)\,\overline{\psi}(\xi)]$ in Eq. (16) is almost linearly related to the the temperature gradient, the fluid velocity can be expected to vary with the fourth power of applied voltage. On the other hand, increased fluid streaming will reduce the temperature rise. Therefore, the fluid velocity is likely to depend on between the second and the fourth power of the applied voltage.

Experimentally, the expected relaxation peak (800 kHz for water at $\sigma =$ 4 mS/m) is found. But also, at low field frequencies, the fluid moves in direction of the travelling electric field and, usually, at a higher velocity. This becomes more pronounced at lower liquid conductivities (ethanol). In such liquids and at low frequencies or high field strength at the interface double layer (with a characteristic dimension of the Debye length) forward streaming can be induced (Ehrlich and Melcher, 1982). These effects were not incorporated in the above considerations. In the higher frequency range, the predicted backward streaming is observed and the dependency of pumping speed on applied voltage fits the theoretical estimations (Fig. 11B). At high voltages the streaming above the electrode array becomes turbulent. Large, stable whirls are formed that collect small particles from the liquid (Fig. 11C). The form and rotation speed of these whirls does not depend on the particle density or properties. The velocity in the whirls is much higher (up to several mm/s) than in the channel outside the electrode array (up to several hundred µm/s). Since almost all particles are collected in the whirls this enables the creation of a micro filter.

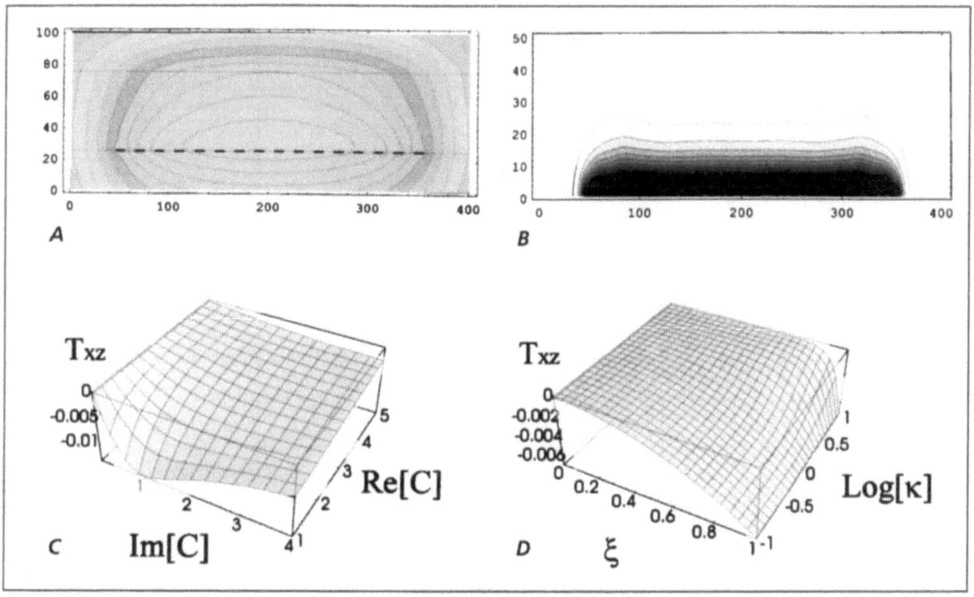

Figure 10 *Analysis of a travelling wave electrode structure. A) Shows the temperature distribution in a section through a channel in a 16 electrode structure. The liquid glass interfaces are symbolised by horizontal lines and the electrodes by thin black bars. Red colour corresponds to high temperatures. Units are given in 2 µm. Calculations were carried out on a larger array (500 µm × 200 µm). At the border the temperature was fixed at 20°C. Maximum temperature in the central part rises to about 24° at an applied voltage 10 V. Electrodes were fed with travelling fields at a frequency of 1 MHz. Ohmic and thermal conductivity as well as permittivity of the liquid were assumed to be functions of the temperature:*

$\sigma = \sigma_0 \times (1 + 0.022 \times (T - T_0)) \quad \sigma_0 = 4\ mS/m \quad T_0 = 20°$
$\varepsilon = 78.54 \times (1 - (T - 25°) \times (4.6 \times 10^{-3} - 8.86 \times 10^{-6} (T - 25°))$
$\lambda = 0.555 - 2.71 \times 10^{-5} \sqrt{T} - T(9.93 \times 10^{-5} - T \times (6.27 \times 10^{-5} - 4.9286 \times 10^{-7}\, T))$

B) Contour plot of the mean square value of the electric field in the liquid channel (section of A)).
C) T_{xz} is the time averaged component of the Maxwell stress tensor responsible for pumping. It depends upon the real and imaginary parts of C calculated from the frequency and the dielectric properties of the liquid:

$$C = \frac{(\sigma + i\omega\varepsilon)|_{\xi=0}}{(\sigma + i\omega\varepsilon)|_{\xi=1} - (\sigma + i\omega\varepsilon)|_{\xi=0}}\ .$$

This plot shows the case at a fixed normalised height $\xi = 0.25$ of the channel height and normalised wavelength $\kappa = 1$. Variations in the average temperature or the temperature gradient (dielectric properties) in the liquid correspond to lines in the Re-Im-plane. Changing ω between 0 and infinity results in half circles starting and ending on the real axis (no acting torque). Due to the flatness of the surface, the fluid velocity shows one slightly asymmetric peak in the frequency spectrum (and as this and Fig. 10 D imply, only backward pumping occurs). D) T_{xz} component of the Maxwell stress tensor plotted against height in the channel and normalised wavelength (at fixed dielectric properties $C = 32 + 58\,i$ of the liquid, this value correspond to water at 4 mS/m , a field frequency of 1 MHz and a temperature gradient of 1°). In C) and D) T_{xz} is given in units of $0.5\ \kappa\ \varepsilon\ (V/d)^2$ (see colour plate 13)

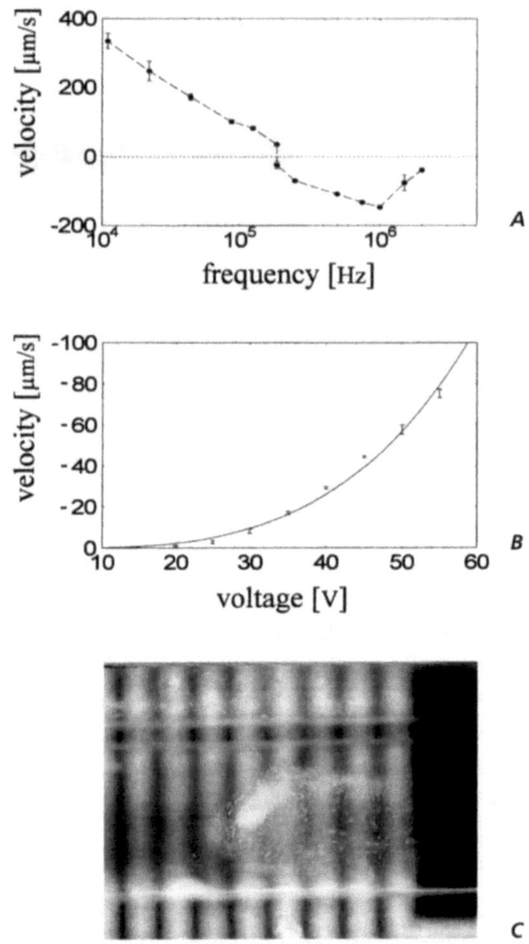

Figure 11 *Pumping behaviour of travelling wave electrode structures. Twelve finger electrodes are covered with an channel forming glass plate (channel height 100 µm, width 300 µm). The channel is filled with a suspension of 3.4 µm latex beads (conductivity 4 mS/m) and the electrodes driven with travelling signals from 15–55* V_{ptp} *at 1 MHz. The fluid is pumped in the direction opposite to electric field propagation. A large whirl forms at the end of the electrode structure and collects nearly all the latex beads. The average streaming velocity in the channel was, therefore, determined 300 µm away from the electrode array. A) Fluid velocity as function of the applied field frequency. B) Fluid velocity as function of the driving voltage measured with a laser scattering method (see Prüger et al., 1997). Each measuring point correspond to 10 measurements, the full line represents the fit:* 6.44×10^{-5} *voltage*$^{3.5}$*. C) Complex streaming patterns at higher voltages*

18.4 Discussion

We have presented the current possibilities that are offered by the use of micro electrode systems and negative DEP for the manipulation of micron and sub micron particles. They can be trapped, enriched from a liquid, transported, characterised and separated according to their geometry, size and/or passive dielectric properties. Similarly, the liquid can be dielectrically analysed if the particle properties are known.

A general difficulty is to select and manipulate one distinct particle from a suspension or to confine only a single particle in a cage or trap, even if freely movable electrodes could be used. These tasks are, however, easily fulfilled by the laser tweezers. Therefore, both techniques are compared in Table 2.

Since dielectrophoretic and optical forces hardly interfere, both methods can be combined (Fig. 12). This should be of great importance for biotechnological applications. In addition, a force calibration for laser tweezers can be established (for details see Fuhr et al., 1998).

Compared with laser tweezers, the negative DEP-tools have the advantage that the objects are forced to regions of low field strength. Besides the heating which can be compensated by lowering the ambient temperature, the main

Table 2 Comparison of DEP and laser beam traps for biological objects

	Negative DEP trap	Single beam optical trap
Field type	Electric	Electro-magnetic
Field frequency	$10^3-5\times10^8$ Hz, depending on particles and liquid	$3-6\times10^{14}$ Hz
Field polarisation	Linear or rotating	Linear or circular
Typical field strength	10^4-10^6 V/m	10^5-10^6 V/m
Stable point	At or close to field minimum	Just beyond the focus (field maximum)
Minimum particle size	$\cong 0.05$ µm	$\cong 0.025$ µm (Ashkin et al., 1986)
Particle properties compared to medium	Less polarisable	More polarisable
Orientation of long particles	Parallel or perpendicular to E-field	Parallel to E-field
Cell chaining forces	Strong	Weak
Aggregate formation	Readily	Possible

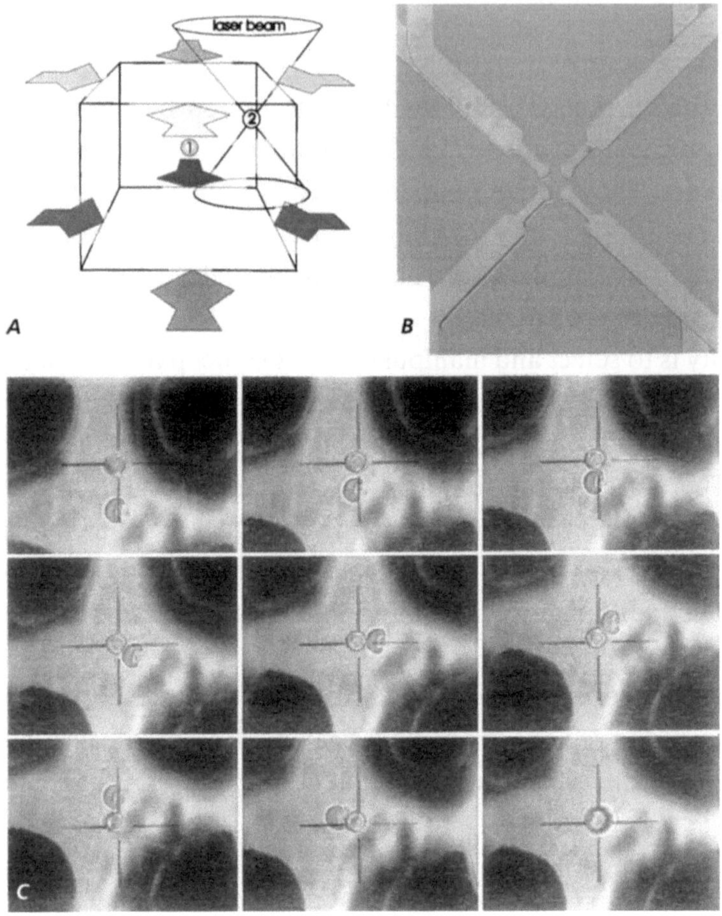

Figure 12 *Combination of field cages and laser tweezers. A) Schematically drawing of a field cage consisting of eight electrodes in two quadrupole layers (top and bottom) in combination with a focused laser beam. Particle (1) is trapped in the centre of the field cage, particle (2) in the focus of a laser tweezers. B) Photograph of an octupole field cage with electrode spacing of 40 μm*

problem for living cells in ac electric fields consists in membrane loading. This may damage cells, especially large cells at high field strengths. Such damage may be reversible but, ideally, manipulation should not damage cells at all, particularly in the case of long-term trapping. To avoid the induction of high additional transmembrane potential field frequencies above 10 MHz should be used. Then even high field strengths (up to 50 kV/m) are well tolerated (Fuhr et al., 1994c). Small cells such as bacteria and yeast (Markx et al., 1994) and viruses can be successfully handled at low frequencies. We were able to cultivate yeast cells in growth medium at 2 MHz and field strengths up to 100 kV/m in the structure shown in Figure 2B. No differences in division rate compared to control were found.

A further alternative for particle confinement consists in the use of standing ultrasonic waves.

Circularly placed ultrasonic sources lead to closed sound minima (see e.g. Schram, 1991). Difficulties arise again from the loading of cells in sound maxima and from non-selectivity. Frequencies of interest are several kHz up to some MHz due to the short wavelength of sound in aqueous solutions.

18.5 Conclusions

Keeping in mind the pros and cons of DEP-microtools and laser tweezers it is most tempting to combine them. For this, optically transparent electrodes are preferred (Fiedler et al., 1997). As an example, one can think of an array of field cages used for long-term confinement and study of cells which are loaded and unloaded using laser tweezers. Dielectric and optical forces can even be applied in parallel. This could enhance the trapping of sub micron particles and macromolecules. Since cells can be accurately handled (e.g. defined rotated or lifted) in a cage, this allows the investigation of freely suspended cells with high precision spectroscopic methods. For monitoring extremely low concentrations (down to 10^{-15} M) a feedback coupled electrophoretic trap and the fluorescence correlation spectroscopy (FCS) were combined (Eigen and Riegler, 1994). In a combination of dielectrophoretic traps with the FCS-method no feedback control for particle centring would be necessary. In addition the discussed enrichment effect in these structures could be exploited.

The Brownian behaviour of small particles can also be utilised for induction of directed transport. The following Langevin equation

$$\dot{\vec{r}}_i(t) = \zeta(t)\,\frac{\vec{\nabla} U}{\gamma} + \sqrt{\frac{2kT}{\gamma}}\,\vec{\xi}(t) \qquad (17)$$

with a (asymmetric periodic) ratchet potential U defines a class of Brownian pumps (for an overview see Hänggi et al., 1998). In the simplest case the so-called "flashing-ratchet" ζ is a time periodic two state function (with the values 1 and 0). This, simulates a periodic switching on and off of the ratchet potential and gives rise to an average net particle flow illustrated in Figure 13.

Using positive DEP and ac driving of planar Christmas tree-like electrodes, a flashing ratchet pump has already been realised (Rousselet et al., 1994). Considering the disadvantages of attracting particles to the electrode tips it is worth trying to collect them by negative DEP which requires a three-dimensional electrode design. An example structure and the resulting ratchet potential in a sub unit and near an electrode plane is shown in Figure 14. To obtain closed traps, rotating electrode driving has to be used. With small distances between the electrode planes, the particles are likely to be collected within the two closed egg-shaped contour lines both in bottom and upper plane (right picture). At larger separation (more than twice the electrode pitch in a plane) particles are likely to be entrapped in free suspension.

Experimentally, 1 µm latex beads were collected at the bottom and top planes of the ratchet structure using rotating electric fields (Fig. 15 A) in the theoretically expected positions. However, in a typical sub unit in each electrode plane only one particle cloud is formed. Typically, an alternating pattern of darker and lighter (lower/upper electrode plane) spots appears. Strong dielectric interaction forces between particles in one electrode plane favours the forming of only one aggregate per sub unit and electrode plane (Fig. 15 B). The two electrode planes (each consisting of 10 electrodes) where mounted over each other in such a way that in each plane one electrode has no face-to-face counterpart. This gives rise to an asymmetry perpendicular to the electrode alignment. This and the particle aggregation should explain the observed alternating pattern. For ac driving, the latex beads are aligned as chains in the mid-plane and central to the electrodes. The particle chains move along their orientation indicating a liquid streaming under ac conditions (in the same direction and with a velocity of several 10 µm/s). This demonstrates a typical behaviour of ac-driven asymmetric electrode structures. They tend to produce liquid streamings (Müller et al., 1996). For rotating driving of our structure no streaming was found.

Since the average net flow in a Brownian pump depends exponentially on the diffusion coefficient of the particles (see e.g., Rousselet et al., 1994), the application for a macromolecule or virus particle fractionating device is promising.

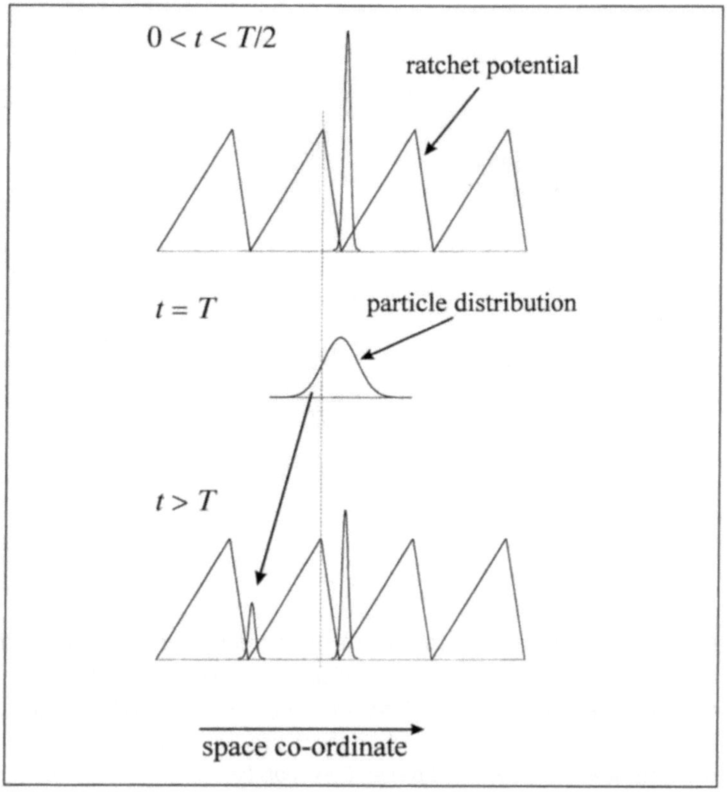

Figure 13 *A noise induced transport mechanism in periodic ratchet potentials. In a ratchet potential Brownian particles will distribute near the potential minima as shown in the upper figure for a single peak distribution. A time interval of free diffusion leads to a symmetric distribution sketched in the middle. When the field is reapplied (lower figure), parts of the single peak distribution will now be trapped in the left minimum. Periodically switching the field on and off (with period T) will, therefore, result in net current to the left*

Microparticle (and even macromolecule) concentrators and separators are likely to be developed. Resolution can be enhanced using cascades of micro structures, which can be easily produced by semiconductor technology. Techniques like dielectric field flow fractionating may become practicable due to the steep gradients that can be realised in micro structures.

Ultra micro electrodes can additionally be used either for chemical liquid characterisation (e.g., by voltammetry, see for example Aoki, 1990) or even for changing the liquid properties using electrode reactions occurring in chopped dc fields (Fiedler et al., 1995b). It is possible to create defined concentration

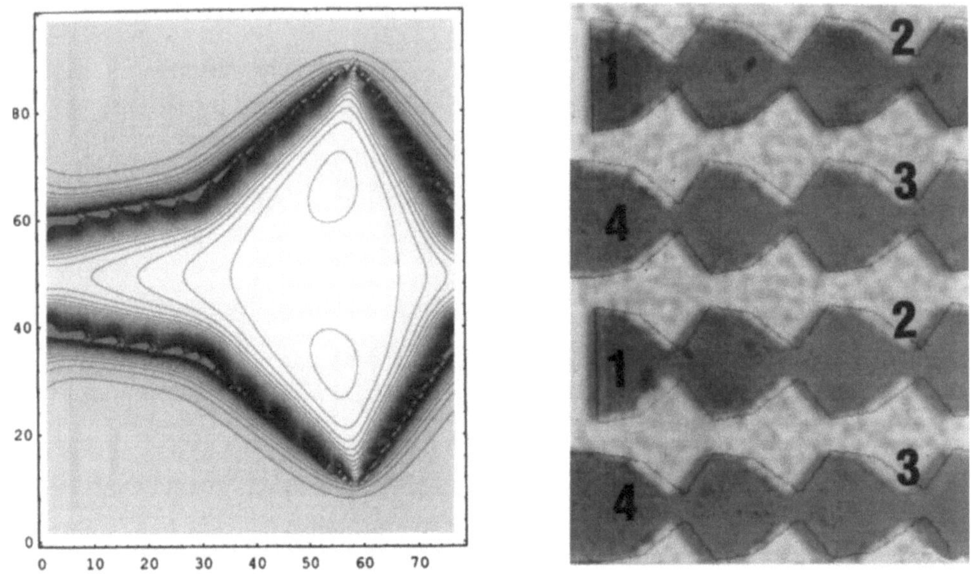

Figure 14 *A three-dimensional electrode structure for Brownian pumping. Left: Two electrode planes were mounted face to face in a distance of 40 μm. Both electrode planes were structured by a laser ablation technique. The lower one was made from a Pt/Ti-sputtered glass substrate (gray coloured) and the upper one from optically transparent ITO. The photo shows a section of structure focused on the upper plane. There is a slight displacement between the planes. Dark spots between the electrodes are metal residue from ablation. To achieve a rotating electric field, the lower electrodes were driven with phases according to the left numbers (1/4) and the upper electrodes with phases 2 and 3 respectively (see Fig. 2). For ac drive there was a phase shift of 180° between neighbours in both planes. The minimum electrode gap in one plane is 25 μm (bar 30 μm). Right: Numerically determined distribution of the mean square field strength of one segment in a plane just below the upper electrodes. Dark regions near the edges of the electrodes corresponds to high values of the field strength, light to lower ones*

gradients on a micron scale. This can find applications for chemistry in micro compartments.

Dielectric spectroscopy will increasingly be used in cellular biology. It has now the ability to measure dispersions of cell interior. This has not only be found in animal cells (e.g., red blood cells as discussed above and fibroblasts, Fuhr et al., 1994c) but also in plant cells (Schnelle et al., 1997). It should open up an additional non-invasive possibility to study cell organelles and even proteins *in vivo*.

The basic dielectric micro tools to built complete closed micro systems are already reality. From the further integration and combination with other techniques such as laser tweezers spectacular developments are to be expected.

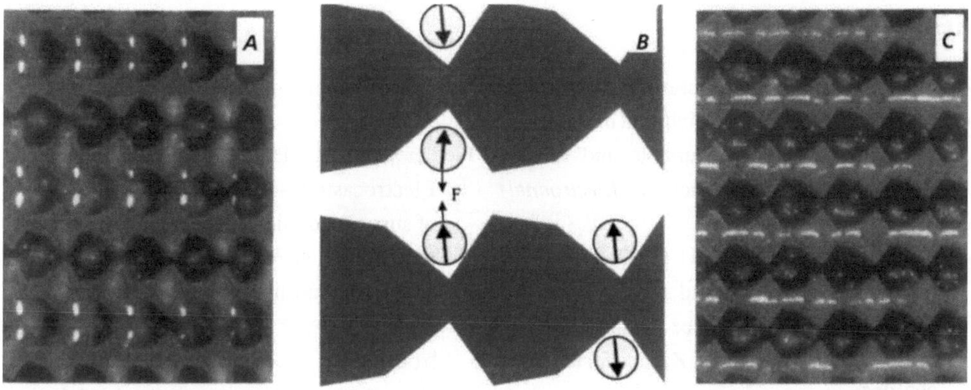

Figure 15 *A three dimensional electrode structure for Brownian pumping. A) Collection of latex beads using rotating electric fields. A suspension (water at 2 mS/m) of fluorescently labelled 1 µm latex spheres (Molecular Probes) fills the structure of Figure 12. The electrodes where driven with rotating electric fields (2 MHz, 9 V_{ptp}). Light spots correspond to the fluorescence signal of collected particles near the field minimum of the upper plane and the halos to those in the lower plane. B) The influence of particle interaction forces on the observed pattern. The polarisation of particles is symbolised by thick arrows and the interaction force (F) by thin arrows. C) Collection of latex beads using ac electric fields. Same conditions as in Figure 14A but with ac-drive of the electrodes. Particles are pushed between the electrodes in to a height of 20 µm where they form lines (see colour plate 14)*

Acknowledgements

We thank Dr. B. Wagner from Fraunhofer-Institut für Siliziumtechnologie (ISiT-Itzehoe) and Dr. S.G. Shirley for the fabrication of the micro structures. Dr. S.G. Shirley's critical reading of the manuscript is kindly acknowledged. We would further like to thank Dr. S. Fiedler for the photographs shown in figures 7B-D and P. Eppmann and Dr. J. Gimsa for the measurements presented in Figure 11B. Mr. K. Ludwig's cooperation in virus experiments is grateful acknowledged. We thank Dr. S. Monajembashi and Prof. K.-O. Greulich (IMB, Jena) for common experiments on the combination of field cages and laser tweezers. This work was supported by the German BMBF/VDI (ELMA-ZELL 16 SV 657).

References

Aoki K (1990) Theory of stationary current-potential curves at interdigitated microarray electrodes for quasi-reversible and totally irreversible electrode reactions. *Electroanal* 2: 229–233

Arnold WM, Zimmermann U (1982) Rotating-field-induced rotation and measurement of the membrane capacitance of single mesophyll cells of *Arena sativa. Z Naturforsch* 37 c: 908–915

Ashkin A (1978) Trapping of atoms by resonance radiation pressure. *Phys Rev Lett* 40: 729–733

Ashkin A, Dziedzic JM, Bjorkholm JE, Chu S (1986) Observation of a single-beam gradient force optical trap for dielectric particles. *Opt Lett* 11: 288–290

Becker FF, Wang X-B, Huang Y, Pethig R, Vykoukal J, Gascoyne PRC (1995) Separation of human breast cancer cells from blood by differential dielectric affinity. *Proc Natl Acad Sci USA* 92: 860–864

Cornell EA (1995) Observation of Bose-Einstein-Condensation in a dilute atomic vapor. *Science* 269/5221: 198–201

Chen CS, Pohl HA (1974) Biological dielectrophoresis: the behavior of lone cells in a nonuniform electric field. *Ann NY Acad Sci* 238: 176–185

Debye P, Debye PP, Eckstein BH (1954 a) Dielectric high-frequency method for molecular weight determinations. *Phys Rev* 94: 1412–1413

Debye P, Debye PP, Eckstein BH, Barber WA, Arquette GJ (1954 b) Experiments on polymer solution in inhomogeneous electrical fields. *J Chem Phys* 22: 152–153

Ehrlich RM, Melcher JR (1982) Bipolar model for traveling-wave induced nonequibrilibium double-layer steaming in insulating liquids. *Phys Fluids* 25/10: 1785–1793

Eigen M, Rigler R (1994) Sorting single molecules: Application to diagnostics and evolutionary biotechnology. *Proc Natl Acad Sci USA* 91: 5740–5747

Fiedler S, Schnelle Th, Wagner B, Fuhr G (1995a) Electrocasting – formation and structuring of suspended microbodies using a.c. generated field cages. *Microsyst Techn* 2: 1–7

Fiedler S, Hagedorn R, Schnelle Th, Richter E, Wagner B, Fuhr G (1995b) Diffusional Electrotitration: Generation of pH gradients over arrays of ultramicroelectrodes detected by fluorescence. *Anal Chem* 67/5: 820–828

Fiedler S, Shirley SG, Schnelle Th (1997) Dielectrophoretic field cages made from Indium Tin oxide. Trapping of submicron particles within transparent electrode structures. *Sens Mat* 9/3: 141–148

Fuhr G, Hagedorn R, Müller T, Wagner B, Benecke W (1991) Linear motion of dielectric particles and living cells in microfabricated structures induced by traveling electric fields. Proc IEEE – MEMS, 91, Japan, Nara 259–264

Fuhr G, Arnold WM, Hagedorn R, Müller T, Benecke W, Wagner B, Zimmermann U (1992) Levitation, holding and rotation of cells within traps made by high-frequency fields. *Biochim Biophys Acta* 1108: 215–223

Fuhr G, Fiedler S, Müller T, Schnelle Th, Glasser H, Lisec T, Wagner B (1994 a) Particle micromanipulator consisting of two orthogonal channels with travelling wave electrode structures. *Sensors and Actuators A* 41–42: 230–239

Fuhr G, Müller T, Schnelle Th, Hagedorn R, Voigt A, Fiedler S, Arnold WM, Zimmermann U, Wagner B, Heuberger A (1994 b) Radio-frequency microtools for particle and live cell manipulation. *Naturwissenschaften* 81: 528–535

Fuhr G, Glasser H, Müller T, Schnelle Th (1994c) Cell manipulation and cultivation under a.c. electric field influence in highly conductive culture media. *Biochim Biophys Acta* 1201: 353–360

Fuhr G, Schnelle Th, Müller T, Hitzler H, Monajembashi S, Greulich K-O (1998) Force measurements of optical tweezers in electro-optical cages. *J Appl Phys A* 67: 385–390

Fuhr G, Schnelle Th, Hagedorn R, Shirley SG (1995) Dielectrophoretic field cages: technique for cell, virus and macromolecule handling. *Cellular Engineering* 1: 47–57

Gimsa J, Müller T, Schnelle Th, Fuhr G (1996) Dielectric spectroscopy of single human erythrocytes at physiological ionic strength: Dispersion of the cytoplasm. *Biophys J* 71: 495–506

Green N, Morgan H, Wilkinson CDW (1995) Dielectrophoresis of virus particles, *Proc St Andrews Meeting SEB*, UK 77

Hänggi P, Bartussek R (1998) Brownian rectifiers: How to convert Brownian motion into directed transport. In: J Parisi, SC Müller, W Zimmermann (eds): *Nonlinear physics of complex systems – current status and future trends,* Springer Series: Lecture notes in physics. Springer, Berlin, *in press*

Hu Y, Glass JL, Griffith AE (1994) Observation of electrohydrodynamic instabilities in aqueous colloidal suspensions. *J Chem Phys* 100 (6): 4674–4682

Huang Y, Hölzel R, Pethig R, Wang X-B (1992) Differences in the AC electrodynamics of viable and non-viable yeast cells determined through combined dielectrophoresis and electrorotation studies. *Phys Med Biol* 37: 1499–1517

Jones TB (1995) Electromechanics of particles. Cambridge University Press, Cambridge

Jones TB, Bliss GW (1977) Bubble dielectrophoresis. *J App Phys* 48: 1412–1417

Jones TB, Washizu M (1996) Multipolar dielectrophoretic and electrorotation theory. *J Electrostat* 37: 121–134

Kaler KVIS, Xie J-P, Jones TB, Paul R (1992) Dual-frequency dielectrophoretic levitation of Canola protoplasts. *Biophys J* 63: 58–69

Masuda S, Washizu M, Iwadare M (1987) Separation of small particles suspended in liquid by nonuniform traveling field. *IEEE Transactions on Industry Applications*, IA–23: 474–480

Markx GH, Talary MS, Pethig R (1994) Separation of viable and non-viable yeast using dielectrophoresis. *J Biotech* 32: 29–37

Meyer G, Zoephel S, Rieder KH (1996) Manipulation of atoms and molecules with low temperature scanning tunneling microscope. *Appl Phys A* 63(6): 557–564

Melcher JR (1966) Traveling-wave induced electroconvection. *Physics of Fluids* 9: 1548–1555

Melcher JR, Firebaugh MS (1967) Traveling-wave bulk electroconvection induced across a temperature gradient. *The Phys of Fluids*, 19/6: 1178–1185

Miura N, Asaka N, Shinyashiki N, Mashimo S (1994) Microwave dielectric study on bound water of globule proteins in aqueous solution. *Biopolymers* 34: 357–364

Monroe C, Meekhof DM, King BE, Wineland DJ (1996) A "Schrödinger Cat" Superposition State of an Atom. *Science* 272: 1131–1136

Müller T, Arnold WM, Schnelle Th, Hagedorn R, Fuhr G, Zimmermann U (1993a) A traveling-wave micropump for aqueous solutions. Comparison of 1 g and µg results. *Electrophoresis* 14: 764–772

Müller T, Küchler L, Fuhr G, Schnelle Th, Sokirko A (1993) Dielektrische Einzelzellspektroskopie an Pollen verschiedener Waldbaumarten – Charakterisierung der Pollenvitalität. *Silvae Genetica* 42/6: 311–322

Müller T, Gerardino AM, Schnelle Th, Shirley SG, Bordoni F, De Gasperis G, Leoni R, Fuhr G (1996) Trapping of micrometre and sub-micrometre particles by high frequency electric fields and hydrodynamic forces. *J Phys D: Appl Phys* 29: 340–349

Muth E (1927) Über die Erscheinung der Perlschnurkettenbildung von Emulsionspartikelchen unter Einwirkung eines Wechselfelds. *Kolloid Z* 41: 97–102

Pastushenko VPh, Kuzmin PI, Chizmadshev YuA (1985) Dielectrophoresis and electrorotation: a unified theory of spherically symmetrical cells. *Studia Biophysica* 110: 51–57

Paul W (1990) Elektromagnetische Käfige für geladene und neutrale Teilchen. *Phys Bl* 46: 227–236

Paul W, Raether M (1955) Das elektrische Massenfilter. *Z Physik* 140: 262–273

Penning FM (1936) Die Glimmentladung bei niedrigem Druck zwischen koaxialen Zylindern in einem axialen Magnetfeld. *Physica* 3: 873–894

Pethig R, Kell DB (1987) The passive electrical properties of biological systems: their significance in physiology, biophysics and biotechnology. *Phys Med Biol* 32: 933–970

Pethig R, Huang Y, Wang X-B, Burt JPH (1992) Positive and negative dielectrophoretic collection of colloidal particles using interdigitated castellated microelectrodes. *J Phys D, Appl Phys* 24: 881–888

Pohl HA, Hawk I (1966) Separation of living and dead cells by dielectrophoresis. *Science* 152: 647–649

Pohl HA, Kaler KVIS, Pollack K (1981) The continuous dielectrophoretic separation of microorganisms. *J BioL Phys* 9: 67–86

Prüger B, Eppmann P, Donath E, Gimsa J (1997) Measurement of inherent particle properties by dynamic light scattering – introducing electrorotational light scattering (ERLS). *Biophys J* 72: 1414–1424

Rousselet J, Salome L, Ajdari A, Prost J (1994) Directional motion of Brownian particles induced by a periodic asymmetric potential. *Nature* 370: 446–448

Saito M, Schwan, HP (1960) The time constants of pearl-chain formation, Proc 4th Ann Triervice Conf Biol Eff Microwave Radiation. 1: 85–97

Schnelle Th, Glasser H, Fuhr G (1997) An optoelectronic technique for automatic detection of electrorotational spectra of single cells. *Cell Eng* 2/2: 33–41

Schnelle Th, Hagedorn R, Fuhr G, Fiedler S, Müller T (1993) Three-dimensional electric field traps for manipulation of cells – calculation and experimental verification. *Biochim Biophys Acta* 1157: 127–140

Schnelle Th, Müller T, Voigt A, Reimer K, Wagner B, Fuhr G (1996) Adhesion inhibited surfaces. Coated and uncoated interdigitated electrode arrays in the micrometer and submicrometer range. *Langmuir* 12: 801– 809

Schram CJ (1991) Manipulation of particles in an acoustic field. *Adv. Sonochemistry* 2: 293–322

Strittmatter W (1988) Numerische Behandlung von stochastischen dynamischen Systemen, thesis, University of Freiburg, Germany

Washizu M, Kurosawa O, Arai I, Suzuki S, Shimamoto N (1993) Applications of electrostatic stretch-and-positioning of DNA, Conference Record IEEE/IAS Annual Meeting, Toronto Oct.: 1629–1637

Washizu M (1991) Handling of biological molecules and membranes in microfabricated structures. In: Karube, I. (ed), Automation in Biotechnology, Elsevier, Amsterdam, 113–125

Washizu M, Jones TB (1996) Dielectrophoretic interaction of two spherical particles calculated by equivalent multipole-moment method, IEEE Transact. Ind Appl 32/2: 233–242

Washizu M, Suzuki S, Kurosawa O, Nishizaka T, Shinohara T (1994) Molecular dielectrophoresis of biopolymers, IEEE Transact Ind Appl 30/4: 835–843

Weiler W (1893) Zur Darstellung elektrischer Kraftlinien. *Z Phys Chem Unterricht* 6: 194–195

Zimmermann U, Neil GA (1996) Electromanipulation of cells. CRC Press, Boca Raton

19 Optical trapping and manipulation

Karl Otto Greulich

19.1 Introduction

A number of applications of microsystem technology require manipulation with high accuracy. Microreactors and microdiagnostic devices can be easily constructed with millimetre accuracy [1], but micrometre precision appears also to be realistic (see for example Fig. 3 below). Also, the generation of chip-arrayed libraries may be facilitated by the use of spatially directed photoaddressable reactions. In addition, for techniques such as microliquid handling or handling of individual beads in bead libraries a sort of ultrafine tweezers is at least helpful, if not mandatory.

In all these cases manipulation techniques based on the interaction of light with matter can be employed. Many processes in biotechnology occur at surfaces, for example the catalysis by surface bound enzyme molecules or the hybridization between probe molecules and target molecules. In protein patterning, antibodies are bound to surfaces which previously have been structured by photolithographic means. Also, one can envision the use of complete cells as micro-bioreactors. By directed positioning of cells at surfaces or by modification of the surface itself biotechnological processes can be modified or adapted to a desired target function.

The positioning of a cell at a surface can be achieved with the help of optical tweezers, or, almost synonymously, an optical trap [2, 3]. This is a continuous infrared laser (often a NdYAG laser working at a wavelength of 1064 nm) with moderate power of a few 100 mW which is coupled to a microscope and focused to the diffraction limit. Using physical effects such as light pressure or gradient forces, microscopic objects can be fixed in the focus of this laser and thus can be held as if an ultrafine pair of tweezers were used. By moving the focus relative to the surface cells, microbeads and other microscopic objects can be moved and transported in the visual field of a microscope.

In contrast to the optical tweezers which are designed to interact with the object only by physical forces of light, a laser microbeam uses a pulsed laser with high transient (peak) power. Such a laser can be used to manipulate optically sensitive chemical reactions. When such a laser is focused to high power density (power per area or intensity) it can be used to precisely ablate biological material, or weld or drill holes with submicrometre accuracy.

The purpose of the present contribution to this book is to stimulate ideas about how to use a laser microbeam combined with optical tweezers [4–6] in microsystem technology and to give a glimpse into the wide field of applications of this versatile microtool.

19.2 Light pressure and gradient forces: The basis for optical tweezers

Light is best described by an electromagnetic wave with the wavelength λ. The wavelength of green light is about 500 nm (0.5 µm), red is 600 nm. Wavelengths below 400 nm are invisible (ultraviolet). For example, UVB has a wavelength around 300 nm. Wavelengths above 800 nm are also invisible (near infrared). High quality mechanical workpieces can have an accuracy of that size and thin plastic or aluminium foils may also be a few wavelengths thick.

When light interacts with matter it can exchange portions of energy which are related to its wavelength λ:

$$E = h\,c/\lambda \quad \text{or in other terms} \quad E = h/\nu \tag{1}$$

where $c = 3 \times 10^{10}$ cm/s (1 s from Earth to the Moon) is the vacuum velocity of light, h is the Planck constant (6.6×10^{-27} g cm^2/s). The frequency is $\nu = c/\lambda$. It is this portion of energy which is meant when one speaks of a photon. The shape of such a portion of energy is not known and therefore the picture of the photon as a strict particle is misleading. The mass-like property of light can be introduced via probably the best known formula of physics

$$E = m\,c^2 \quad \text{or rearranged} \quad m = E/c^2 \tag{2}$$

which implies that energy is always correlated with a mass m, irrespective of any strict particle like structure.

Multiplication of the mass (E/c^2, Eq. (2)) with the speed c gives the momentum = E/c carried by light. Again, no strict particle-like property has to be invoked. Since pressure P is generally defined as momentum per time t and area S one obtains

$$P = E/(t\,S\,c) \qquad (3)$$

E/t is the power W and W/S = E/(t S) is the intensity I. Eq. (3) can then be simplified

$$P = I/c \qquad (4)$$

This is the pressure which is exerted by light of intensity I when it is falling on an object. For precise calculations Eq. (4) has to be somewhat modified, but this does not affect the principal understanding of light pressure. By dividing both sides of the light pressure Eq. (4) through S one obtains the formula for calculation of the light force:

$$F = W/c \qquad (5)$$

A 1 Watt laser falling on a black object exerts a force of 3×10^{-9} Newton. When this radiation is focused for example onto a bacterium with a volume of 1 femtolitre or a mass of 1 picogram, it can accelerate it over a short distance with approx. 30000 fold graviational acceleration.

The force related to this light pressure may be used to balance a cell in the microscope against gravity and thereby to fix it spatially. An additional effect, however, makes life even simpler: due to gradient forces, a dielectric biological object such as a cell or a subcellular particle is pulled into the focus of the laser, even when it has to be pulled against the direction of motion of the light. In some sense this process is comparable with a boat sailing against the wind. Gradient forces are, for example dependent on the refraction indices of the object and of the medium, and on the beam quality of the focused light. Since so far no way has been found to calculate such forces exactly, one usually quantifies gradient forces semiempirically. A quality factor Q is introduced into Eqs (4) or (5) which then become

$$P = Q \times I/c \qquad (6)$$

and

$$F = Q \times W/c \qquad (7)$$

There are several detailed studies available where the value of Q is determined for different experimental situations [7–9]. Gradient forces can act axially, i.e. against the direction of light propagation and perpendicularly. Both forces can cooperate to pull an object against light pressure into the focus. For 1 µm silica microbeads, the axial value for Q is of the order of 0.05, the transversal value is approx. 0.15 (or 15% of the value for pure light pressure, for which Q is per definition 1 or 100%). The corresponding values for 20 µm polystyrene microbeads are 0.1 and up to 0.4, i.e. 10% and up to 40% of the light pressure value. As a rule of thumb, one can say that gradient forces of optimally focused light pull an object into the focus with a force corresponding to a few per cent of the force which one would calculate using the simple light force formula (5). The lateral light forces can amount to up to 40%.

Since for some types of experiments these Q values are still not sufficiently accurate, sophisticated calibration procedures have been developed to get exact values for the forces exerted by light. An overview on such calibration procedures can be found in [10].

19.3 Generation of heat in the laser microbeam and its fast dilution

In addition to momentum (responsible for pressure and forces) light carries energy, as was easily seen in Eq. (1). This is a mixed blessing. Unfortunately, upon interaction with matter, light can transfer its energy to the interacting matter. Quite often, the matter is heated up by absorption and thermal dissipation of the light energy. Particularly when light is used in optical tweezers such a heating is an unwanted effect since one may damage the object to be trapped. Detailed studies have found a pronounced wavelength dependence. When infrared light with a wavelength of 1064 nm is used, the effect is, however, small. For example, continuous infrared irradiation with 100 mW at 1064 nm causes a temperature increase of

1.45 ± 0.15°C in liposomes

and

1.15 ± 0.25°C in CHO cells

(for details see [11]). The reason for the difference is that the delicate balance of heating by the light and flow of heat into the environment is dependent on material constants such as thermal conductivity, which are different for liposomes and for living cells.

Things become quite different with the laser microbeam. As will be seen immediately, extreme local temperatures are generated by focused laser pulses. At first glance the laser microbeam appears to be just a special form of laser microprocessing with lens-focused lasers, as is known from industrial applications. A detailed quantitative discussion shows, however, that the use of a microscope objective instead of a simple lens introduces a new quality of laser processing. In both types of microprocessing often a pulsed UV laser is used. In standard UV laser microprocessing a pulse energy of 10 millijoule at a pulse duration of 10 nanoseconds (typically from an excimer laser) focused to a diameter of 10 µm generates a power density of 10^{12} Watt per cm^2. A similar power density can be achieved when a pulse of 25 microjoule is focused to the theoretical (diffraction) limit, which is of the order of 0.5 µm. Since 25 microjoules can be generated by simple nitrogen lasers, this is a much cheaper way to generate very high power density. But in addition, since nitrogen lasers are available with a pulse duration of 0.5 nanoseconds (instead of the nanoseconds of excimer lasers) an even smaller total pulse energy is sufficient to achieve 10^{12} Watt per cm^2. Table 1 summarizes this.

The thermal interaction of such a laser pulse with matter can be calculated with the Stefan-Boltzmann law. At its face value, this law relates the temperature T of a black body with the intensity I of the radiation which is emitted by this body

$$I = \sigma T^4 \tag{8}$$

Table 1 Comparison of standard UV laser microprocessing and microbeam processing. A power density of 10^{12} Watt per cm^2 can be generated in different ways

By focusing to a diameter of	With a pulse energy	With a pulse duration	Laser used
10 µm (lens)	10 millijoule	10 ns	excimer
0.5 µm (microscope)	1.25 microjoule	0.5 ns	nitrogen

where the Stefan-Boltzmann constant is $\sigma = 5.67 \times 10^{-12}$ Watt cm^{-2} K^{-4}. In turn it is assumed that an intensity I generates a temperature

$$T = (I/\sigma)^{1/4} \tag{9}$$

Note that Eq. (9) is just a rearranged form of Eq. (8).

When the radiation is fully absorbed, i.e. when the penetration depth is only a few micrometres (as is the case for UV radiation) this assumption is suitable at least for an estimate of the order of magnitude of the generated temperature. As shown by Eq. (9) the temperature depends on the intensity by a fourth root law. That means, that even when only 1:1000 of the total energy is really absorbed, the temperature will only be half an order of magnitude lower.

For an intensity of 10^{12} Watt per cm^2 one calculates a temperature of 4 million degrees. This means that even a nitrogen laser can generate physical plasma. In this state of matter the temperature is not well defined. Therefore plasma physicists prefer to give the energy of the free electrons in this plasma instead of a temperature. For 5×10^{12} Watt per cm^2 the electron energy is approx. 300 electron Volt (eV), i.e. they have a speed as if they had been accelerated by a potential of 300 Volt. Such electrons can already generate soft x-rays (for comparison: medical x-rays have an energy of a few 10000 eV). Generation of soft x-rays may become important for the interpretation of some experiments, although absorption by biological tissue is low and therefore only minor damage to the biological sample has to be expected. More critical is the heat itself: millions of degrees are not what one expects to be a temperature suitable for work with biological material. However, the heat pulse is also very short and, according to Table 1, the total energy which generates the heat is only a few microjoules.

Let us thus estimate how this heat is diluted: Due to the large temperature gradients one cannot use the (laminar) laws of heat diffusion. It is rather a shock wave which expands with the speed of sound (in aqueous environment as it prevails in many biological objects 1 km per second or 1 micrometre per nanosecond) in all three dimensions of space. The consequence for biological material is surprising and shows the striking difference between standard laser processing and UV laser microbeams. In order to cool down from 4 million degrees to room temperature, the radius of the spot must expand by a factor of 50 (i.e. the volume expands by a factor of 125000). In standard microprocessing with an initial radius of 10 µm or more, this is the case after 500 nanoseconds. In the 0.5 µm microbeam this time is only 25 nanoseconds. As far as it is known today, the latter time is probably shorter than a protein and other biological macromolecules can denature [12]. As a consequence, biological tissue is only denatured and ablated where the laser pulse hits directly. Secondary effects are probably small and caused by stray light and the above-mentioned soft x-rays. In conclusion, the laser microbeam allows very precise microprocessing with comparably small secondary damage.

19.4 How to build laser microbeams and optical tweezers

The effects described on p. 454 and p. 456 can, in principle, be generated with any sort of light. In practise, however, lasers and focusing close to the theoretically possible limit are required. This is the reason why the advent of lasers caused a boom in micromanipulation by light, particularly in biological applications. The construction of a laser microbeam with optical tweezers is comparably simple.

- Use a conventional fluorescence microscope
- Replace standard optics with UV optics
- Add two semitransparent mirrors in the fluorescence illumination path
- Couple, via the mirrors, a pulsed UV laser and a continuous IR laser, into the fluorescence illumination path and focus them via the objective.

One has to be careful about some details. For example the laser beams have to be adapted for ideal focusing.

A large variety of lasers is available. Quite often an inexpensive nitrogen laser (approx. $ 9000) with a working wavelength of 337.1 nm and a pulse duration of approx. 1 nanosecond is used for the microbeam. For the optical tweezers, Nd YAG lasers (wavelength 1064 nm) prevail, but cheaper semiconductor lasers (wavelengths between 760 and 980 nm) are becoming increasingly popular, particularly since knowledge on the biophysics of interaction of submicrometre wavelengths is accumulating. Figure 1 shows such an apparatus.

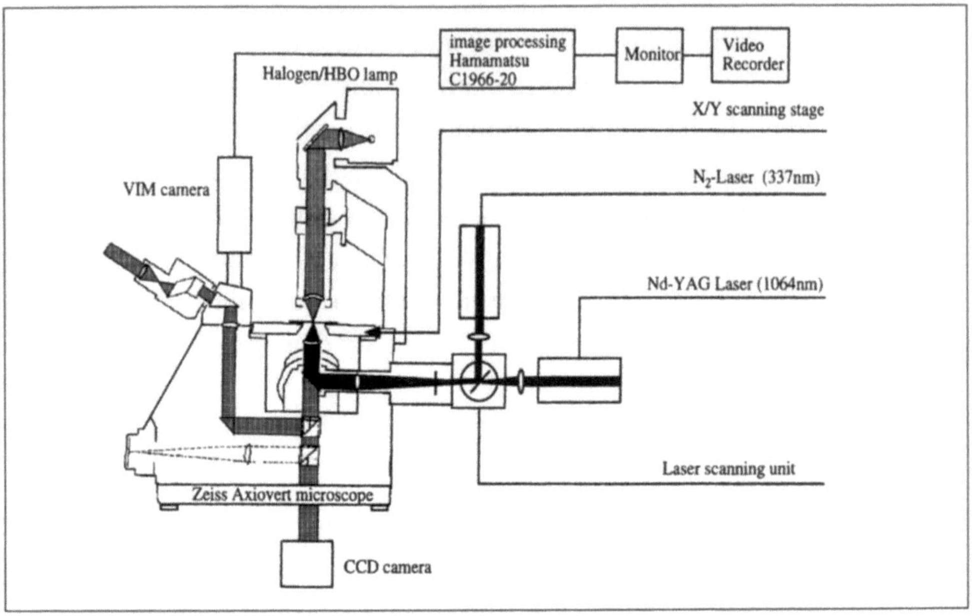

Figure 1 *The equipment: In this version a pulsed ultraviolet laser as microbeam and a continuous NDYAG laser for optical trapping are coupled into the microscope via the epi-fluorescence illumination path. More detailed descriptions can be found for example in references [4–6]*

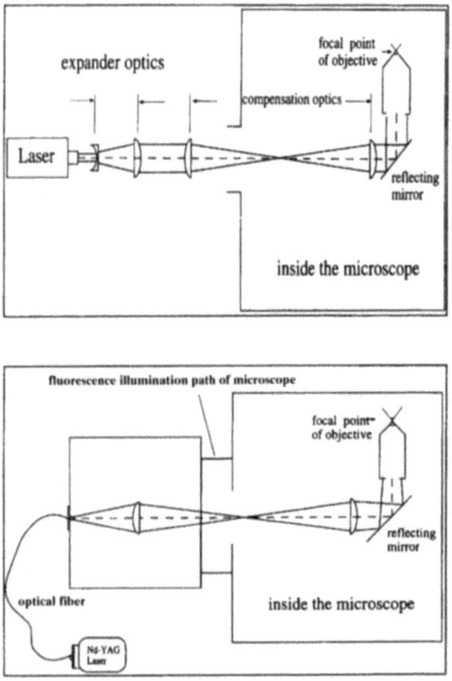

Figure 2 *Details of the coupling unit: The expander optics "prepare" the beam for optimal focusing and is required for any type of laser microbeam trap. The compensation optics are only required in those microscopes which have already a lens built in the illumination path*

In order to achieve optimal effects, the lasers have to be coupled properly into the microscope, mostly via the fluorescence epi-illumination path. One has to take care that the objective-pupil is fully illuminated. Therefore, the diameters of the laser beams often have to be expanded by a combination of diverging lens and a collimating lens (Fig. 2 top). Instead of the diverging lens a flexible light guide can be used (Fig. 2 bottom) which makes the system vibration resistant and allows to work with a larger variety of particles. On the other hand, the lens system provides better collimation when extreme focusing or very high power densities in a microbeam are required.

19.5 A first application: Patterning surfaces

The most simple and yet, in the context of microsystem technology, quite useful application of the laser microbeam is the modification of surfaces. Figure 3 (courtesy Klaus Kessler, Physical Chemistry Institute, University of Heidel-

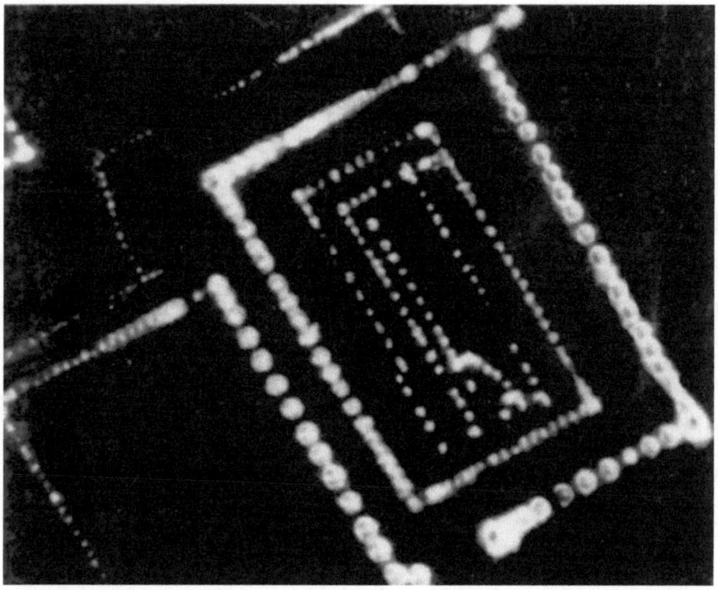

Figure 3 *Patterning a surface of a microchip substrate with a UV laser microbeam*

berg, Germany) shows how a UV laser microbeam can write a structure on a microchip substrate. In principle, each spot is a microvessel with a depth of a few hundred nanometres. The diameter in the outer area is large, since high intensities have been used, and becomes smaller in the inner areas of the pattern since the microbeam has been attenuated. The smaller diameters are approx. 200 ... 300 nm, i.e. close to the limits of visibility.

Since microbeam patterning is a writing technique and not an imaging technique, no imaging masks have to be generated. Direct writing is particularly useful when small numbers of patterns are required. Writing of the pattern in Figure 3 took approx. 2 min. This technique is also useful when masks for imaging photolithography are to be generated.

19.6 Optical trapping and manipulation of individual DNA molecules

The DNA molecule is probably the most prominent molecular individuum we know. Optical trapping and manipulation allows contact free handling, fixing and moving particles under physiological conditions [13]. Individual DNA molecules are bound via a streptavidin-biotin bridge to microbeads and an electrostatic field is applied. By trapping the bead with the optical tweezers and applying the electric field the DNA molecule can be rotated and stretched in the direction of the field's positive electrode (Fig. 4, scale bar 30 μm). The full length of a totally stretched λ-phage DNA molecule (48 kb) would be about 15 μm. The molecule in Figure 4a is obviously not stretched to its maximum length. A further increase of the electric field does not stretch the molecule further but the bead with the attached DNA is pulled out of the focus and the DNA seems to collapse back onto the bead. It was never possible to tear the DNA molecule off from the bead just by electric forces, i.e. the force is not stronger than the bonds between streptavidin-biotin or biotin-DNA, which is in the order of 100 to 200 piconewtons (see Tab. 2 below).

The next step is to observe the enzymatic cutting (restriction) of an individual molecule and to visualize the reaction under a fluorescence microscope. During trapping the bead and stretching the DNA molecule in an electric field the enzyme is still active. In the case of restriction of a single DNA molecule

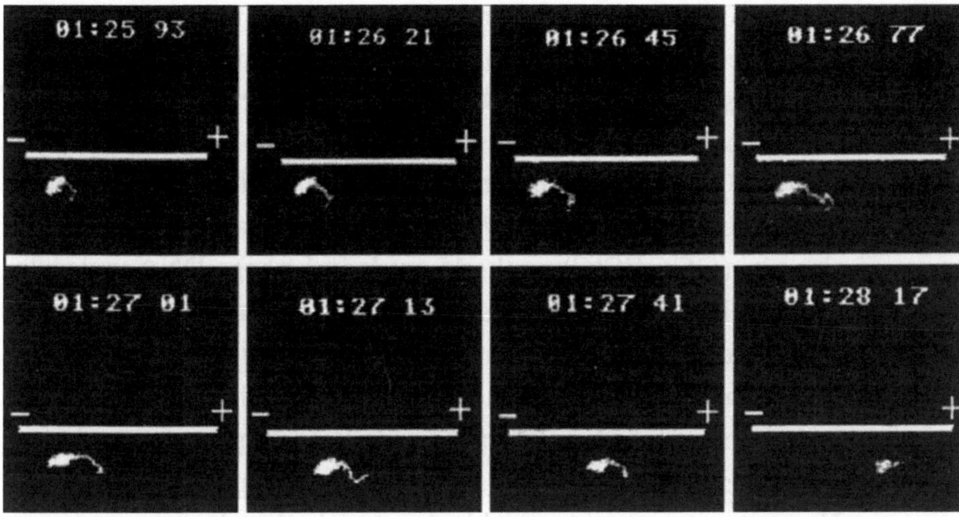

Figure 4 a Top, from left to right: *with an increasing electric field the DNA is increasingly stretched.* Bottom, from left to right: *After reaching 5 V/cm (01:27:01) the bead and molecule are pulled out of the focus and the DNA collapses back onto the bead*

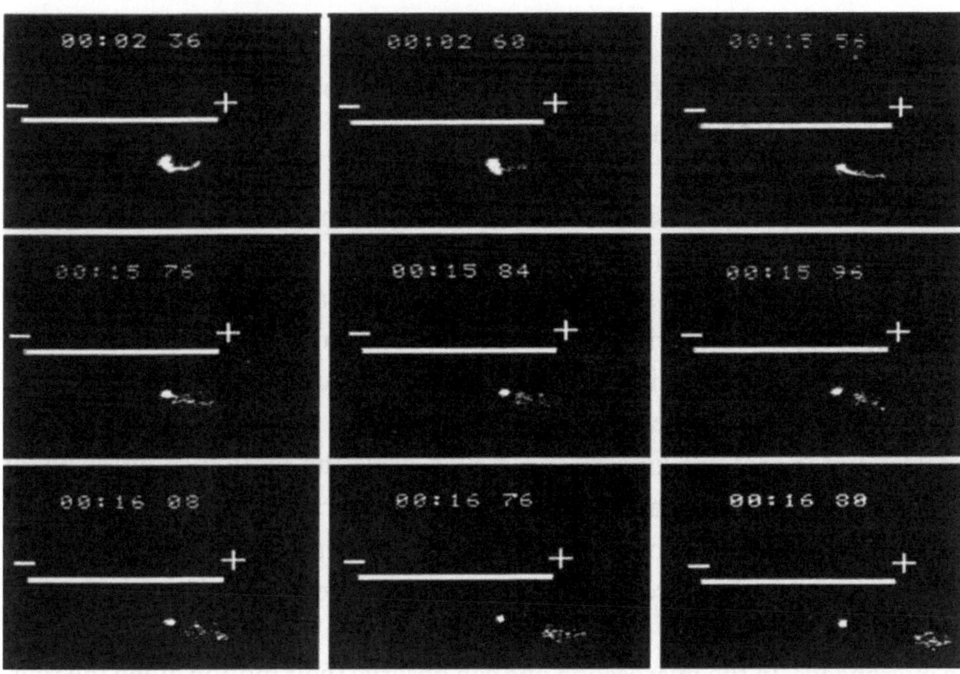

Figure 4 b *After 15.76 s the restriction by ApaI is visible. One part of the molecule is pulled away and leaves the focus plane, therefore it becomes defocused, while the rest is still bound to the bead and remains in the trap. Scale bar 30 µm*

there is only a "yes or no" answer; that means the yield of the reaction is either 0% or 100%. Therefore the starting point of the restriction reaction has to be controlled. For this purpose the Mg^{2+}-ions, required for the endonuclease activity, are complexed. The caged compound of Mg^{2+} with DM-nitrophen can be destroyed by UV-light (360 nm), the ions are liberated, the enzyme is activated and the reaction is started. With the endonuclease ApaI with only one restriction site (GGGCC!C) on the 15 µm long λ-phage DNA molecule, two fragments should be visible after a successful restriction. The length of the fragments should be 11 µm and 4 µm at maximum stretching.

While one fragment remains bound to the bead, the other part was pulled away in the electric field. In Figure 4b the restriction can be observed after 15.76 s. The bead and the remaining fragment are still trapped by the optical tweezers. However, it was not possible to stretch the shorter remaining fragment. Probably the DNA is collapsed onto the bead because of the free Mg^{2+}-ions being present after UV photolysis of the caged compound.

19.7 Measuring forces in individual motor proteins by optical tweezers

With a technique quite similar to that used with individual DNA molecules motor proteins can be investigated. The motor proteins are interesting for microsystem technology since they convert chemical into mechanical energy with unprecedented efficiency and since one can envision their use as motors in microsystems. Of particular interest is the measurement of forces generated by single motor proteins.

The motor proteins kinesin and dynein drive the motion of intracellular structures along microtubuli, which act as a sort of rail. Better known is probably the interplay of myosin with actin in the generation of muscle force. In order to measure the forces in such single molecule interactions, the "rails," microtubuli or actin, are fixed in space. The motor proteins are often bound to microbeads. In the microscope one can observe their motion along the rails. This motion can be stopped by the transversal force (see discussion of transversal Q factor on p. 454 (2)). The Q factor for these experiments has to be determined by sophisticated calibration measurement. Once it is known,

Table 2 Forces between single molecules

Hydrogen bond	300 pN
Avidin and biotin bond	160 pN
Protein protein interaction, dissociation constant of 10^{-5}	11–22 pN
Protein protein interaction, dissociation constant 10^{-9}	74–3300 pN
Force to break DNA	480 pN
Elastic structural transition of DNA	65–70 pN
Straightening DNA kinks	6 pN
Motility force of a sperm cell	40 pN
Traction force of a locomoting cell	45000 pN
Optical tweezers 1 Watt in object plane against light pressure (close to focus)	150–500 pN
Perpendicular to beam propagation	1000 pN

pN = Piconewton = 10^{-12} Newton
1 Newton = 10^5 dyn

force measurements are possible using the fact that, according to Eq. (7), the force is linearly dependent on the laser power.

One way to measure the force is to determine the laser power required to stop the motion (escape force method), but meanwhile there are much more sophisticated techniques available (reviewed in [10]). Kinesin forces have been measured to amount to 3–6 piconewton per molecule [14, 15]. The best studied interaction is that of myosin with actin. Here forces from 1.8 to 4 pN were found with optical tweezers [16, 17]. In order to give an impression of what such forces quantiatively mean, Table 2 summarizes some other known forces between single molecules.

19.8 How to isolate individual DNA molecules: Laser microdissection

The experiments described in the two previous chapters allowed to get very detailed information on individual molecules. However, in those experiments the molecules under investigation were prepared in bulk and one individual was selected for investigation, similarly as individual molecules are prepared and selected for electron or scanning probe microscopies.

Even more interesting would be when the preparation of one molecule could be achieved on a single molecule basis, since this would, for example, allow the study of DNA from a single preselected cell. The technique for such a preparation does not yet exist, but important details are already available and a microsystem for such a preparation can at least be envisioned.

For example, it is possible to isolate a single DNA molecule from a human chromosome: This technique uses the fact that a human chromosome does contain only one (immediately after cell division) or two (before the next cell division) DNA molecules. They are complexed with histone and other proteins and are visible in a light microscope. When such a chromosome is microdissected either with a glass microtool or with a laser microbeam, one can isolate one single piece of a DNA molecule from a single cell, i.e. one has access to DNA in its full molecular individuality. Figure 5 shows how a chromosome is microdissected with a UV laser microbeam [18, 19].

So far, a chromosome segment has still to be amplified enzymatically and analyzed by classical molecular biological methods. A complex microsystem which not only allows handling of individual molecules and of bead, but also controls the relevant single molecule reactions will be required for a preparation method which allows the isolation of individual DNA molecules from individual cells.

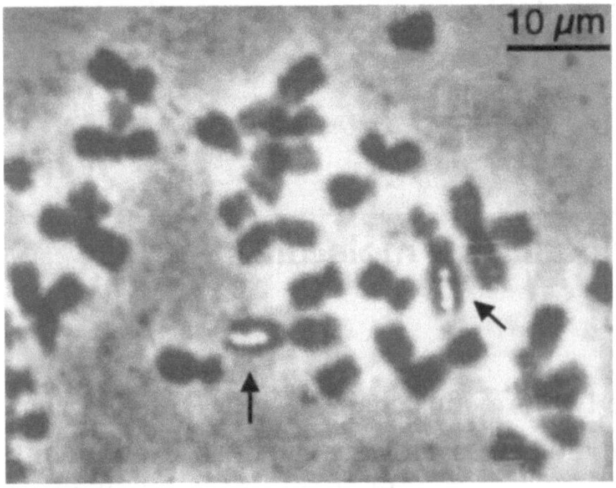

Figure 5 *Laser microdissection of a chromosome along the chromosome`s long axis. Dissection perpendicular to this axis is also possible, for excample to generate DNA probes for gene diagnosis*

19.9 Laser microbeam induced gene transfer into plant cells

Gene transfer generally implies the perforation of cell membranes in order to allow the gene (the DNA molecule) to get into the cytoplasm. Laser microbeams are attractive for microinjecting foreign genetic material into *plant cells* since the hard wall of many plant cell types cannot be penetrated by glass microcapillaries, which might be an alternative tool for direct gene transfer. On the contrary, microperforation with a laser microbeam is almost trivial and can be employed virtually without any practise. Once the equipment is established, microperforation is simple and the experimenter can concentrate on the cell biological aspects of gene transfer.

The first successful laser induced gene transfer into animal cells has been reported in 1984 [20], (noteworthy, in a journal devoted to applied physics). The laser technique is particularly useful when individual cell have to be genetically modified, for example selected cells of a tissue. Gene transfer can be enhanced by establishing an osmotic gradient from outside to inside of a cell. Membranes opened in this way reseal at temperatures of 20°C within a time period of less than 5 s. One day after treatment 40% of single plant cells and 25–30% of pollen grains not only survive but continue to grow. Genes encoded by the introduced DNA are expressed transiently as well as incorporated into the genome. Newly introduced traits can be passed through meiosis into the daughter generation. Using the laser technique, rapeseed and tobacco have already been genetically modified [21].

19.10 Laser induced cell fusion

When the laser microbeam is slightly defocused it will not longer burn holes into the membranes of mammalian cells or cut subcellar structures. The power density will, however, still be sufficient to disturb the cell membrane in a similar way as it would be disturbed when it were touched by a micromechanical tool. If two cells are in contact with each other, they may be fused by a short series of laser pulses. The contact may be established either by adhesive forces or due to high cell density on a microscope slide or by specific coupling via a bridging molecule system such as avidin/biotin. So far, plant protoplasts and different types of immune cells [22, 23] have been fused with each other. In the latter case the contact was enhanced using the optical tweezers, and thereby the fusion yield could be significantly improved. The mechanism of laser induced cell fusion is not yet clear, although it appears reasonable to assume that surface tension is important.

The fusion of plant cell protoplasts is more difficult since, for each type of plant cell, different physicochemical conditions are required and it takes some time to optimize them. Particularly critical is the adjustment of osmotic pressure of the environment, which is quite often achieved by the use of mannitol. Usually, protoplasts are prepared from plant leaves by digestion of the cell wall with a suitable enzyme cocktail. The protoplasts are transferred onto a cover slide at high concentration. Individual pairs of protoplasts stick together due to the hydrophobicity of their membranes. If this is not the case, contact can be established by pushing protoplasts toward each other with the optical tweezers. In the latter case, protoplasts can be well selected, for example according to their size or according to the number of chloroplasts. In order to induce protoplast fusion, the laser microbeam is slightly defocused and directed to the contact area between two protoplasts. Usually single pulses or short series of 10–20 pulses are sufficient to induce the fusion which is completed after a few seconds up to 30 s, depending on temperature and mannitol concentration.

19.11 Preparation of plant cell membranes

A recently presented application of laser microbeams is the preparation of free plant membranes. When the root hair tips of *Medicago sativae* are cut with a laser microbeam, the protoplast is expelled from the root hair, thus exposing the membrane to the environment. Figure 6 shows the result of these experiments. The original experiment and the physicochemical condition have been reported by Kurkdjian et al. [24]. Similar experiments using other plants have been reported by de Boer and colleagues [25].

When intracellular viscoelastic properties of the cytoplasma have to be measured in unopened cells, optical tweezers are the tool of choice. In cells of

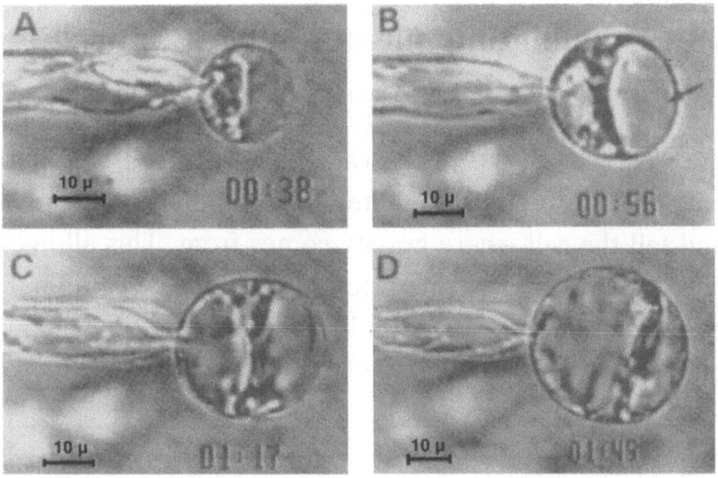

Figure 6 *Preparing plant cell membrane: The tip of a root hair is cut off by a laser microbeam. The protoplast is expelled by internal pressure (turgor)*

Pyrocystis noctilucae subcellular organelles can be picked up by the tweezers, which were previously directed into the cell interior, and moved away from their original position, for example into the free space of a vacuole [26]. Since the organelles remain in contact with their original position via a filament of membrane, they will snap back as soon as the optical tweezers are switched off. Figure 7 shows how optical tweezers can work in the interior of cells without opening them.

The speed of the redirection movement depends on the viscoelasticity of the cell interior and thus, by measuring the speed of redirection, relative viscoeelasticites of different cells can be compared.

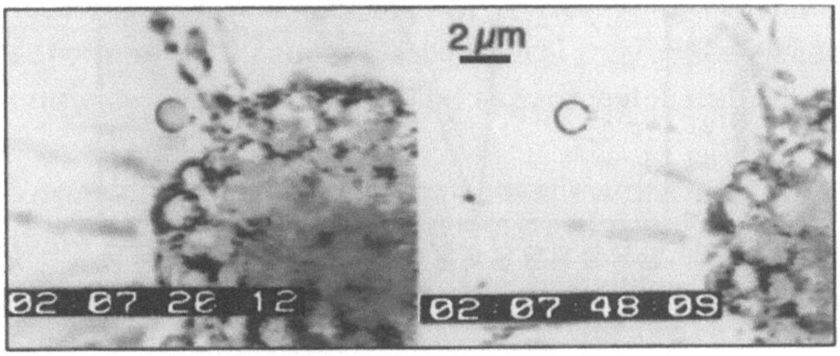

Figure 7 *Moving subcellular structures by optical tweezers in unopened cells of Pyrocystics noctilucae*

19.12 Simulating microgravity of optical tweezers

Finally, optical tweezers have been used to simulate microgravity effects in the alga *Chara* [27]. In the rhizoids of this cell type barium sulfate crystals enveloped by membrane material (statoliths) fall on gravity sensing receptors and tell the cell where gravity comes from. This allows the rhizoid to grow "downwards". When the statoliths are relocated within the unopened rhizoid by optical tweezers, the later lose their orientation and grow in a desoriented manner.

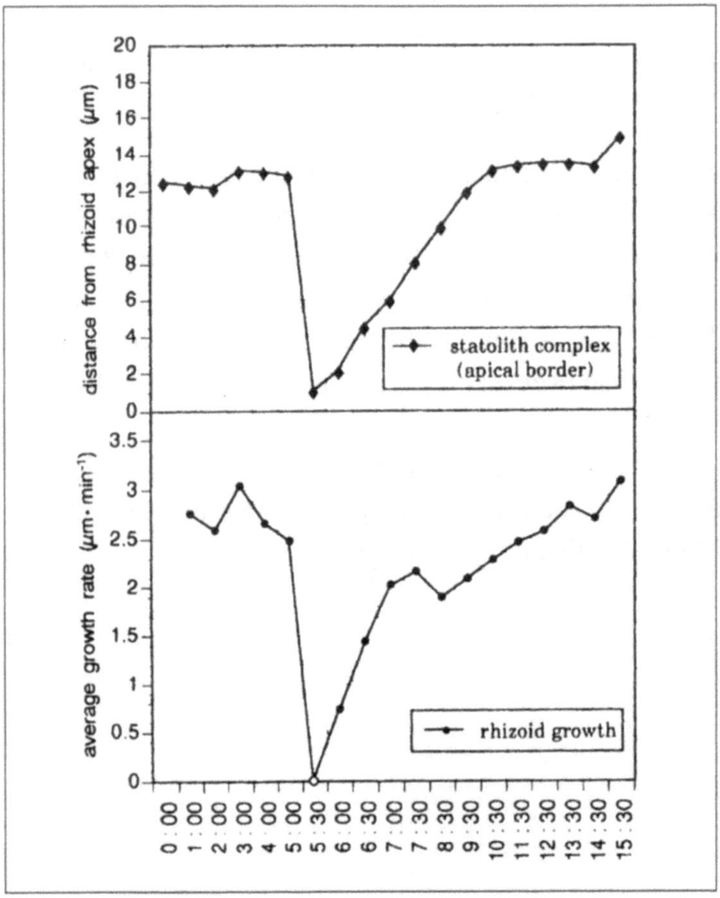

Figure 8 Top: *displacement of Chara statolithes from their original position as a function of time.* Bottom: *Growth rate under the same condition*

This is approximately what one would have expected. The surprise comes when the velocity of growth of the rhizoid is quantitatively evaluated as a function of displacement of the statoliths. Figure 8 shows that the growth velocity is reduced by almost a factor of 10 when the sensing of gravity is artificially inhibited by interference with the optical tweezers.

A speculative conclusion from this experiment is that Chara's growth behaviour has been optimized during evolution for Earth's gravity and that it would grow slower under other gravity conditions.

19.13 Lasers in *in vitro* fertilization

In a previous section cell fusion was induced by disturbing cell membranes with a laser pulse. A process where cell fusion occurs naturally is fertilization. During human fertilization the sperm cell has to penetrate a viscous envelope (the zona pellucida) around the egg cell. Occasionally, this zona is too rigid to be penetrated and infertility is the consequence. *In vitro* fertilization may help, and laser microtools may support this therapy: Laser zona drilling, one variant of *in vitro* fertilization combines perforation of the zona pellucida by a laser microbeam (zona drilling) with optical trapping of the sperm cells, i.e. a laser microbeam trap is used. It is a non-contact method in the sense that no micromechanical tools are required. By drilling a micrometre sized hole into the zona pellucida, the viscuous barrier can be opened. Sperm cells approaching a thus treated egg cell by chance will be attracted and can slip through the hole into the interior of it, where the normal processes of fertilization continue. In order to increase the chance of finding the channel through the zona, the sperm may be caught by the optical trap/optical tweezers and led to the entrance of the channel. The experimental strategy is depicted in Figure 9.

Basic experiments using mouse egg and sperm cells show that with laser support particularly at low sperm density, the rate of successful *in vitro* fertilization is 58% as compared to 33% for the non-laser technique. Thus it is evident that zona drilling with the laser microbeam increases the chance of fertilization. For a long time it was, however, not clear if optical tweezers have an additional beneficial effect. This was studied using bovine oocytes [28].

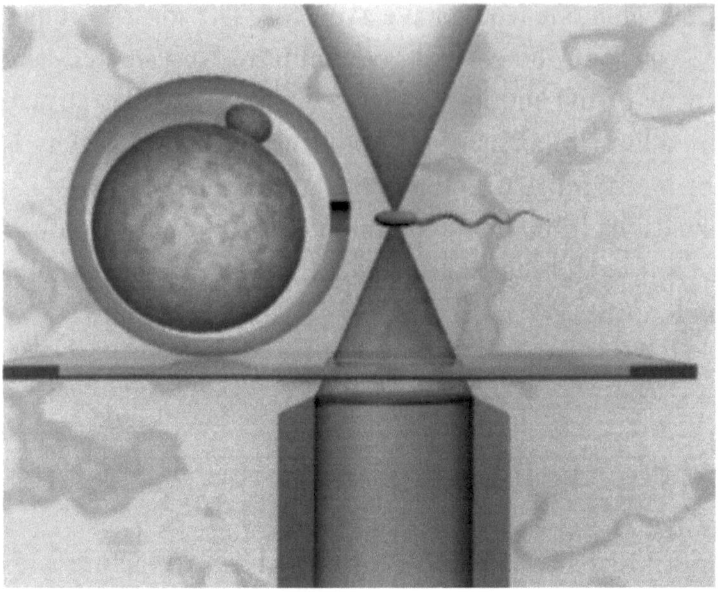

Figure 9 *Principle of laser supported in vitro fertilization: The zona pellucida has already been perforated and the optical tweezers/trap are used to lead the sperm cell towards the entrance of the perforation channel (Reprinted with permission from Dr. K. Schütze, Klinikum Harlaching, München and Dr. G. Kerlen, IWF Göttingen)*

A small additional chance of fertilization was reported when optical tweezers were used to complement laser microbeam treatment.

In many countries the application of laser microtools in human *in vitro* fertilization is not allowed. Whenever the technique was used clinically, it showed advantages as compared to other IVF approaches. However, a final statement on its usefulness can at this time, not be given.

This chapter contains passages taken from [29].

References

1 Epstein AH, Senturia SD (1997) Macro Power from micro machinery. *Science* 276: 1211

2 Ashkin A (1970) Acceleration and trapping of particles by radiation pressure. *Phys Rev Lett* 24: 156–159

3 Ashkin A, Dziedzic JM (1987) Optical trapping and manipulation of viruses and bacteria. *Science* 235: 1517–1520

4 Greulich KO, Bauder U, Monajembashi S, Ponelies N, Seeger S, Wolfrum J (1989) UV

Laser Mikrostrahl und optische Pinzette (UV laser microbeam and optical tweezers). *Labor* 2000: 36–42

5 Greulich KO, Weber G (1992) The light microscope on its way from an analytical to a preparative tool. *J Microscopy* 167(2): 127–151

6 Weber G, Greulich KO (1992) Manipulation of cells, organelles and genomes by laser microbeam and optical trap. *Int Review of Cytology* 133: 1–41

7 Ashkin A (1992) Forces of a single-beam gradient laser trap on a dielectric sphere in the ray optics regime. *Biophys J, Vol* 61: 569–582

8 Wright WH, Sonek GJ, Berns MW (1994) Parametric study of the forces on microspheres held by optical tweezers. *Applied Optics* 33: 1735–1748

9 Simmons RM (1996) Force calibration in optical trapping. *Biophys J* 70: 1813–1822

10 Svoboda K, Block SM (1994) Biological applications of optical forces. *Ann Rev Biomol Struct* 23: 247–285

11 Liu Y, Cheng DK, Sonek GJ, Berns MW, Chapman CF, Tromberg BJ (1995) Evidence for localized cell heating induced by IR optical tweezers. *Biophys J* 68 (5): 2137–2144

12 Williams S, Causgrove TP, Gilmanshin R, Fang KS, Callender RH, Woodruff WH, Dyer RB (1996) Fast events in protein folding: Helix melting and formation in a small peptide. *Biochemistry* 35: 691–697

13 Hoyer C, Monajembashi S, Greulich KO (1996) Laser manipulation and UV induced single molecule reactions of individual DNA molecules. *J Biotech* 52 (2): 65–73

14 Kuo SC, Sheetz MP (1993) Force of single kinesin molecules measured with optical tweezers. *Science* 260: 232–234

15 Svoboda K, Block SM (1994) Force and velocity measurement for single kinesin molecules. *Cell* 77: 773–784

16 Finer JF, Simmons RM, Spudich JA (1994) Single molecule mechanics: piconewton forces and nanometre steps. *Nature* 368: 113–119

17 Molloy JE, Burns JE, Sparrow JC, Tregear RT, Kendrick-Jones J, White DCS (1995) Single molecule mechanics of heavy meromyosin and S1 interacting with rabbit or *Drosophila* actins using optical tweezers. *Biophys J* 68: 298a–305a

18 Monajembashi S, Cremer C, Cremer T, Wolfrum J, Greulich KO (1986) Microdissection of chromosomes by a laser microbeam. *Exp Cell Research* 167: 262

19 Monajembashi S, Hoyer C, Greulich KO (1997) Laser Microbeams and optical tweezers convert the microscope into a versatile microtool. *Microscopy and Analysis* 97 (1): 7–9

20 Tsukakoshi M, Kurata S, Nomiya Y, Ikawa Y, Kasuya T (1984) A novel method of DNA transfection by laser microbeam cell surgery. *Appl Phys B* 35: 135–140

21 Weber G, Monajembashi S, Greulich KO, Wolfrum J (1988) Genetic manipulation of plant cells and organelles with a microfocussed laser beam. *Plant Cell, Tissue and Organ Culture* 12: 219

22 Wiegand R, Weber G, Zimmermann K, Monajembashi S, Wolfrum J, Greulich KO (1987) Laser induced fusion of mammalian cells and plant protoplasts. *J Cell Science* 88: 145

23 Berns MW, Wright WH, Wiegand-Steubing R (1991) Laser microbeam as a tool in cell biology. *Int Rev Cytol* 129: 1–44

24 Kurkdjian A, Leitz G, Manigault P, Harim A, Greulich KO (1993) Non-enzymatic access to the plasma membrane of Medicago root hairs by laser microsurgery. *J Cell Science* 105: 263–268

25 de Boer AH, van Duijn B, Giesberg P, Wegner L, Obermeyer G, Köhler W, Linz KW (1994) Laser microsurgery: A versatile tool in plant (electro) physiology. *Protoplasma* 178: 1–10

26 Leitz G, Greulich KO, Schnepf E (1994) Laser microsurgery and optical trapping in the marine dinophyte Pyrocystis noctiluca. *Botan Acta* 107: 90–94
27 Leitz G, Schnepf E, Greulich KO (1995) Micromanipulation of statoliths in gravity sensing chara rhizoids by optical tweezers. *Planta* 197 (2): 278–288
28 Clement-Sengewald A, Schütze K, Ashkin A, Palma GA, Kerlen G, Brem G (1996) Fertilization of bovine oocytes induced solely with combined laser microbeam and optical tweezers. *J of Assisted Reprod and Genetics* 13 (1): 259–265
29 Greulich KO (1999) Micromanipulation by light. The laser microbeam and optical tweezers. Birkhäuser Verlag, Switzerland

20 Computer modeling of protein, nucleic acid, and drug structures

Jürgen Sühnel

20.1 Introduction

Microsystem technology is devoted to the handling of small amounts of chemicals and biochemicals. It can assist and is in fact a necessary requirement for all types of automatic processes including optimization. In addition to the technological aspects a basic understanding of the molecular structures to be processed is a *sine qua non* for the successful automatization in general and for the optimization of desired functions in particular. Computer modeling of biopolymer and drug structures has already proven to be helpful and becomes even more important if large numbers of structures are to be processed. Molecular modeling can also contribute to the design of nanostructures (Russell et al., 1997; Stupp et al., 1997). This aspect is beyond the scope of this contribution, however. Therefore, this chapter is focused on the application of computational and theoretical approaches to the structure modeling of proteins, nucleic acids, and drugs. It covers a relatively broad subject and cannot be very specific or comprehensive. Rather, it is intended as an introductory reading to modeling techniques and their application. More in-depth information should easily be possible both *via* the references and the sampling of web resources.

In physics and chemistry there has been a very fruitful interplay between theory and experiment. In biology, however, we come across a different situation. So far, the major approach of attacking problems was primarily experimental. One of the reasons is certainly the greater complexity of biopolymers, not to mention cells or complete organisms. Currently, however, this situation seems to be changing. Biomedical research projects are generating data at an enormous pace, the various genome projects being the major source. To give an example, the European Bioinformatics Institute, which is maintaining the EMBL and SwissProt sequence databases receives each minute a new

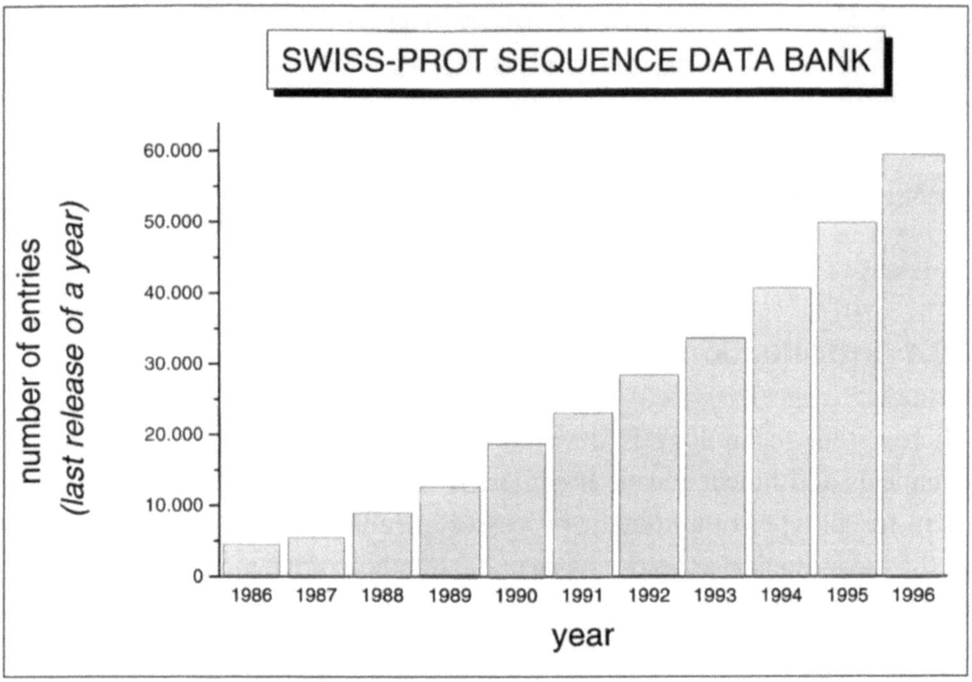

Figure 1 *Number of entries deposited at the Swiss-Prot protein sequence data bank. There is more than one release per year. The data given correspond to each year's last release*

sequence, 24 hours per day and 365 days per year (see Fig. 1). However, the genome projects are not the only source of this data explosion. Combinatorial and evolutionary approaches are another one and a further example comes from structural biology. Between 1973 and 1989 the number of new entries released per year by the Protein Data Bank was below 100 structures. This growth rate has dramatically increased especially since 1991, leading to more than 1200 new structures in 1996, which corresponds to slightly more than three structures per day (see Fig. 2).

Simultaneously, the development of new network information systems has led to a phenomenal growth of the world-wide computer network Internet (Schatz and Hardin, 1994). In fact, the Internet or World-Wide Web (WWW) is one of the basic requirements for the effective organization of large-scale international genome projects, for example. Moreover, without the web it would be absolutely impossible to make accessible to the scientific community the huge amounts of data generated by current biomedical research.

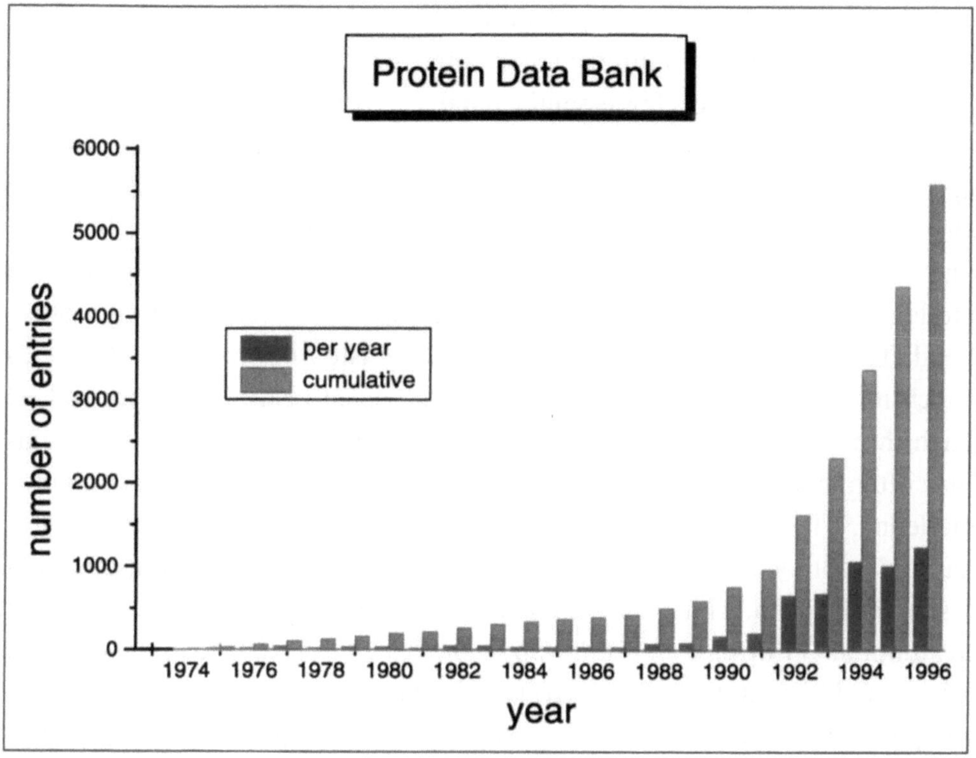

Figure 2 *Number of entries deposited at the Protein Data Bank*

In face of these developments and challenges a new discipline has emerged, bioinformatics, which covers all aspects of acquisition, processing, storage, distribution, analysis, and interpretation of biological data (Murray-Rust, 1994; Benton, 1996; Lyall, 1996; Ruediger, 1997). In a first step, and this is currently happening, these data are classified and searched for regularities. This can lead to a formulation of new rules or laws and they in turn can be used to build scientific theories. Currently, bioinformatics is primarily related to data from genome projects, high-throughput screening and structural biology. In a next step, however, these data will be used to study more complex phenomena adopting such methods like system analysis, for example (Palsson, 1997).

Our impression is that currently biology moves slowly towards a greater awareness on the importance of theoretical concepts. Whether or not this should be called a paradigm shift is not at all important (Gilbert, 1991).

Information on sequences and structures of biopolymers stored in freely accessible databases is of utmost importance for all types of modeling efforts. Therefore, we describe a few essential databases first. We proceed with modeling methods and end up with a brief discussion of the application of these approaches to proteins, nucleic acids, and small molecules. From the point of view of optimization the great challenge is to define interfaces between known experimental facts and theoretical knowledge including predictive modeling efforts on the one hand, and the experimental setup on the other hand. This can lead to automatic computer-assisted optimization strategies which may be much more effective than the usual brute-force approaches. High-throughput screening, antisense techniques and combinatorial chemistry often linked with genomics are expected to gain benefit from computer modeling. This point is discussed in more detail in the chapter on combinatorial approaches. In addition to the references a sampling of useful web resources is given, which should enable interested readers to get easy access to additional information. If the web resources are referred to in the text this is indicated by a citation of the type [wxx], where xx is a number. The web site compilation can also be used as stand-alone information. The journals *Current Opinion in Structural Biology* and *Current Opinion in Biotechnology* provide yearly updates of many aspects mentioned in this contribution. The series *Reviews in Computational Chemistry* edited by K.B. Lipkowitz and D.B. Boyd (VCH Publishers) contains both reviews on modeling methods and on applications and is thus a highly recommended additional reading [w42].

20.2 Databases

The general agreement that the overwhelming part of biological data generated is made freely available to the scientific community is one of the primary reasons for the fast development we currently see in biology. Database information is a necessary prerequisite for modeling of biopolymers and small molecules and is thus discussed first. With the increase of both size and number of databases there is, however, a pressing need for improved quality control and interoperability (Karp, 1996).

Structural information on biological macromolecules and on small molecules of biological interest is an essential requirement for our understanding of the function of biologically active agents and for a deliberate variation of this function by rational or evolutionary approaches.

The Cambridge Structural Database (CSD) is the unique source of structural information from small-molecule organic and organometallic crystal structures determined by x-ray and neutron diffraction techniques (w1). On October 1998, CSD held 164732 entries with three-dimensional (3D) coordinates. The overwhelming majority of these structures was determined by x-ray diffraction and only less than 1% were resolved using neutrons (Allen and Kennard, 1993).

The KLOTHO database maintained at the Washington University Institute for Biomedical Computing provides structural information on slightly more than 400 small molecules of biochemical relevance [w2]. The database includes images and coordinate files for interactive work with molecular graphics programs.

The Protein Data Bank (PDB) at Brookhaven National Laboratory has collected coordinate files of proteins, nucleic acids and carbohydrates for about 25 years (Bernstein et al., 1977; w3). The number of entries released per year has slightly increased from 10 in 1973 to 89 in 1989. Progress in recombinant DNA technology and RNA synthesis, x-ray and nuclear magnetic resonance (NMR) instrumentation and computer and software technology has led to an increasing rate of accumulation of new structures since 1990. Currently (January 6, 1999), the PDB holds 8974 coordinate entries including 8329 proteins, 633 nucleic acids and 12 carbohydrates.

The Nucleic Acid Database (NDB) provides information on 3D nucleic acid structures (Berman et al., 1992; w4). The great majority of nucleic acid structures can be obtained both from NDB and PDB. NDB provides various additional x-ray structures which are not available from the PDB. On the other hand, contrary to the NDB the PDB also includes NMR and theoretical structures.

In addition to the structures determined at atomic resolution recent developments in electron microscopy enable structure determinations at much higher resolution than before. This is especially important for macromolecular assemblages (DeRosier and Harrison, 1997). One example is the ribosome, where cryoelectron microscopy studies have led to models of the *Esche-*

richia coli 70S ribosome at 20–25 Å resolution (Frank, 1997). The very near future will see a further improved resolution of microscopy techniques below 10 Å and a fruitful interplay with x-ray and NMR studies on subunits and with modeling attempts (Chiu and Schmid, 1997). In a kind of three-dimensional jigsaw puzzle this will lead to a further refinement and improvement of the low-resolution models for molecular assemblages (Brimacombe, 1995).

Internet-based image archives of biological macromolecules are a useful addition to the usage of molecular graphics software. Automatically generated images of all PDB entries are provided by the PDB (w3) and by the Molecules R US utility (w5). The advantage of these image collections is their completeness. However, the automatic generation procedure restricts the information content of the images, even though Molecules R US allows for a certain interactive selection. High-quality images, including mono and stereo representations which were manually generated, are available from the Swiss-3D-Image collection (Peitsch et al., 1995; w6) and from the Image Library of Biological Macromolecules (Sühnel, 1996, 1997; w7). The disadvantage of the latter databases is, of course, that they only provide images of a small fraction of the structures known.

In spite of the increasing rate of determination of 3D biopolymer structures the gap between the number of known sequences and structures is increasing. The current release 36 of the annotated protein sequence database SWISS-PROT contains 76803 entries. This number should be compared to the 8329 resolved 3D protein structures, which means that the structures of slightly more than 10% of sequenced proteins are known (see Figs 1 and 2). Almost the same figures are obtained for individual genomes. For only 475 out of the 6195 yeast proteins the structures are known, which corresponds to a fraction of 7.7%. In addition, there seems to be a kind of culture clash between the sequence and structure worlds which is primarily due to the greater complexity of 3D structure data. If theoretical methods were able to bridge the sequence-structure gap this would be a major breakthrough. Sequence databases including the genome databases are an extremely important working tool for modeling purposes of any type. As was already mentioned, SWISS-PROT is an annotated database of protein sequences (Bairoch and Apweiler, 1997; w9). The ENTREZ information system at the National Center for Biotechnology Information provides a very attractive combination of nucleotide, protein and genome databases including the bibliographic

PubMed retrieval system (Schuler et al, 1996; w8). The Sequence Retrieval System SRS is a very efficient network browser for accessing a wide variety of sequence data banks (Etzold et al., 1996; w10). The sampling of web resources includes further links to databases on the human, yeast, and various microbial genomes (w11–w14).

20.3 Modeling methods

Visualization plays an important role in structure determination, modeling, understanding biopolymer structures and possibly deriving new insights and is therefore briefly mentioned first.

From a theoretical point of view the most satisfying modeling approach would be to start out from basic physicochemical principles and finally end up with correct structures. This method is currently not feasible. A more successful alternative is knowledge-based methods which combine statistical information and empirical rules to predict the 3D structures of biopolymers. Nevertheless, the *ab initio* modeling of biopolymer structures remains an important challenge and the mutual interplay between the two methods can be fruitful for both of them.

Visualization

From the very beginning of structural biology visualization was essential for determining and understanding structures. Model building with metal plates and other materials has played a crucial role for the DNA double helix structure proposed by Watson and Crick in 1953. Further, it is reported that the first model of hemoglobin built by M. Perutz and colleagues in the 1960s took up 16 square feet of floor space (Hall, 1995). Since the early 1980s interactive computer graphics has greatly facilitated and improved the visualization of biopolymer structures. Currently, a great variety of molecular graphics programs is available for different computer platforms (w18–w25). The usual approach for the visualization of known structures is to retrieve the coordinate file from one of the structure databases and then to use a molecular graphics package. The recent developments in the World-Wide Web enable one to transfer images or, in general, multimedia information over the Internet very easily. Therefore, Internet-based image archives of biopolymer structures have been set up (Peitsch et al., 1995; Sühnel, 1996; w5–w7). They are a useful addition to the usage of molecular graphics software, can contribute to a better dissemination of visual structural information on biopolymers, and are of particular importance for researchers outside structural biology and for educational purposes.

In 1995, the new The Virtual Reality Modeling Language (VRML) format was defined (Brickmann and Vollhardt, 1996; Sühnel, 1996). It is essentially a 3D image format supplemented by network tools. Contrary to the static images, it enables one to interact with the 3D image objects. Of course, this can be done much better using molecular graphics packages. However, VRML viewers are standard parts of current web browsers. This means that one can interact with the 3D biopolymer images without having access to a molecular graphics software. The very same format can also be used in conferencing systems appropriate for collaborative work on 3D objects.

Quantum chemistry

Broadly speaking, there are two basically different approaches for the physicochemical modeling of chemical structures, namely the quantum-mechanical or quantum-chemical and the molecular mechanical approaches.

Quantum mechanics requires the solution of the Schrödiger equation

$$H\Psi = E\Psi \tag{1}$$

where H is the Hamiltonian, an order to derive the kinetic and potential energy of the system (Boyd, 1990). The squared wavefunction $\Psi^2(x, y, z)$ is proportional to the probability of finding electron density in a given region of space described by the spatial particle coordinates and the eigenvalue E represents the energy of the system in a particular state. For practical calculations on almost all molecules various approximations must be made. They include neglect of relativistic effects, time independence of the Hamiltonian, the Born-Oppenheimer or adiabatic approximation and the orbital approximation. Born-Oppenheimer approximation means that the electronic wavefunction is calculated assuming fixed nuclei. This assumption is the prerequisite for the formulation of the concept of the potential energy hypersurface.

Within the orbital approximation the wavefunction can be written as a linear combination of orbital functions each dependent on only one electron. Adopting all these approximations one ends up with a simplified version of the Schrödinger equation called Hartree-Fock self-consistent field equation. Quantum-chemical calculations performed on this level are described as *ab initio* methods. Introducing further approximations like the consideration of valence electrons only, neglect of particular types of integrals and using a very simple basis set leads to the *semiempirical* methods. The difference between the exact energy and the Hartree-Fock *ab initio* energy is called correlation energy. There are basically two methods for taking into account the correlation energy, at least in part, many-body perturbation energy and configuration interaction. Most of the current *ab initio* quantum chemistry packages include options for taking into account the correlation energy (w26–w28). Finally, it should be noted that density functional theory is a less computationally demanding alternative to the conventional calculations (St-Amant, 1996; Bartolotti, Flurchick, 1996).

Improved computing facilities have made possible the application of more sophisticated quantum-chemical methods to larger systems. Nevertheless, the quantum-chemical approach is currently restricted to molecular structures with no more than 200 or 300 atoms. Therefore, for biopolymer structures quantum-chemical calculations are not feasible currently. So the application of *ab initio* quantum-chemical methods is limited to studies on building blocks, like active sites in proteins or base pairs in nucleic acids (Sponer et al., 1996). *Ab initio* quantum-chemical methods are also used for the exploration of the conformational space of small molecules and for the calculation of charge distributions which are required for the derivation of parameters for molecular mechanics force fields.

Even though the application of *ab initio* quantum chemical approaches has become more feasible in recent years semiempirical approaches are still in use (Zerner, 1991; w29). They enable one to treat larger systems or to perform more comprehensive studies of exploring the conformational space of structures, for example.

An interesting development is to treat parts of a system with the hopefully more accurate quantum-chemical methods and the remaining part with molecular mechanics (Liu et al., 1996).

Molecular mechanics

Molecular mechanics means that the dependence of the total energy of a molecular system on the atom coordinates is described by an empirical expression which contains a variety of parameters.

A widely used expression for the empirical description of the dependence of the total energy of the molecular system on the atom coordinates is

$$
\begin{aligned}
E(R) = {} & (1/2)\sum K_b(b-b_0)^2 && \text{(bonds)} \\
& (1/2)\sum K_\Theta(\Theta-\Theta_0)^2 && \text{(bond angles)} \\
& (1/2)\sum K_\varphi[1+\cos(n\varphi-\delta] && \text{(torsional angles)} \\
& (1/2)\sum [(A/r^{12})\text{-}(B/r^6)+(q_1q_2/Dr)] && \text{(non bonded and electrostatic interaction)}
\end{aligned} \qquad (2)
$$

The energy E is a function of the Cartesian coordinate vector R from which the internal coordinates used in the expression can be calculated. They include the bond distances b, the bond angles Θ, the dihedral or torsional angles φ and the distance r between nonbonded atoms.

The bond and bond angle expressions describe the energy change associated with a displacement from the equilibrium positions b_0 or Θ_0 adopting a harmonic approximation with the force constants K_b and K_Θ as proportionality factors. Torsions around bonds are described by a periodic potential and the last term takes into account the van der Waals and Coulomb interaction between all nonbonded atom pairs at the distance r. The first two expressions in this term give together the well-known Lennard-Jones 6–12 potential which yields an energy minimum for the sum of the van der Waals radii. Finally, the electrostatic interaction is modeled by a Coulomb term containing the charges q and an effective dielectric function D.

The parameters of molecular mechanics potentials are determined by taking into account experimental data and results of quantum-chemical studies on small molecules. The basic assumption is then that the force fields derived in this way can be carried over to biological macromolecules. Depending on the systems studied and on the specific approaches adopted the results are more or less reliable. One should realize, however, that with increasing availability of computer power substantial improvements of this approach can be expected.

The current force fields are often referred to as "second generation" as they contain revisions of original versions. Force fields which can be applied to biopolymers include AMBER (Cornell et al., 1995; w30), CHARMM (Brooks et al., 1983; w31) and GROMOS (Van Gunsteren et al., 1996; w32).

Molecular mechanics calculations are widely used if the quantum-chemical approach cannot be adopted and a dynamical treatment is not necessary or possible (Weber, Harrison, 1996).

Molecular dynamics

Biopolymers are not static systems but show an inherent dynamics. In a few cases it has even been claimed that a particular rigidity or flexibility may be relevant for biological activity. Time-dependent processes in biological macromolecules can be treated with molecular dynamics simulations (Karplus and Petsko, 1990; van Gunsteren and Mark, 1992). The molecular dynamics method is based on the statistical mechanics of many-body systems and on the (classical) dynamics of particles on potential energy surfaces.

The dynamical treatment requires the solution of Newtons's equations of motion for all atoms of the biopolymer and of the surrounding solvent. As for the static treatment of biopolymer structures a quantum approach is completely unfeasible. Therefore, one has to use empirical force fields like the one described in Eq. (2). In addition to the force field problems the dynamics pose further major challenges to this approach. They refer primarily to the size of biological macromolecules and to the large range of time scales relevant to biological processes.

Currently, the most direct comparison of molecular dynamics simulations with experimental data refers to the isotropic temperature factors (B-factors) obtained from x-ray diffraction data. These factors are a measure of the mean square displacements of atoms. It has been shown that the simulation yields reasonable results for this property but not for the fluctuations in the separation of pairs of atoms (Caspar, 1995). This may possibly indicate a sampling problem, which could be overcome with longer simulations.

Application of molecular dynamics simulations include structure determination in NMR spectroscopy and structure refinement in x-ray studies, the simulation of macromolecule motions including functionally relevant motions and free energy calculations (Karplus and Petsko, 1990).

Recent methodological advances in molecular dynamics simulations of biological systems refer to improved force fields, more precise definitions of the statistical ensembles used and the particle mesh Ewald summation (Brooks, 1995; Louise-May et al., 1996). The latter approach represents a more exact method to calculate electrostatic interactions as compared to the conventional cutoff procedure. With the increasing computer power and further methodological developments we can expect that the reliability of molecular dynamics simulations of biological systems will be substantially improved.

Knowledge-based potentials

With the rapidly increasing number of experimental protein structures available there have been attempts to derive information on inter-residue or more recently on inter-atomic interactions from experimental protein structures (Sippl, 1995; Jernigan and Bahat, 1996; Moult, 1997). This approach complements the molecular mechanics force fields described above.

The application of knowledge-based potentials starts out from the Boltzmann relation for a population of protein molecules. It is usually defined as

$$P(\Delta G_i) = \exp(-\Delta G_i/kT)/\Sigma_j \exp(-\Delta G_j/kT) \qquad (3)$$

where $P(\Delta G_i)$ is the probability of finding the system in state i with a free energy ΔG_i and the summation is over all possible states. The potential energy E of the molecular mechanics force fields is related to the free energy ΔG by the relation

$$\Delta G = \Delta E + T\Delta S \qquad (4)$$

where T is the absolute temperature and S is the entropy of the system.

Knowledge-based or statistical or database potentials can be viewed according to statistical mechanics, and in this sense they should allow for the experimental determination of free energy of particular types of interaction in biopolymers. On the other hand, they can be treated in a purely statistical way, which requires no physical assumptions at all. The latter approach can be applied in a straightforward manner if, for example, correct and incorrect biopolymer structures are to be distinguished (Sippl, 1993; Luthy et al., 1992). There is, however, a current debate about whether the free energies derived in this manner have the same properties as free energies used in physics. If so this would enable one to derive free energies of particular interaction types from experimental data, which is a really fascinating possibility (Sippl et al., 1996). This, however, has been questioned by Thomas and Dill (1996). One of their basic arguments is that interactions are not independent. Nevertheless, statistical potentials have been surprisingly successful in structure prediction (Moult, 1996). Finally, it should be noted that, like molecular mechanics or molecular dynamics results which depend on the parametrization used, the knowledge-based potentials rely heavily on the quality of the database in terms of resolution, sample size and possible errors. Furthermore, due to the very nature of the approach it cannot be applied to address specific questions in single biopolymer structures.

Comparative modeling by homology

Comparative modeling or modeling by homology is possible if the sequence of the target protein is related to the sequence of a protein whose 3D structure is known (Moult, 1996; Sanchez and Sali, 1997). This approach works because small sequence variations have only minor effects on the 3D fold. One basic question is, of course, how a significant sequence relationship is defined. The current experience is that a sequence identity of 40% or higher allows for an automatic modeling procedure with very good results. More challenging is the region between 25% and 40% identity. Sequences with a sequence identity below 20% cannot be used with this approach. Comparative modeling includes the sequence alignment, selection of the core backbone structure, loop and sidechain modeling and finally, structure refinement. There are basically three approaches for core modeling. Rigid body assembly uses core regions, and possible loops and sidechains as well from other structures. Segment matching uses a database of short protein structure segments and additional geometrical and/or energetical rules. Modeling by satisfaction of spatial constraints adopts distance geometry approaches widely used in NMR spectroscopy, for example, or other optimization techniques. One major problem of sequences below 30% sequence identity seems to be related to the quality of the alignment.

Nevertheless, for a sequence identity above 40% comparative modeling is the method of choice. It yields correct backbone folds with possibly inaccurate loop regions. A disadvantage of this approach is, of course, that no unique new folds can be predicted.

Fold recognition/Threading

We have learned in recent years that structures are more conserved than sequences. In other words, sequences with no detectable homology may nevertheless have similar folds. In these cases fold recognition or threading approaches can be applied (Torda, 1997). An amino acid sequence is compared to a library of protein folds by calculating some kind of score. This requires, of course, sequence-structure alignment which is necessarily of an heuristic nature. Fold recognition is a relatively new method. Most work is still focused on methodological improvements. However, the first applications were already published (Madej et al., 1995).

Ab initio predictions

If neither comparative modeling nor fold recognition approaches can be applied, *ab initio* prediction methods are all that remain (Moult, 1996). These methods make no direct use of structure databases. As far as secondary structure prediction is concerned this is the most traditional form of structure prediction (Barton, 1995). However, knowledge-based approaches and direct numerical simulations adopting lattice methods or molecular dynamics can also be applied (Defay and Cohen, 1995). In a certain sense *ab initio* prediction is the intellectually most challenging approach. On the other hand, it is certainly correct to claim that it is currently the least successful method. The accuracy of secondary structure prediction is usually measured in terms of a correct assignment of residues as alpha helix, beta sheet or other conformations. The success rate for using only one sequence seems to be limited to an accuracy of about 60%. More recently, methods have been developed which include evolutionary information on proteins. This has increased the prediction accuracy above 70% (Barton, 1995; Rost and Sander, 1996). Contrary to the proteins the very same phylogenetic or comparative sequence analysis approach has been successfully applied over the years for RNA secondary structure prediction and more recently for the identification of tertiary interactions as well (Konings and Gutell, 1995; Gautheret et al., 1995). Another approach widely used in RNA secondary structure prediction calculates structures of minimum free energy (Zuker and Jacobson, 1995).

Ab initio prediction methods remain a challenging field of research even though currently the results obtained are seldom reliable enough to be used in drug design, for example.

3D-QSAR and structure based design

Quantitative structure activity relationships (QSAR) were the first attempt to put the search for new drugs on a more quantitative basis and to enable predictions (Green and Marshall, 1995). In this approach various molecular parameters like the partition coefficient between octanol and water were correlated with biological activity. The next step was to take into account information on the molecular structure by means of indices describing the molecular structures or structure changes by means of one net index. Finally, 3D-QSAR uses a more detailed information of the potential drug structures (Kubinyi, 1993). The most widely used 3D-QSAR method seems to be the CoMFA approach (comparative molecular field analysis) (Cramer et al., 1988). These methods take into account the 3D structure of drugs but do not necessarily require information on the receptor structure. If, however, the target structure is available 3D-QSAR combined with structure-based design should lead to improvements in lead generation and optimization. Structure-based drug design starts out from the known 3D structure of the target and tries to find optimum ligands (Verlinde and Hol, 1994). Figure 3 shows an example from a modeling project, where a peptidic and a non-peptidic inhibitor are fitted into the binding site of the corresponding HIV protease. A few success stories have already reported (Edwards et al., 1996). However, most of the examples known are improvements of known lead structures. An especially challenging research area is the *de novo* design of ligands which

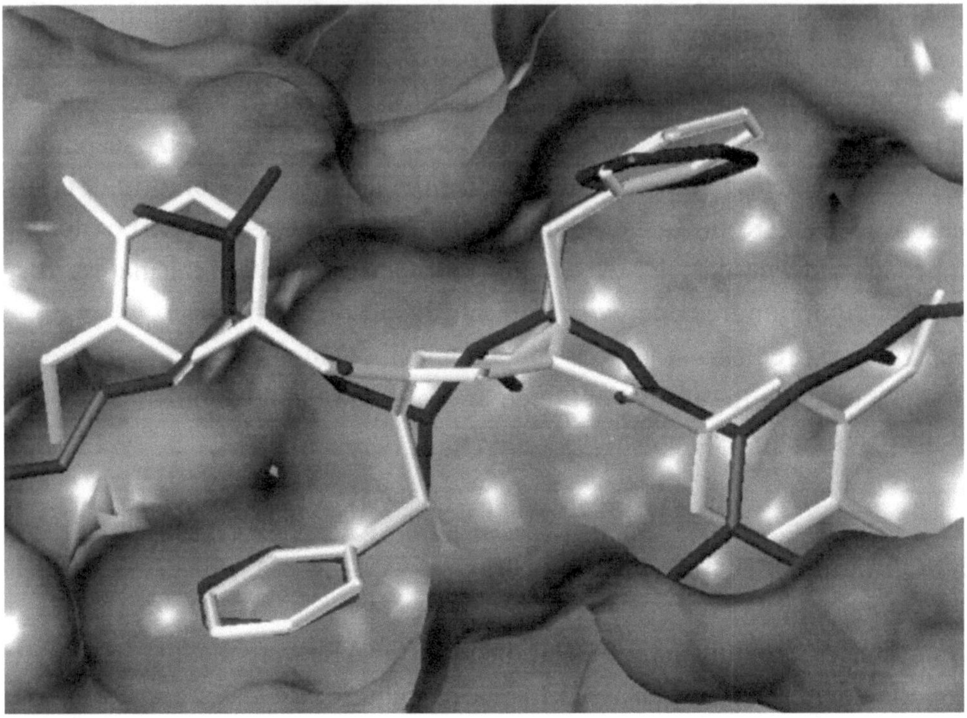

Figure 3 *Superposition of a non-peptidic (dark) and a peptidic HIV-protease inhibitor (bright). The image was generated by a superposition of the protein parts of the HIV-1 protease complexes with A-74704 (PDB code: 9hvp; Erickson et al., 1990) and with A-98881 (PDB code: 1pro; Sham et al., 1996). It is obvious that the orientation and the sterical demands of both inhibitors are very similar. The image was generated with SYBYL, Tripos, Inc. and provided by A. Hillisch, Molekularbiologie, Institut für Molekulare Biotechnologie, Jena*

uses the 3D target structure alone without taking into account prior information on other ligands (Böhm, 1996; Bamborough and Cohen, 1996). Ligands identified in this way are new lead structures which have to be subjected to further lead optimization.

Both for structure-based drug design, where the target structure is known, and for pharmacophore design, where simply some information on the geometric and electronic properties of the active agent is available, searching of 3D databases is a valuable tool. Algorithms which generate 3D structures of small molecules from simple connectivity information have played an important role because they have enabled the conversion of large corporate and commercial chemical structure databases into 3D format (Sadowski, 1997).

Manual versus automatic modeling

Automatic modeling techniques can easily cope with a large number of structures and their results are not dependent on subjective decisions. On the other hand, human intervention may improve the quality of the results significantly. One of these examples is the evaluation of secondary structure prediction data (Barton, 1995). Manual modeling has been especially successful in 3D RNA structure prediction. A database of preformed structure modules is assembled interactively taking into account biochemical and other data which contain indirect structural information (Michel and Westhof, 1990; Cate et al., 1996).

Therefore, it can be concluded that both approaches have their own strengths and weaknesses and the application of both of them is appropriate for specific questions.

Combinatorial techniques

Combinatorial chemistry is comprised of methods which generate very rapidly large numbers of compounds which can be tested as drug candidates, for example (Gallop et al., 1994; Blondelle and Houghten, 1996; Hogan, 1997). Usually, the compounds synthesized are run through a separate screening scheme. In other cases (SELEX, phage display) the approach is reiterative which means that compound generation and screening are combined within one step (Gold, 1995).

It is immediately obvious that combinatorial chemistry needs an effective system of information management. Even though the rapid generation of extremely large numbers of molecules has become feasible by combinatorial chemistry it is nevertheless time-consuming and expensive. Therefore, it would be extremely useful if information already available could be taken into account to develop focused libraries. This information could substantially reduce the number of compounds to be tested.

One basic issue which has to be addressed is a quantitative description of molecular diversity. A possible application is, for example, the selection of a highly diverse subset from a large pool of molecules without having information on the target structure (Chapman, 1996). Other approaches use structural constraints provided by the target to restrict the size of the combinatorial library (Murray et al., 1997).

In view of the fact that combinatorial approaches are probably the only means to keep pace with the rate at which genome sequencing generates new potential drug targets, the development of new techniques which make the combinatorial approaches faster and more effective can hardly be overestimated.

The impact of the world-wide web

The importance of web sites for storing and accessing biological information was already pointed out. In addition to the conventional database query tools there is an increasing number of web sites where one can process own data. Examples are protein or RNA secondary structure prediction and multiple sequence alignments (w37, w43). Moreover, the new VRML format enables the collaborative work on three-dimensional structures over the net (Brickmann and Vollhardt, 1996; Sühnel, 1996). The web tools can either work as mail services or in an interactive manner. Internet search engines enable easy access to relevant information when entering new fields. Text-based information is gradually replaced by multimedia documents (Schatz, 1997).

The world-wide web has already changed the communication of science and will open up new possibilities in the future.

20.4 Structures

Proteins

The prediction of protein 3D structure from sequence information remains probably the largest unsolved problem in structural biology. Myriads of papers have been written on this subject and the pessimistic point of view is that there is no light at the end of the tunnel. Upon closer inspection, however, it turns out that there is progress at least. The most direct way of testing the possible success of structure predictions is to predict structures for which experimental work is in progress. Therefore, two structure prediction contests were performed in 1994 and 1996 (Lattman, 1995; Moult, 1996; w41). A detailed and authorative report on the contest can be found in articles focusing on comparative modeling (Mosimann et al., 1995), fold recognition (threading) (Lemer at al., 1995) and *ab initio* prediction (Defay and Cohen, 1995). More information on the second contest can be obtained from the corresponding website (w41). It is, of course, very difficult to draw any general conclusions on the state of the art of protein structure prediction. Approximately, 25–30% of new sequences deposited at sequence databases and about 7–9% from genome projects have sufficient sequence homology to apply comparative modeling approaches. In these cases a useful though approximate 3D structure can be obtained. The cases for which fold recognition approaches can be applied is difficult to evaluate. For the remaining part of structures *ab initio* prediction is the only possible method. For secondary structure prediction a success rate between 50 and 80% is usually assumed. The predictions may be more reliable if evolutionary information is taken into account and if possible hydrophobicity patterns are carefully inspected. How to assemble the secondary structure elements to a 3D fold remains still in most cases a mystery. So far, the results obtained from *ab initio* predictions cannot be used in drug design, for example.

Another major goal of modeling efforts for proteins is the prediction of ligands starting out from a known 3D target structure (Bamborough and Cohen, 1996). These efforts include the prediction of the binding site and ligand orientation and ranking of a series of ligands according to their binding affinity. Even though this goal seems to be more easy to reach, success stories are rare (Strynadka et al., 1996). One additional difficulty with protein-ligand docking is possible structure changes upon binding. The classical key and lock principle works in some cases. Other examples exhibit an induced fit. So far, there is almost no possibility to predict from the free structure if key and lock or induced fit will be operating for a particular example. Ligand-protein docking was included in the second protein structure prediction contest (Moult, 1996).

Nucleic acids

The interest in structure determination and structure prediction is primarily focused on proteins. As already mentioned, less than 10% of the known 3D biopolymer structures are nucleic acids. One the other hand, DNA structure determination has come of age. In the DNA field the most interesting contributions to the current understanding of structure comes probably from investigations on unusual conformations or structural motifs, like triple helices, quadruplexes, telemore-like structures and DNA-RNA chimeras (Joshua-Tor and Sussmann, 1993; Wahl and Sundaralingam, 1995).

The current computational approaches are sufficient for a modeling of static and dynamical DNA structures which are in overall agreement with experimental data (Louise-May et al., 1996). An important step forward came from the application of Ewald summation techniques which no longer require unphysical truncations of the longe-range electrostatic forces.

Studies on the principles of DNA-drug recognition have been performed over the years (Neidle, 1997). Even though the possible reasons for effective binding to DNA and for sequence specificity seem to be known in principle, their importance may vary from ligand to ligand. To date, no simple rule for drug-DNA binding has emerged. Figure 4 is taken from a modeling project where a drug is moved along the DNA minor groove.

There is a similar situation for DNA-protein interaction (Choo and Klug, 1997). It seems to be clear now that no universal protein-nucleic acid recognition code will be found. However, there are attempts to derive general rules which can be called recognition code for particular families of protein-nucleic acid complexes like zinc fingers, for example (Choo and Klug, 1997).

In addition to these studies, which are primarily focused on relatively short DNA structures of 10, 12 or 14 base pairs long, the properties of long DNA strands are interesting as well. One interesting background is the three-dimensional organisation of large genomes and their structural response to the binding of proteins (Olson, 1996).

Especially exciting is the development in the RNA field. Since the discovery that RNA cannot only encode genetic information but is also able to catalyze chemical reactions there is a renewed interest in RNA structures. Occasionally, it is even asked if ribozyme engineering may be easier than protein engineering (Breaker, 1996; Cech, 1992; Wedel, 1996). For about 20 years, tRNA was the only larger RNA structure whose 3D coordinates were known. This situation has changed within recent years. Currently, 3D structural infor-

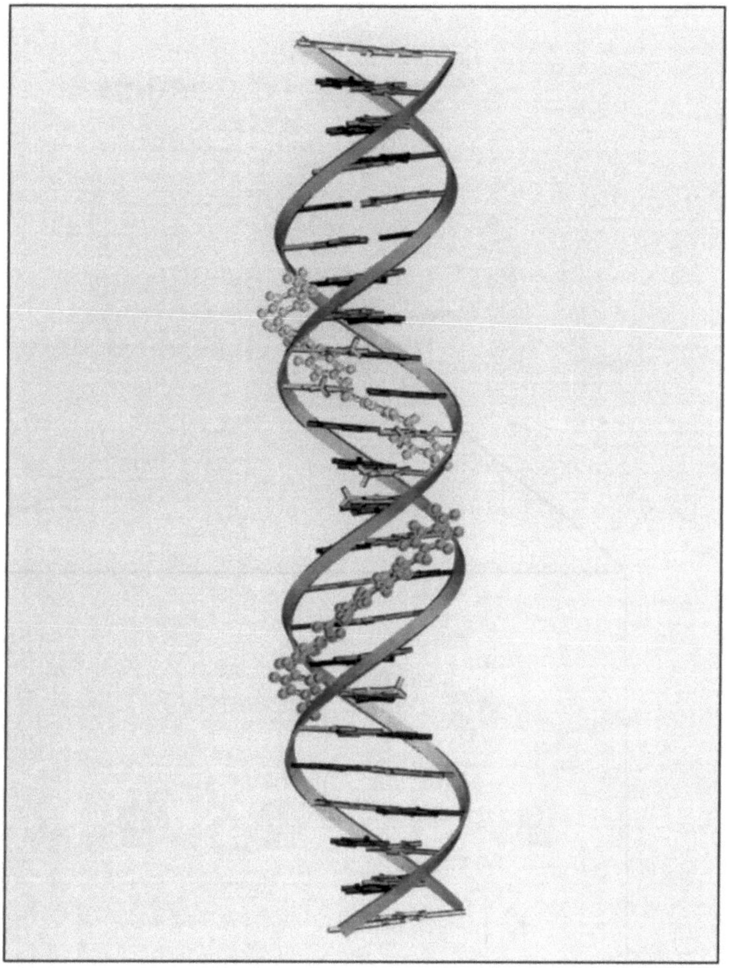

Figure 4 *Moving a ligand along the minor groove of a standard B-DNA. This image is from a modeling project aimed at a better understanding of sequence-specific DNA-drug recognition. The coordinates were kindly provided by P. Slickers, Biocomputing, Institut für Molekulare Biotechnologie, Jena. The image was generated with SETOR (Evans, 1993)*

mation on pseudoknots, hammerhead ribozymes and of a domain of the group I intron ribozyme has become available. This information is supplemented by RNA-protein and RNA-aptamer complexes (Feigon et al., 1996; Uphoff et al., 1996). Aptamers are DNA or RNA oligonucleotides that have been selected *in vitro* for specific binding to a target molecule. Figure 5 shows a ribbon drawing of the group I intron domain with a hydrogen bond

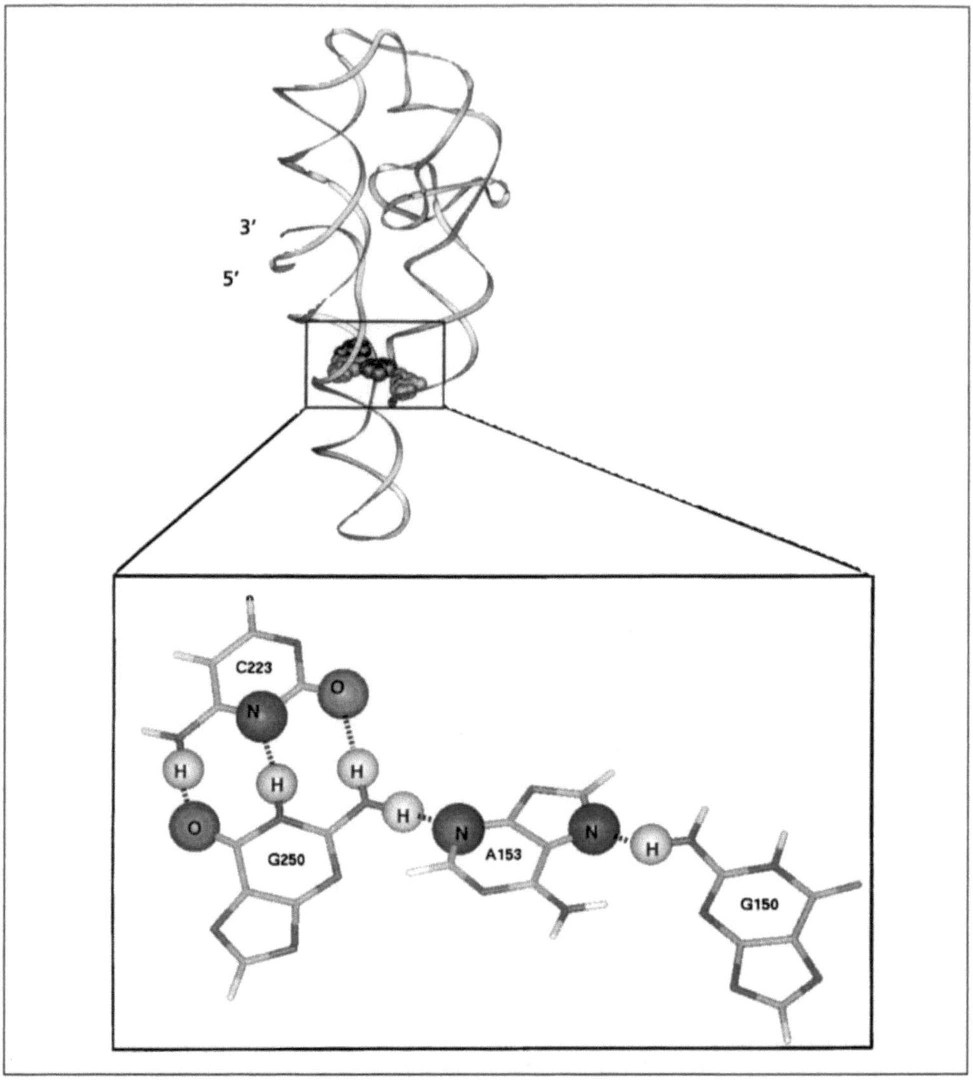

Figure 5 *Ribbon drawing of the structure of a group I ribozyme domain with a tetra-nucleotide base-base hydrogen bond interaction highlighted (PDB code: 1gid; Cate et al., 1996). The image was generated with InsightII, Molecular Simulations, Inc. and SETOR (Evans, 1993)*

polynucleotide interaction highlighted, and in Figure 6 an RNA aptamer structure is shown. We now have much more experimental information on RNA, which will certainly catalyze studies on structure prediction. From the structural point of view one of the basic questions is how RNA manages to fold into relatively complex structures which are obviously necessary for the catalytic

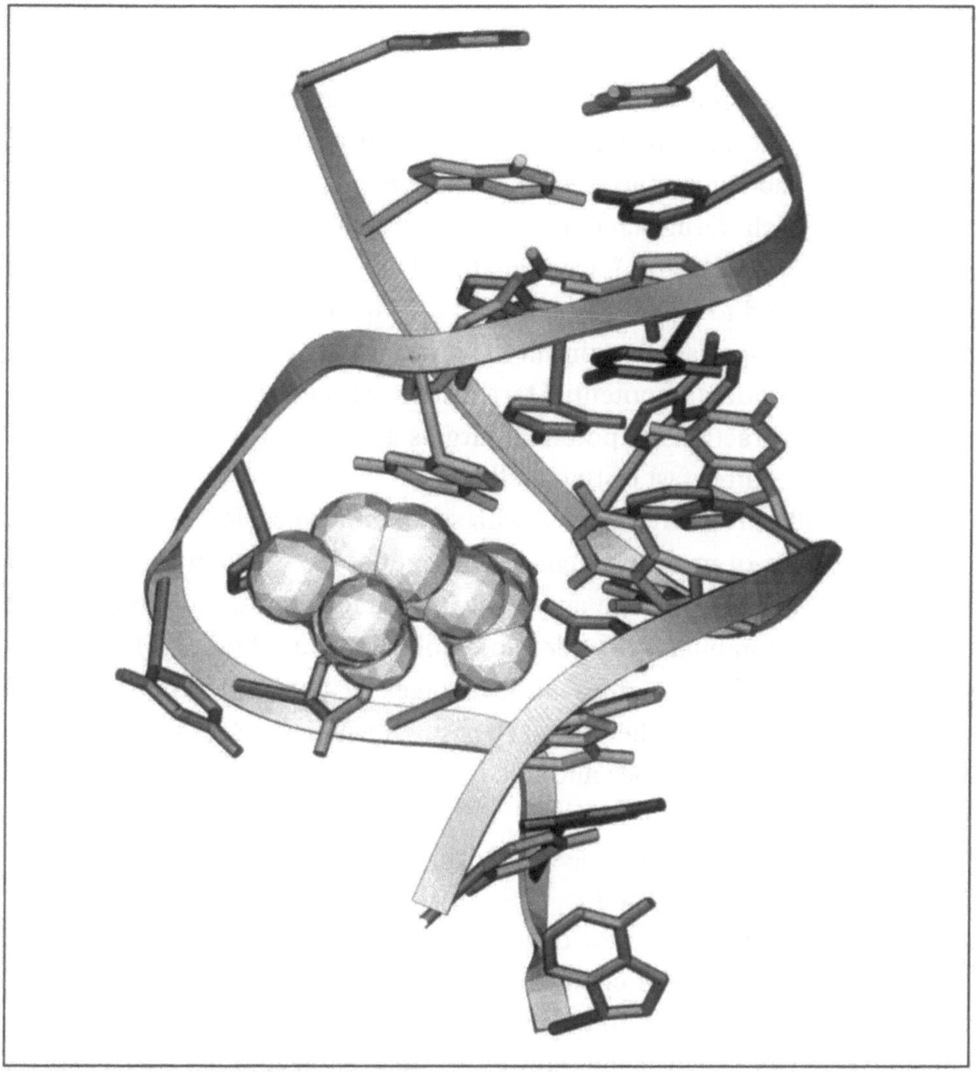

Figure 6 *Structure of an RNA aptamer complexed with citrulline (PDB code: 1kod; Yang et al., 1996). The image was generated with SETOR (Evans, 1993)*

properties, for example, having only a repertoire of four relatively similar nucleotides as compared to the 20 amino acids.

The usual view is that RNA structure is more protein-like than DNA. The aptamer approach is ideally suited to test this hypothesis by selecting DNA and RNA oligonucleotides for the very same target.

Drugs

The traditional approach to drug discovery has been to synthesize a number of substances, test them in an appropriate biological assay, identify a lead and optimize the lead compound. Recent developments have dramatically modified this approach. Drug screening has become target-based (receptors, enzymes, transcription factors). Combinatorial chemistry and high-throughput screening have led to an increased rate of testing drug candidates. This latter development is especially important in view of the enormous speed at which genome projects identify potential targets. One should realize that so far it is not clear which fractions of potential targets are real targets. To answer this question also requires fast testing methods.

Traditionally, drugs are equated with small molecules. Recently, however, the development of macromolecular therapeutics has come of age (Cho and Juliano, 1996). They include antisense oligonucleotides, ribozymes, gene therapeutics, recombinant peptides and monoclonal antibodies. Macromolecule therapeutics have various disadvantages in terms of production and target delivery. On the other hand, they might offer a target specificity not known from small molecules and this is exactly the reason why so many researchers are interested in this field.

Modeling approaches have assisted drug discovery over the years (Guida, 1994). They are important for both small molecule and macromolecule drugs. The most relevant new development in the field of drug discovery is probably combinatorial chemistry. Combinatorial chemistry started out with peptide libraries. Due to the well-known disadvantages of macromolecular therapeutics the focus has shifted to small-molecule libraries (Hogan, 1997).

20.5 Conclusions

The accurate prediction of biopolymer structures from the sequence alone is a problem which remains to be resolved. However, structure prediction and modeling methods have gained immensely both from the rapidly growing sequence and structure databases and from the increasing availability of computer power. Therefore, computer-assisted approaches become more and more important and successful in strategies for optimizing biological functions.

References

Allen FH, Kennard O (1993) 3D search and research using the Cambridge Structural Database. *Chemical Design Automation News* 8: 31–37

Bairoch A, Apweiler R (1997) The SWISS-PROT protein sequence data bank and its supplement TrEMBL. *Nucl Acids Res* 25: 31–36

Bartolotti LJ, Flurchick K (1996) An introduction to density functional theory. In: KB Lipkowitz, DB Boyd (eds): *Reviews in computational chemistry*. VCH Publishers, Inc, New York, Vol. 7: 187–216

Barton GJ (1995) Protein secondary structure prediction. *Curr Opinion Struct Biol* 5: 372–376

Bamborough P, Cohen FE (1996) Modeling protein-ligand complexes. *Curr Opinion Struct Biol* 6: 236–241

Benton D (1996) Bioinformatics – principles and potential of a new multidisciplinary tool. *Trends Biotechnol* 14: 261–273

Berman HM, Olson WK, Beveridge DI, Westbrook J, Gelbin A, Demeny T, Hsieh SH, Srinivasan AR, Schneider B (1992) The nucleic acid database. A comprehensive relational database of three-dimensional structures of nucleic acids. *Biophys J* 63: 751–759

Bernstein FC, Koetzle TF, Williams GJ, Meyer EE, Brice MD, Rodgers JR, Kennard O, Shimanouchi T, Tasumi M (1977) The Protein Data Bank: a computer-based archival file for macromolecule structures. *J Mol Biol* 112: 535–542

Blondelle SE, Houghten RA (1996) Novel antimicrobial compounds identified using synthetic combinatorial library technology. *Trends Biotechnol* 14: 60–65

Böhm H-J (1996) Current computational tools for *de novo* ligand design. *Curr Opinion Biotechnol* 7: 433–436

Boyd DB (1990) Aspects of molecular modeling. In: KB Lipkowitz, DB Boyd (eds): *Reviews in computational chemistry.* VCH Publishers, Inc., New York, Vol. I: 321–354

Brickmann J, Vollhardt H (1996) Virtual reality on the world-wide web: a paradigm shift in molecular modeling? *Trends Biotechnol* 14: 167–172

Brimacombe R (1995) The structure of ribosomal RNA: a three-dimensional jigsaw puzzle. *Eur J Biochem* 230: 365–383

Breaker RR (1996) Are engineered proteins getting competition from RNA? *Curr Opinion Biotechnol* 7: 442–448

Brooks BR, Bruccoleri R, Olafson B, States D, Swaminathan S, Karplus M (1983) CHARMm: a program for macromolecular energy, minimization, and dynamics calculations. *J Comp Chem* 4: 187–193

Brooks III CL (1995) Methodological advances in molecular dynamics simulations of biological systems. *Curr Opinion Struct Biol* 5: 211–215

Caspar DL (1995) Problems in simulating macromolecular movements. *Structure* 3: 327–329

Cate JH, Gooding AR, Podell E, Zhou K, Golden BI, Kundrot CE, Cech TR, Doudna

JA (1996) Crystal structure of a group I ribozyme domain: principles of RNA packing. *Science* 273: 1678–1685

Cech T (1992) Ribozyme engineering. *Curr Opinion Struct Biol* 2: 605–609

Chapman D (1996) The measurement of molecular diversity: A three-dimensional approach. *J Comput-Aided Mol Design* 10: 501–502

Chiu W, Schmid MF (1997) Pushing back the limits of electron cryomicroscopy. *Nature Struct Biol* 4: 331–334

Cho MJ, Juliano R (1996) Macromolecular versus small-molecule therapeutics: drug discovery, development and clinical considerations. *Trends Biotechnol* 14: 153–158

Choo Y, Klug A (1997) Physical basis of a protein-DNA recognition code. *Curr Opinion Struct Biol* 7: 117–125

Cornell WD, Cieplak P, Bayly CI, Gould IR, Merz Jr., KM, Ferguson DM, Spellmeyer DC, Fox T, Caldwell JW, Kollman PA (1995) A second generation forcefield for the simulation of proteins, nucleic acids and organic molecules. *J Am Chem Soc* 117: 5179–5197

Cramer RD III, Patterson DE, Bunce JD (1988) Comparative molecular field analysis (CoMFA). I. Effect of shape on binding of steroids to carrier proteins. *J Am Chem Soc* 110: 5959–5967

Defay T, Cohen FE (1995) Evaluation of current techniques for *ab initio* protein structure prediction. *Proteins* 23: 431–445

DeRosier DJ, Harrison SC (1997) Macromolecular assemblages. Sizing things up. *Curr Opinion Struct Biol* 7: 237–238

Elber R (1996) Novel methods for molecular dynamics simulations. *Curr Opinion Struct Biol* 6: 232–235

Edwards PD, Andisik DW, Strimpler AM, Gomes B, Tuthill PA (1996) Nonpeptidic inhibitors of human neutrophil elastase. 7. Design, synthesis, and *in vitro* activity of a series of pyridopyrimidine trifluoromethyl ketones. *J Med Chem* 39: 1112–1124

Erickson J, Neidhart DJ, VanDrie J, Kempf DJ, Wang XC, Norbeck DW, Plattner JJ, Rittenhouse JW, Turon M, Wideburg N, Kohlbrenner WE, Swimmer R, Helfrich R, Paul DA, Knigge M (1990) Design, activity, and 2.8 Å crystal structure of a C2 symmetric inhibitor complexed with HIV-1 protease. *Science* 249: 527–533

Evans SV (1993) SETOR: hardware-lighted three-dimensional solid model representations of macromolecules. *J Mol Graphics* 11: 134–138

Feigon J, Dieckmann T, Smith FW (1996) Aptamer structures form A to ξ. *Chemistry & Biology* 3: 611–617

Frank J (1997) The ribosome at higher resolution – the donut takes shape. *Curr Opinion Struct Biol* 7: 266–272

Gallop MA, Barrett RW, Dower WJ, Fodor SPA, Gordon EM (1994) Application of combinatorial technologies to drug discovery. Background and peptide combinatorial libraries. *J Med Chem* 37: 1233–1251

Gautheret D, Damberger SH, Gutell RR (1995) Identification of base-triples in RNA using comparative sequence analysis. *J Mol Biol* 248: 27–43

Gilbert W (1991) Towards a paradigm shift in biology. *Nature* 349: 99

Gold L (1995) Oligonucleotides as research, diagnostic and therapeutic agents. *J Biol Chem* 270: 13581–13584

Gold L, Brown D, He Y-Y, Shtatland T, Singer BS, Wu Y (1997) From oligonucleotide shapes to genomic SELEX: Novel biological regulatory loops. *Proc Natl Acad Sci* 94: 59–64

Green SM, Marshall GR (1995) 3D-QSAR: a current perspective. *Trends Pharmacol Sci* 16: 285–291

Guida WC (1994) Software for structure-based drug design. *Curr Opinion Struct Biol* 4: 777–781

Hall SH (1995) Protein images update natural history. *Science* 267: 620–624

Hogan Jr. JC (1997) Combinatorial chemistry in drug discovery. *Nature Biotechnol* 15: 328–330

Jernigan RL, Bahar I (1996) Structure-derived potentials and protein simulations. *Curr Opinion Struct Biol* 6: 195–209

Joshua-Tor L, Sussman JK (1993) The coming of age of DNA crystallography. *Curr Opinion Struct Biol* 3: 323–325

Karp PD (1996) Database links are a foundation for interoperability. *Trends Biotechnol* 14: 273–279

Karplus M, Petsko GA (1990) Molecular dynamics simulations in biology. *Nature* 347: 631–639

Konings DAM, Gutell RR (1995) A comparison of thermodynamic foldings with comparatively derived structures of 16S and 16S like RNAs. *RNA* 1: 559–574

Lattman EE (ed) (1995) Protein structure prediction: a special issue. *Proteins* 23: 295–460

Lemer CM, Rooman MJ, Wodak SJ (1995) Protein structure prediction by threading methods: evaluation of current techniques. *Proteins* 23: 337–355

Lipkowitz KB, Boyd DB (eds): Reviews in Computational Chemistry, VCH Publishers

Liu H, Müller-Plathe F, Van Gunsteren WF (1996) A combined quantum/classical molecular dynamics study of the catalytic mechanism of HIV protease. *J Mol Biol* 261: 454–469

Louise-May S, Auffinger P, Westhof E (1996) Calculations of nucleic acid conformations. *Curr Opinion Struct Biol* 6: 289–298

Luthy R, Bowie JU, Eisenberg D (1992) Assessment of protein models with three-dimensional profiles. *Nature* 356: 83–85

Lyall A (1996) Bioinformatics in the pharmaceutical industry. *Trends Biotechnol* 14: 308–312

Madej T, Boguski MS, Bryant SH (1995) Threading analysis suggests that the obese gene product may be a helical cytokine. *FEBS Lett* 373: 13–18

Michel F, Westhof E (1990) Modeling of the three-dimensional architecture of group I catalytic introns based on comparative equence analysis. *J Mol Biol* 216: 585–610

Mosimanu S, Meleshko R, James MNG (1995) A critical assessment of comparative molecular modeling of tertiary structures of proteins. *Proteins* 23: 301–317

Moult J (1996) The current state of the art in protein structure prediction. *Curr Opinion Biotechnol* 7: 422–427

Moult J (1997) Comparison of database potentials and molecular mechanics force fields. *Curr Opinion Struct Biol* 7: 194–199

Murray CW, Clark DE, Auton TR, Firth MA, Li J, Sykes RA, Waszkowycz B, Young SC (1997) PRO_SELECT: Combining structure-based drug design and combinatorial chemistry for rapid lead discovery. 1. Technology. *J Comput-Aided Mol Design* 11: 193–207

Murray-Rust P (1994) Bioinformatics and drug discovery. *Curr Opinion Biotechnol* 5: 648–653

Neidle S (1997) Crystallographic insights into DNA minor groove recognition by drugs. *Biopolymers* 44: 105–121

Olson WK (1996) Simulating DNA at low resolution. *Curr Opinion Struct Biol* 6: 242–256

Palsson OP (1997) What lies beyond bioinformatics? *Nature Biotechnology* 15: 3–4

Peitsch MC, Wells TN, Stampf DR, Sussman JL (1995) The Swiss-3D-image collection and PDB-browser on the world-wide web. *Trends Biochem Soc* 20: 82–84

Rost B, Sander C (1996) Bridging the protein sequence-structure gap by structure predictions. *Annu Rev Biophys Biomol Struct* 25: 113–136

Ruediger N (1996) Bioinformatics: New frontier call young scientists. *Science* 273: 265

Russell VA, Evans CC, Li W, Ward MD (1997) Nanoporous molecular sandwiches: Pillared two-dimensional hydrogen-bonded networks with adjustable porosity. *Science* 276: 575–579

Sanchez R, Sali A (1997) Advances in comparative protein-structure modeling. *Curr Opinion Struct Biol* 7: 206–214

Sadowski J (1997) A hybrid approach for addressing ring flexibility in 3D database searching. *J Comput-Aided Mol Design* 11: 53–60

Schatz BR, Hardin JB (1994) NCSA Mosaic and the World Wide Web: Global hypermedia protocols for the Internet. *Science* 265: 895–901

Schatz BR (1997) Information retrieval in digital libraries: Bringing search to the net. *Science* 275: 327–334

Schuler GD, Epstein JA, Ohkawa H, Kans JA (1996) Entrez: molecular biology database and retrieval system. *Methods Enzymol* 266: 141–162

Sham HL, Zhao C, Stewart KD, Betebenner DA, Lin S, Park CH, Kong XP, Rosenbrook W Jr, Herrin T, Madigan D, Vasavanonda S, Lyons N, Molla A, Saldivar A, Marsh KC, McDonald E, Wideburg NE, Denissen JF, Robins T, Kempf DJ, Plattner JJ, Norbeck DW (1996) A novel, picomolar inhibitor of human immunodeficiency virus type 1 protease. *J Med Chem* 39: 392–397

Sippl MJ (1993) Recognition of errors in three-dimensional structures of proteins. *Proteins* 17: 355–362

Sippl MJ (1995) Knowledge-based potentials for proteins. *Curr Opinion Struct Biol* 5: 229–235

Sippl MJ, Ortner M, Jaritz M, Lackner P, Flöckner H (1996) Helmholtz free energies of atom pair interactions in proteins. *Fold Des* 1: 289–298

Sponer J, Leszynski J, Hobza P (1996) Hydrogen bonding and stacking of DNA bases: A review of quantum-chemical *ab initio* studies. *J Biomol Struct Dyn* 14: 117–135

St-Amant A (1996) Density functional methods in biomolecular modeling. In: Reviews in Computational Chemistry. (Lipkowitz, KB, Boyd, DB, eds) VCH Publishers, Inc., New York, Vol. 7: 217–259

Strynadka NCJ, Eisenstein M, Katchalski-Katzir E, Shoichet BK, Kuntz ID, Abagyan R, Totrov M, Janin J, Cherfils J, Zimmermann F, Olson A, Duncan B, Rao M, Jackson R, Sternberg M, James MNJ (1996) Molecular docking programs successfully predict the binding of a beta-lactamase inhibitory protein to TEM-1 beta-lactamase. *Nature Struct Biol* 3: 233–239

Stupp SI, LeBonheur V, Walker K, Li LS, Huggins KE, Keser M, Armstutz A (1997) Supramolecular materials: Self-organized nanostructures. *Science* 276: 384–389

Sühnel J (1996) Image library of biological macromolecules. *Comput Appl Biosci* 12: 227–229

Sühnel J (1997) Views of RNA on the World-Wide Web. *Trends Genetics* 13: 206–207

Thomas PD, Dill K (1996) Statistical potentials from protein structures: How accurate are they? *J Mol Biol* 257: 457–469

Torda AE (1997) Perspectives in protein-fold recognition. *Curr Opinion Struct Biol* 7: 200–205

Uphoff KW, Bell SD, Ellington AD (1996) *In vitro* selection of aptamers: the dearth of pure reason. *Curr Opinion Struct Biol* 6: 281–288

Vajda S, Sippl M, Novotny J (1997) Empirical potentials and functions for protein folding and binding. *Curr Opinion Struct Biol* 7: 222–228

Van Gunsteren WF, Mark AE (1992) On the interpretation of biochemical data by molecular dynamics computer simulation. *Eur J Biochem* 204: 947–961

Van Gunsteren WF, Billeter SR, Eising AA, Hünenberger PH, Krüger P, Mark AE, Scott WRP, Tironi IG (1996) Biomolecular simulation: The GROMOS 96 manual and users guide. vdf Hochschulverlag AG, ETH Zürich; BIOMOS b.v., Zürich, Groningen

Verlinde CLJM, Hol WGJ (1994) Structure based drug design: progress, results and challenges. *Structure* 2: 577–587

Wahl MC, Sundaralingam M (1995) New crystal structures of nucleic acids and their complexes. *Curr Opinion Struct Biol* 5: 282–295

Watson JD, Crick FCH (1953) A structure for deoxyribose nucleic acid. *Nature* 171: 737

Weber IT, Harrison RW (1996) Molecular mechanics calculations on HIV-1 protease with peptide substrates correlate with experimental data. *Protein Eng* 6: 679–690

Wedel AB (1996) Fishing the best pool for novel ribozymes. *Trends Biotechnol* 14: 459–465

Yang Y, Kochoyan M, Burgstaller P, Westhof E, Famulok M (1996) Structural basis of ligand discrimination by two related aptamers resolved by NMR spectroscopy. *Science* 272: 1343–1347

Zerner MC (1991) Semiempirical molecular orbital methods. In: Reviews in Computational Chemistry (Lipkowitz, KB, Boyd, DB, eds) VCH Publishers, Inc., New York, Vol. II, 313–365

Zuker M, Jacobson AB (1995) Well-determined regions in RNA secondary structure prediction: analysis of small subunit ribosomal RNA. *Nucleic Acids Res* 23: 2791–2798

Sampling of useful web resourcees

Structure databases

(w1) Cambridge Crystallographic Data Centre
(http://csdvx2.ccdc.cam.ac.uk/)

(w2) KLOTHO: Biochemical Compounds Database
(http://www.ibc.wustl.edu/klotho/)

(w3) Protein Data Bank
(http://www.pdb.bnl.gov/; http://www2.ebi.ac.uk/pdb/)

(w4) Nucleic Acid Database
(http://ndbserver.rutgers.edu:80/; http://ndbserver.ebi.ac.uk:5700/NDB/)

Image archives

(w5) Molecules R US
(http://molbio.info.nih.gov/cgi-bin/pdb)

(w6) Swiss-3D-Image
(http://expasy.hcuge.ch/sw3d/sw3d-top.html)

(w7) Image Library of Biological Macromolecules
(http://www.imb-jena.de/IMAGE.html)

Sequence and genome databases

(w8) ENTREZ database
(http://www.ncbi.nlm.nih.gov/Entrez/)

(w9) SWISS-PROT protein sequence database
(http://expasy.hcuge.ch/sprot/sprot-top.html)

(w10) Sequence Retrieval System – SRS
(http://www.ebi.ac.uk/queries/queries.html)
(w11) Gene Map of the Human Genome
(http://www.ncbi.nlm.nih.gov/SCIENCE96/)
(w12) TIGR Human Gene Index
(http://www.tigr.org/tdb/hgi/hgi.html)
(w13) TIGR Microbial Database
(http://www.tigr.org/tdb/mdb/mdb.html)
(w14) Yeast Genome Access
(http://speedy.mips.biochem.mpg.de/mips/yeast/)

Software collections

(w15) EBI BioCatalog
(http://www.ebi.ac.uk/biocat/biocat.html)
(w16) Network Science Software Lists
(http://www.netsci.org/Resources/Software/top.html)
(w17) QCPE – Quantum Chemistry Program Exchange
(http://ccl.osc.edu/ccl/qcpe/QCPE/)

Public domain molecular graphics programs

(w18) RasMol
(http://www.umass.edu/microbio/rasmol/)
(w19) WebLab Viewer
(http://www.msi.com/weblab/viewer/)

Commercial molecular graphics and molecular modeling software

(w20) AMBER (Oxford MolecularGroup)
(http://www.oxmol.co.uk/prods/amber/)
(w21) InsightII (Molecular Simulations, Inc.)
(http://www.msi.com/solutions/products/insight/)
(w22) QUANTA (Molecular Simulations, Inc.)
(http://www.msi.com/solutions/products/quanta/)
(w23) SCULPT (Interactive Simulations, Inc.)
(http://www.intsim.com/)
(w24) SYBYL (Tripos, Inc.)
(http://www.tripos.com/products/sybylsummary.html)
(w25) VAMP (Oxford Molecular Group)
(http://www.oxmol.co.uk/prods/vamp/)

Ab initio quantum chemistry programs

(w26) GAMESS
(http://www.msg.ameslab.gov/GAMESS/GAMESS.html)

(w27) GAUSSIAN
(http://www.gaussian.com)
(w28) SPARTAN
(http://www.wavefun.com/software/features/features_spartan.html)

Semiempirical quantum chemistry programs

(w29) MOPAC
(http://home.att.net/~mrmopac/)

Molecular dynamics packages

(w30) AMBER
(http://www.amber.ucsf.edu/amber/amber.html)
(w31) CHARMM
(http://yuri.harvard.edu/charmm/CHARMM-docs.html)
(w32) GROMOS96
(http://igc.ethz.ch/gromos/)

Internet courses

(w33) The NIH guide to molecular modeling
(http://cmm.info.nih.gov/modeling/gateway.html)
(w34) Virtual School of Natural Sciences – Biocomputing Division
(http://www.techfak.uni-bielefeld.de/bcd/ForAll/welcome.html)
(w35) Virtual School of Molecular Sciences
(http://www.vsms.nottingham.ac.uk/vsms/)

Other

(w36) DALI server for 3D protein structure comparison
(http://croma.ebi.ac.uk/dali/)
(w37) PredictProtein server for secondary structure prediction
(http://www.embl-heidelberg.de/predictprotein/predictprotein.html)
(w38) Protein structure classification – CATH
(http://www.biochem.ucl.ac.uk/bsm/cath/)
(w39) Structural classification of proteins – SCOP
(http://scop.mrc-lmb.cam.ac.uk/scop/)
(w40) SWISS-MODEL (Automated knowledge-based protein modeling server)
(http://expasy.hcuge.ch/swissmod/)
(w41) Critical assessment of techniques for protein structure prediction
(http://predict.sanger.ac.uk/casp2/)
(w42) Reviews in Computational Chemistry
(http://chem.iupui.edu/~boyd/rcc.html)
(w43) RNA secondary structure prediction
(http://mfold1.wustl.edu/~mfold/ma/form1.cgi)

21 Optimizing structure and function relationship of nucleic acid molecules

Michael Famulok, Andres Jäschke and Stefan Wölfl

21.1 Biomolecular interaction

Interactions between two or more molecules are central to all living systems, as they provide the basis for biochemical reactions and regulatory processes. Complex reactions require a precise control in the interaction of the partners and, given the relevance of these interactions, it is of high importance to understand the structural and chemical mechanisms that enable specific molecular recognition.

In recent years, continously increasing numbers of biomolecular structures were elucidated. New high resolution structures are published every month, but still our understanding of structure function relationships is limited. In particular, more detailed information is required on structural variations that are within the frame of biomolecular function. This requires to obtain both structural and functional data on a large number of related molecules, in order to investigate the functional consequences of structural alterations.

The two basic biological macro-molecules, proteins and nucleic acids are both polymers made of monomeric building blocks. Proteins are built of 20 amino acids and nucleic acids are built of four nucleotides. A small protein (peptide) of 20 amino acids could be built in 20^{20} (10^{26}) different sequences. This makes it impossible to practically provide all potential sequence permutations. Even nucleic acids built of only four nucleotides (nt) soon reach very large numbers of different molecules. For a 20 nt long molecule 4^{20} ($\sim10^{12}$) different sequences and for a 40 nt long nucleic acid 4^{40} ($\sim10^{24}$) different sequences are possible.

How can the optimal structures for specific interactions be found in the laboratory, when the shear number of molecules vastly exhausts the available means? Of course, for most biological molecules the *in vivo* partner molecules should be found in their natural environment. Still, if various interactions are utilized, a rare interaction partner could be hidden among more abundant interacting molecules.

Furthermore, optimal ligands that are not utilized in the natural system are of high interest. They have the potential to provide new insights into the constraints of biomolecular structures and functions and may be used to develop new molecules for medical applications. Molecules that interact with cell surface markers, with molecules of signaling pathways or with transcription factors, can be used to elucidate the interaction interfaces or may even be used directly in the diagnosis or therapy of a related disease.

In chemistry, new molecules based on peptides or nucleic acids that provide specific binding sites for the reaction partners or for transition state analogues may provide a selective catalytic surface that enhances chemical reactions and even provides stereoselectivity. Thus, the demand for molecules with specific affinities is underlining the need for new nano-technology approaches to enhance the chances to identify optimized molecules.

21.2 Microstructured arrays and/or *in vitro* biological systems

The most intriguing gain using miniaturized devices is that a large number of reactions can be performed in parallel. For example 10^8 different molecular species can be immobilized on an array of 1 cm^2 using just 1 μm^2 spots for each species. The scale of reduction will be only limited by the detection system, the size of the immobilized molecule by the reaction of interest, e.g. the partner molecules in a (bio-)molecular interaction, and by the techniques available to create a diverse library of molecules on a given array. Although the number of different molecular species is very high, it is still quite limited with respect to the number of all different molecular species possible.

Using the means of molecular biology, i.e. enzymes such as polymerases or ligases as well as cloning and expression systems, various *in vitro* selection schemes can be designed. This led to the development of strategies based on nucleic acids (Ellington and Szostak, 1990; Joyce, 1989; Tuerk and Gold, 1990), phage particles (Scott and Smith, 1990) or other display systems that use nucleic acid coupled expressed peptides (Mattheakis et al., 1994). In all

Table 1 Comparison between synthetic combinatorial arrays and in vitro biological systems

Combinatorial arrays	Biological system
– Any chemical molecule	– Only peptides or nucleic acids
– Diversity is limited by the size of the array	– Diversity is limited by total amount of molecules
– Direct detection of interaction is required	– Indirect detection of interaction by amplification
	– *In vitro* mutation, evolution to optimize properties

cases, molecules that perform a given function will be amplified selectively. In the following we will focus on the use of nucleic acid libraries and their potential for self-optimization.

21.3 *In vitro* selection of nucleic acids

To carry out selection-amplification reactions from a large pool of random nucleic acid sequences, elements of known sequence must be included. In the basic design, the randomized sequence is flanked by two defined sequence elements. These two elements provide annealing sites for complementary primer molecules that allow amplification by PCR. This basic setup can be used in selections with RNA, single or double stranded DNA or with modified nucleic acid molecules, as long as they are compatible with the enzymatic reactions required. Figure 1 outlines the steps of an *in vitro* selection experiment with RNA. First, DNA containing the defined flanking sequences and an internal random sequence element is synthesized by standard phosphoramidite chemistry. Synthesis of 1 µmol DNA can provide up to 10^{17} different molecules. From this single stranded template double stranded DNA is prepared using two amplification primers complementary to the flanking sequences. One of the two primer molecules contains the sequence motif of the T7 RNA polymerase promoter. The double stranded DNA molecules serve as

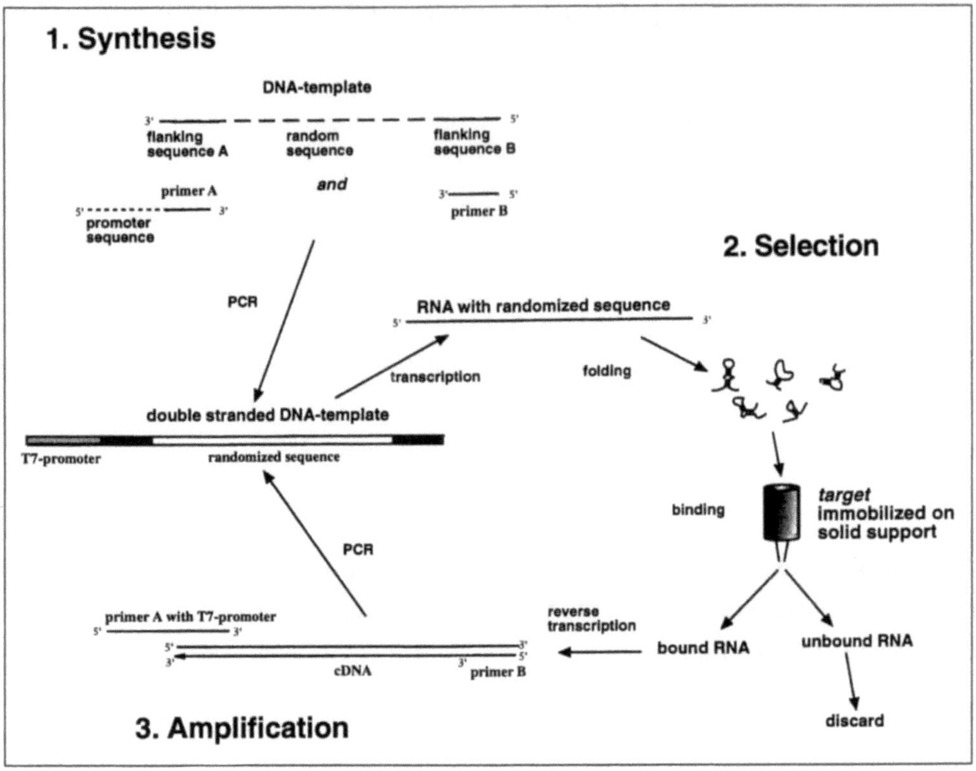

Figure 1 *Graphic outline of the in vitro selection protocol. The initial synthesis of the nucleic acid pool is followed by the cyclic selection process*

templates for the synthesis of single stranded RNA molecules. After sequence specific folding this pool of RNA molecules will provide a variety of shapes that can be selected based on a specific property like affinity for a target molecule. The selected RNA molecules are then re-amplified using reverse transcription and PCR resulting in a new pool of double stranded DNA. The complexity of this pool is smaller than the original pool but is enriched with molecules of the desired property.

Repetition of this selection cycle results in further enrichment of molecules that fit the desired property. Increase in stringency during the selection process should favour optimized molecules. After several rounds of selection the resulting molecules will be cloned, sequenced and assayed for their properties.

An alternative way for amplification that avoids the use of PCR is the isothermal amplification or self-sustained sequence replication (3SR) (Guatelli

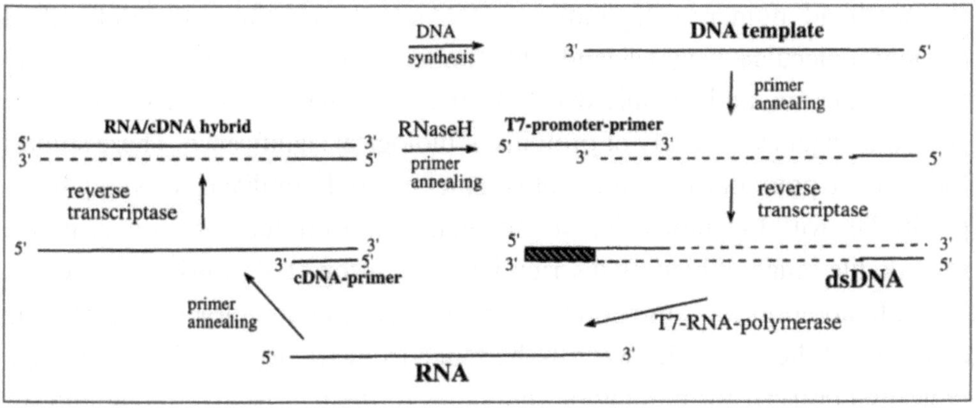

Figure 2 *The 3SR protocol employs a continuous exchange between a single stranded RNA form, usually the active state, and a double stranded DNA form, usually the replicative (template) state*

et al., 1990). Here amplification is carried out by an alternation between the RNA and the DNA cycle. With reverse transcriptase and RNaseH, double stranded cDNA is synthesized from single stranded RNA that contains a T7-promoter. The third enzyme, T7 RNA polymerase, then synthesizes RNA from this template that in turn serves as a new template for reverse transcription (Fig. 2). This technique allows continuous amplification of RNA molecules.

21.4 Selection criteria

The above described procedures allow the systematic investigation of libraries with complexities of up to 10^{16} different RNA species (Bartel and Szostak, 1993). Moreover, similar strategies have been applied to DNA (Ellington and Szostak, 1992) and modified nucleic acid libraries. These strategies can be applied to answer questions for binding properties, substrate properties and catalytic activity.

Binding properties

The search for specifically binding molecules has been performed against targets of various molecular complexity (Osborne and Ellington, 1997). On the one hand, numerous high affinity RNAs (aptamers) have been identified for low molecular weight ligands, like amino acids, nucleotides, biologically active cofactors, and organic dyes. On the other hand, aptamers have been generated against a variety of proteins of biological significance. These studies focus on the potential application of aptamers as tools for diagnostics and therapy. Studies with low molecular weight ligands showed that aptamers display a remarkable range of binding discrimination (Jenison et al., 1994; Geiger et al., 1996). In addition, various peptides were used as selection targets. The binding constants of the molecules were in the range from 10^{-8} to 10^{-6}; the same range typically observed for binding of antibodies to their respective target and for several other biomolecular interactions. Already available aptamers against pharmacologically relevant targets pose the question of bioavailability and efficacy. The problem of stability in biological fluids has been solved in different ways, like introduction of chemical modifications, conjugation and mirror image design (Heidenreich et al., 1993; Jellinek et al., 1995; Klussmann et al., 1996).

Substrate properties

The specificity of aminoacylation of tRNAs has been studied using *in vitro* selection approaches. The methodology is somewhat different from the above described binding studies (Fig. 3a). For example, a partially randomized tRNA pool was incubated with an aminoacyl-tRNA-synthetase and the corresponding amino acid. The aminoacylated species were then specifically derivatized and separated from the unreacted tRNAs. At least two different approaches have been described (Pütz et al., 1997; Sampson and Saks, 1996). Similar strategies can be applied in principal to numerous RNA modifying enzymes. In contrast to classical mutation studies involving the parallel preparation of individual mutants, these combinatorial methods allow the simultaneous screening of libraries with almost exhaustive complexity.

Catalytic activities

The search for catalytic activity poses a new category of difficulty, as a true catalyst is per definition required to leave a reaction unchanged. Therefore catalysis has to be coupled with some other principle that leads to a selectable

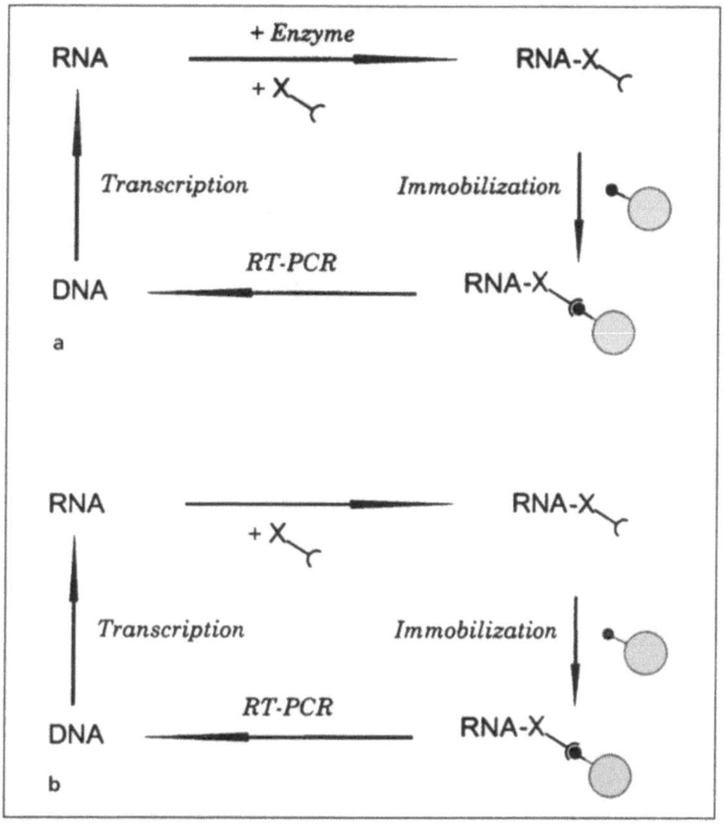

Figure 3 *Variations in the protocol allow to address various questions. (a) The RNA molecule is the substrate for an enzymatic reaction, e.g. aminoacyl transferase. (b) The RNA molecule provides the catalytic activity*

change in the catalytic molecule. This is normally achieved by self-modification, i.e., one and the same molecule acts as both catalyst and substrate (Fig. 3b). For example, Lorsch and Szostak identified RNA molecules that were able to transfer the terminal thiophosphate of a thio analog of ATP (ATP-γS) to its own 5′-end (Lorsch and Szostak, 1994). On incubation with ATP-γS some molecules of the randomized RNA pool got attached to the thiophosphate. These molecules were separated from unreacted RNA species by coupling to an activated thiol matrix.

The isolated molecules were then enzymatically amplified and the whole scheme repeated several times. The activity of the selected molecules was in the range of 0.2 min^{-1} (k_{cat}). Similar reaction schemes have been successfully

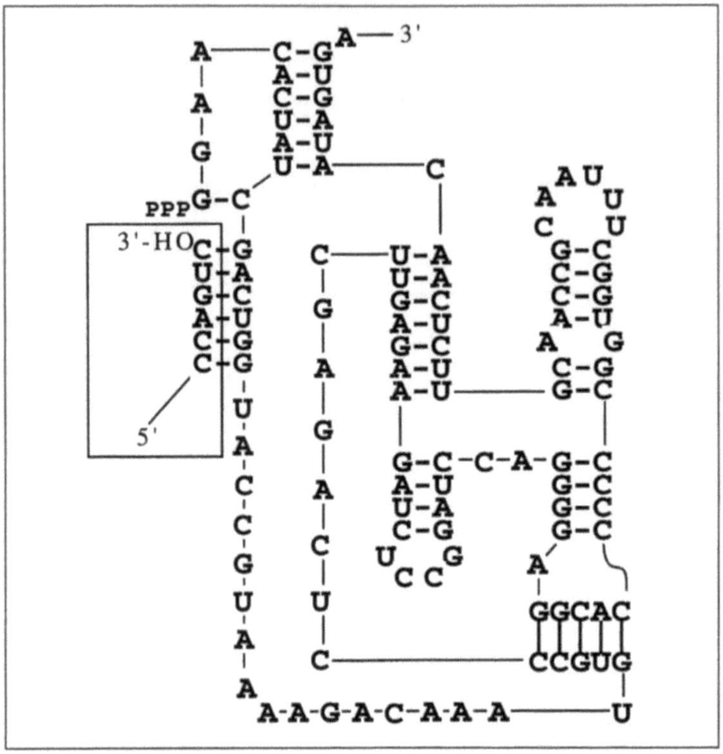

Figure 4 *Structure of the class I ligase (Eckland et al., 1995). The ligation substrate (3′-site) aneals with six complementary nucleotides, 12 positions away from the 5′-site on the ribozyme. The structure of the ribozyme is defined by several complementary domains that can form double helices*

applied to aminoacylation, alkylation, ligation, transesterification and other reactions (see Appendix).

The second general approach for the selection of catalysts is an indirect two-step approach. In the first step aptamers are raised against immobilized transition state analogs (TSAs) of the reaction under investigation. The identified species are then screened for catalytic activity. While the selection of aptamers against TSAs was successful in several examples, only in a few cases catalytically active molecules were found (Conn et al., 1996; Li and Sen, 1997; Prudent et al., 1994).

21.5 Counter selection

In vitro selection offers a striking difference to all other combinatorial or immunological approaches, as it allows the experimentator to emphasize or de-emphasize certain positions in a target molecule for interaction. The way this can be achieved is counter selection. For example, RNAs bound to the immobilized target (Fig. 1) are first washed with a chemically similar compound and then eluted with the free target. Thereby RNA species are amplified that can discriminate between the two similar compounds.

Using the approach of selection and counter selection, Jenison and coworkers selected aptamers that tightly bind to theophylline but do not recognize caffeine (Jenison et al., 1994). The two molecules are bound with a 10000 fold difference in binding affinity. This finding was particularly interesting since theophylline and caffeine differ only in one methyl group. Because of the small binding surface, the overall affinity of the aptamer is only in the micromolar range. Still, the oligonucleotide will selectively recognize theophylline even in the presence of a large excess of caffeine.

21.6 *In vitro* evolution

In principle, biological evolution can be reduced to the three features: selection, amplification and mutation. *In vitro* selection as described above is an ideal methodology to study biomolecular evolution, if the element of mutation is introduced into the protocol. The strategies described so far only allowed the isolation and amplification of RNA molecules that were already present in the initial RNA pool. As pools with long randomized domains only contain a tiny fraction of all theoretical possible sequences (theoretical complexity 4^n; practical complexity 10^{13} to 10^{15}) the selected species typically represent suboptimal structures. Random *in vitro* mutagenesis enables to introduce new sequences at any point. Combined with selection pressure, only the best performing variants will be propagated. Another round of mutation will increase the chance to find molecules with optimized properties.

In vitro evolution of a ligase ribozyme

Bartel and Szostak used a library of 1.6×10^{15} different RNA molecules, completely randomized at 220 base positions, to isolate ribozymes with ligase activity. This length of randomization was chosen to increase the probability of isolating catalytically active sequences (Bartel and Szostak, 1993). Bartel and Szostak immobilized their highly complex RNA pool non-covalently on agarose beads to avoid precipitation of the RNA, and incubated the immobilized RNA with a short substrate oligonucleotide to be ligated to the 5′-end of catalytically active ribozymes present in the pool. After four rounds of selection and amplification, 3% of the pool became ligated to the substrate within 1 h of incubation. After cycle 4, additional rounds of selection were carried out using mutagenic PCR in the amplification step in order to obtain ribozymes with enhanced catalytic activity with respect to the ones from cycle 4. This strategy led to an enhancement of the ligation rate from 2×10^{-2} h^{-1} to 100 h^{-1}. The unselected pool had ligation rates of 2×10^{-6}, which is the background activity of the uncatalyzed reaction. The relative abundance of active catalysts was measured to be in the range of one sequence in 3×10^{14}. This study represented the first example of the *in vitro* selection of catalytic RNAs from a completely randomized RNA-pool which had no relationship to any sequence previously invented by nature.

The characterization of individual sequences from this selection experiment revealed three different structural classes of ligases from which only one, the class I ligase, generates a 3′–5′ phosphodiester bond at the ligation site (Ekland and Bartel, 1995; Ekland et al., 1996) (Fig. 4). The catalytic domain of the class I ligase was localized by partially randomizing a deletion construct of this ribozyme at 172 of its 186 relevant positions. By carrying out four cycles of a selection scheme that was slightly modified from the previous one (Fig. 6), RNA-ligases were isolated that self-ligated 700 times faster than the original class I ligase sequence. This optimized version of the class I ribozyme which differs at 10 positions from the original sequence has a k_{cat} of 100 min^{-1}, a value comparable to that of the protein enzyme ligase. The analysis also revealed that the class I ligase constitutes a highly complex nucleic acid; the probability of isolating it from the starting 1.6×10^{15} sequence pool was calculated to be 5×10^{-4} (Ekland et al., 1996). Either the isolation of this ribozyme from this pool was an extraordinarily lucky event, or, more likely, a

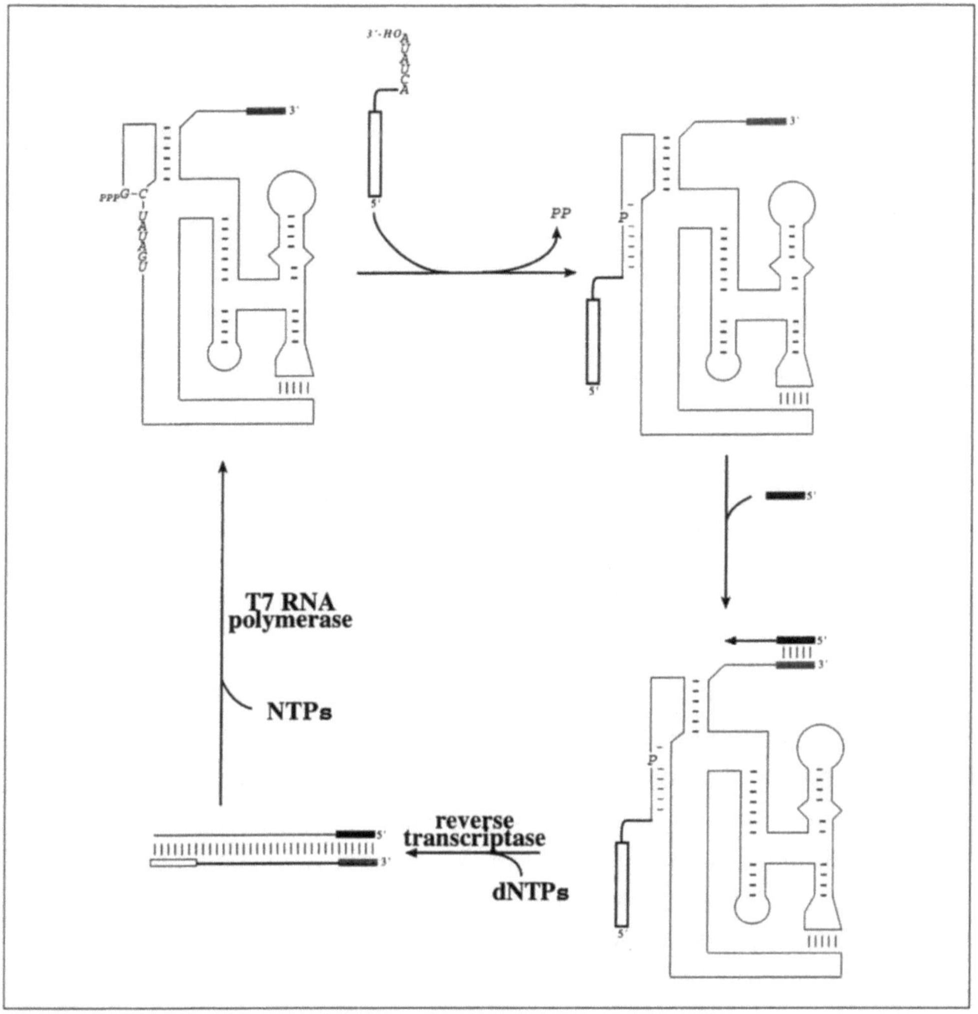

Figure 5 *Selection cycle for the optimization of the ligase molecule. Ligation is required to incorporate the T7-RNA-promoter sequence in the RNA molecule. Only reverse transcription from a ligated molecule results in a new double stranded DNA template for T7-RNA-polymerase. Under this condition the amplification rate depends on the efficiency of the ligation reaction*

great many different large RNA structures exist which are capable of catalyzing ligation reactions of the type studied.

Ekland and Bartel then took one step further in creating a plausible scenario for an "RNA world" in which ribozymes catalyzing the replication of RNA are postulated. They demonstrated that the class I ribozyme is capable

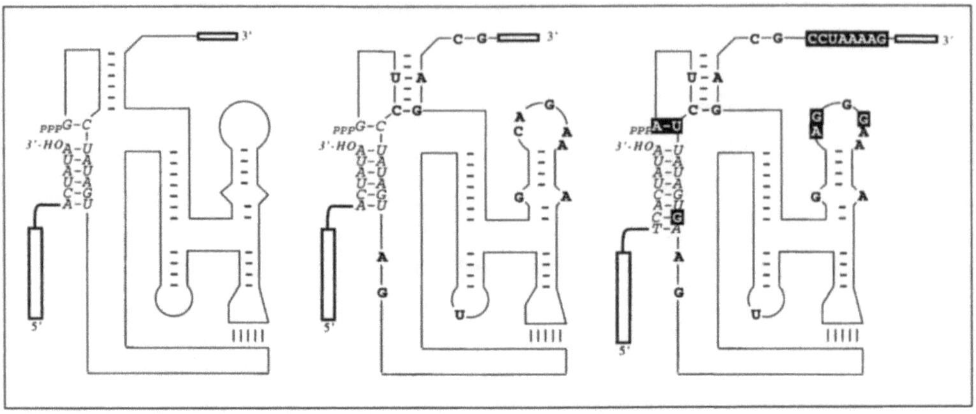

Figure 6 *Results from various optimization experiments for the ligase molecule (Wright and Joyce, 1997). From left to right: the structure of the class I ligase (Fig. 4, Eckland et al., 1995); example of a ligase molecules resulting from the stepwise evolution protocol; example of a ligase isolated after continuous evolution*

of extending a separate RNA primer by one nucleotide in the presence of a template oligonucleotide and nucleoside triphosphates (Ekland and Bartel, 1996). The polymerization reaction exhibits a remarkable template directed fidelity. Mismatched nucleotides are added 1000-fold less efficiently. By linking the primer covalently to the ribozyme, the reaction could be expanded to the addition of three nucleotides. By designing the primer in a way that it was able to slip onto the template, even six nucleotides could be added to the primer in a template directed way.

Using the class I ligase for continous evolution

A modified version of the class I ligase which contained a different substrate hybridization site was evolved by Wright and Joyce and used in a continuous evolution system (Wright and Joyce, 1997). The hexameric substrate hybridization site of the original class I ligase, 5′-GACUGG-3′ had to be changed to 5′-UAUAGU-3′ in order to make it complementary to a substrate which corresponded to the sequence of the T7-promotor. This switch in substrate specificity was necessary because the continuous evolution scheme was designed to evolve ribozymes which were capable of ligating a T7-promotor sequence onto their 5′-end with very high efficiency in the presence of reverse transcriptase, the 3′-primer, dNTPs, NTPs, and T7-RNA polymerase. In this way, a competing situation was generated in which only those ribozymes were replicated by T7-polymerase that had ligated the T7-promotor onto their 5′-end *before* reverse transcriptase had generated too much sense DNA from the ribozyme-template to make catalysis impossible.

The change in substrate specificity of the class I ligase, however, resulted in a 1000-fold reduced catalytic activity. Therefore, the new class I ligase construct had to be evolved in 15 cycles of stepwise *in vitro* evolution to improve the ligation rate (Tsang and Joyce, 1996). In the stepwise evolution protocol, the ribozymes were incubated with the DNA-3′-r(N_1)-substrate in the absence of the replicating enzymes with gradually decreasing the incubation time in each cycle. Then, the ligation products were purified, and amplified by reverse transcription and T7-transcription. After the stepwise evolution the ribozyme still had to be taught to accept a DNA-substrate containing four 3′-terminal ribooligonucleotides by performing an additional 100 cycles of rapid evolution. In the rapid evolution scheme the ribozymes were allowed to perform the ligation reaction for 5 min with a DNA-3′-r(N_4)-substrate before they were transferred to a replication mixture containing reverse transcriptase, 3′-primer, dNTPs, T7-RNA polymerase, and NTPs to allow for selective amplification.

At the end, ribozymes emerged which were not only capable of performing fast enough catalysis with the modified substrate hybridization sequence, but also of accepting a DNA-3′-r(N_4)-substrate. A ribozyme which was mutated at 17 positions compared to the starting sequence was used to start the continuous evolution reaction.

Continuous evolution was performed by doing 100 serial transfers into reaction vessels containing a mix of substrate, 3′-primer, dNTPs, NTPs, reverse transcriptase, and T7-RNA polymerase. The incubation time between each transfer was 60 min. In this way, approximately 300 cycles of catalysis and amplification were achieved within less than 2 days with a net amplification of roughly 2^{1000}, or 3×10^{298}. The improvement of catalytic efficiency of an evolved variant ribozyme was 14000-fold better than the starting sequence. It was mutated at 15 positions compared to the sequence which was used as the input in the continuous evolution reaction (Fig. 6). With their system, Wright and Joyce have developed a continuous evolution scheme similar to the Qβ-replicase system used by Spiegelman (1967), with the difference that the actual catalytic step is carried out by a ribozyme. The amplification reaction, however, still depends on the "helper proteins" reverse transcriptase and T7-RNA polymerase. The continuous evolution reaction might also be used to develop a ribozyme with RNA polymerase activity, which, perhaps, someday makes the use of a proteinaceous RNA-polymerase obsolete. This suggestion, remarked by Wright and Joyce in their paper, is supported by the observation that the ribozyme, when devoid of the 5′-triphosphate, can add a single ATP onto the 3′-end of a substrate which lacks the 3′-terminal adenylate.

Why did Wright and Joyce spend all the effort of pre-evolving the class I ligase to become a good catalyst which accepts the modified substrate? Could one not expect that the continuous evolution experiment would simply restore the catalytic power that was lost due to the changes in the substrate and substrate annealing site within a few generations? The answer is "no": Breaker and Joyce had shown in previous studies that if a ribozyme with only poor catalytic ability is used as a starting point, continuous evolution will produce "selfish" RNAs, so-called "mini monsters", or "RNA Z" (Breaker et al., 1994; Breaker and Joyce, 1994). These short parasital RNAs quickly appeared during the experiment because they were able to fold back and self-prime their own cDNA synthesis. From those RNA Z hairpins in which the sequence constituted a T7 promoter site, T7 RNA polymerase was able to transcribe RNA which completed the cycle.

21.7 Conclusions

The above described examples show the optimization of molecules that perform a self-ligation reaction. Only the successful catalysis of the ligation reaction leads to the formation of molecules that possess all the components required to initiate a new round of "replication". To use this evolutionary concept to optimize molecules that perform other functions (see Appendix) several adaptations are required. Each mutation/amplification step must be coupled with a specific enrichment of the molecules that show the desired function. This can be acheived with transfer strategies that limit the transfer from one round of mutation/amplification to the next to molecules of the desired activity. Furthermore, the specificity of the selection can be adjusted through changes in the kinetic parameters of the selection reaction. For example, the reaction time and the concentration of the interacting partners can be reduced to increase the stringency of the selection.

Molecules optimized following these criteria will provide new examples of molecular recognition. To understand the constraints of the sequence structure space, a detailed knowledge of the resulting 3D structure is required. In particular, the structures of the recognition interfaces of closely related molecules are of high interest. What are the parameters in the aptamer molecule that define the interaction? So far, only a limited number of 3D structures of nucleic acid molecules that were obtained by *in vitro* selection had been solved. A summary of currently available 3D structures of nucleic acid aptamers was recently published (Feigon et al., 1996).

Driven by the need of new active molecules in fields as diverse as from chemical catalysis to medical therapy, aptamers are sought in many research and screening projects. In chemistry, the focus is on catalytic molecules that enable stereoselective chemical reactions similar to those performed by enzymes (Prudent et al., 1994). In this way new substrate specificities may be obtained. Another application will be provided by molecules that show a high selectivity for a specific ligand. Such molecules can be used to design new detection devices based on individual molecular recognition. Aptamers that bind to physiologically active molecules, e.g. vascular endothelial growth factor VEGF (Jellinek et al., 1994), may be employed as a diagnostic device *in vitro* and *in vivo*. If physiological functions can be inhibited in a highly specific way, such a molecule may even be used as a tool in medical therapy.

In some cases a combination of various active domains may be required to perform a well defined multistep process. As in nature, optimized domains may be combined in a complex, multienzyme-like structure. Here a very simple strategy based on the simple basepairing of reverse complementary nucleic acid strands can be followed. To each active domain one or more sequence tags are added. Hybrid formation of the reverse complementary tags then leads to a spatial organization of the domains of interest (Seeman, 1996). As a result, a large number of individual nucleic acid molecules can be assembled, and after formation this structure will perform as one large multi-domain molecule. This selforganization based on reverse complementary sequence tags even works when the tags are linked to non DNA units. In such a way nanoparticles were associated into complex arrays coupled by DNA tags (Mirkin et al., 1996).

This brings optimization of biomolecules and general nanotechnology in close relation. For a variety of applications, we imagine that microstructured arrays and biomolecular libraries in solution will be combined to result in highly specific optimization processes. In one strategy, a variety of target molecules and many related molecules will be arranged on microstructured arrays using microprinting, *in situ* synthesis or related technologies (Ermantraut et al., this volume). These arrays will be incubated with the complex randomized nucleic acid library. After incubation only the molecules that bind at the matrix spots with the target will be recovered and amplified. In this way, library molecules are confronted with an immense variation of potential target molecules. This results is a highly parallel integrated selection and counter selection process. Driven by reaction or binding kinetics, molecules with high discrimination properties in molecular recognition should be discovered.

The integration of *in vitro* selection and evolution technologies with well defined microtechnology devices will provide many useful tools for the discovery of highly specific interaction partners or active molecules. In the future, this could provide a basic, economically and ecologically effective method for the expanding field of molecular biotechnology.

References

Bartel DP, Szostak JW (1993) Isolation of new ribozymes from a large pool of random sequences. *Science* 261: 1411–1418

Breaker RR, Banerji A, Joyce GF (1994) Continuous *in vitro* evolution of bacteriophage RNA polymerase promoters. *Biochem* 33: 11980–11986

Breaker RR, Joyce GF (1994) Emergence of a replicating species from an *in vitro* RNA evolution reaction. *Proc Natl Acad Sci USA* 91: 6093–6097

Conn MM, Prudent JR, Schultz PG (1996) Porphyrin metalation catalyzed by a small RNA molecule. *J Am Chem Soc* 118: 7012–7013

Ekland EH, Bartel DP (1996) RNA-catalyzed RNA polymerization using nucleoside triphosphates. *Nature* 382: 373–376

Ekland EH, Bartel DP (1995) The secondary structure and sequence optimization of an RNA ligase ribozyme. *Nucleic Acids Res* 23: 3231–3238

Ekland EH, Szostak JW, Bartel DP (1996) Structurally complex and highly active RNA ligases derived from random RNA sequences. *Science* 269: 364–370

Ellington AD, Szostak JW (1990) *In vitro* selection of RNA molecules that bind specific ligands. *Nature* 346: 818–822

Ellington AD, Szostak JW (1992) Selection *in vitro* of single-stranded DNA molecules that fold into specific ligand-binding structures. *Nature* 355: 850–852

Feigon J, Dieckmann T, Smith FW (1996) Aptamer structures from A to zeta. *Chem Biol* 3: 611–617

Geiger A, Burgstaller P, von der Eltz H, Roeder A, Famulok M (1996) RNA aptamers that bind L-arginine with sub-micromolar dissociation constants and high enantioselectivity. *Nucleic Acids Res* 24: 1029–1036

Guatelli JC, Whitfield KM, Kwoh DY, Barringer KJ, Richman DD, Gingeras TR (1990) Isothermal, *in vitro* amplification of nucleic acids by a multienzyme reaction modeled after retroviral replication [published erratum appears in *Proc Natl Acad Sci* USA 1990 Oct; 87(19):7797]. *Proc Natl Acad Sci USA* 87: 1874–1878

Heidenreich O, Pieken W, Eckstein F (1993) Chemically modified RNA: approaches and applications. *Faseb J* 7: 90–96

Jellinek D, Green LS, Bell C, Janjic N (1994) Inhibition of receptor binding by high-affinity RNA ligands to vascular endothelial growth factor. *Biochemistry* 33: 10450–10456

Jellinek D, Green LS, Bell C, Lynott CK, Gill N, Vargeese C, Kirschenheuter G, McGee DP, Abesinghe P, Pieken WA et al (1995) Potent 2′-amino-2′-deoxypyrimidine RNA inhibitors of basic fibroblast growth factor. *Biochemistry* 34: 11363–11372

Jenison RD, Gill SC, Pardi A, Polisky B (1994) High-resolution molecular discrimination by RNA. *Science* 263: 1425–1429

Joyce GF (1989) Amplification, mutation and selection of catalytic RNA. *Gene* 82: 83–87

Klussmann S, Nolte A, Bald R, Erdmann VA, Fürste JP (1996) Mirror-image RNA that binds D-adenosine. *Nature Biotechnology* 14: 1112–1115

Li Y, Sen D (1997) Toward an efficient DNAzyme. *Biochemistry* 36: 5589–5599

Lorsch JR, Szostak JW (1994) *In vitro* evolution of new ribozymes with polynucleotide kinase activity. *Nature* 371: 31–36

Mattheakis LC, Bhatt RR, Dower WJ (1994) An *in vitro* polysome display system for identifying ligands from very large peptide libraries. *Proc Natl Acad Sci* USA 91: 9022–9026

Mirkin CA, Letsinger RL, Mucic RC, Storhoff JJ (1996) A DNA-based method for rationally assembling nanoparticles into macroscopic materials. *Nature* 382: 607–609

Osborne SE, Ellington AD (1997) *Chemical Rev* 97: 349–370

Prudent JR, Uno T, Schultz PG (1994) Expanding the scope of RNA catalysis. *Science* 264: 1924–1927

Pütz J, Wientges J, Sissler M, Giege R, Florentz C, Schwienhorst A (1997) Rapid selection of aminoacyl-tRNAs based on biotinylation of alpha-NH2 group of charged amino acids. *Nucleic Acids Res* 25: 1862–1863

Sampson JR, Saks ME (1996) Selection of aminoacylated tRNAs from RNA libraries having randomized acceptor stem sequences: using old dogs to perform new tricks. *Methods Enzymol* 267: 384–410

Scott JK, Smith GP (1990) Searching for peptide ligands with an epitope library. *Science* 249: 386–390

Seeman NC (1996) The design and engineering of nucleic acid nanoscale assemblies. *Curr Opin Struct Biol* 6: 519–526

Spiegelman S (1967) An *in vitro* analysis of a replicating molecule. *Am Sci* 55: 221–264

Tsang J, Joyce, GF (1996) *In vitro* evolution of randomized ribozymes. *Methods Enzymol* 267: 410–426

Tuerk C, Gold L (1990) Systematic evolution of ligands by exponential enrichment: RNA ligands to bacteriophage T4 DNA polymerase. *Science* 249: 505–510

Wright MC, Joyce GF (1997) Continuous *in vitro* evolution of catalytic function. *Science* 276: 614–617

Selection of published nucleic acid aptamers

Target	kD	Reference
A: RNA		
Cibacron blue 3G-A reactive blue 4 and other dyes	<500 µM	Elington and Szostak (1990) *Nature* 346: 818–822
T4 DNA polymerase		Tuerk and Gold (1990) *Science* 249: 505–510
HIV Rev protein		Giver et al (1993) *Nucleic Acids Res* 21: 5509–5516
ATP	≈ 14 µM – 0.7 µM	Sassanafar and Szostak (1993) *Nature* 364: 550–553
Human α-thrombin		Kubik et al (1994) *Nucleic Acids Res* 22: 2619–2626
Prot. kinase C		Conrad et al (1994) *J Biol Chem* 269: 32051–32054
Rev. transcriptase		Chen and Gold (1994) *Biochemistry* 33: 8746–8756
VEGFactor		Jellinek et al (1994) *Biochemistry* 33: 10450–10456
Cyanocobalamin	≈ 10^{-7}	Lorsch and Szostak (1994) *Biochemistry* 33: 973–982
Caffeine, theophylline	≈ 10^{-7}	Jenison et al (1994) *Science* 263: 1425–1429
HIV Rev protein	≈ 6 nM – 0.8 nM	Jensen et al (1995) *Proc Natl Acad Sci USA* 92: 12220–12224
HIV-1 integrase	≈ 10^{-6} – 10^{-8}	Allen et al (1995) *Virology* 209: 327–336
L22 ribosomal protein		Dobbelstein and Schenk (1995) *J Virol* 69: 8027–8934
NGF		Binkley et al (1995) *Nucleic Acids Res* 23: 3198–3205
Tachykinin substance P	≈ 10^{-7}	Nieuwlandt et al (1995) *Biochemistry* 34: 5651–5659
Tax		Tian et al (1995) *RNA* 1: 317–326
HNE		Smith et al (1995) *Chemistry and Biology* 2: 741–750
Neomycin	≈ 10^{-7}	Wallis et al (1995) *Chemistry and Biology* 2: 543–552
Aminoglycoside antibiotics	≈ 10^{-5}	Wang and Rando (1995) *Chemistry and Biology* 2: 281–290
L-arginine, citrulline	≈ 10^{-5}	Burgstaller et al (1995) *Nucleic Acids* Res 23: 4769–4776
L-arginine	≈ $3.3 \cdot 10^{-7}$	Geiger et al (1996) *Nucleic Acids Res* 24: 1029–1036
Chloramphenicol	≈ 25 –65 µM	Burke et al. (1997) *Chemistry and Biology* 4: 833-843
HIV Gag protein	≈ 1–10 nM	Lochrie et al. (1997) *Nucleic Acids Res* 25: 2902–2910
SelB (elongation factor of *E. coli*)		Klug et al. (1997) *Proc Natl Acad Sci USA* 94: 6676–81
Prion protein PrP		Weiss et al. (1997) *Virology* 71: 8790–7
Xanthine/guanine		Kiga et al. (1998) *Nucleic Acids Res* 26: 1755–60
Coenzyme A		Burke and Hoffman (1998) *Biochemistry* 37: 4653–63

Target	kD	Reference
B: DNA		
Cibacron blue 3G-A reactive blue 4 and other dyes	≈ 46 µM – 33 µM	Ellington and Szostak (1992) *Nature* 355: 850–852
Thrombin	≈ 200 nM – 25 nM	Bocke et al (1992) *Nature* 355: 564–566
Adenosine/ATP	≈ 10^{-6}	Huizenga et al (1995) *Biochemistry* 34: 656–665
RT of HIV-1	≈ 10^{-9}	Schneider et al (1995) *Biochemistry* 34: 9599–9610
Porphyrins		Li et al (1996) *Biochemistry* 35: 6911–6922
IgE (Fc-e-receptor I)	≈ 10^{-8}	Wiegand et al (1996) *J Immunol* 157: 221–230
L-selectin		O'Connell et al (1996) *Proc Natl Acad Sci USA* 93: 5883–5887
HNE		Davis et al (1996) *Nucleic Acids Res* 24: 702–706
C: 2'-amino-pyrimidine RNA		
Human neutrophil elastase (HNE)	≈ 30 nM – 10 nM	Lin et al (1994) *Nucleic Acids Res* 22: 5229–5234
VEGFactor	≈ 2.4 nM	Green et al (1995) *Chemistry and Biology* 2: 683–695
Basic FGF	≈ 350 nM	Jellinek et al (1995) *Biochemistry* 34: 11363–11372
IgE (Fc-e-receptor I)	≈ 30 nM, ≈ 35 nM	Wiegand et al (1996) *J Immunol* 157: 221–230
D: mirror-image RNA		
D/L-adenosine	≈ 1 µM	Klußmann et al (1996) *Nature Biotechnology* 14: 1112–1115
D/L-arginine		Nolte et al (1996) *Nature Biotechnology* 14: 1116–9
Vasopressin		Williams et al. (1997) *Proc Natl Acad Sci USA* 94: 11285–11290

Selection of published nucleic acid catalysts (ribozymes)

(This table does not include molecules derived from natural ribozymes by optimization)

Reaction	Reference
Biphenyl isomerization	Prudent et al (1994) *Science* 264: 1924–1927
Kinase nucleic acid phosphorylation	Lorsch and Szostak (1994) *Nature* 371: 31–36
Alkylation (self)	Wilson and Szostak (1995) *Nature* 374: 777–782
Phosphodiesterbond formation Metalloenzyme Zn2+/Cu2+	Cuenoud and Szostak (1995) *Nature* 375: 611–614
RNA ligase	Ekland et al. (1995) *Science* 269: 364–370
Porphyrin metalation	Conn et al (1996) *J Am Chem Soc* 118: 7012–7013
RNA polymerization	Ekland and Bartel (1996) *Nature* 382: 373–376
Amino acid transfer	Lohse and Szostak (1996) *Nature* 381: 442–444
Peptidyl-transferase	Zhang and Cech (1997) *Nature* 390: 96–100
Carbon-carbon bond	Tarasow et al. (1997) *Science* 389: 54–57
Amide synthase	Wiegand et al. (1997) *Chemistry and Biology* 4: 675–683
Ester transferase	Jenne and Famulok (1998) *Chemistry and Biology* 5: 23–34
N-glycosidic bond	Unrau and Bartel (1998) *Nature* 395: 260–263

Structures of nucleic acid aptamers

Molecule	Reference
Thrombin-DNA aptamer	Macaya et al (1993) *Proc Natl Acad Sci USA* 90: 3745–3749 Wang et al (1993) *Biochemistry* 32: 1899–1904
FMN-RNA aptamer	Fan et al (1996) *J Mol Biol* 258 (3): 480–500
Arginine-RNA aptamer Citrulline-RNA aptamer	Yang et al (1996) *Science* 272: 1343–1347
ATP-RNA aptamer	Dieckmann et al (1996) *RNA* 2: 628–640
AMP-RNA aptamer	Jiang et al (1996) *Nature* 382: 183–186
HIV-1 rev peptide-RNA aptamer	Ye et al (1996) *Nat Struct Biol* 3: 1026–1033
AMP-DNA aptamer	Lin and Patel (1997) *Chemistry and Biology* 4: 817–832
MS2-RNA aptamer	Rowsell et al. (1998) *Nat Struct Biol* 5:970–975
Tobramycin-RNA aptamer	Jiang and Patel (1998) *Nat Struct Biol* 5: 769–74

22 Optimisation of molecular function

Christian V. Forst

22.1 Introduction

The search for optimal biopolymer sequences with a desired function seems to be hopeless at first glance. There are a astronomically high number (i.e. 4^n RNA molecules and 20^n for peptides and proteins) of different sequences of chain length n. How will it ever be possible to find the appropriate ones? Systematic search is instantaneously caught in unfavorable combinations and is doomed to failure. Even if knowledge on rational design would be sufficient (which is not at the current state) to predict spatial conformations from sequence data, the next step relating functions to structures is still an unsolved problem in molecular biology. On the other hand selection of biomolecules has been successful even if *a priori* chances are highly unfavorable: Bartel and Szostak [3] succeeded to select RNA with catalytic ligase activity for RNA molecules aligned to a template from a sample of 1.6×10^{15} different molecules containing a randomised domain of $n = 220$ bases (yielding a total number of $4^{220} = 2.75 \times 10^{132}$ different sequences). Starting with an incredibly small fraction of only 10^{-117} of the total sequence library the enzymatic activity was increased more than a millionfold in ten selection cycles. Why and how can such a small initial population be successful? An attempt to solve this puzzle will be made in this contribution.

Evolutionary strategies based on Darwin's principle of (i) creation of genetic diversity, (ii) selection of the best suited genotypes, and (iii) amplification of the selected molecules are apparently very successful. In order to conceive evolutionary experiments one would nevertheless like to solve following non-trivial problems:

- Configuration of initial population – uniform randomly distributed, quasispecies distribution, clustered starting population with distinct distributions

- "Doped libraries" – Here a bias in the starting population is introduced
- Error-rates – What is the optimal error rate to choose? Are strategies with alternating low and high error rates more successful than models with constant error rates?
- "Limit" of hill climbing – If dependencies between subsystems in the system arise one has to switch to complex dynamics.

Since no reference is given to particular sequences the problems here will have answers of a statistical nature.

We first give an overview on coarse graining of both RNA and protein folding landscapes in the next section. In section on p. 535 we will report on generic properties of folding landscapes. The section on p. 542 will present evolutionary dynamics as optimisation strategies. Here questions regarding the above-mentioned problems will be discussed. In section on p. 546 an attempt of a synthesis between theoretical concepts and experiments is made.

22.2 A redundant relationship

The function of biomolecules, especially peptides and nucleic acids, is predetermined significantly by their tertiary structure in space. Active residues of these molecules are kept in precise position by a huge spatially organized framework of interacting residues and backbone. As conserved active residues in, for example, catalytically active sites are, as flexible is the structural framework. Here complete motives can be omitted maintaining (almost) unperturbed functionality. Thus a relevant structure of biopolymers in a given context is seldom described with atomic resolution. In order to detect phylogenetic relations, e.g., structures of proteins are often considered to be similar when polypeptide backbones coincide roughly. A large fraction of amino-acid residues can be exchanged without changing these coarse-grained structures that are apparently relevant in an evolutionary context. Rost and Sander investigated that 25% pairwise sequence identity of residues are sufficient for folding into the same structure [58]. That is 75% sequence dissimilarity (in best cases) is compatible with conserved structures.

Similar results are known for RNA structures. Here an adequate coarse-graining is represented by the secondary structure. It commonly is understood as a list of Watson-Crick (A = U and G ≡ C) and Wobble (G – U) base-pairs which are compatible with unknotted and pseudo-knot-free two-dimensional graphs (for a precise formal definition we refer to Waterman [75]). The relevance of RNA secondary-structures for biomolecular function is significantly reflected in viral life cycles [76]. Replication of RNA molecules in the $Q\beta$-system depends exclusively on the structural feature of a hairpin at the 5′-end [8]. Especially kinetics of RNA replication by $Q\beta$-replicase and the dependence of structural features have been studied [9]. A different system – which is exensively examined – is the internal initiation of translation for specific +-strand RNA viruses. This so-called IRES-region (I*nternal* R*ibosomal* E*ntry* S*ite*) – a highly structured region close to the 5′-end of the virus genome – is responsible for the success of the genome translation in the host cell [55].

Mathematically the relationship between genotypes and phenotypes is described as (non-necessarily invertible) mapping from sequence space Q_α^n into a suitable space of phenotypes. It is simplified by partitioning it in partial maps:

$$Q_\alpha^n \xrightarrow{f} S_n \xrightarrow{g} F_n \xrightarrow{\varphi} \phi. \tag{1}$$

For distinguished structure-families S_n (such as RNA secondary structures and proteins to some extent) sequence – structure maps denoted by f are well studied objects and generic properties are known. Only to a certain extent is similar knowledge available for structure – function maps g and function – fitness relationships φ (F and ϕ indicate function and fitness resp.). Both functions g and φ are commonly concatenated as a single mapping.

RNA secondary structures

RNA secondary structures and the induced sequence – structure relationship are a suitable and generic description for genotype – phenotype mappings which are important in molecular evolutionary biology. RNA secondary structures represent a type of coarse-graining of biopolymer structures. They commonly are understood as a list of Watson-Crick (A=U and G≡C) and Wobble (G–U) base-pairs which are compatible with unknotted and pseudo-knot-free two-dimensional graphs (for a precise formal definition we refer to Waterman [75]).

Defining secondary structures independently of chemical or physical restrictions yields a general description based on contacts with respect to arbitrary alphabetes A with arbitrary pairing rules Π. A *pairing rule* Π on A is given as a set of pairs of letters from the given alphabet (i.e. AU, UA, GC, CG, GU, UG for natural RNA-molecules with alphabet A, U, G, C). This concept can easily be extended to a general description of biopolymer structures via contact maps [46]. Similar to secondary structures a general *contact structure* c is determined by a *set of contacts* of c omitting the trivial contacts due to adjacent letters in the succession of the sequence.

A relevant concept in studying sequence – structure relation is how sequences have to be composed to fulfil necessary conditions for folding into a desired structure. In the following we define *compatibility* of a sequence to a given structure: A sequence x is said to be *compatible* to a structure s if all base-pairs required by s can be provided by x_i and $x_j \in x$ with respect to the pairing rule Π for each base pair (Fig. 1). C(s) is the set of all sequences which are compatible to structure s. The number of compatible sequences is readily computed for secondary structures (with n_u unpaired bases and n_p base pairs this evaluates to $4^{n_u} \cdot 6^{n_p}$).

Proteins

Compared to DNA and RNA a lot more work has been done on proteins, on protein folding and on their energy landscapes. For a description of kinetic protein folding with their extremely large numbers of possible pathways leading through similar but distinct states, the notion of a *statistical energy landscape* has been introduced [11]. This landscape view of structure formation in proteins provides explanations for following phenomena:

- for folding behavior like a two-state first-order phase transition,
- for the existence of metastable collapsed and only partially folded states known as *molten globules*,
- and for the curved Arrhenius plots found in laboratory measurements and in discrete lattice simulations.

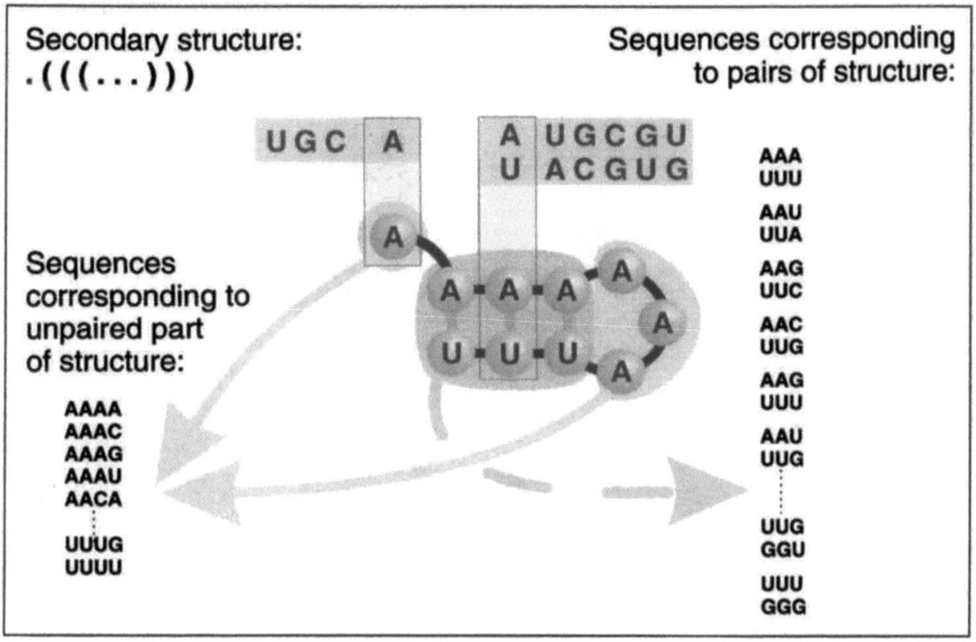

Figure 1 *Compatible sequences with respect to a fixed secondary structure. A sequence is called* compatible *with a given secondary structure if for all base pairs in the structure there are pairs of matching bases in the sequence. Sequences compatible to a structure* do not *fold in general into this structure. However the structure will always be found as a result of suboptimal folding*

Lattice simulations are a suitable description of coarse grained protein structures. Similar to RNA secondary structures, lattice models represent structures in terms of contact matrices. A significant difference between these two abstractions of structure representations is that computational results for lattice models face difficulties when related to experimental structure predictions. Nevertheless, lattice heteropolymer models by Ken Dill and co-workers [14, 15, 18] and by other groups [11, 12, 67] are successful in studying, e.g., the uniqueness of ground states and questions of kinetic folding.

The above-mentioned paradigm of a statistical energy landscape provides a solution of the well known Levinthal paradox [48]: How can a protein search the cosmologically large number of unfolded configurations in order to find the native folded structure in biologically relevant time? In other words, a huge number of initial conformations is reduced hierarchically yielding fewer and fewer intermediate states of increasing hierarchy until the set of trajectory converges into the native state (Fig. 2).

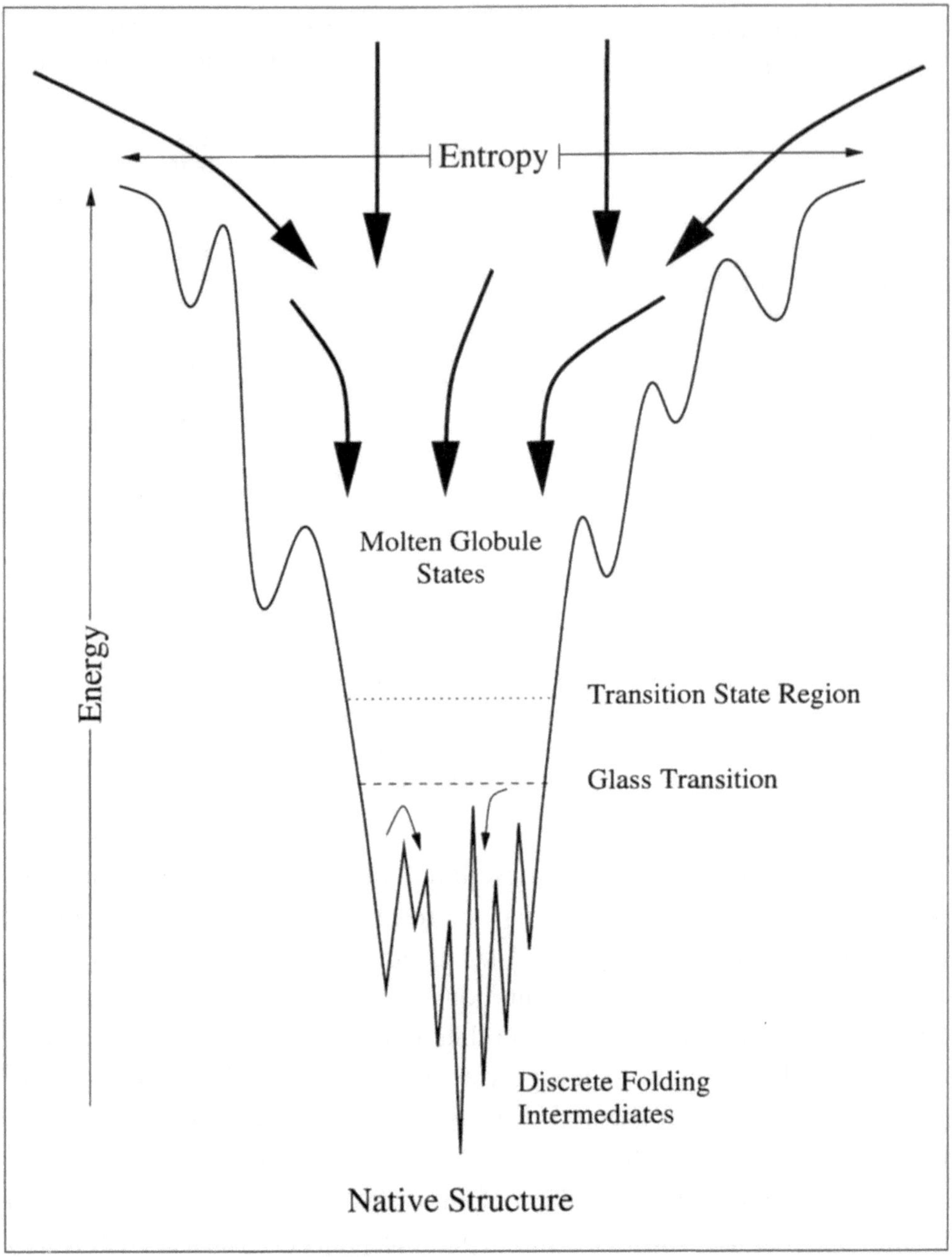

Figure 2 *"Funnelling" trajectories of protein folding towards the native states. Funnelling provides a solution of the Levinthal paradox [48]. Proteins have extremely large numbers of conformations. Thus random coils would not be able to fold into native states in reasonable times unless there are folding pathways guiding molecules into states of minimal energy*

A common problem in structure prediction is the following: The current insight into regularities of biopolymer structures is not sufficient to allow reliable determination of molecular structures from known sequences. For a review on the present state of the art in protein structure prediction see [58] (regarding RNA tertiary structures see [1, 47]). Investigations on biopolymers have revealed a huge amount of empirical knowledge providing detailed insight into molecular structures and mechanisms of molecular-biological systems. A common approach is to rely on sequence homologies to derive identical or at least similar structures. Similar structures are considered to be a very strong hint for identical biological functions. Examples, however, are also known for proteins with divergent sequences but with almost structural identical active sites. Thus these bio-molecules have practically the same enzymatic functions. The next chapter will deal with both knowledge based methods for structure prediction and sequence – structure relationships with partly neutral properties.

22.3 Statistics on biopolymer structures

First steps towards optimisation of biomolecules are a detailed study of the underlying sequence – function relationship. This is taken for granted in the case of rational design. Nevertheless, this knowledge is essential for evolutionary strategies to "optimise the optimisation". Thus a high effort has been performed to deduce generic properties of biopolymer folding landscapes and to relate structural features to function.

Common and rare structures

By application of combinatoric methods to RNA secondary structures [75] an asymptotic expression for the number of possible (pseudo-knot free) structures which can be formed by sequences of chain length n has been derived [37, 65]

$$|S_{\mathrm{n}}| \approx 1.4848 \cdot n^{3/2} \cdot 1.8488^{\mathrm{n}}. \tag{2}$$

This formula is based on two assumptions: (i) the minimum stack length is two base pairs and (ii) the minimal size of hairpin loops is three bases. The number of sequences evaluate to 4^n and 2^n for natural RNA sequences and for sequences with binary alphabet (AU- and GC- only sequences) resp. In the evolutionary relevant cases for common structures (*common* will be defined in *Definition 1*) there will be a large number of sequences folding into the same secondary structure s. These fundamental properties are not only restricted to RNA molecules. For proteins the number of sequences is enormous. For n residues there are 20^n sequences. On the other hand, the repertoire of stable native folds seems to be highly restricted or even vanishingly small [16]. For example, frequently made observations indicate that seemingly unrelated sequences have essentially the same fold [38, 52, 53].

Returning to RNA shows that not all acceptable secondary structures are actually formed as minimum free energy structures. The number of stable secondary structures $|\widehat{S_n}|$ was calculated by exhaustive folding [33, 34] of all GC-only sequences with chain length up to $n = 30$ (Tab. 1). About 20% to 50% of acceptable structures are obtained by folding. An estimate of the dependency of $|\widehat{S_n}|$ on n from exhaustive folding data is $|\widehat{S_n}| \propto 1.65^n$. A classification of secondary structures relevant in an evolutionary context is a classification in common and rare structures:

Definition 1. (*c.f.* [33]) A structure s is called common if it is formed by more sequences than the average structure:

$$|f^{-1}(s)| \geq \alpha^n / |\widehat{S_n}|, \qquad (3)$$

whereas α denotes the alphabet-size.

Two important general properties are implications of the above made definition of common structures [33, 34]: (i) common structures only present a small fraction of all structures and this fraction decreases with increasing chain length, and (ii) almost all sequences fold in common structures in the limit of infinite chain length. Thus for sufficiently long RNA molecules almost all sequences fold into a small fraction of structures. The effective ratio of sequences to structures is larger than calculated from Eq. (3) since only common structures play a role in molecular evolution.

Computational studies of protein folding landscapes similar to explorations in the RNA world are more difficult due to the unequivocal complexity of

Table 1 Common secondary structures of GC-only sequences

	Number of sequences		Number of structures	GC *		
n	4^n	2^n	$\lvert S_n \rvert$	$\lvert \widehat{S_{GC}} \rvert$	R_c	n_c
7	16384	128	6	2	1	120
10	1.5×10^6	1024	22	11	4	859
15	1.07×10^9	322768	258	116	43	28935
20	1.10×10^{12}	1.05×10^6	3613	1610	286	902918
25	1.13×10^{15}	3.36×10^7	55848	18590	2869	30745861
30	1.15×10^{18}	1.07×10^9	917665	218820	22718	999508805

* The total number of minimum free energy secondary structures formed by GC-only sequences is denoted by $\lvert \widehat{S_{GC}} \rvert$, R_c is the rank of the least frequentn common structure and thus is tantamount to the number of common structures, and n_c is the number of sequences folding into common structures.

protein folding, and by the fact that there is no biophysically meaningful and computationally simple coarse description of protein structures (protein secondary structures are referred to local features that may or may not be present but do no capture the global organization of the molecule). Hence one has to restrict oneself to a less ambitious approach based on *inverse folding* only [2]. The problem on deciding whether a given sequence x folds in a structure s (i.e. $x \in f^{-1}(s) : f^{-1}(s)$ is the preimage of the mapping in sequence-space with respect to a fixed structure s) is less demanding than predicting the unknown protein structure of a given amino acid sequence. The former can be investigated by inverse folding techniques [10, 20].

In the following we shortly present an approach pursued by Babajide et al. [2] using a set of *knowledge-based potentials of mean force* that have been derived by Sippl and co-workers [35, 68]. Recent studies demonstrated that the energy of the native structure (i.e. the putative ground state) of a sequence x can be estimated from the energy values of x in its configuration space [4, 10, 30–32, 35, 69, 70]. This allows the construction of an energy scale by which conformations of different sequences can be compared. As a measure for the quality of fitting sequence x in structure s the *z-score* has been defined [13]:

$$z(x, s) = \frac{\Phi(x, s) - \bar{\Phi}(x)}{\sigma_\Phi(x)} \tag{4}$$

Here $\Phi(x, s)$ is a potential function evaluating the energy of x when folded into s which is defined by spatial coordinates of its C^α and C^β atoms, resp. $\bar{\Phi}(x)$ is the average energy of x in all configurations in a database and $\sigma_\Phi(x)$ is the standard deviation of the corresponding distribution. Empirically, native folds have z-scores in a narrow characteristic range. Hence one may assume that x is a member of $f^{-1}(s)$ if the z-score of x in conformation s is in the native range [13]. Obviously, only native structures that are already in the database can be explored by this method. Thus inverse folding of protein structures is been translated into an optimisation problem in the set of sequences.

One interesting question arising in the context of inverse folding is: "What is the minimal set of amino-acids providing reasonable folds?". In literature arguments have been stated on the dominating influence of hydrophobic versus hydrophilic amino-acid pattern (HP-pattern) on the protein structure [19, 42, 72]. The success for predicting wild-type like z-scores depends essentially on the chosen combination of amino acids. For example, the two-letter alphabet AD gives very poor results while LS or DL yield z-scores. ADGL as a proposed candidate for a primordial set of amino acids [51] yield good sequences since DL is already sufficient.

The most striking similarity in sequence-structure maps of RNA molecules and proteins is the existence of few common and many rare structures. Comparing log-rank/log-frequency plots of GC-sequences of length $n = 27$ with the corresponding plot derived from HP-lattice proteins indicates a strong similarity (see [33, 49] and Fig. 3). Such distributions can be characterised by a generalised Zipf's law [78]. The interpretation of this data shows that for GC-sequences of length $n = 30$ more than 93% of all sequences fold into only 10% structures. Extrapolation to longer chain lengths indicates an increasing fraction of sequences folding into a decreasing percentage of sequences. Implications for evolutionary optimisation are evident: Populations live in a space of common structures or phenotypes. Rare phenotypes are extremely hard to find in random searches and thus play no role in evolution.

Neutral networks

It has been frequently observed (and already mentioned in the previous chapter) that unrelated protein sequences fold into the same structure [38, 52,

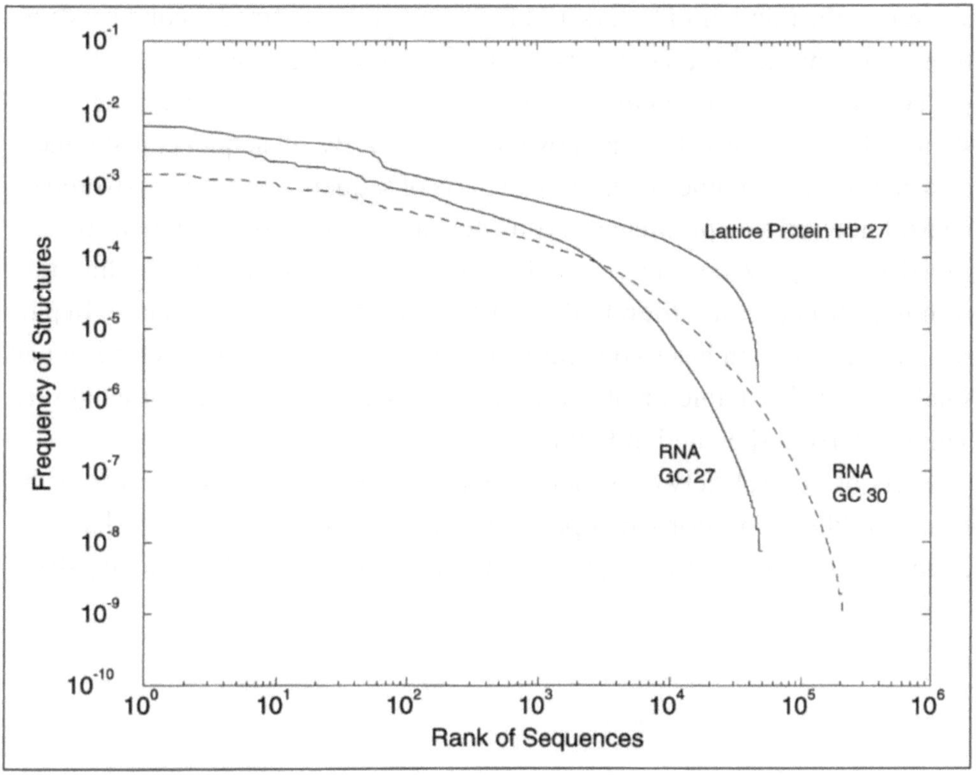

Figure 3 *The distribution of frequencies of RNA secondary structures and lattice model proteins. The diagram shows the distribution of preimage size of sequence-structure maps for GC-sequences of length 27 and 30 [33] compared with the corresponding plot for HP-lattice proteins of length 27 [49]*

53]. Similar results have been reported for RNA viruses which show a large degree of sequence variation while sharing many conserved features in their secondary structures (i.e. see [77]). Whether these may have originated from a common ancestor or whether they must be the result of convergent evolution depends on the topology of the so-called *neutral networks* (for a definition see below) in sequence space. Another well known example is represented by the clover-leaf secondary structure of tRNAs: The sequences of different tRNAs have very little structure homology but nevertheless fold into the same secondary structure motif [23]:

First let us formulate some definitions: A *neutral network* $\Gamma_n(s)$ with respect to s is defined as a set of all sequences folding into a given structure s. Q_α^n denotes the generalised hypercube of dimension n over an alphabet A of size α (i.e. the number of letters in A is α), and $s \in S_n$ is a fixed secondary structure. Mathematically $\Gamma_n(s)$ refers to the induced subgraph of $f_n^{-1}(s)$ in $\mathbf{C}(s)$ ($f_n^{-1}(s)$ indicates the *preimage* of a fixed structure s w.r.t. the mapping f_n). A sketch of these embeddings is shown in Figure 4. By subsequenct assignment of structures to sequences (almost) all parts of sequence-space are covered by neutral networks as preimages. As a consequence of this assignment the so-called *shape space conjecture* can be observed [65]: Sequences folding into common structures are found (almost) everywhere in sequence space. In particular only a region has to be searched at maximum which diameter is much smaller than the diameter of the total sequence-space (i.e. chain length n). This property is sketched in Figure 5.

Parts of neutral networks can be explored by either neutral walk or by inverse folding [36]. For a complete exploration, up to now, only exhaustive enumeration leads to success [33]. Neutral networks of RNA secondary struc-

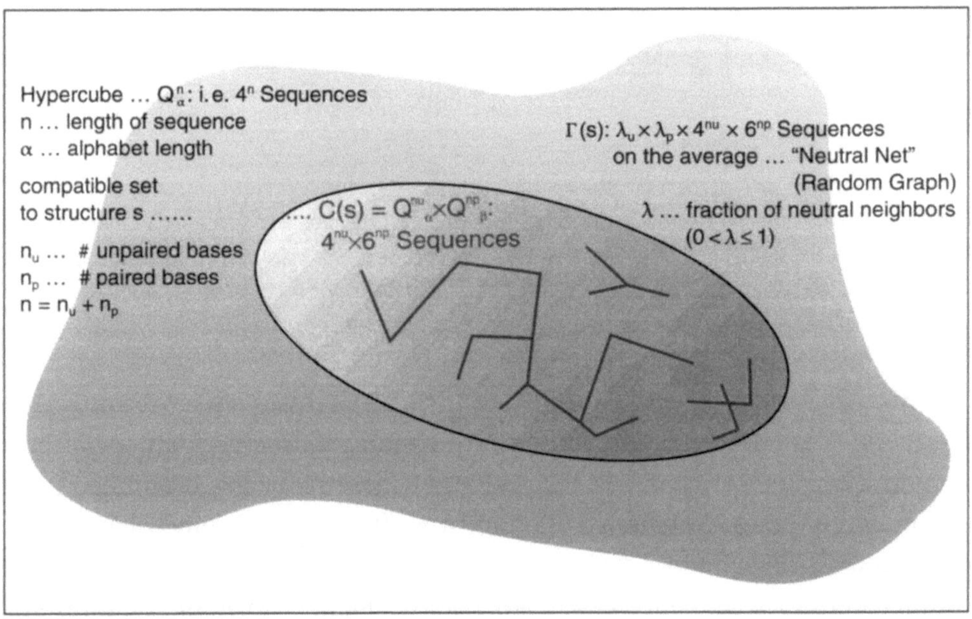

Figure 4 *Sketch of a neutral network $\Gamma(s)$ (shown as solid line graph) embedded in the set of sequences compatible to structure s (i.e. $C(s)$ – indicated as oval) which itself is embedded in sequence space Q^n_α (realized as shaded background). n_p and n_u are the number of base-pairs and the number of unpaired residues corresponding to the fixed secondary structure s. α is the alphabet size (for polynucleotides $\alpha = 4$), β corresponds to the alphabet size of the "paired alphabet" (for base-pairs observed in RNA secondary structures $\beta = 6$). The fraction of neutrality of a given sequence is expressed in λ_u and λ_p for unpaired and paired regions resp.*

tures to a first approximation can be modelled as random graphs [57], with a fixed expected number of neutral neighbours per structure. We have tested some predictions from the random graph model on the neutral networks which arise if minimum free energy is the criterion for structure assignment [29]. On minimum free energy networks, the ease of accessibility of alternative structures is critical for the expected number of neutral neighbours of a sequence. It can be shown that stable, well defined structures lead to networks which conform to the r.g. model: they are dense and locally connected. Indeed these networks are near-regular graphs, with the expected neutrality of a given position being nearly independent of the sequence. If, however, a structure is not very well defined, then it depends on the actual sequence which alternative structures are most easily (or at all) accessible (in terms of the r.g. model, different sequences on the network belong to different *intersections* with the neutral networks of other struc-

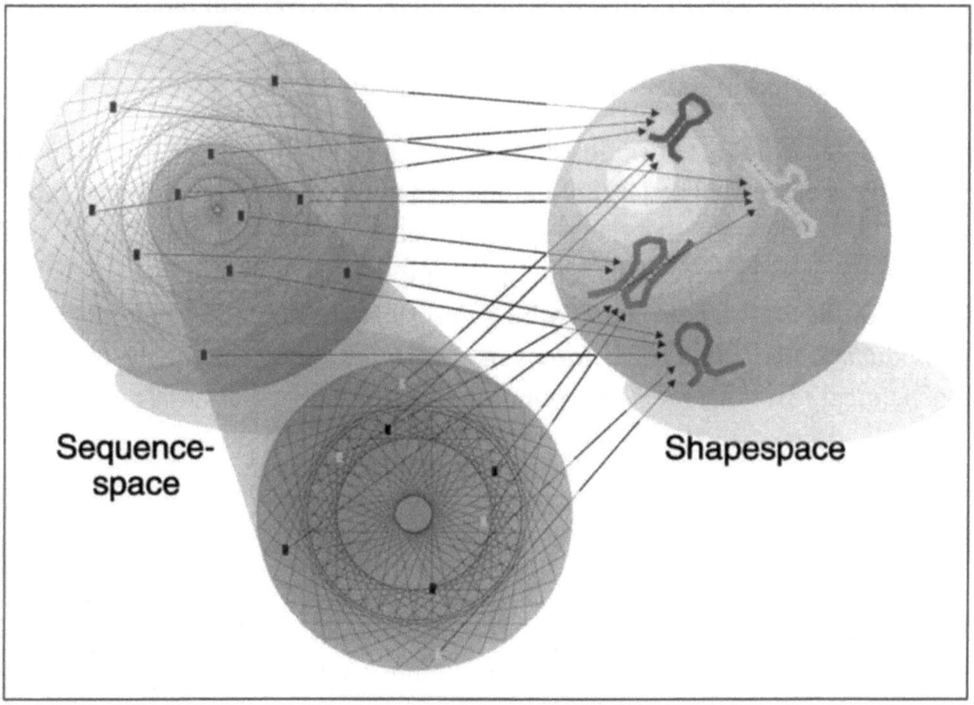

Figure 5 *A sketch of an RNA sequence to secondary structure map. In close vicinity of any random sequence on can find mutants which fold in (almost) all common structures*

tures). Accordingly, these structures exhibit a broad distribution of the number of neutral neighbours per sequence.

22.4 Evolutionary dynamics

A concept

Evolutionary dynamics itself is a highly complex process. Therefore we omit additional difficulties in considering spatiotemporal patterns and introduce a comprehensive model which tries to account for most of the relevant features of molecular evolution. Peter Schuster proposed an interaction of three processes described in three different abstract metric spaces [63] as essential building blocks of *evolutionary dynamics*:

- the *sequence space* of genotypes being DNA or RNA sequences,
- the *shape space* of phenotypes, and
- the *concentration space* of biochemical reaction kinetics.

A Darwinian scenario of optimising species of a population by evolutionary processes is easily realized as hill-climbing in a high-dimensional fitness landscape. The underlying dynamics of this process is linear (in most cases) and approaches a stationary state. One quite important class of dynamics is the so-called *quasispecies dynamics* which describes a population of replicating individuals under mutational forces in a constant environment. In terms of biochemical reactions the corresponding reaction-equations are as follows:

$$I_i \xrightarrow{f[s(I_i)] \cdot W_{ij}} I_i + I_j \qquad (5)$$

$$I_i \xrightarrow{\Phi} \theta$$

We denote I_i, $i = 1, \ldots, n$ as reacting species with a fixed phenotype $s(I_i)$, and W_{ij} as stochastic matrix indicating the probability of reproducing I_j by replicating I_i. $s(I_i)$ corresponds to the structure (or phenotype) of the individual, and $f(s)$ is the fitness of phenotype s. Thus the top reaction-equation of Eq. (5) describes an autocatalytic, error-prone replication of I_i. The bottom equation refers to an unspecific dilution flux maintaining the total numbers of individuals constant in the system. Applying chemical reaction-kinetics yields the following selection-mutation equation, originally formulated by Eigen [21] in a simpler representation:

$$\dot{x}_i = x_i[f(s(I_i))\,W_{ii} - \Phi(\mathbf{x})] + \sum_{j \neq i} k_j W_{ij}\, x_i \qquad i = 1, \ldots, n. \tag{6}$$

Eigen in his original paper described evolution in molecular-biological systems without explicit usage of a genotype-phenotype mapping. Thus he assigned constant reaction rates $k_i \equiv f(s(I_i))$ for each genotype. He focused especially on dynamics on the so-called *single-peak landscape* as mean field approach, where a single genotype has superior fitness (high reaction rate k_i implying fast replication) upon all other genotypes with equal but lower fitness. In the past both deterministic and stochastic models of quasispecies upon single-peak landscapes have been studied [23, 50, 54, 62, 66, 73]. An essential result is the report of a so-called *error-threshold*. This threshold is the maximal error-rate for the system where the organization of the population in a cloud around a master-sequence (the quasispecies) is replaced by a random distribution of individuals all over sequence space. It can be understood as *phase-transition* from an ordered and organized phase to a disordered, random phase.

Neutral evolution and adaptation

Introducing a genotype – phenotype mapping motivated by sequence – structure relations of RNA molecules yields an expansion of the quasispecies dynamics with totally new properties. First studies of quasispecies-dynamics on RNA secondary-structure folding-landscape have been followed by Fontana et al. [26, 27]. Forst et al. performed evolutionary dynamics on *single-shape landscapes* where shape refers to a fixed secondary structure [28]. Analogous to the approach of a single-peak landscape Forst et al. classified the sequence-space in fit sequences (sequences which are mapped in the distinct structure) and non-fit sequences (all other sequences). This implies a classification of an evolving population in masters (fit individuals) and non-masters (non-fit individuals). Starting with low error rates the population is localised around a non-moving master as the quasi-species in the single-peak landscape. At a distinct error-rate (the error-threshold for a single-peak landscape) the population starts moving and drifts on a neutral network analogue to a diffusion-process. For even higher error rates the population breaks into small clusters with lifetime obeying a power law (large clusters have long lifetime, small clusters have short lifetime) [40]. After a sufficiently high error-rate a so-called *phenotypic error-threshold* can be observed. Here the population is no longer able to conserve the information of the phenotype but diffuses randomly distributed all over sequence-space. Analytical expression of stationary distributions of a population in this landscape and of diffusion constant has been reported [28]. Derivations and proofs of these formulae can be found elsewhere [56]. For infinite chain lengths the following implicit formula for the critical error-rate p can be deduced:

$$[1-(1-p)^{n_u}]\,\lambda_u(1-p)^{2n_p} + (1-p)^{n_u}\,\lambda_p\,\Phi(p) + [1-(1-p)^{n_u}]\,\lambda_u\lambda_p\Phi(p) + (1-p)^n = \frac{1}{\sigma} \tag{7}$$

with $\Phi(p) \stackrel{\text{def}}{=} [(\frac{p^2}{\alpha-1} + (1-p)^2)^{n_p} - (1-p)^{2n_p}]$. Positive, real roots of Eq. (7) for p indicate critical error-rates, where the error-threshold occurs. For an explanation of variables see the legend of Figure 4.

As a consequence of neutral networks, a population seeking the global optimum is likely to find its goal (c.f. [64]). Figure 6 shows different dynamical behavior of an optimising population for landscapes without and with selective neutrality. Landscapes without partly neutral properties lead an evolving population to the nearest local optimum. In landscapes with selective neutrality a population reaches a local optimum but is not doomed to stay there forever. Instead, the population drifts neutrally in genotype space, changes neutral networks by transitions occasionally and looks for a better place. Only in the eyes of the impatient experimentalist is this local, so-called *virtual optimum* real. After finding better blessed grounds the population heads towards new optimal phenotypes.

In this context, an interesting question to ask is: "Is there an optimal error-rate p for which an evolving population evolves fastest?" A naive answer would be to choose p close to the error-threshold. In this scenario a population searches far away from safe grounds. Thus a "good" phenotype cannot be maintained sufficiently for such high error-rates. Keeping good phenotypes

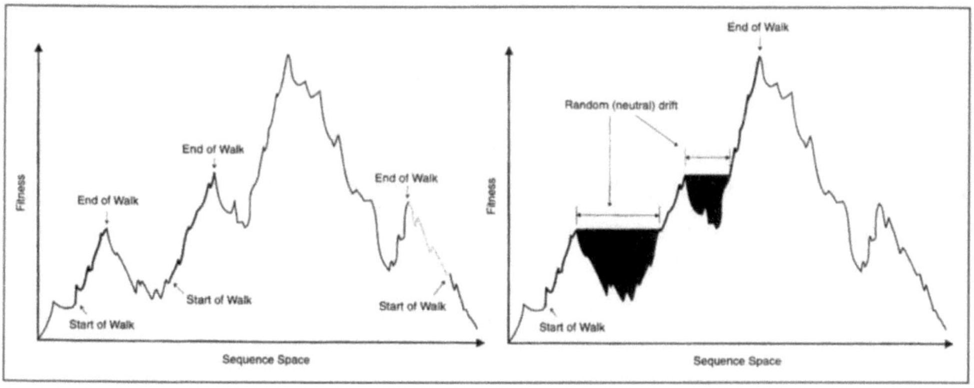

Figure 6 *Optimisation on Fitness landscapes: The difference of optimization in landscapes without and with selective neutrality is shown. On landscapes without partly neutral properties optimization reaches a local optimum and stays there. Having selective neutrality optimization does not have to stop at local optima, but is able to move neutrally in sequence space, thus finding new opportunities to optimise*

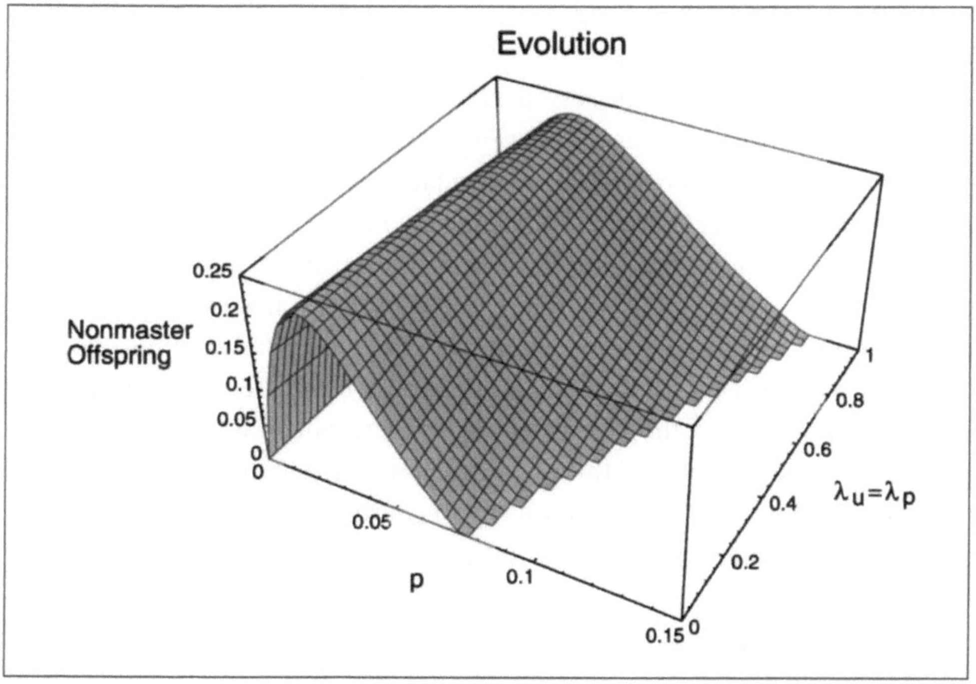

Figure 7 *Offspring of population with a phenotype different to the parental phenotype. Surprisingly, the maximum of the surface for p (for λ arbitrarily chosen but fixed) is hardly dependent on λ. This implies the evaluation of an optimal p-value almost independent of the topology of the given sequence – structure map*

as established "bases" to search for better ones is only feasible for lower p-values. Here both the maintenance of a good phenotype and the production of a reasonable high quantity of mutants, searching for better phenotypes is performed. Figure 7 shows the dependency of the fraction of offspring generated from a fixed parental phenotype on the error-rate p and the neutrality λ of the underlying neutral network. Note that the maximum of the surface for p is hardly dependent on λ.

22.5 Models and machines

Eigen and Gardiner proposed the application of RNA-based replication assays to solve problems in biotechnology by means of evolutionary adaptation through Darwin's selection principle [22]. Two years later Kauffman suggested to start from libraries of random sequences and to synthesize biopolymers with desired catalytic properties through large-scale screening based on recombinant DNA techniques and selection methods [43]. Meanwhile, both concepts have become reality: several research groups apply evolutionary concept to produce biomolecules with new properties (for reviews see [24, 41, 44]).

A typical strategy for optimizing molecular propertis is sketched in Figure 8. As outlined in the previous chapter, experiments are carried out at the level of populations of molecules. Replication of molecules is used as an amplification factor. By either artificially increased mutation rates or by partial randomisation of the initial sequence pool variation is introduced into populations. These two techniques differ with respect to the selection procedure. The first approach to the optimisation problem is suitable for "batch experiments": The idea behind this procedure is to encode the desired functions into the selection constraint. Several reports of the success in applying directed evolution techniques to biochemical problems are found in literature [3, 7, 17, 25, 39, 71, 74].

But often it will be very difficult or even impossible to encode a desired function implicitly into the selection constraint. Then spatial separation of individual genotypes and massively parallel screening provide a solution [5, 6]. This technique, however, requires highly sophisticated equipment which is currently in a final phase of development [60, 61]. Compartments are created

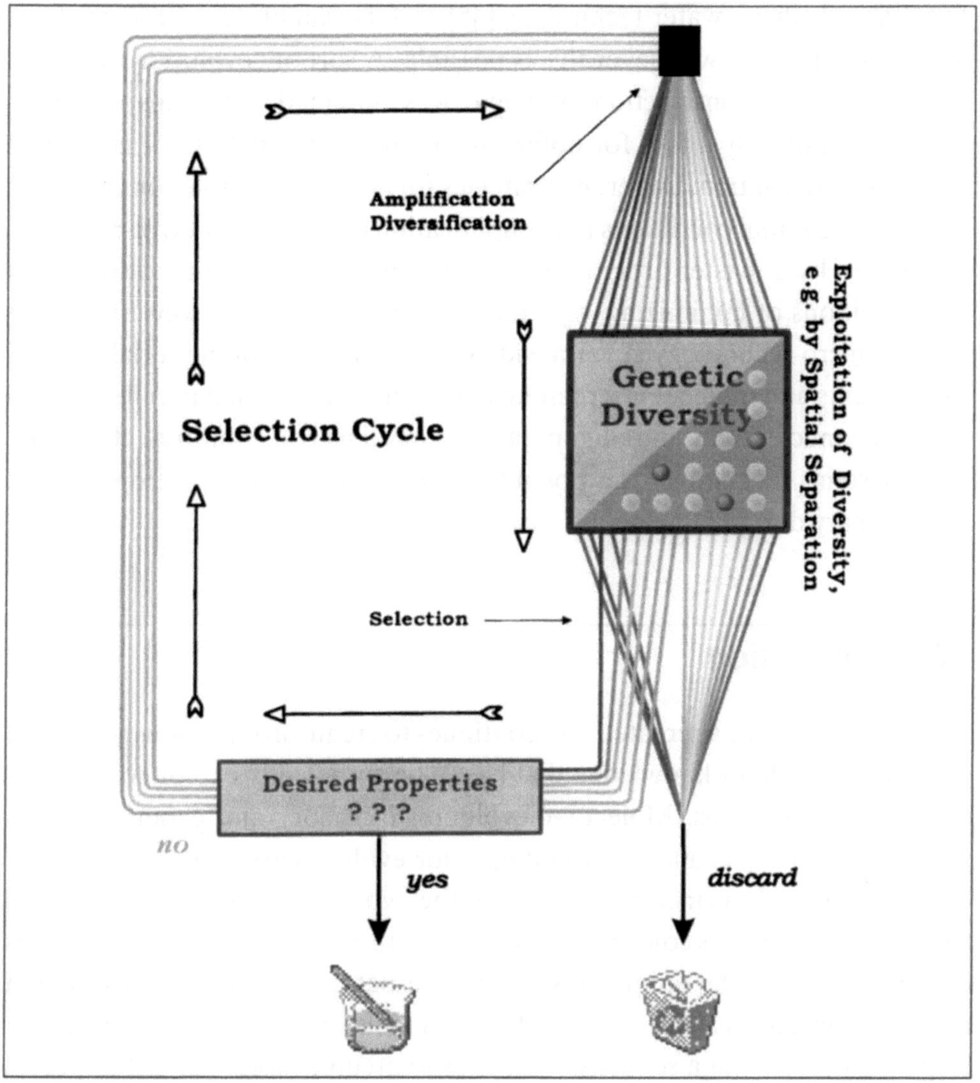

Figure 8 *The principle of an evolutionary strategy in molecular optimisation: In a series of selection cycles predetermined molecular properties and functions of biopolymers or other macromolecules are optimised iteratively. Each cycle consists of the following three phases: (i) amplification, (ii) diversification, and (iii) selection. Cycles are repeated until the desired result has been obtained, or no further improvement on consecutive selection rounds was observed*

by means of silicon wafer technology [45] and the samples are processed and analysed in parallel whenever this is possible. A sort of "evolution machine" with a modular construction principle was designed [59]. This evolution machine is not only a tool for optimizing biopolymers, but also it is itself an object of optimisation. Different strategies for accessing samples on the wafer, for choosing initial conditions and amplification protocols, for error-rates, etc. are chosen by a master program to optimize biopolymer optimisation under the conditions of minimizing, e.g., time and costs. Problems solvable by this kind of micro-technology cover a wide range from elaborating serial transfer experiments and molecular screening to computer controlled parallel synthesis of oligonucleotides and oligopeptides as well as combinatorial chemicals. Microsystems technology is a possible approach for future developments in evolutionary design of proteins *in vitro*.

22.6 Conclusion

Molecular biology has created and continues to create an enormously rich and fast growing field of knowledge that has to be sufficiently exploited for technological applications. Molecular evolution in theory and experiment can provide new insights and methodologies for evolutionary strategies in molecular evolution. In principle, the idea of the evolutionary approach is copied from nature: Darwin's concept of natural selection in laboratory systems. This concept possesses the unique advantage that detailed structural information on target molecules is not required for the success of the optimization. Three steps are repeated in a sequential manner to archive molecules with optimal properties:

- diversification through error-prone copying of biopolymers,
- selection of useful candidates from molecular populations according to predefined criteria, and
- amplification of selected mutants.

This process is continued until the desired result has been achieved or no further improvements are observed.

In order to be successful in creating and optimising biomolecules with an efficiency suitable for technological applications, further development of a theory of molecular evolution is required. A comprehensive knowledge on evolutionary dynamics and on structural biology is essential for a practical success. Most results on genotype – phenotype maps presented here were derived from coarse-grained structure representation, such as RNA secondary structures, random RNA-structure models, and lattice protein models. Generic properties such as the ratio of sequences to (compatible) structures and the existence of extended neutral networks turn out to be universal for biopolymers. Extending the "zoo" of sequence – structure relationships by protein structure prediction based on knowledge-based potentials of mean force completes the picture.

Studying the interplay between adaption and neutral evolution provides essential insights in the mechanisms of evolutionary dynamic. Neutral networks play a key-role in this scenario. They supply a powerful medium through which evolution can become really efficient. Adaptive walks of populations, generally ending in a nearby local optimum, are supplemented by random drift on neutral networks. Periods of neutral diffusion end when the population visit regions of better phenotypes. By subsequent phases of adaptive walks and neutral drift the global optimum is approached, provided that the neutral networks are sufficiently large. Tunable parameters in this scenario are, for example, composition of initial population, error-rate, optimisation strategy, constraints on the fitness landscape, etc., which are altered in order to optimize under the condition of external constraints such as time and costs. An "evolution machine" as technological tool for optimizing molecular function would have to consider all these aspects to be successful.

Acknowledgments

I thank Prof. Peter Schuster and Dr. Peter Stadler for fruitful discussions, helpful hints, and for providing data for Figure 3. The results presented here are derived from joint work with Dr. Andreas Schober and Dipl.-Biol. Ulrike Göbel, many thanks are due them. I also would like to thank Prof. Peter Hammerstein for the kind invitation and the community of the *Innovationskolleg Theoretische Biologie* at the Humboldt University of Berlin, Germany, for the pleasant stay.

References

1 Allain F-T, Varanin G (1997) How accurately and precisely can RNA structures be determined by NMR? *J Mol Biol* 267: 338–351

2 Babajide A, Hofacker IL, Sippl MJ, Stadler PF (1996) Neutral networks in protein space: A computational study based on knowledge-based potentials of mean force. *Folding & Design* 96: 12–085

3 Bartel DP, Szostak JW (1993) Isolation of new ribozymes from a large pool of random sequences. *Science* 261 (5127): 1411

4 Bauer A, Beyer A (1994) An improved pair potential to recognize native protein folds. *Proteins* 18: 254–261

5 Bauer G (1990) *Biochemische Verwirklichung und Analyse von kontrollierten Evolutionsexperimenten mit RNA-Qasispezies* in vitro. Doctoral thesis, Universität Braunschweig, Germany

6 Bauer GJ, McCaskill JS, Otten H (1989) Traveling waves of *in vitro* evolving RNA. *Proc Natl Acad Sci USA* 86: 7937–7941

7 Beaudry AA, Joyce Jul GF (1992) Directed evolution of an RNA-Enzyme. *Science* 257: 635–641

8 Biebricher CK (1983) Darvinian selection of self-replicating RNA molecules. *Evolutionary Biology* 16: 1–52

9 Biebricher CK, Eigen M (1988) Kinetics of RNA replication by Qβ-replicase. In: E Domingo, J Holland, P Ahlquist (eds): *RNA directed virus replication Vol. I of RNA Gentics*. CRC Press, Boca Raton, FL, 1–21

10 Bowie JU, Luthy R, Eisenberg D (1991) A method to identify protein sequences that fold into a known three-dimensional structure. *Science* 253: 164–170

11 Bryngelson J, Onuchic J, Socci N, Wolynes P (1995) Funnels, pathways, and the energy landscape of protein folding: A synthesis. *Proteins: Structure, Function and Genetics* 21: 167– 195

12 Camacho CJ, Thirumalai D (1993) Minimum energy compact structure of random sequences of heteropolymers. *Phys Rev Lett* 71(15): 2505–2508

13 Casari G, Sippl MJ (1992) Structure-derived hydrophobic potentials – hydrophobic potentials derived from X-ray structures of globular proteins is able to indentify native folds. *J Mol Biol* 224: 725–732

14 Chan H, Dill K (1991) Sequence space soup. *J Chem Phys* 95: 3775–3787

15 Chan H, Dill K (1996) Comparing folding codes for proteins and polymers. *Proteins* 24: 335–344

16 Chothia C (1992) Proteins. One thousand families for the molecular biologist. *Nature* 357: 543–544

17 Devlin J, Pangaiban L, Devlin P (1990) Random peptide libraries: A source of specific protein binding molecules. *Science* 249: 404–406

18 Dill K, Bromberg S, Yue K, Fiebig K, Yeo D, Thomas P, Chan H (1995) Principles of protein folding: A perspective from simple exact models. *Prot Sci* 4: 561–602

19 Dill KA, Bromberg S, Yue K, Fiebig KM, Yeo DP, Thomas PD, Chan HS (1995) Principles of protein folding: a perspective from simple exact models. *Prot Sci* 4: 561–602

20 Drexler KE (1981) Molecular engineering: An approach to the development of general capabilities for molecular manipulation. *Proc Natl Acad Sci* 78: 5275–5278

21 Eigen M (1971) Selforganization of matter and the evolution of biological macromolecules. *Die Naturwissenschaften* 10: 465–523

22 Eigen M, Gardiner W (1984) Evolutionary molecular engineering based on RNA replication. *Pure & Appl Chem* 56: 967–978

23 Eigen M, McCaskill J, Schuster P (1988) Molecular Quasi-Species. *Journal of Physical Chemistry* 92: 6881–6891

24 Ellington AD (1997) Ribozymes in wonderland. *Science* 276 (5312): 546

25 Ellington AD, Szostak JW (1990) *In vitro* selection of RNA molecules that bind specific ligands. *Nature* 346: 818–822

26 Fontana W, Schnabl W, Schuster P (1989) Physical aspects of evolutionary optimization and adaption. *Physical Review A* 40 (6): 3301–3321

27 Fontana W, Schuster P (1987) A computer model of evolutionary optimization. *Biophysical Chemistry* 26: 123–147

28 Forst CV, Reidys C, Weber J (1995) Evolutionary dynamics and optimization: Neutral Networks as model-landscape for RNA secondary-structure folding-landscapes. In: F Morán, A Moreno, J Merelo, P Chacón (eds): *Advances in Artificial Life*, vol 929 of *Lecture Notes in Artificial Intelligence*, p 128–147, Berlin, Heidelberg, New York. ECAL '95, Springer, Santa Fe Preprint 95-10-94

29 Göbel U, Forst CV, Schuster P (1997) Structural constraints and neutrality in RNA. In: R Hofestädt (ed): *LNCS/LNAI Proceedings of GCB96*, Lecture Notes in Computer Science, Berlin, Heidelberg, New York. Springer-Verlag

30 Godzik A, Kolinski A, Skolnick J (1992) Topology fingerprint approach to the inverse folding problem. *J Mol Biol* 227: 227–238

31 Goldstein R, Luthey-Schulten Z, Wolynes P (1992) Protein tertiary structure recognition using optimized Hamiltonians with local interaction. *Proc Natl Acad Sci USA* 89: 9029–9033

32 Grossman T, Farber R, Lapedes A (1995) Neural net representations of empirical protein potentials. *Ismb* 3: 154–61

33 Grüner W, Giegerich R, Strothmann D, Reidys C, Weber J, Hofacker IL, Stadler PF, Schuster P (1996) Analysis of RNA sequence structure maps by exhaustive enumeration I. Neutral networks. *Monatsh Chem* 127: 355–374

34 Grüner W, Giegerich R, Strothmann D, Reidys C, Weber J, Hofacker IL, Stadler PF, Schuster P (1996) Analysis of RNA sequence structure maps by exhaustive enumeration II. Structures of neutral networks and shape space covering. *Monatsh Chem* 127: 375–389

35 Hendlich M, Lackner P, Weitckus S, Floeckner H, Froschauer R, Gottsbacher K, Casari G, Sippl MJ (1990) Identification of native protein folds amongst a large number of incorrect models – the calculation of low energy conformations from potentials of mean force. *J Mol Biol* 216: 167–180

36 Hofacker IL, Fontana W, Stadler PF, Bonhoeffer S, Tacker M, Schuster P (1994) Fast folding and comparison of RNA secondary structures. *Monatshefte f Chemie* 125 (2): 167–188

37 Hofacker IL, Schuster P, Stadler P (1998) Combinatorics of RNA secondary structures. *SIAM J Disc Math* 88: 207–237

38 Holm L, Sander C (1997) Dali/FSSP classification of three-dimensional protein folds. *Nucl Acids Res* 25: 231–234

39 Horwitz M, Loeb L (1986) Promotors selected from random DNA sequences. *Proc Natl Acad Sci USA* 83: 7405–7409

40 Huynen MA, Stadler PF, Fontana W (1996) Smoothness within ruggedness: The role of neutrality in adaption. *Proc Natl Acad Sci* 93: 397–401

41 Joyce GF (1992) Directed molecular evolution. *Sci Am* 267 (6): 48–55

42 Kamtekar S, Schiffer JM, Xiong H, Babik JM, Hecht MH (1993) Protein design by binary patterning of polar and nonpolar amino acids. *Science* 262: 1680–1685

43 Kauffman SA (1986) Autocatalytic Sets of Proteins. *J Theor Biol* 119: 1–24

44 Kauffman SA (1992) Applied Molecular Evolution. *J Theor Biol* 157: 1–7

45 Köhler JM, Pechmann R, Scharper A, Schober A, Jovin TM, Thürk M, Schwienhorst A (1995) Micromechanical elements for the detection of molecules and molecular design. *Microsystems Technol*, 202–208

46 Kopp S, Reidys CM, Schuster P (1996) Exploration of artificial landscapes based on random graphs. In: F Schweitzer (ed): *Self-Organizati-*

on of Complex Structures: From Inividual to Collective Behavior, Gordon and Breach, London, UK

47 Leclerc F, Srinivasan J, Cedergren R (1997) Predicting RNA structures: The model of the RNA element binding Rev meets the NMR structure. *Folding & Design* 2: 141–147

48 Levinthal C (1969) How to fold graciously. In: P Debrunner, J Tsbiris, E Munck (eds): *Mössbauer Spectroscopy in biological Systems*, pp 22–24, Urbana, IL. Proceedings of a Meeting Held at Allerton House, Monticelli, IL, University of Illinois Press

49 Li H, Helling R, Tang C, Wingreen N (1996) Emergence of preferred structures in a simple model of protein folding. *Science* 273: 666–669

50 McCaskill JS (1984) A localization threshold for macromolecular quasispecies from continuously distributed replication rates. *J Chem Phys* 80: 5194–5202

51 Miller S, Orgel L (1974) *The Origin of Life on the Earth*. Prentice Hall

52 Murzin AG (1994) New protein folds. *Curr Opin Struct Biol* 4: 441–449

53 Murzin AG (1996) Structural classification of proteins: new superfamilies. *Curr Opin Struct Biol* 6: 386–394

54 Nowak M, Schuster P (1989) Error tresholds of replication in finite populations, mutation frequencies and the onset of Muller's ratchet. *Journal of theoretical Biology* 137: 375–395

55 Pelletier J, Sonenberg N (1988) Internal initiation of translation of eukaryotic mRNA directed by a sequence derived from poliovirus RNA. *Nature* 334: 320–325

56 Reidys C (1995) *Neutral Networks of RNA Secondary Structures*. PhD thesis, Friedrich Schiller Universität, Jena

57 Reidys C, Stadler PF, Schuster P (1997) Generic properties of combinatory maps and neutral networks of RNA secondary structures. *Bull Math Biol* 59(2): 339–397

58 Rost B, Sander C (1996) Bridging the protein sequence-structure gap by structure predictions. *Ann Rev Biophys* 25: 113–136

59 Schober A, Schwienhorst A, Köhler J, Fuchs M, Günther R, Thürk M (1995) Microsystems for independent parallel chemical and biological processing. *Microsystems Technol*, p 168– 172

60 Schober, A., Thürk, M Eigen, M (1993) Optimization by Hierarchical Mutant Production. *Biol Cybern* 60: 493–501

61 Schober A, Walter N, Tangen U, Strunk G, Ederhof T, Daprich MEJ (1995) Multichannel PCR and serial transfer machine as a future tool in evolutionary biotechnology. *Bio Techniques* 18(4): 652–658

62 Schuster P (1989) Optimization and complexity in molecular biology and physics. In: PJ Plath (ed): *Optimal Structures in Heterogenous Reaction Systems*. Springer-Verlag: Springer Series in Synergetics

63 Schuster P (1995) Artificial life and molecular evolutionary biology. In: F Morán, A Moreno, J Merelo, P Chacón (eds): *Advances in Artificial Life*, Lecture Notes in Artificial Intelligence, Berlin, Heidelberg, New York. ECAL '95, Springer

64 Schuster P (1997) Landscapes and molecular evolution. *Physica D* 107: 351–356

65 Schuster P, Fontana W, Stadler PF, Hofacker IL (1994) From sequences to shapes and back: A case study in RNA secondary structures. *Proc Roy Soc (London) B* 255: 279–284

66 Schuster P, Swetina J (1988) Stationary mutant distributions and evolutionary optimization. *Bull Math Biol* 50: 635

67 Shakhnovich E, Abkevich V, Ptisyn O (1996) Conserved residues and the mechanism of protein folding. *Nature* 379: 96–98

68 Sippl MJ (1990) Calculation of conformational ensembles from potentials of mean force – an approach to the knowledge-based prediction of local structures in globular proteins. *J Mol Biol* 213: 859–883

69 Sippl MJ (1993) Boltzmann's principle, knowledge-based mean fields and protein folding. an approach to the computational determination of protein structures. *J Computer-Aided Molec Design* 7: 473–501

70 Sippl MJ (1993) Recognition of errors in three-dimensional structures of proteins. *Proteins* 17: 355–362. URL: http://lore.came.sbg.ac.at/Extern/software/Prosa/prosa.html

71 Soumillion P, Jespters L, Bouchet J, Marchand-Brynaert J, Winter G, Fastrez J (1994) Selection of β-lactamase on filamentous bacteriophage by catalytic activity. *J Mol Biol* 237: 415–422

72 Sun S, Brem R, Chan HS, Dill KA (1995) Designing amino acid sequences to fold with good hydrophobic cores. *Protein Eng* 8: 1205–1213

73 Swetina J, Schuster P (1982) A model for polynucleotide replication. *Biophys Chem* 16: 329–345

74 Tuerk C, Gold L (1990) Systematic evolution of ligands by exponential enrichment: RNA ligands to bacteriophage T4 DNA polymerase. *Science* 249: 505

75 Waterman MS (1978) Secondary structure of single-stranded nucleic acids. *Studies on foundations and combinatorics, Advances in mathematics supplementary studies,* Academic Press NY, 1: 167–212

76 Weissmann C (1974) The making of a phage. *FEBS Letters* (Suppl), 40: S10–S12

77 Zell R, Stelzner A (1997) Application of genome sequence information to the classification of bovine enteroviruses: the importance of 5′- and 3′-nontranslated regions. *Virus Research,* 51: 213–229

78 Zipf G (1949) *Human Behaviour and the Principle of Least Effort.* Addison-Wesley, Reading (Mass.)

23 Appendix

Udo Luhmann

Selected patent rights

The following table is a compilation of patents and patent applications (summarized as "patent rights") dealing with the use of microsystem technology for biomolecular problems. Due to the huge number of patent applications filed and patents issued recently in this technical field it was unavoidable to make a more or less willful choice and to select a limited number as representative examples of relevant patent rights.

The table is arranged according to applicants (or owner of the patent rights). The cited patent rights are German (DE), European (EP) or US ones or are world patent applications (WO). It is to be noted that any patent right actually cited may only be one member of a great patent family consisting of equivalent patent rights in other countries. The filing date is arranged in the order year/month/day, i.e. 970123 means January 23, 1997.

Nr.	Applicant of Patent or Patent Application	Inventor (s)	Title	No. of Patent or Patent Application	Filing Date
1.	Affymax Technologies N.V., NL	Anderson, Rolfe C. Lipschutz, Robert J. Rava, Richard P. Fodor, Stephen P.A.	Integrated nucleic acid diagnostic device	WO 97/02357	970123
2.	Affymax Technologies N.V., NL	Besemer, Donald M. Goss, Virginia W. Winkler, James L.	Bioarray chip reaction apparatus and its manufacture	WO 95/33846	951214
3.	Affymax Technologies N.V., NL	Chee, Mark Cronin, Maureen T. Fodor, Stephen P.A. Gingeras, Thomas R. Huang, X.C. Hubbell, Earl A. Lipschutz, Robert J. Lobban, P.E. Miyada, C.G. Morris, M.S. Shah, N. Sheldon, E.L.	Arrays of nucleic acid probes on biological chips	WO 95/11995	950504
4.	Affymax Technologies N.V., NL	Fodor, Stephen P.A. Lipschutz, Robert J. Huang, X. Jevons, Luis C.	Hybridization and sequencing of nucleic acids	WO 95/00530	950105
5.	Affymax Technologies N.V., NL	McGall, Glenn H. Fodor, Stephen P.A.	Spatially-addressable immobilization of	WO 93/22680	931111

		Sheldon, Edward L.	oligonucleotides and other biological polymers on surfaces		
6.	Affymax Technologies N.V., NL	Winkler, James L. Fodor, Stephen P.A. Buchko, Christopher J. Ross, Debra A. Aldwin, Lois Modlin, Douglas N.	Combinatorial strategies for polymer synthesis	WO 93/09668	930527
7.	Affymax Technologies N.V., NL	Schatz, Peter J. Cull, Millard G. Miller, Jeff F. Stemmer, Willem P.	Peptide library and screening method	WO 93/08278	93 0429
8.	Affymax Technologies N.V., NL	Dower, William J. Fodor, Stephen P.A.	Sequencing of surface immobilized polymers utilizing microfluorescence detection	WO 92/10587	920625
9.	Affymax Technologies N.V., NL	Fodor, Stephen P.A. Stryer, Lubert Winkler, James L. Holmes, Christopher W. Solas, Dennis W.	Very large scale immobilized polymer synthesis	WO 92/0092	920625
10.	Affymax Technologies N.V., NL	Barrett, Ronald W. Pirrung, Michael C. Stryer, Lubert Holmes, Christopher P. Sundberg, Steven A.	Spatially-addressable immobilization of anti-ligands on surfaces	WO 91/07087	910530

Nr.	Applicant of Patent or Patent Application	Inventor (s)	Title	No. of Patent or Patent Application	Filing Date
11.	Affymax Technologies N.V., NL	Pirrung, Michael C. Read, J. L. Fodor, Stephen P.A. Stryer, Lubert	Very large scale immobilized peptide synthesis	WO 90/15070	901213
12.	Affymax Technologies N.V., NL	Chee, Mark Cronin, Maureen T. Fodor, Stephen P.A. Gingeras, Thomas R. Huang, X.C. Hubbell, E.A. Lipschutz, R.J. Lobban, P.E. Miyada, C.G. Morris, M.S. Shah, N. Sheldon, E.L.	Arrays of nucleic acid probes on biological chips	EP 730663	960911
13.	Affymax Technologies N.V., NL	Pease, R. Fabian Fodor, Stephen P.A. McGall, Glenn Goss, Virginia Goldberg, Martin J. Stryer, Lubert Rava, Richard P. Winkler, James L.	Printing molecular library arrays	EP 728520	960828
14.	Affymax Technologies N.V., NL	Sugarman, J.H. Rava, Richard P.	Synthesizing and screening molecular	EP 726906	960821

		Kedar, Haim Dower, William J. Barrett, Ronald W. Gallop, Mark A. Needels, Michael C.	diversity		
15.	Affymax Technologies N.V., NL	Lockhart, David J. Chee, Mark S. Vetter, Dirk Diggelmann, Martin	Nucleic acid library arrays, methods for synthesizing them and methods for sequencing and sample screening using them	EP 721016	960710
16.	Affymax Technologies N.V., NL	Fodor, Stephen P.A. Lipschutz, Robert Huang, Xiaohua Jevons, Luis C.	Hybridization and sequencing of nucleic acids	EP 705271	960410
17.	Affymax Technologies N.V., NL	Winkler, James L. Fodor, Stephen P.A. Buchko, Christopher J. Ross, Debra A. Aldwin, Lois Modlin, Douglas N.	Combinatorial strategies for polymer synthesis	EP 624059	941117
18.	Affymax Technologies N.V., NL		Very large scale immobilized peptide synthesis	EP 619321	941012
19.	Affymax Technologies N.V., NL	Schatz, Peter J. Cull, Millard G. Miller, Jeff F. Stemmer, Willem P.	Peptide library and screening method	EP 610448	940817

Nr.	Applicant of Patent or Patent Application	Inventor (s)	Title	No. of Patent or Patent Application	Filing Date
20.	Affymax Technologies N.V., NL	Dower, William J. Barrrett, Ronald W. Gallop, Mark A. Needels, Michael C.	Method of synthesizing diverse collections of oligomers	EP 604552	940706
21.	Affymax Technologies N.V., NL	Fodor, Stephen P.A. Solas, Dennis W. Dower, William J.	Sequencing by hybridization of a target nucleic acid to a matrix of defined oligonucleotides	EP 562047	930929
22.	Affymax Technologies N.V., NL	Fodor, Stephen P.A. Stryer, Lubert Winkler, James L. Holmes, Christopher W. Solas, Dennis W.	Very large scale immobilized polymer synthesis	EP 562025	930929
23.	Affymax Technologies N.V., NL	Barrett, Ronald W. Pirrung, Michael C. Stryer, Lubert Holmes, Christopher P. Sundberg, Steven A.	Spatially-addressable immobilization of anti-ligands on surfaces	EP 502060	920909
24.	Affymax Technologies N.V., NL	Pirrung, Miachael C. Read, J.L. Fodor, Stephen P.A. Stryer, Lubert	Very large scale immobilized peptide synthesis	EP 476014	920325
25.	Affymax Technologies N.V., NL	Dower, William J. Fodor, Stephen P.A.	Sequencing of surface immobilized polymers utilizing microfluorescence detection	US 5,547,839	960820

26.	Affymax Technologies N.V., NL	Rava, Richard Fodor, Stephen P.A. Trulson, Mark	Methods for making a device for concurrently processing multiple biological chip assays	US 5,545,531	960813
27.	Affymax Technologies N.V., NL	Fodor, Stephen P.A. Pirrung, Michael C. Read, J.L. Stryer, Lubert	Synthesis and screening of immobilized oligonucleotide arrays	US 5,510,270	960423
28.	Affymax Technologies N.V., NL	Fodor, Stephen P.A. Stryer, Lubert Winkler, James L. Holmes, Christopher W. Solas, Dennis W.	Photolabile nucleoside and peptide protecting groups	US 5,489,678	960206
29.	Affymax Technologies N.V., NL	Barrett, Ronald W. Pirrung, Michael C. Stryer, Lubert Holmes, Christopher P. Sundberg, Steven A.	Spatially-addressable immobilization of anti-ligands on surfaces	US 5,482,867	960109
30.	Affymax Technologies N.V., NL	McGall, H. Fodor, Stephen P.A. Sheldon, Edward L.	Spatially-addressable immobilization of oligonucleotides and other biological polymers on surfaces	US 5,412,087	950502
31.	Affymax Technologies N.V., NL	Pirrung, Michael C. Read, J.L. Fodor, Stephen P.A. Stryer, Lubert	Large scale photolithographic solid phase synthesis of an array of polymers	US 5,405,783	950411

Nr.	Applicant of Patent or Patent Application	Inventor (s)	Title	No. of Patent or Patent Application	Filing Date
32.	Affymax Technologies N.V., NL	Fodor, Stephen P.A. Mazzola, Laura T.	Method and apparatus for measuring binding affinity	US 5,324,633	940628
33.	Affymax Technologies N.V., NL	Barrett, Ronald W. Pirrung, Michael C. Stryer, Lubert Holmes, Christopher P. Sundberg, Steven A.	Spatially-addressable immobilzation of anti-ligands on surfaces	US 5,252,743	931012
34.	Affymax Technologies N.V., NL	Pirrung, Michael C. Read, J.L. Fodor, Stephen P.A. Stryer, Lubert	Large scale photolithographic solid phase synthesis of polypeptides and receptor binding screening thereof	US 5,143,854	920901
1.	Trustees of the University of Pennsylvania, US	Wilding, Peter Kricka, Larry J.	Mesoscale polynucleotide amplification devices	WO 96/15269	960523
2.	Trustees of the University of Pennsylvania, US	Kricka, Larry J. Wilding, Peter	Mesoscale devices for analysis of motile cells	WO 96/14933	960523
3.	Trustees of the University of Pennsylvania, US	Wilding, Peter Kricka, Larry J.	Polynucleotide amplification analysis using a microfabricated device	WO 93/22058	931111
4.	Trustees of the University of Pennsylvania, US	Wilding, Peter Kricka, Larry J.	Mesoscale polynucleotide amplification devices	EP 739423	961030

5.	Trustees of the University of Pennsylvania, US	Wilding, Peter Kricka, Larry J.	Mesoscale sample preparation device and systems for determination and processing of analytes	EP 739240	961030
6.	Trustees of the University of Pennsylvania, US	Kricka, Larry J. Wilding, Peter	Microfabricated sperm handling devices	EP 639223	950222
7.	Trustees of the University of Pennsylvania, US	Wilding, Peter Kricka, Larry J.	Polynucleotide amplification analysis using a microfabricated device	EP 637999	950215
8.	Trustees of the University of Pennsylvania, US	Wilding, Peter Kricka, Larry J. Zemel, J.N.	Fluid handling in microfabricated analytical devices	EP 637998	930215
9.	Trustees of the University of Pennsylvania, US	Wilding, Peter Kricka, Larry J.	Analysis based on flow restriction	EP 637997	950215
10.	Trustees of the University of Pennsylvania, US	Wilding, Peter Kricka, Larry J. Zemel, J.N.	Microfabricated detection structures	EP 637996	950215
11.	Trustees of the University of Pennsylvania, US	Wilding, Peter Kricka, Larry J.	Mesoscale polynucleotide amplification devices	US 5,587,128	961224
12.	Trustees of the University of Pennsylvania, US	Wilding, Peter Kricka, Larry J.	Mesoscale polynucleotide amplification device and method	US 5,498,392	960312

Nr.	Applicant of Patent or Patent Application	Inventor (s)	Title	No. of Patent or Patent Application	Filing Date
13.	Trustees of the University of Pennsylvania, US	Wilding, Peter Kricka, Larry J. Zemel, J.N.	Fluid handling in mesoscale analytical devices	US 5,304,487	940419
1.	Evotec Biosystems GmbH, DE	Günther, Rolf	Method and device for determining substance-specific parameters of one or a plurality of molecules by correlation-spectroscopy	WO 96/13744	960509
2.	Evotec Biosystems GmbH, DE	Eigen, Manfred Rigler, Rudolf Henco, Karsten	Process and device for selectively extracting components from complex mixtures	WO 95/35492	951228
3.	Evotec Biosystems GmbH, DE	Eigen, Manfred Henco, Karsten Schober, Andreas Schwienhorst, Andreas Köhler, Michael Thürk, Marcel Günther, Rolf Döring Michael	Sample holder and its use	WO 95/01559	950112
4.	Evotec Biosystems GmbH, DE	Rigler, Rudolf Eigen, Manfred Henco, Karsten	Method and device for assessing the suitability of biopolymers	WO 94/16313	940721

5.	Evotec Biosystems GmbH, DE	Eigen, Manfred Henco, Karsten Schober, Andreas Schwienhorst, Andreas Köhler, Michael Thürk, Marcel Günther, Rolf Döring Michael	Sample holder and its use.	EP 706646	960417
6.	Evotec Biosystems GmbH, DE	Rigler, Rudolf Eigen, Manfred Henco, Karsten	Method and device for assessing the suitability of biopolymers	EP 679251	951102
7.	Evotec Biosystems GmbH, DE	Henco, Carsten Eigen, Manfred Lindemann, Björn Schwienhorst, Andreas	Evolutive vaccination	EP 667917	950823
1.	Diagen Institut für Molekular-Biologische Diagnost., DE	Henco, Karsten Eigen, Manfred	Support material for simultaneous binding of genotypic and phenotypic substances	WO 94/10572	940511
2.	Diagen Institut für Molekular-Biologische Diagnost., DE	Eigen, Manfred Henco, Karsten Mc Caskill, John	Process for separating substances from dilute solutions and suspensions	WO 94/10564	940511
3.	Diagen Institut für Molekular-Biologische Diagnost., DE	Henco, Karsten Eigen, Manfred Riesner, Detlev	Process for determining (*in vitro*) amplified nucleic acids	WO 93/16194	930819
4.	Diagen Institut für Molekular-Biologische Diagnost., DE	Henco, Karsten Eigen, Manfred	Method for preparing new biopolymers	WO 92/18645	921029

Nr.	Applicant of Patent or Patent Application	Inventor (s)	Title	No. of Patent or Patent Application	Filing Date
5.	Diagen Institut für Molekular-Biologische Diagnost., DE	Rigler, Rudolf Eigen, Manfred Henco, Karsten	Method and apparatus for evaluating the fitness of biopolymers	DE 4301005	940721
1.	Fraunhofer-Gesellschaft e.V.	Fuhr, Günther Voigt, Andreas Hagedorn, Rolf Schnelle, Thomas Hornung, Jan Müller, Torsten Fiedler, Stefan Glasser, Henning Wagner, Bernd	Shaping of microparticles in electric-field cages	WO 95/23020	950831
2.	Fraunhofer-Gesellschaft e.V.	Fuhr, Günther Voigt, Andreas Hagedorn, Rolf Lisec, Thomas Müller, Torsten Wagner, Bernd	Ultra-miniaturised surface structures with controllable adhesion	WO 95/17258	950629
3.	Fraunhofer-Gesellschaft e.V.	Fuhr, Günther Voigt, Andreas Hagedorn, Rolf Schnelle, Thomas Hornung, Jan Müller, Torsten Fiedler, Stefan	Shaping of microparticles in electric-field cages	EP 746408	961211

		Glasser, Henning Wagner, Bernd			
4.	Fraunhofer-Gesellschaft e.V.	Fuhr, Günther Voigt, Andreas Hagedorn, Rolf Lisec, Thomas Müller, Torsten Wagner, Bernd	Ultra-miniaturised surface structures with controllable adhesion	EP 735923	961009
5.	Fraunhofer-Gesellschaft e.V.	Benecke, Wolfgang Wagner, Bernd Hagedorn, Rolf Fuhr, Günther Müller, Torsten	Apparatus for separating mixtures of microscopic small dielectric particles dispersed in a fluid or a gel	EP 96103459	920819
1.	Ciba Geigy AG, CH;	Oroszlan, Peter Erbacher, Christoph Duveneck, Gert L. Verpoorte, Elisabeth	Flow cell	WO 97/01087	970109
2.	Ciba Geigy AG, CH	Rink, Hans Vetter, Dirk Gercken, Bertold Felder, Eduard	Process for the production of combinatorial compound libraries	WO 96/30392	961003
3.	Ciba Geigy AG, CH	Ensing, Kees Oroszlan, Peter Paulus, Aran Effenhauser, Carlo S.	Device and method for combined bioaffinity assay and electrophoretic separation	EP 671626	950913

Nr.	Applicant of Patent or Patent Application	Inventor (s)	Title	No. of Patent or Patent Application	Filing Date
4.	Ciba Geigy AG, CH	Manz, Andreas Effenhauser, Carlo S.	Apparatus and method for the electrophoretical separation of mixtures of fluid substances	EP 653631	950517
1.	Howitz, Steffen Bürger, Mario Wegener, Thomas	Howitz, Steffen Bürger, Mario Wegener, Thomas	Electrically controllable micro-pipette	WO 96/24040	960808
2.	Howitz, Steffen Pham, Minh T.	Howitz, Steffen Pham, Minh T.	Fluid micro-diode	WO 95/22696	950824
3.	Forschungszentrum Rossendorf e.V., DE	Howitz, Steffen Bürger, Mario Wegener, Thomas	Electrically controlled micro-pipette	EP 725267	960807
4.	Forschungszentrum Rossendorf e.V., DE	Howitz, Steffen Pham, Minh T. Fiehn, Hendrik	Microcapillary with inbuilt chemical microsensors and method for its manufacture	EP 633468	950111
1.	Nanogen, Inc., USA	Heller, Michael J. Tu, Eugene Montgomery, Donald D. Butler, William F.	Automated molecular biological diagnostic system including generator, microelectronic relay system, and electrodes	WO 96/07917	960314

2.	Nanogen, Inc., USA	Heller, Michael J. Tu, Eugene Evans, Glen A. Sosnowski, Ronald G.	Self-addressable self-assembling microelectronic systems and devices for molecular biological analysis and diagnostics	WO 96/01836	960125
3.	Nanogen, Inc., USA	Heller, Michael J. Tu, Eugene	Self-addressable self-assembling microelectronic systems and devices for molecular biological analysis and diagnostics	WO 95/12808	950511
1.	Selectide Corporation	Krchnak, Viktor Lebl, Michal Seligmann, Bruce	Apparatus and method for multiple synthesis of organic compounds on polymer support	WO 96/22157	960725
2.	Selectide Corporation	Lebl, Michal	Combinatorial libraries having a predetermined frequency of each species of test compound	WO 96/18903	960620
3.	Selectide Corporation	Lebl, Michal Lam, Kit S. Salmon, Sydney E. Krchnak, Viktor Sepetov, Nikolai Kocis, Peter	Topologically segregated, encoded solid phase libraries	EP 705279	960410
1.	The Regents of the University of California, US	Northrup, M. Allen Mariella, Raymond P. Carrano, Anthony V. Balch, Joseph W.	Silicon-based sleeve devices for chemical reactions	WO 97/00726	970109

Nr.	Applicant of Patent or Patent Application	Inventor (s)	Title	No. of Patent or Patent Application	Filing Date
2.	The Regents of the University of California, US	Northrup, M. Allen White, Richard M.	Microfabricated reactor	WO 94/05414	940317
3.	The Regents of the University of California, US	Northrup, M. Allen White, Richard M.	Microfabricated reactor	EP 711200	960515
1.	Affymetrix, Inc. US	Anderson, Rolfe C. Lipschutz, Robert J. Rava, Richard P. Fodor, Stephen P.A.	Integrated nucleic acid diagnostic device	WO 97/02357	970123
2.	Affymetrix, Inc., CA		Surface-bound, unimolecular, double-stranded DNA	US 5,556,752	960917
1.	Beckmann Instruments, Inc.	Caskey, Charles T. Matson, Robert S. Coassin, Peter J Rampal, Jang B.	Oligonucleotide repeat arrays	WO 95/30774	951116
2.	Beckmann Research Institute of the City of Hope, US	Corbett, John Michael Reed, Kenneth C. Riggs, Arthur D.	Device and method for the automated cycling of solutions between two or more temperatures	WO 92/13967	920820
1.	IPHT e.V., DE	Baier, Volker Bodner, Ulrich Dillner, Ulrich Köhler, Michael Poser, Siegfried Schimkat, Dieter Schulz, Torsten	Miniaturisierter Mehrkammer-Thermocycler (Miniaturized multi chamber thermocycler)	DE 19519015	960905

2.	IPHT e.V., DE	Keßler, Ernst Köhler, Johann Michael Steinhage, Gudrun	Probenaufnehmer und Sensor für die Scanning-Kalorimetrie (Sample collector and sensor for scanning calorimetry)	DE 4429067	940817
1.	Pharmacia Biosensor Ab, SE	Hansson,Thord Sjlander, Stefan	Valve, especially for fluid handling bodies with microflow channels	WO 95/07425	950316
2.	Pharmacia Biosensor Ab, SE	Ekström, Björn Jacobson, G. Öhman, O. Sjödin, H.	Microfluidic structure and process for its manufacture	WO 91/16966	911114
	Abbott Laboratories, USA	Stroupe, Stephen D.	Scanning probe microscopy immunoassay and test kit	WO 92/15709	920917
	Arizona Board of Regents, USA; City University of New York	Lindsay, Stuart M. Philipp, Manfred	Scanning probe microscopy method for visualizing the base sequence of nucleic acids	EP 410618	910130
	Biometra biomedizinische Analytik GmbH, DE	Baier, Volker Bodner, Ulrich Dillner, Ulrich Köhler, Michael Poser, Siegfried Schimkat, Dieter	Miniaturisierter Fluß-Thermocycler (Miniaturized flow thermocycler)	DE 4435107	940930

Nr.	Applicant of Patent or Patent Application	Inventor (s)	Title	No. of Patent or Patent Application	Filing Date
	California Institute of Technology, USA	Baldeschwieler, John D. Baselt, David Unger, Mark A. O'Connor, Stephen D.	Probes for sensing and manipulating microscopic environments and structures	WO 96/38705	961205
	H & N Instruments, Inc. US	Nishioka, Gary M.	Synthesis of chain chemical compounds	WO 93/02992	930218
	International Business Machines Corp., USA	Binning, Gerd K. Haeberle, Walter Rohrer, Heinrich Smith, Douglas P. E.	Fine positioning apparatus with atomic resolution	WO 96/07074	960307
	Leland Stanford Junior University, USA	Mathies, Richard A. Peck, Konan Stryer, Lubert	High sensitivity fluorescent single particle and single molecule detection apparatus and method	WO 90/14589	901129
	Max-Planck-Gesellschaft Berlin	Eigen, Manfred Rigler, Rudolf	Detection of nucleic acid sequences at very low concentrations using multiple probes with fluorescent labels	EP 731173	960911
	Messerschmitt-Bölkow-Blohm, DE	Kroy, Walter, Seidel, Helmut Dette, Eduard Deimel, Peter Binder, Florian Hilpert, Reinhold	Mikromechanische Struktur (Micromechanical structure)	DE 39 15920	890516

Ro Institut Za Molekularnu Genetiku i Geneticko Inzenjerstvo. Beograd	Drmanac, Radoje T. Crkvenjakov, Radomir B.	Process for determination of a complete or partial contents of very short sequences in the samples of nucleic acids connected to the discrete particles of microscopic size by hybridization with oligonucleotide probes	EP 392546	901017
Sequenom, Inc.	Köster, Hubert Tang, Kai Fu, Dong-Jing Siegirt, Carsten Little, Daniel P. Higgins, G.S. Braun, Andreas Darnhofer-Demar, Brigitte Jurinke, Christian	DNA diagnostics based on mass spectrometry	WO 96/29431	960926
Sigrist, Hans, CH Klingler-Dabral, Vibhuti Dolder, Max Wegmüller, Bernhard	Sigrist, Hans Klingler-Dabral, Vibhuti Dolder, Max Wegmüller, Bernhard	Method for the light-induced immobilization of biomolecules on chemically "inert" surfaces	WO 91/16425	911031
Sri International, US	Tsien, Roger Y. Ross, Pepi Fahnestock, Margaret Johnston, Allan J.	DNA sequencing	WO 91/06678	910516

Nr.	Applicant of Patent or Patent Application	Inventor (s)	Title	No. of Patent or Patent Application	Filing Date
	Trustees of Boston University	Cantor, Charles R. Koster, Hubert Smith, Cassandra Fu, Dong-Jing	Solid phase sequencing of biopolymers	WO 96/32504	961017
	Unilever, PLC, GB/NL	Izzard, Martin Wilding, Peter Lane, S.J. Patrick, A.J. Hammond, K.A.	Moisture barrier and its preparation	EP 471558	920219
	United States Dept. of Commerce, USA	Cavicchi, Richard Semancik, Stephen Suehle, John S. Gaitan, Michael	Application of microsubstrates for materials processing	WO 94/10821	940511

Index

actin 464
advanced silicon etching (ASE) plasma etching 37
algorithm 258
alkyl silane 103
aminoacylation 512
amplification 509, 515
amplified primer extension (APEX) 338
anisotropic wet etching 35
anodic bonding 93
antibacterial compound 225
antinoistic compound 225
aptamer 512
arrayed primer extension (APEX) method 387
arraying of libraries 215
autocalibrating procedure 336
automation 22

BAC 200
basepairing 521
bead-binding assay 270
binary encoding 271
binding constant 512
bioinformatics 477
biomolecular function 507
biomolecular interaction 507
biotechnological application 331
biotechnology 453
bonds, class of 5
Borofloat 94
BRCA1 gene 383
Brownian pump 445

caged compound 464
Cambridge Structural Database (CSD) 479
carry-over effect 53
catalytic activity 512
chamber resonance 427
Chara 470
chemical tagging 226
CHO cell 456
chromosome 466
Clausius-Mosotti-factor 423
closed-loop system 70
combinatorial chemistry 120, 268, 311, 491
combinatorial explosion 3
combinatorial synthesis 120
compartment 76
complementary DNA (cDNA) 189, 201
– array 384
– library 189, 209
– synthesis 211, 212
contact angle 100, 102
contact free handling 462
contig 190
contiguous stacking hybridization 378
continuous evolution 518
coordinative bond 5
coordinative interaction 6
cos site 196
cosmid 196
– library 203
counter selection 515
covalent bond 5
customized manifold of a silicon glass construction 40
cutting (restriction) of an individual molecule 462
cystic fibrosis transmembrane conductance regulator (CTFR) gene 382

database 218
decoding 270
diagnostics of pathogens 338
dielectric biological object 455
dielectric dispersion 427
dielectrophoresis 417
dielectrophoretic spectroscopy 427
diffraction limit 453
diffusion 79
- time 336
dispensing 53
DNA 17, 189, 479
–, recombinant 225
–, scanning force microscopy (SFM) imaging 357
–, single 462
- amplification, rapid 179
- chip 372
- library 201
- molecule 462
- polymerase extension 397
- preparation 203
- sequencing 311
dosing system for placing biological cells 48
drop-on-demand micropump 64
drug 498
dry etching 37
dynamic mixer 67
dynamic viscosity, η^c, of the cytoplasm 346
dynein 464

Edman degradation, automated 226
egg cell 471
electric field 462
electrocaloric flow sensor 67
electrohydrodynamic pumping 419
electron microscopy 479
electroporation 200, 214
encoding 270
endonuclease activity 464
endonuclease ApaI 464
ENTREZ 480
entropy 9
enzymatic reaction, analysis of real-time 344
error rate 530
error-threshold 543
evanescent wave 379
evolution 11, 471
–, early 12
–, *in vitro* 515, 516, 519
- machine 548
evolutionary concept 520
evolutionary dynamics 530
evolutionary process 13
evolutionary strategy 121, 529
expression vector 211

farnesyl-protein transferase (FPTase) 226
- inhibitor 226
FEM simulation 64
fertilization 471
–, *in vitro* 472
field cage 432
filamentous bacteriophage 225
finite difference method 424
flow sensor 65
flow-injection analysis 62
fluid diode 63
fluid handling 120
fluidic switching element 59
fluorescence 22
–, specific 334
- microscope 459
fluorescence correlation spectometry (FCS) 332
–, inside living cells 348
focused library 491
fold recognition 488
folding landscape 530
force between single molecules 465
force field 485
freedom of rotation 6

gene expression study 383
gene transfer 467
genome, structure of 13
genome size 10
genome subdivision 11
genomic library 189
genotype-phenotype relationship 531

glass 32
– microtool 466
glass-silicon nanolitre plate 55
gold thiol chemistry 104, 117
gradient force 453
gravitational acceleration 455

H-bridge bond 6
heatable pipette 58
hierarchy of lifetime 8
hierarchy of molecular interaction 4
hierarchy of structures 6
hierarchy of topologies 7
high density colony membrane 200
homogenous assay 333
hybridization 18, 217
– array, high density 216–218
– -based array 381
hydrogen 5
hydrophilizing 102, 116
hydrophobizing 102, 116

imaging mask 462
immobilization support 256
individual bead 266
individual molecule 465
infertility 471
information 9
intermolecular bonds, scanning force microscopy (SFM) study 362
interspersed repeat sequence (IRS) 199
– IRS-polymerase chain reaction (PCR) 199
iterative deconvolution 226

kinesin 464
kinetically-balanced mixture 227
knowledge-based potential 486

laser induced cell fusion 467, 468
laser induced fluorescence (LIF) 311
laser induced gene transfer 467
laser microbeam 454
laser tweezers 443
lattice protein 533, 539
level of biological complexity 13
levels, number of 121
library 255
library-microchip assembly 226
ligand 5
ligase 516
light 454
– guide 461
– pressure 453
liposome 456
liquid evaporation 112
liquid handling 76, 112, 116
long-time stability 8

macromolecular therapeutics 498
manifold 32
mass spectometry 226
measurement of biochemical interaction 331
Medicago sativae 468
membrane 85, 88, 97
membrane receptor 339
metaphase chromosome, scanning force microscopy (SFM) imaging 365
metrifield resin 266
micro compartment 78
micro compartment array 81
micro filter 440
micro lens array 94
micro motor 432
micro sieve 85, 88
microbead 456
microcapillary electrophoresis 62
microchannel 32
microdispensing 255
microejection pump 46
–, high-flow-rate 48
microfluidic component 31
microfluidics 31
microgravity 470
microinjector 62
microliquid handling 31, 453
microlitre plate 52
microlitre range 43
micromeniscus 64

micromixer 32, 67
microperforation 467
micropipette with heater 57
micropump 32
microreactor 453
microrestrictor 32
microscope 453
microsensor 32
microsystem 31
microtool 454
microtubulus 464
microvalve 32, 60
microvessel 462
miniaturisation 23, 258
mirror image design 512
mismatch discrimination 377
mixing ratio 70
modeling by homology 487
modified nucleic acid library 511
molecular aggregation 334
molecular diffusion 334
molecular diversity 2, 4, 6
molecular dynamics 485
molecular evolution 12, 531, 536
molecular mechanics 484
molecular modeling 475
molecular recognition 521
motor protein 464
mRNA preparation 210
mtDNA 383
multichannel dispenser 50
multichannel pipette 55
multiple peptide synthesis 225
mutagenesis, *in vitro* 515
mutation 515
myosin 464

N-(9-fluorenyl)methoxycarbonyl (Fmoc) amino acid 227
nanolitre plate 52
nanolitre range 43
nanoplotter 52
nanotiterplate 75
Nd YAG laser 459
needle pipette 54
neutral network 538
Newton 455
nitrogen laser 457
node strength 11
nuclear magnetic resonance (NMR) 479
nucleic acid 337, 493, 507
– aptamer 525, 527
– array, production of 255
– -based sequence amplification (NASBA) 338
– catalyst 526
– library 509
Nucleic Acid Database (NDB) 479

oligonucleotide, amino-modified 376
oligonucleotide array 384
oncogene 226
one-bead-one-compound library 269
optical trap 453
optical tweezers 453
optimization 10
ordering parameter 13
organization level 6
orientation-dependent etching 81

P1 199
p53 389
PAC 200
packaging extract 196
packaging, *in vitro* 208
parallel processing 120
parallel single bead synthesis 272
parallelization 75, 120
parasital RNA 520
partial digest 206
particle-particle interaction 436
patenting 555
–, analysis of motile cells 562
–, arrays of nucleic acid probes 556, 558
–, assessing the suitability of biopolymers 564, 565
–, automated cycling of solutions 570
–, chip reaction apparatus 556
–, combinatorial compound libraries 567

–, combinatorial libraries 569
–, detection of nucleic acid sequences 572
–, determining substance-specific parameters by correlation-spectroscopy 564
–, diagnostic device 556
–, diagnostic system 568
–, diagnostics 569, 573
–, electrophoretical separation 568
–, evolutive vaccination 565
–, fine positioning apparatus 572
–, flow thermocycler 571
–, fluid handling 563, 564
–, fluid micro-diode 568
–, fluorescent single molecule detection 572
–, hybridization 556, 559, 573
–, immobilization of antiligands 560-562
–, immobilization of biological polymers 556, 561
–, light-induced immobilization of biomolecules 573
–, measuring binding affinity 562
–, microcapillary with inbuilt microsensors 568
–, microfabricated reactor 570
–, microfluidic structure 571
–, microfluorescence detection 557, 560
–, micro-pipette 568
–, multi chamber thermocycler 570
–, multiple synthesis of organic compounds 569
–, nucleic acid diagnostic device 570
–, nucleotide amplification 562, 563
–, peptide library 557, 559
–, peptide synthesis 558-560
–, photolabile protecting groups 561
–, photolithographic solid phase synthesis 561, 562
–, polymer synthesis 557, 560
–, preparing biopolymers 565
–, printing molecular library arrays 558
–, sample preparation device 563
–, scanning calorimetry 571
–, scanning probe microscopy 571
–, screening 557–560, 562
–, self-assembling microelectronic systems 569
–, sensing and manipulating microscopic structurs 572
–, separating mixtures of dielectric particles 567
–, sequencing 556–561, 573, 574
–, solid-phase libraries 569
–, synthesizing of oligomers 560
penetration depth 458
peptide library 225
peptidomimetic inhibitor 227
λ-phage 195
– DNA molecule 462
phenotypic error-threshold 544
photosensitive glass 95
pico titer plate 99
picolitre range 43
piconewton 462
piezoelectronic dispensing 23
pipette 53
pipetting 53
Planck constant 454
plant cell 467
plasma 458
plasmid 194
polydimethylsiloxane (PDMS) 94, 117
polymer 97
polymerase chain reaction (PCR) 179, 199, 338, 509, 510
– pool 200
population, artificial 14
positional scanning 226
primary library 190
printing 116
probability 9
prokaryote 12
protein 459, 492, 507
Protein Data Bank (PDB) 479
protein/DNA binding assay 339
pulsed field gel electrophoresis (PFGE) 200

Qβ replicase 520
quality factor Q 455
quantum chemistry 482

radiofrequency encoding 226
random graph 541
rapeseed 467

ratchet potential 446
reference library 218
replacement vector 196
resin loading 267
resist 257
reverse transcriptase 511, 519
reverse transcription 510
Reynolds number 67
ribozyme 516, 518, 519, 526
RNA 479
–, high affinity 512
– pool 515
–, randomized 513
– preparation 210
– secondary structure 531
– world 517
RNaseH 511
robotics 20

scanning force microscopy (SFM) 353
–, cantilever 356
–, constant force mode 354
–, force-distance curve 360
–, image processing 358
–, single molecule detection 356
–, tapping mode 354
–, tip 354
–, volume determination 358
second messenger 342
selection 515
–, *in vitro* 508, 509
– cycle 510
self assembled monolayer (SAM) 104
self-modification 513
self-optimization 509
self-organization 522
self-sustained sequence replication 510
separation of particles 437
sequence, randomized 509
sequence space 13, 531
sequencing by hybridization (SBH) 372
serial transfer 122
shape space conjecture 540
shock wave 458
Si_3N_4 85
Si_3N_4/SiO_2-membrane 86
silicon 32, 77, 81
silicon deep etching 81, 87
single molecule detection (SMD) 311
single molecule sensitivity 332
SiO_2 85
SiO_2/Si_3N_4-membrane 98
solid-phase library 225
solid-phase synthesis 257, 266
solution-phase library 225
solvation 266
spectrum 426
sperm cell 471
split synthesis 268
stamp 258
stamping 117
static mixer 67
Stefan-Boltzmann constant 458
Stefan-Boltzmann law 457
structure formation, rules of 6
structure-based drug design 490
sublibrary 229
submicrometer accuracy 454
sub-micron particle 435
submolecular structuration 8
substance library 14
surface modification 100, 103
SWISS-PROT 480
synthetic peptide combinatorial library 225

T7 RNA polymerase 509, 511, 519
target of selection 10
technology of manufacturing microfluidic components 34
temperature control 105
therapy of infertility 471
thermal conductivity 457
thermal force 425
thermo control 106
thin film 107
three-dimensional micromachining 33
three-dimensional quantitative structure activity relationship (3D-QSAR) 489

three-dimensional state 8
tip 355
tobacco 467
topology 4
μ-total analysis system μ-TAS) 39
transcription factor 338
transition state analog (TSA) 514
trap 429
travelling-wave dielectrophoresis (TWD) 418
tRNA 12

ultrasonic levitator 52
ultrathin-walled microwell plate 179
universal chip 392
uracil-N-glycosylase 378
UV laser microprocessing 457

vector 189, 205
viscoelasticity 469
visualization 481

wet masking 258
wetting 100, 105, 116, 256
World-Wide Web (WWW) 476

x-ray 458, 479

yeast artificial chromosome (YAC) 197

MIX
Papier aus verantwortungsvollen Quellen
Paper from responsible sources
FSC® C105338

If you have any concerns about our products,
you can contact us on
ProductSafety@springernature.com

In case Publisher is established outside the EU,
the EU authorized representative is:
Springer Nature Customer Service Center GmbH
Europaplatz 3, 69115 Heidelberg, Germany

Printed by Libri Plureos GmbH
in Hamburg, Germany